Templated Organic Synthesis

Related Titles from WILEY-VCH:

F. Diederich / P. J. Stang (eds.)
Metal-catalyzed Cross-coupling Reactions
1997. XXII. 518 pages with approx 1000 figures and 20 tables
Hardcover. ISBN 3-527-29421-X

N. V. Gerbeleu / V. B. Arion / J. Burgess (eds.)
Template Synthesis of Macrocyclic Compounds
1999. IX. 574 pages with over 300 figures and 10 tables
Hardcover. ISBN 3-527-29559-3

J.-P. Sauvage / C. Dietrich-Buchecker (eds.)
Catenanes, Rotaxanes and Knots
A Journey through the World of Molecular Topology
1999. X. 406 pages with over 300 figures
Hardcover. ISBN 3-527-29572-0

J. Mulzer / H. Waldmann (eds.)
Organic Synthesis Highlights III
1998. X. 412 pages with 302 figures
Softcover. ISBN 3-527-29500-3

Templated Organic Synthesis

Edited by François Diederich and Peter J. Stang

WILEY-VCH

Weinheim · New York · Chichester · Brisbane · Singapore · Toronto

Prof. Dr. F. Diederich
Laboratorium für Organische Chemie
Eidgenössische Technische Hochschule
ETH-Zentrum
Universitätstrasse 16
CH-8092 Zürich
Switzerland

Prof. Dr. P. J. Stang
Department of Chemistry
University of Utah
Salt Lake City, UT 84112
USA

Library of Congress Card No. applied for.

A catalogue record for this book is available from the British Library.

Deutsche Bibliothek Cataloguing-in-Publication Data:
A catalogue record for this book is available from Der Deutschen Bibliothek

© WILEY-VCH Verlag GmbH, D-69469 Weinheim (Federal Republic of Germany), 2000

Printed on acid-free and chlorine-free paper.

Composition, Printing and Bookbinding: Konrad Triltsch, D-97070 Würzburg
Printed in the Federal Republic of Germany

Preface

Templated synthesis is a timely research area where molecular and supramolecular sciences meet. Moreover, it represents an emerging interface between chemistry, biology, and materials sciences. This multi-author monograph includes authoritative contributions from ten leading research laboratories involved in the development of novel template-assisted organic synthesis at this interface. It demonstrates how non-covalent and covalent templates are successfully applied to control the rate and the regio- and stereoselectivity of molecular and supramolecular organic reactions.

In view of the broad, often indiscriminate use of the words "templates" and "tethers", the scope of the monograph requires careful definition. According to *Webster's Dictionary*, a template is "a gauge, pattern, or mold (as a thin plate or board) used as a guide to the form of a piece being made". The same source defines a tether as "something (as a rope or chain), by which an animal is fastened so that it can range only within a set radius". "Mold" is perhaps the best single word to define a template in the frame of this monograph. Since clear definitions are important, it is appropriate that Chapter 1 provides definitions and describes the roles of templates in organic synthesis.

Either non-covalent or covalent templates need to be absent in the original substrate as well as in the product or, at least, should be removable after having accomplished their task. Tethers should be effective in the transition states of reactions and, by acting as a mold, control the rate and the regio- and stereochemistry. Thus the monograph focuses on reactivity rather than on static structural chemistry. It is intended to promote the use of intermolecular non-covalent interactions, besides steric discrimination, to control reactivity and selectivity. Its content was selected to provide useful template- or tether-based methodology to both expert and novice practitioners in both the molecular and supramolecular sciences.

This book is focused on organic synthesis and reactivity, naturally combining and merging molecular and supramolecular aspects. It includes one contribution (Chapter 5) on templated synthesis of biological polymers and their mimics. A profound discussion of biological templated reactions or inorganic templated synthesis (such as biomineralization) is beyond the scope of this monograph. Likewise, templates (better: scaffolds) in medicinal chemistry are excluded. Thus, carbohydrates or other central scaffolds (or cores), from which functional groups for protein recognition diverge, have sometimes been named as templates. Furthermore, the concept of reversibility, i.e., the removal of a non-covalent or covalent tether after the function of providing a "mold" has been accomplished, is important. This eliminates a variety of supramolecular self-assembly processes as well as "structural templates" such as those used to organize four-helix bundle proteins.

Although all metal-catalyzed reactions such as the Pauson–Khand reaction, Co-catalyzed cyclotrimerization, and metal-catalyzed cross-coupling reactions are templated reactions in which the metal center acts as template, they are excluded since they have been the subject of the previous monographs, *Modern Acetylene Chemistry* and *Metal-catalyzed Cross-coupling Reactions* from the same editors and publishers as this book.

The requirement for ultimate removal of the covalent template precludes examples in which the reaction has been changed from an intermolecular to an intramolecular one by connecting the reaction partners by some kind of permanent tether which remains in the molecule. Even with this restriction, organic synthesis provides an extremely rich variety of reversible templating and tethering, as illustrated in Chapter 10; here, we would only like to mention the "silicon connection" chemistry introduced by G. Stork. Claimed template effects sometimes deserve critical mechanistic examination, and this is nicely illustrated in Chapter 9.

The increasingly successful use of intermolecular interactions by non-covalent and covalent molecular and polymeric tethers provides an important link between organic synthesis and supramolecular chemistry in this monograph. Synthesis in Nature abundantly takes advantage of non-covalent bonding and templating and, therefore, the chemistry described herein can be rightly qualified as biomimetic. Emphasis is placed upon key developments and important advances, which are illustrated with attractive and useful examples. Areas covered range from molecular imprinting (Chapter 2), molecular recognition and supramolecular self-assembly (Chapters 3 and 4), biomimetic catalysis (Chapter 6), templated synthesis of biopolymers and unnatural oligomers and polymers (Chapters 1, 5, and 8), regio- and stereoselective synthesis of covalent fullerene derivatives (Chapter 7), templated macrocyclizations (Chapters 1 and 9), to the use of temporary connections in organic synthesis (Chapter 10). Carefully chosen references guide the reader through the extensive literature. A useful feature is the inclusion of key synthetic protocols, in experimental format, in six chapters. We are hopeful that the monograph will stimulate the further development of efficient and exciting new templated synthesis methodology in the molecular and supramolecular sciences.

ETH Zürich   François Diederich
University of Utah   Peter J. Stang
Salt Lake City
May 10, 1999

List of Contributors

Jennifer R. Allen
Department of Chemistry
Box 1822, Station B
Vanderbilt University
Nashville, TN 37235
USA

Harry L. Anderson
Department of Chemistry
Oxford University
Dyson Perrins Laboratory
South Parks Road
Oxford OX 13QY, UK

Sally Anderson
Sharp Laboratories of Europe LTD
Edmund Halley Road
Oxford Science Park
Oxford OX4 46A, UK

Ronald Breslow
Department of Chemistry
Columbia University
New York, New York 10027
USA

Liam R. Cox
School of Chemistry
University of Birmingham
Edgbaston, Birmingham B15 2TT
UK

Prof. François Diederich
Laboratorium für Organische Chemie
ETH Zürich
Universitätstrasse 16, CH-8092 Zürich
Switzerland

Ken S. Feldman
Department of Chemistry
The Pennsylvania State University
University Park, Pennsylvania 16802
USA

Alois Fürstner
Max-Planck-Institut für Kohlenforschung
D-45470 Mülheim/Ruhr
Germany

Yahaloma Gat
Searle and Jones Chemical Laboratories
5735 Ellis Avenue
University of Chicago
Chicago, Illinois 60637
USA

Roland Kessinger
Laboratorium für Organische Chemie
ETH Zürich
Universitätstrasse 16, CH-8092 Zürich
Switzerland

Steven V. Ley
Department of Chemistry
University of Cambridge
Lensfield Road, Cambridge CB2 1EW
UK

David G. Lynn
Searle and Jones Chemical Laboratories
5735 Ellis Avenue
University of Chicago
Chicago, Illinois 60637
USA

Darren Makeiff
Department of Chemistry
University of British Columbia
2036 Main Mall
Vancouver, BC
Canada V6T 1Z1

Ned A. Porter
Department of Chemistry
Box 1822, Station B
Vanderbilt University
Nashville, TN 37235
USA

Françisco M. Raymo
Department of Chemistry
and Biochemistry
University of California, Los Angeles
405 Hilgard Avenue
Los Angeles, California 90095-1569
USA

John C. Sherman
Department of Chemistry
University of British Columbia
2036 Main Mall
Vancouver, BC
Canada V6T 1Z1

J. Fraser Stoddart
Department of Chemistry
and Biochemistry
University of California, Los Angeles
405 Hilgard Avenue
Los Angeles, California 90095-1569
USA

Günter Wulff
Institute of Organic Chemistry
and Macromolecular Chemistry
Heinrich-Heine-University of Düsseldorf
Universitätstrasse 1, D-40225 Düsseldorf
Germany

List of Abbreviations

A	acceptor; adenine
A^{bz}	N^6-benzoyladenine
acac	acetylacetonate
ACT	addition, cyclization, transfer
AIBN	2,2′-azobis(isobutyronitrile)
ATPH	aluminum tris(2,6-diphenylphenoxide)
BINOL	1,1′-binaphthalene-2,2′-diol
Bn	benzyl
Boc, BOC	*tert*-butyloxycarbonyl
BOM	benzyloxymethyl
BSA	*N,O*-bis(trimethylsilyl)acetamide
CAN	cerium(IV) ammonium nitrate
Cbz	benzyloxycarbonyl
CD	circular dichroism
CDB	charged hydrogen bond
COD	1,5-cyclooctadiene
Cp	cyclopentadienyl
CSA	camphorsulfonic acid
CTAB	cetyl triethylammonium bromide
Cy	cyclohexyl
D	donor
DBU	1,8-diazabicyclo[5.4.0]undec-7-ene
DCC	*N,N*′-dicyclohexylcarbodiimide
DCE	1,2-dichloroethane
DDQ	2,3-dichloro-5,6-dicyanobenzoquinone
d.e.	diastereomeric excess
DIBALH	diisobutylaluminum hydride
DICI	*N,N*′-diisopropylcarbodiimide
DIPEA	diisopropylethylamine
DMA	*N,N*-dimethylacetamide; 9,10-dimethylanthracene (Chapter 7)
DMAD	dimethyl acetylenedicarboxylate
DMAP	4-(dimethylamino)pyridine
DME	1,2-dimethoxyethane
DMF	*N,N*-dimethylformamide
DMSO	dimethyl sulfoxide
DMTr	4,4′-dimethoxytriphenylmethyl
DMTST	dimethyl(methylthio)sulfonium trifluoromethanesulfonate

DNA	deoxyribonucleic acid
DNOE	difference nuclear Orerhauser effect
L-DOPA	3-hydroxy-L-tyrosine
dppe	bis(diphenylphosphino)ethane
DSC	differential scanning calorimetry
DTBP	2,6-di-*tert*-butylpyridine
EDTA	ethylenediaminetetraacetic acid
e.e.	enantiomeric excess
EM	effective molarity
Fmoc	(9*H*-fluoren-9-ylmethoxy)carbonyl
GC	gas chromatography
GDS	guest determining step
HETCOR	{^{1}H-^{13}C}heteronuclear shift correlation
HMDS	hexamethyldisilazane
HMPA	hexamethylphosphoric triamide
HOBT	1-hydroxy-1*H*-benzotriazole
HPLC	high-performance liquid chromatography; high-pressure liquid chromatography (Chapter 8)
h-RNA	homo-RNA
IDCP	iodonium collidine perchlorate
IDCT	iodonium collidine triflate
IMDA	intramolecular Diels–Alder
i-PrOH	isopropanol
KHMDS	potassium hexamethyldisilazide
LDA	lithium diisopropylamide
LNA	locked-nucleic acid
L-selectride	lithium tri-*sec*-butylborohydride
LUMO	lowest unoccupied molecular orbital
MC	Monte Carlo
*m*CPBA	*m*-chloroperoxybenzoic acid
MMA	methyl methacrylate
MMTR	4-methoxytriphenylmethyl
MOM	methoxymethyl
ms	molecular sieves
NBD	norbornadiene
NBS	*N*-bromosuccinimide
NIS	*N*-iodosuccinimide
NFP	*N*-formylpiperidine
NMO	*N*-methylmorpholine *N*-oxide
NMP	*N*-methyl-2-pyrrolidinone
OPNA	oxypeptide-RNA
OTf	trifluoromethane sulfonate
PCC	pyridinium chlorochromate
PDC	pyridinium dichromate

PEG	poly(ethylene glycol)
phen	1,10-phenanthroline
Phth	phthalyl
Piv	pivaloyl
PM3	parametric method 3
PMB	*p*-methoxybenzyl
pMMA	poly(methyl methacrylate)
PNA	peptide-nucleic acid
Pr	propyl
pRNA	pyranosyl-RNA
*p*TSA	*p*-toluenesulfonic acid
py	pyridine
Ra-Ni	Raney nickel
RCM	ring-closing metathesis
RNA	ribonucleic acid
RORCM	ring-opening and ring-closing metathesis
rSNA	dimethylene sulfone RNA
rt	room temperature
RVC	reticulated vitreous carbon
SDDS	sodium dodecylsulfate
T	thymine
TBAF	tetrabutylammonium fluoride
TBDMS	*tert*-butyldimethylsilyl
TBDPS	*tert*-butyldiphenylsilyl
TBS	*tert*-butyldimethylsilyl
Tf	trifluoromethane sulfonyl
TFA	trifluoroacetic acid
THF	tetrahydrofuran
THP	3,4,5,6-tetrahydro-2*H*-pyran-2-yl
TLC	thin layer chromatography
TMP	tetramethylpiperidine
TMS	trimethylsilyl
TMSI	trimethylsilyl iodide
Tr	triphenylmethyl
Ts	*p*-toluenesulfonyl
TsOH	*p*-toluenesulfonic acid
T. S.	transition state
TSA	*p*-toluenesulfonic acid

Contents

1   **Templates in Organic Synthesis: Definitions and Roles**

Sally Anderson, Harry L. Anderson

1.1   Introduction – Early Templates   1

1.2   The Definition of a Molecular Template   4

1.3   Roles of Templates   5
1.3.1   Thermodynamic and Kinetic Templates   5
1.3.2   Covalent and Non-covalent Template-Substrate Interactions   8
1.3.3   Topology of Reaction   9
1.3.3.1 Cyclization templates   10
1.3.3.2 Linear templates   14
1.3.3.3 Interweaving templates   18
1.3.4   Scavenger Templates   19
1.3.5   Negative Templates   20

1.4   Measuring Template Effects   21
1.4.1   Qualitative Detection of Template Effects   21
1.4.2   Quantification of Kinetic Template Effects in Terms of Effective
        Molarity, Substrate Affinity, and Maximum Rate Enhancement . . .   22
1.4.2.1 Linear templates   23
1.4.2.2 Quantitative analysis of template effects in tethered reactions   . . .   29
1.4.2.3 Cyclization templates   29

1.5   Conclusion   33
        Appendix 1a: Equations for Figure 1-5   34
        Appendix 1b: Equations for Figure 1-10   34
        References   35

2   **Templated Synthesis of Polymers – Molecularly Imprinted
         Materials for Recognition and Catalysis**

Günter Wulff

2.1   Introduction   39

2.2   Preparation of Optically Active Linear Vinyl Polymers
        by Templated Synthesis   40

2.3   Exact Placement of Functional Groups on the Surfaces of Rigid
          Polymeric Materials Using Template Molecules   42

2.4   Molecular Imprinting in Polymeric Materials Using Template
          Molecules   45
2.4.1   The Principle   45
2.4.2   The Optimization of the Structure of the Polymer Network   50
2.4.3   The Role of the Binding-site Interactions   52
2.4.4   Chiroptical Properties of the Crosslinked Polymers   57
2.4.5   Chromatography Using Molecularly Imprinted Polymers   58
2.4.6   Catalysis With Molecularly Imprinted Polymers   60
2.4.7   Outlook   64

2.5   Experimental Procedures   65
2.5.1   Polymer from Scheme 2-5   65
2.5.1.1   Preparation of template monomer 7 [34]   65
2.5.1.2   Preparation of the polymer [35]   66
2.5.2   Polymer from Scheme 2-6   66
2.5.2.1   Thermally initiated polymerization [36]   66
2.5.2.2   Photochemically initiated polymerization [50]   66
2.5.3   Polymer from Scheme 2-9 [136, 137]   67
2.5.3.1   N-Ethyl-4-vinylbenzamide   67
2.5.3.2   N-Ethyl-4-vinylbenzocarboximide acid ethyl ester   67
2.5.3.3   N,N'-Diethyl-4-vinylbenzamidine (10a)   67
2.5.4   Preparation of the polymer [112]   68
          References   68

3   Templated Synthesis of Catenanes and Rotaxanes

Françisco M. Raymo, J. Fraser Stoddart

3.1   Introduction   75

3.2   Metal-Templated Syntheses   76

3.3   Hydrogen Bonding-assisted Syntheses   78

3.4   Hydrophobically Driven Syntheses   81

3.5   Aromatic Templates   84

3.6   Dialkylammonium-containing Rotaxanes   91

3.7   Conclusions   95

3.8   Experimental Procedures   95
3.8.1   [2]Catenane 4 [13]   95
3.8.2   [2]Catenane 12 [16]   96

3.8.3	[2]Catenane 43 [31]	96
3.8.4	[2]Rotaxane 51 [38]	96
3.8.5	[2]Rotaxane 56 [38]	97
3.8.6	[2]Rotaxane 68 [45]	97
	References	97

4   Templated Synthesis of Carceplexes, Hemicarceplexes, and Capsules

Darren Makeiff, John C. Sherman

4.1	Introduction	105
4.2	Carceplexes	106
4.2.1	The First Soluble Carceplex	106
4.2.1.1	Template ratios in the formation of an acetal-bridged carceplex	106
4.2.1.2	Formation of a charged hydrogen bonded complex	107
4.2.1.3	Mechanism of formation for an acetal-bridged carceplex	108
4.2.2	Large Carceplexes from Cyclic Arrays of Cavitands	109
4.2.2.1	Synthesis of a trimer carceplex containing three DMF molecules	110
4.2.2.2	Synthesis of a bis(carceplex)	110
4.2.3	Benzylthia-bridged Carceplex	111
4.2.4	Calix[4]arene–Cavitand Hybrid Carceplexes	112
4.2.5	Metal-bridged Carceplexes	113
4.3	Hemicarceplexes	114
4.3.1	Template Effects in the Formation of a Trismethylene-Bridged Hemicarceplex	114
4.3.2	Hemicarceplexes Containing Four Slotted Portals	115
4.4	Capsules	118
4.4.1	Capsules Composed of Cavitands Linked via Covalent Bonds or Charged Hydrogen Bonds	118
4.4.2	Rebek's "Tennis Balls", "Softballs", and "Wiffle Balls"	119
4.4.2.1	Template effects of solvent in the synthesis of "softball" dimers	122
4.4.3	Other Hydrogen Bonded Capsules	123
4.4.4	Capsules that do not Involve Hydrogen Bonding	125
4.4.5	Other Capsules	126
4.5	Conclusions	127
	References	127

5 **Template-Directed Ligation:**
Towards the Synthesis of Sequence Specific Polymers

Yahaloma Gat, David G. Lynn

5.1 Introduction 133

5.2 Template-directed Ligation: Minimalist Scheme 134

5.3 Catalytic RNA and DNA 136

5.4 Molecular Recognition 137
5.4.1 Sugar Substitution 137
5.4.2 Phosphate Substitution 138
5.4.3 Neutral Acyclic Backbones 142
5.4.4 Peptide–Peptide Association 144

5.5 Ligation 146

5.6 Product Dissociation 149

5.7 Summary 152
References 153

6 **Biomimetic Reactions Directed by Templates**
and Removable Tethers

Ronald Breslow

6.1 Introduction 159

6.2 Biomimetic Reactions Using Covalently Linked Tethers
and Templates 160
6.2.1 Photochemical Functionalizations by Tethered Benzophenones . . . 160
6.2.2 Free-radical Halogenations by Tethered Reagents 163
6.2.3 Free-radical Reactions Directed by Tethered Templates –
the Radical Relay Mechanism 164
6.2.4 Selective Intramolecular Epoxidations Directed
by Removable Tethers 171

6.3 Selective Reactions Directed by Non-covalently Linked Templates 172
6.3.1 Selective Aromatic Substitution Directed
by Cyclodextrin Complexing 172
6.3.2 Photochemical Functionalizations by Complexed Benzophenones
and Chlorinations Directed by Ion-paired Templates 174
6.3.3 Oxidations Directed by Metalloporphyrin
and Metallosalen Templates 177
6.3.4 Other Reactions Catalyzed by Coordinated Template Catalysts . . . 183
References 185

7 **Regio- and Stereoselective Multiple Functionalization of Fullerenes**

François Diederich, Roland Kessinger

7.1 Introduction 189

7.2 Anthracenes as Reversible Covalent Templates 189

7.3 Tether-directed Remote Functionalizations of C_{60} 192
7.3.1 The First Tether-directed Multiple Functionalization of C_{60}:
Formation of Higher Adducts with Novel Addition Patterns 193
7.3.2 The Bingel Macrocyclization 196
7.3.2.1 Formation of *cis*-2 bis-adducts 200
7.3.2.2 Formation of *cis*-3 bis-adducts 202
7.3.2.3 Other bis-functionalization patterns 204
7.3.3 Formation of Bis-adducts by Double Diels–Alder Addition
of Tethered Bis(buta-1,3-diene)s to C_{60} 205
7.3.4 Double [3+2] Cycloaddition of Tethered Vinylcarbenes 209
7.3.5 Double [3+2] Cycloadditions of Tethered Bis-azides 210

7.4 Conclusions 211

7.5 Experimental Procedures 212
7.5.1 Preparation of Hexakis-adduct 11 using DMA as a Template
(Scheme 7-2) [20] 212
7.5.2 Preparation of Bis-adduct 24 by Reaction
of Methano[60]fullerenecarboxylic Acid 21
with the Tether–Reactive-group Conjugate 22 (Scheme 7-4) [47] . 212
7.5.3 Preparation of Tris-adduct 28 by Addition of the Anchor–Tether–
Reactive-group Conjugate 27 to C_{60} (Scheme 7-4) [47] 213
7.5.4 Formation of Tetrakis-adduct 20 by Removal of the Tether
in Hexakis-adduct 30 and Transesterification (Scheme 7-5) [2] . . . 213
7.5.5 Bingel Macrocyclization: Synthesis of *cis*-2 Bis-adduct 42 Starting
from Benzene-1,2-dimethanol (Scheme 7-8) [25] 213
7.5.6 Enantioselective Synthesis of *cis*-3 Bis adduct (^fC)-2
by Bingel Macrocyclization of (*S,S*)-66 with C_{60} Followed
by Transesterification (Scheme 7-11) [25] 214
References 214

8 **Template Controlled Oligomerizations**

Ken S. Feldman, Ned A. Porter, Jennifer R. Allen

8.1 Introduction 219

8.2 Controlled Oligomerizations with Tethered Monomers 221

8.2.1   Templates for *n*=2 Telomers   221
8.2.2   Templates for *n*=4 Telomers   223
8.2.3   Ether-based Templates   228

8.3   Controlled Oligomerizations with a Polynorbornane-based Template   230
8.3.1   Template Design and Synthesis   231
8.3.2   Measuring the Initiator–Terminator Gap in Molecular Terms   235
8.3.3   Model Studies for Optimizing Terminator Performance   236
8.3.4   Template Controlled Oligomerization Studies   237
8.3.5   Mechanistic Studies: Probing Bimolecular Termination
         and Solvent Effects   239

8.4   Concluding Remarks   243

8.5   Experimental Procedures   244
8.5.1   (2*R*,4*S*)- and (2*S*,4*S*)-2-(5-Bromopentyl)-2-methyl-3-acryloyl-
         4-*tert*-butyl oxazolidine (23)   244
8.5.2   1,2-Bis-[3,4-bis-(5-(2*R*-methyl-3-acryloyl-4*S*-*tert*-butyl-
         2-oxazolidinyl)-pentyloxy)]-ethane (24)   244
8.5.3   Typical conditions for ACT reaction (2.5^A/80^I/200^{Sn})   245
8.5.4   Telomer assay   245
8.5.5   Synthesis of Template Diester 48   245
8.5.6   Controlled Oligomerization of Methyl Methacrylate
         with Template 63   246
         References   247

**9   Templated or Not Templated, That is the Question:
       Analysis of Some Ring Closure Reactions**

Alois Fürstner

9.1   Introduction   249

9.2   Possible Pitfalls   251

9.3   Titanium-induced Intramolecular Carbonyl Coupling Reactions . .   253

9.4   Zinc versus Samarium Mediated Reformatsky Reactions   258

9.5   Macrocycle Syntheses by Ring Closing Metathesis (RCM)   259

9.6   Macrocycles via Tsuji–Trost Allylation   265

9.7   Summary   268

9.8   Experimental Procedures   269
9.8.1   Synthesis of 15,16-Diphenyl-1,4,7,10-tetraoxa(10.2)[20]-
         paracyclophan-15-ene (14, Scheme 9-8) by an Intramolecular
         McMurry Coupling under "Salt Free" Conditions [19]   269

9.8.2   Synthesis of Oxacyclohexadec-11-en-2-one (22, Scheme 9-14)
         via RCM [32]   269
9.8.3   Palladium-catalyzed Alkenyl Epoxide Cyclization:
         Synthesis of Macrocycle 39 (Scheme 9-22) [42]   269
         References   270

10   Use of the Temporary Connection in Organic Synthesis

         Liam R. Cox, Steven V. Ley

10.1   Introduction   275

10.2   Cycloaddition Reactions   277
10.2.1   [4+2] The Diels–Alder Reaction   277
10.2.1.1 Type I IMDA reactions   278
10.2.1.2 Type II IMDA reactions   295
10.2.2   Other Cycloadditions   299
10.2.2.2 An efficient synthesis of (–)-detoxinine using a tandem [4+2]/[3+2]
         reaction   300
10.2.2.3 Sulfur tethers in [5+2] cycloaddition   303
10.2.2.4 Azomethine ylides   304

10.3   Temporary Tethering Strategies in Radical Cyclization Reactions .   307
10.3.1   Introduction   307
10.3.2   Silyl Ethers Containing the Radical Precursor   308
10.3.2.1 The α-(bromomethyl)dimethylsilyl ether in radical cyclizations . .   308
10.3.2.2 Other (haloalkyl)dimethylsilyl ethers   318
10.3.2.3 Dimethylsilyl ethers possessing alkenyl and aryl radical precursors   321
10.3.3   Silyl Ethers Containing the Radical Acceptor   323
10.3.4   Silyl Acetals   327
10.3.5   Acetal Tethers in Radical Cyclizations   331
10.3.6   Tether-directed Radical Cyclization Approaches to the Synthesis
         of *C*-Glycosides   333

10.4   Use of the Temporary Connection in O-Glycosylation
         and Nucleosidation   337
10.4.1   Tether-Directed O-Glycosylation   337
10.4.2   Directed Nucleosidations   354

10.5   Tether-mediated Nucleophile Delivery   356
10.5.1   Synthesis of (+)-Hydantocidin   356
10.5.2   Synthesis of Lincomycin   357
10.5.3   Intramolecular Allylation – Use in the Synthesis
         of Tricyclic β-Lactam Antibiotics   358
10.5.4   Synthesis of Corticosteroids   359
10.5.5   Synthesis of (–)-α-Kainic Acid   361

10.6   Silicon-tethered Ene Cyclization   362

10.7   Intramolecular Hydrosilylation and Related Reactions   364
10.7.1   Hydrosilylation of Olefins   364
10.7.2   Hydrosilylation of Acetylenes   368
10.7.3   Directed Bis-Silylation Reactions   369
10.7.4   Intramolecular Hydrosilylation of Ketones   370

10.8   Olefination   371

10.9   Ullmann Coupling   373

10.10   Use of the Temporary Connection in Carbenoid Chemistry   376

10.11   The Temporary Sulfur Connection   377

10.12   Metathesis   380

10.13   Dötz Benzannulation   381

10.14   Miscellaneous   382

10.15   Concluding Remarks   384

10.16   Experimental Procedures   385
10.16.1   Use of an Acetal Tether in Diels–Alder Reactions:
            IMDA Reaction of Dimethyl Acetal 28 (Scheme 10-9) [15]   385
10.16.1.1   Tether formation   385
10.16.1.2   IMDA reaction and tether cleavage   385
10.16.2   Use of a Boronate Tether in Diels–Alder Reactions: IMDA Reaction
            between Anthrone 50 and Methyl 4-Hydroxybut-2-enoate 3
            (Scheme 10-19) [20a]   386
10.16.2.1   Tether formation and IMDA reaction   386
10.16.2.2   Tether cleavage   386
10.16.3   Use of the (Bromomethyl)dimethylsilyl Ether Group in a Radical
            Cyclization in the Synthesis of Talaromycin A, 140 [54]   386
10.16.3.1   Tether formation   386
10.16.3.2   Radical cyclization and tether cleavage (Scheme 10-47)   386
10.16.4   Use of Acetals in Intramolecular O-glycosylation:
            Synthesis of ethyl 2,3,6-Tri-*O*-benzyl-4-*O*-(3,4,6-tri-*O*-benzyl-
            β-D-mannopyranosyl)-*α*-D-glucopyranoside 246
            (Scheme 10-77) [89c]   387
10.16.4.1   Tether formation   387
10.16.4.2   Glycosylation and tether cleavage   387
            References   387

Subject Index   397

1 Templates in Organic Synthesis: Definitions and Roles

Sally Anderson, Harry L. Anderson

1.1 Introduction – Early Templates

The word "template" originated among mediaeval architects, builders, stonemasons, and craftsmen. A template was a full-scale pattern or mold, made of wood or metal, used as a guide when cutting stone, molding plaster, or building archways [1]. Figure 1.1 shows craftmen using templates in the 13th century [2]. At that time, most craftsmen were illiterate and templates were used to communicate and record design specifications [3]. Templates convey information. They are not permanently incorporated into the structures which they specify. They can be reused. When a template is used to build a stone arch (Figure 1.2), it holds the stones in place while they are cemented together, then it is removed to leave the free-standing structure.

Chemists, the architects and builders of the molecular world, have learnt to use templates in an identical fashion. Molecular templates are control elements which favor the formation of a single product from a reactant, or reactants, that has, or have, the potential to assemble and react in a variety of ways.

Figure 1-1 Thirteenth century stonemasons, from a woodcut of a stained glass window in Chartres Cathedral (France) by E. Viollet-Leduc [2]. The templates used for building arches and stone moldings are depicted above the heads of the craftsmen.

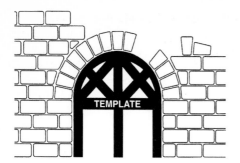

Figure 1-2 The template (in black) is used to hold stone blocks in place while they are cemented together; the template is removed from the finished archway.

Molecular templates are ubiquitous in Nature. DNA is a template for RNA, which in turn is a template for protein. The idea of using molecular templates in synthesis dates from about the time Watson and Crick [4] deduced the structure of double stranded DNA (1953). Biochemistry provided the inspiration for this concept in chemical synthesis, although it was first applied to artificial systems by coordination chemists. The challenge of emulating the exquisite selectivity of biological templates was proclaimed by Todd [5] in 1956, when he suggested that templates might one day be used to control synthesis in the laboratory just as they do in Nature. With hindsight, we can see that chemists had made serendipitous use of templates in organic reactions as early as 1926 [6], but without recognizing the significance, or generality, of the template effect.

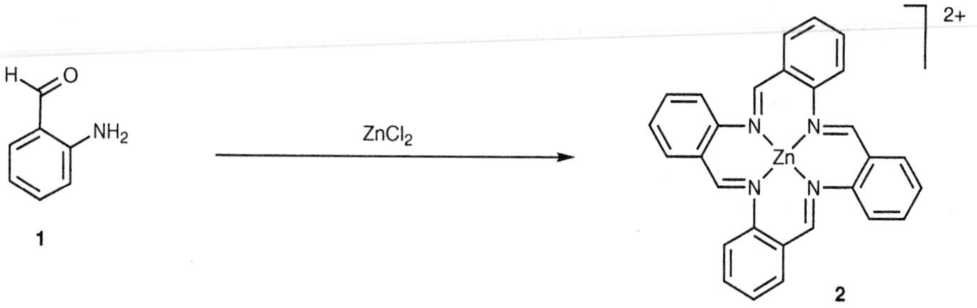

Scheme 1-1 Condensation of 2-aminobenzaldehyde in the presence of zinc(II) chloride leads to the tetrameric macrocycle **2**.

Early work involved the use of metal cations as templates in macrocyclization reactions. Seidel [6] reacted 2-aminobenzaldehyde **1** with anhydrous zinc chloride (Scheme 1-1), but did not recognize that he had prepared the macrocycle **2**, and it was not until 40 years later that Busch [7] and co-workers elucidated the structure of this type of macrocycle and investigated how metal cations template its formation.

In 1932, a dark blue by-product was isolated during the industrial production of phthalimide from phthalic anhydride and ammonia in an iron vessel. This unexpected by-product was eventually identified as iron(II)phthalocyanine: Now we know that the iron had templated the formation of this macrocycle [8]. Curtis investigated the condensation of acetone **4** and 1,2-diaminoethane **3** in the presence of metal cations and showed that the macrocycles **5** and **6** could be prepared both by templated and untemplated processes (Scheme 1-2) [9, 10].

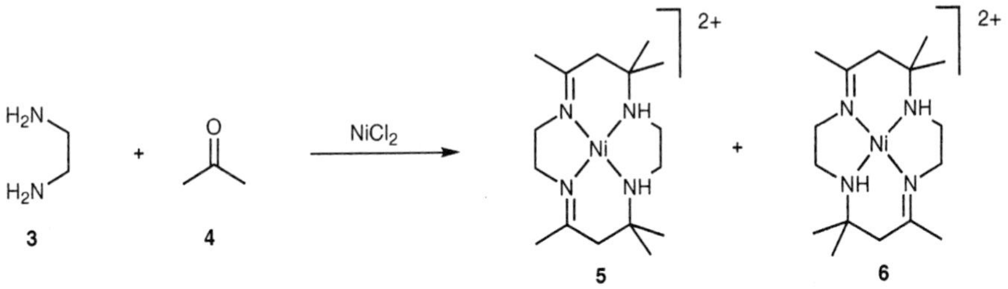

Scheme 1-2   Nickel(II) salts template the formation of tetraazamacrocycles **5** and **6** from 1,2-diaminoethane and acetone.

Another celebrated example of the template effect is the synthesis of crown ethers by Pedersen [11], but it was Busch who first intentionally used templates in synthesis and who first articulated the concept of the "template effect" in the 1960s [12]. Busch used the reaction of a nickel(II) dithiolate complex **7** with 1,2-bis(bromomethyl)benzene **8** to illustrate his ideas (Scheme 1-3) [13]. Once one end of the 1,2-bis(bromomethyl)benzene has reacted with the nickel complex, the nickel template induces the reactive ends of the intermediate **9** to come into close proximity and favors cyclization. The metal template allows the synthesis of a metallated macrocycle **10**; the free ligand cannot be prepared by the reaction of 1,2-bis(bromomethyl)benzene with the unbound thiol (in the absence of a template other cyclic and acyclic products are formed).

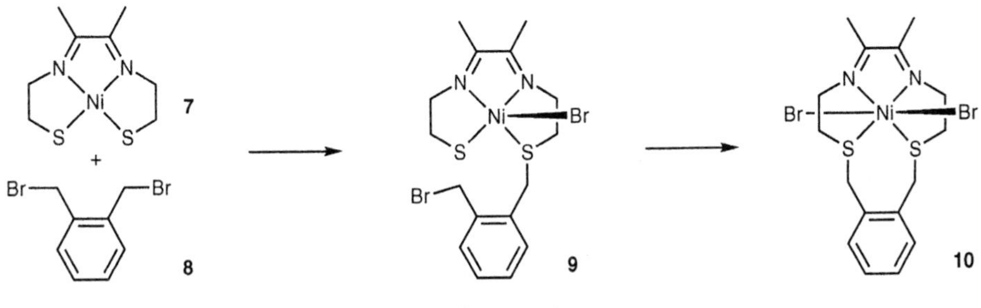

Scheme 1-3   Nickel(II) as a macrocyclization template.

In this chapter, we will show how the concept of template-directed organic synthesis has expanded so that now metal–ligand binding, hydrogen bonding, π–π interactions, and covalent bonding can all be exploited to allow the synthesis of molecules with a remarkable degree of control. Several detailed reviews of molecular templates have been published [14–20]. This chapter explores the definition and classification of molecular templates and discusses how their performance can be quantified and optimized.

1.2   The Definition of a Molecular Template

The word "template" is used widely in chemistry, biochemistry, and materials science; its use has spread dramatically since the 1960s, particularly since 1990. What do we mean by a template in the context of organic synthesis? Busch gave the following definition [21]:

> *A chemical template organizes an assembly of atoms with respect to one or more geometric loci, in order to achieve a particular linking of atoms.*

This very general definition emphasizes the essential features of a template: (a) it organizes an assembly of atoms in a specific spatial arrangement, (b) it favors the formation of a single product where the possibility of forming more than one exists, and (c) it is more than just a platform onto which a structure is built – it must promote attractive interaction between the units which are ordered around it. In the context of organic synthesis this interaction is usually covalent; however, the strength of interaction induced by the template is continuously variable, as discussed later. Reagents and catalysts also influence the course of a reaction; the distinction between a template and a reagent is that a template intervenes in the spatial arrangement of atoms, rather than in the intrinsic chemistry. The template provides instructions for the formation of a single product from a substrate, or substrates, which otherwise have the potential to assemble and react in a variety of ways. Changing the template should result in a different substrate assembly and consequently a different product. In general, after the template has directed the formation of the product, it is removed to yield the template-free product, although sometimes the template becomes inextricably locked into the structure it helps to create.

The use of templates (both molecular and supramolecular) has also become a popular synthetic strategy in materials science. For example, organic cations template the formation of zeolites [22], surfactant mesophases template the synthesis of mesoporous solids [23], track-etched pores in membranes template the formation of metal nanocylinders [24], and colloidal silica particles template the synthesis of photonic crystals [25]. Molecular templates are also used widely in biology and biochemistry, but these enzyme catalyzed processes are outside the scope of this review.

1.3 Roles of Templates

1.3.1 Thermodynamic and Kinetic Templates

Busch classified templates as thermodynamic [13a, 27] or kinetic [13b, 28]. A thermodynamic template shifts the position of equilibrium of a reversible reaction by preferentially binding one product. Kinetic templates operate on irreversible reactions by stabilizing the main transition states leading to the desired product. Kinetic templates almost invariably bind the product more strongly than the starting material, so they also favor the formation of the product thermodynamically. Conversely, thermodynamic templates are likely to accelerate formation of the product by transition state stabilization, so classification of the observed effect depends crucially on the reaction conditions and time scale.

Kinetic templates are more difficult to design and understand but they are ultimately more versatile than thermodynamic templates. A thermodynamic template needs only to bind the desired product more strongly than other species in the reaction mixture. The criteria for an efficient kinetic template are more complex; it needs to bind all the key transition states through which the reaction proceeds, lowering their energies, thus creating a well defined channel in the energy landscape, so that the reaction is prevented from following unwanted paths. By considering how enzymes work, we can see how they resemble templates. Enzymes increase the speed and specificity of a reaction in three ways:

a) Enzymes lower $\Delta H^{\ddagger}$ for a reaction by stabilizing positive and negative charges as they develop during the course of the reaction and by modifying the substrate so as to enhance reactivity.
b) Enzymes lower $\Delta S^{\ddagger}$ for a reaction by increasing the proximity of reactive groups and control the specificity of the reaction by binding substrates in specific orientations.
c) Enzymes achieve catalytic turnover by binding the transition state more strongly than the substrates or the products.

Kinetic templates recognize and bind a species during a reaction in such a way as to favor a particular geometry and orientation of reactive groups, so inducing the reaction to proceed towards a single product. Thus they mimic one important feature of enzymic catalysis: by binding substrates in close proximity, they lower $\Delta S^{\ddagger}$ for a reaction. The most economical template would also be catalytic by showing complementarity only to the transition state, but most templates bind strongly to their products and so exhibit little, or no, catalytic turnover.

Most of the templates discussed in this chapter are kinetic templates. Some of the early examples of metal cation templated macrocyclization studied by Curtis and Busch [7, 9] (Schemes 1-1 and 1-2) are thought to operate under thermodynamic control. The clearest evidence for a thermodynamic template effect comes when the template-free product is not stable under the reaction conditions. For example, treatment of 1,2-dicyanobenzene **11** with boron trichloride or uranyl chloride results in the formation of subphthalocyanine **12** [29a] or superphthalocyanine **13** [29b] complexes, respectively (Scheme 1-4), but when the metals are removed from these macrocycles, they transform into normal phthalocyanine **15**. Reaction of subphthalocyanine **12** with diiminoisoindolines such as **14** provides a useful route to unsymmetrically substituted phthalocyanines [29c–e]. Subphthalocyanines and superphthalocyanines only exist as their boron and uranium complexes, respectively, whereas normal phthalocyanines exhibit a wide variety of metal complexes.

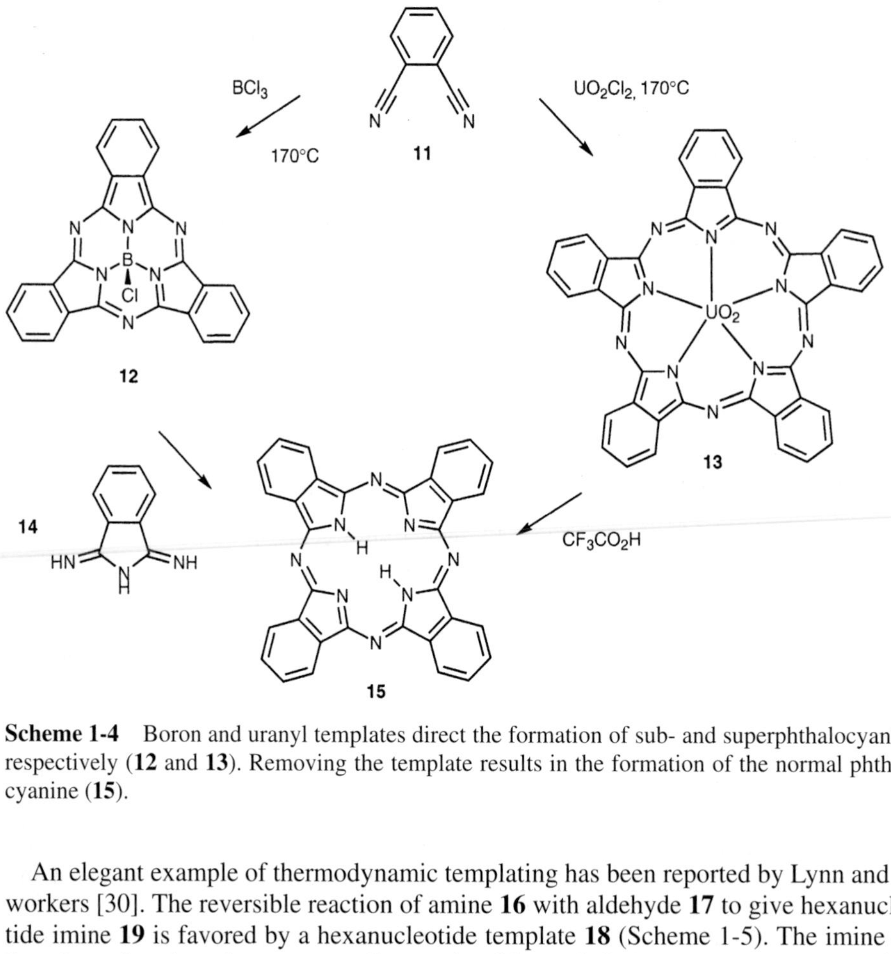

Scheme 1-4   Boron and uranyl templates direct the formation of sub- and superphthalocyanines respectively (**12** and **13**). Removing the template results in the formation of the normal phthalocyanine (**15**).

An elegant example of thermodynamic templating has been reported by Lynn and co-workers [30]. The reversible reaction of amine **16** with aldehyde **17** to give hexanucleotide imine **19** is favored by a hexanucleotide template **18** (Scheme 1-5). The imine can then be reduced to the corresponding amine **20** to "fix" the new hexanucleotide and remove it from the thermodynamic cycle. The hexanucleotide template **18** has a catalytic role in this overall process because, amazingly, it binds **19** about 10^6 times more strongly than **20**, so product inhibition is not a problem. (Systems of this type are described in detail in Chapter 5.)

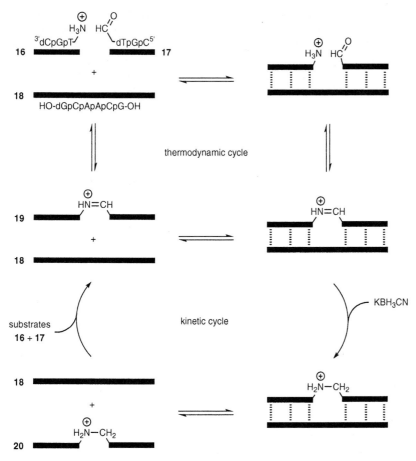

Scheme 1-5 Catalytic and thermodynamic cycles for the DNA catalyzed reductive amination in 0.5 M NaCl at pH 6 and 25 °C.

Another interesting example of a thermodynamic template is adamantanecarboxylate **21** in the reaction shown in Scheme 1-6, discovered by Fujita and co-workers [31]. Cage **24** is only formed in very low yield, as part of an intractable mixture of oligomeric products, when **22** and **23** are mixed. When four equivalents of the template **21** are added, the equilibrium shifts towards **24** and the 1 : 4 complex is the only species observable by ^{1}H-NMR. The cage **24** remains intact after the templates have been removed by acidification and solvent extraction, and is kinetically stable at room temperature.

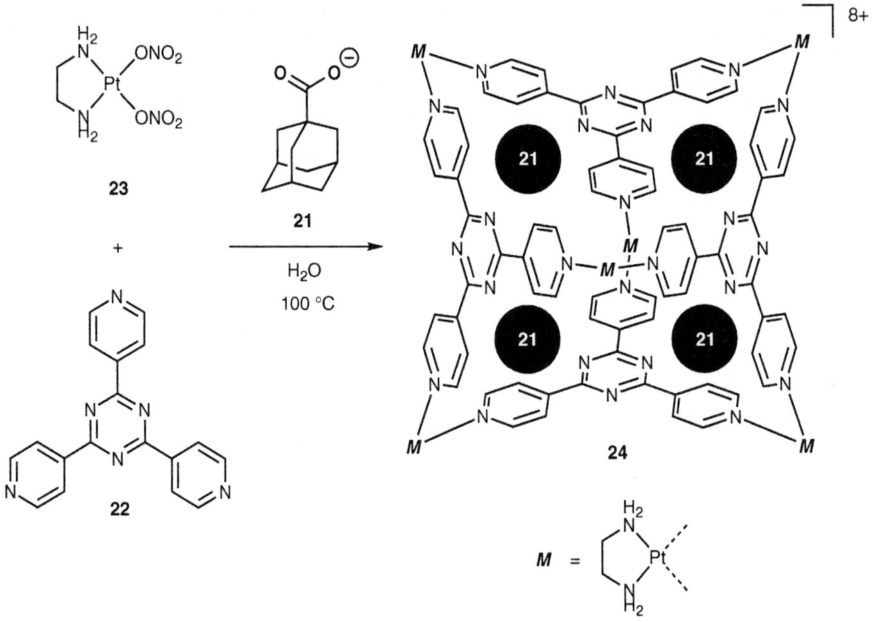

Scheme 1-6   Four adamantanecarboxylates template the formation of cage **24**.

1.3.2   Covalent and Non-covalent Template–Substrate Interactions

Templates can be classified according to the strength of the template–substrate interaction and according to the strength of the subsequent substrate–substrate bonding (Figure 1-3) [16]. Classical templates bind non-covalently to their substrates, and assist in the formation of covalent bonds. A closely related class includes templates like Feldman's [32], top right in Figure 1-3 (discussed in Chapter 8); the only difference between this and the classical templates is the covalent nature of the bond between template and substrates. When the template is a linear chain, fully covalent systems of this type are called tethered reactions (see Chapters 6, 7, and 10). The tether brings groups into close proximity, thus enhancing reactivity. Many of the common reactions of organic synthesis can be viewed as examples of covalent/covalent templating. For example, in Scheme 1-7, the boron atom holds the three alkyl groups together so that they can bond to the carbon of potassium cyanide [33]. The Fischer indole synthesis is another classical example.

The other examples shown in Figure 1-3 involve the template induced formation of non-covalent bonds; this type of process is best described as self-assembly. For example, in Mutter's artificial proteins, several amphiphilic helices are covalently attached to a template, so determining the folding topology of the final globular structure [34]. The self-assembly of surface monolayers is another good example: when aliphatic carboxylic acids form monolayers on a water surface, or long-chain thiols self-assemble on gold,

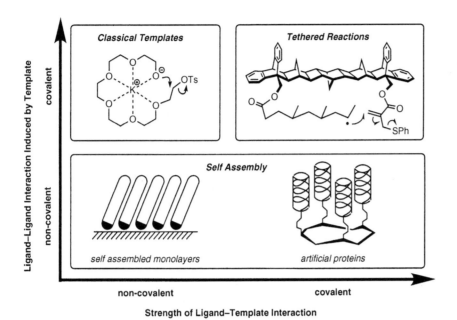

Figure 1-3   Templating and self-assembly.

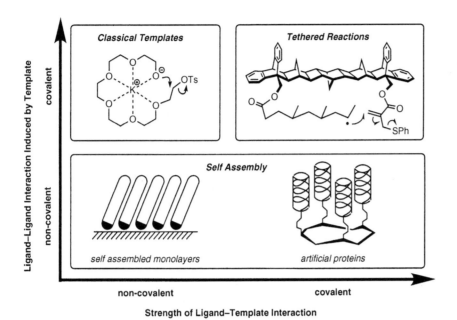

Scheme 1-7   Boron acts as a template, holding the three alkyl groups together, facilitating the reaction with the carbon atom of potassium cyanide.

the surface holds the alkyl chains in the correct orientation for optimal van der Waals interaction between the hydrocarbon chains [35]. A non-covalent interaction between the template and the substrates induces a second non-covalent interaction. These examples show the close relationship between templating and self-assembly [36]. The emphasis of this book is on the use of templates for creating covalent bonds.

1.3.3   Topology of Reaction

Templates can be categorized as "cyclization", "linear", or "interweaving", according to their topology of operation, as shown in Figure 1-4 [16].

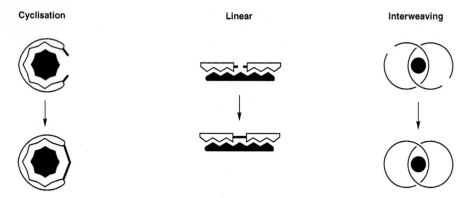

Figure 1-4   Classification of templates according to their topology of operation.

1.3.3.1   Cyclization templates

The simplest type of template directed cyclization is metal cation induced macrocyclization, as discussed in Section 1.1. Mandolini and co-workers [37] have made a detailed
study of the influence of alkali and alkaline earth metal cations on the cyclization of
2-hydroxyphenyl-3,6,9,12-tetraoxa-14-bromotetradecyl ether **28** in methanol and in
dimethyl sulfoxide (Scheme 1-8). They studied different sizes of macrocycle and metal
cation to determine the most effective combination for cyclization; the results from this
study will be summarized in Section 1.4.2.3.

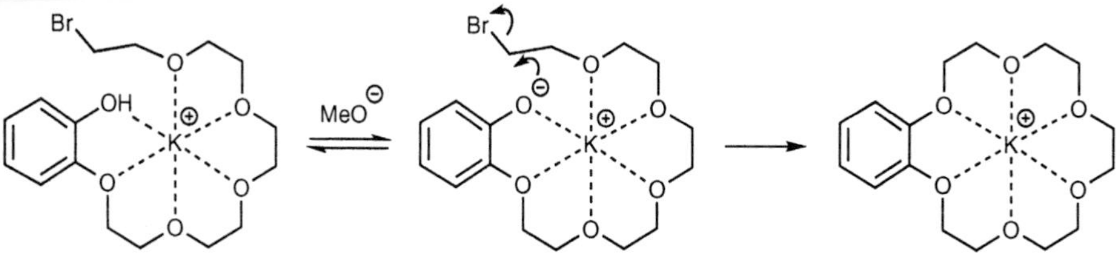

$28 \cdot K^+$ $29 \cdot K^+$ $30 \cdot K^+$

Scheme 1-8   Metal cation template directed formation of benzo[18]crown-6.

Macrocyclization templates are not restricted to metal cations but also include hydrophobic organic units, such as the adamantanecarboxylate templated formation of cage **24**
described by Fujita (Scheme 1-6) [31]. Other examples, which will be described in later
chapters, include Cram and co-workers' use of solvents to template formation of carceplexes: solvents such as dimethylacetamide were found to stabilize the charged intermediate formed during carceplex synthesis. The empty carcerands were never formed

because this would require the intermediate to be unsolvated [38]. Organic charge transfer and $\pi-\pi$ interactions have also been used in template directed synthesis, as in the preparation of Stoddart's [39a−c] macrocycle **35** (Scheme 1-9) (see Chapter 3). The initial dication **31** reacts with one end of the 1,4-bis(bromomethyl)benzene **32** to yield a tricationic intermediate **34**. $\pi-\pi$ Interactions between the electron-deficient 4,4′-bipyridinium unit and the electron rich hydroquinone of the template **33**, and hydrogen bonding between the polyether chains of the template and CH_2 groups in **34**, result in rapid formation of a 1:1 complex. The complex induces the reactive ends of the trication to come into close proximity, enhancing intramolecular coupling [39b]. The template can be removed by solvent extraction, providing an efficient route to the free cyclophane **35**. This system has been extended to achieve the templated synthesis of catenanes and rotaxanes, as will be mentioned later.

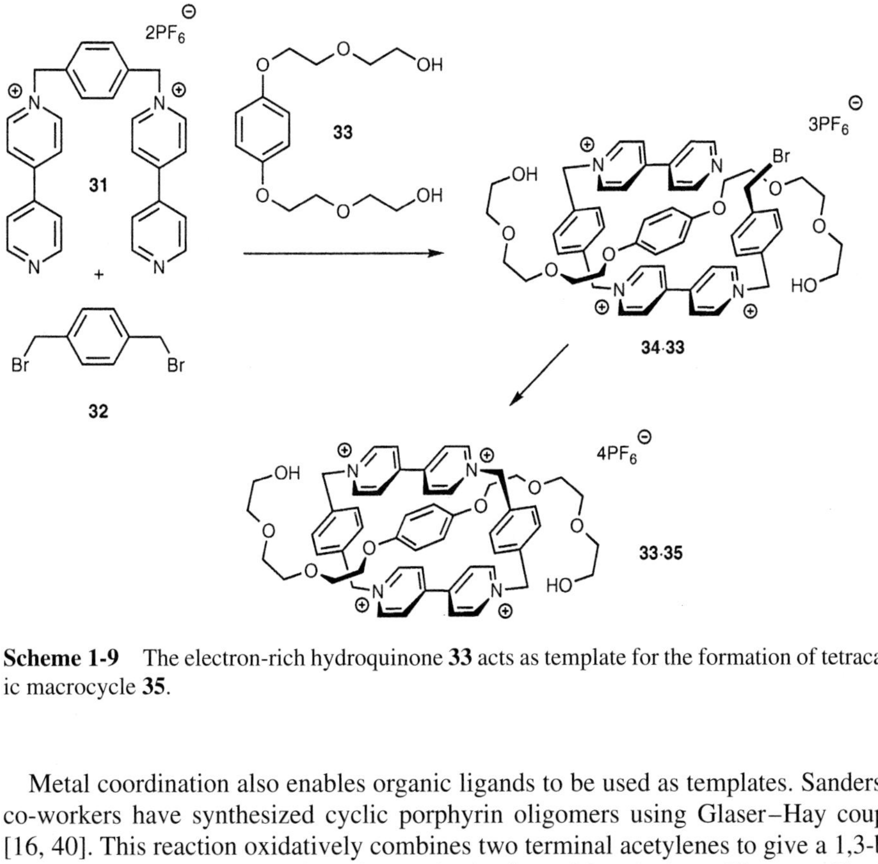

Scheme 1-9 The electron-rich hydroquinone **33** acts as template for the formation of tetracationic macrocycle **35**.

Metal coordination also enables organic ligands to be used as templates. Sanders and co-workers have synthesized cyclic porphyrin oligomers using Glaser−Hay coupling [16, 40]. This reaction oxidatively combines two terminal acetylenes to give a 1,3-butadiyne link. The syntheses of cyclic porphyrin dimer **38** and trimer **39** were effectively templated by 4,4′-bipyridine **37** and tripyridyltriazine **22**, respectively (Scheme 1-10).

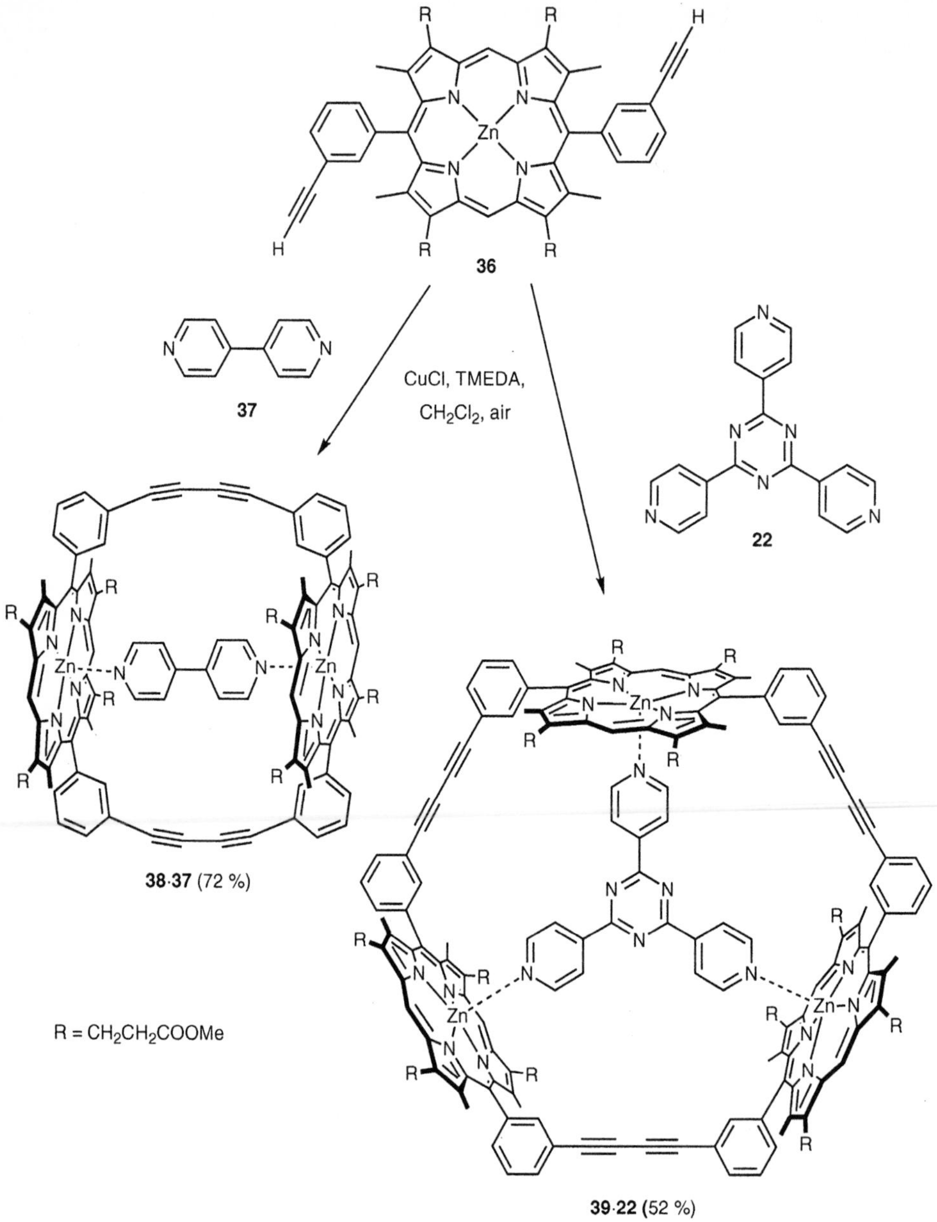

Scheme 1-10   Template directed synthesis of cyclic porphyrin dimer **38** and cyclic porphyrin trimer **39**.

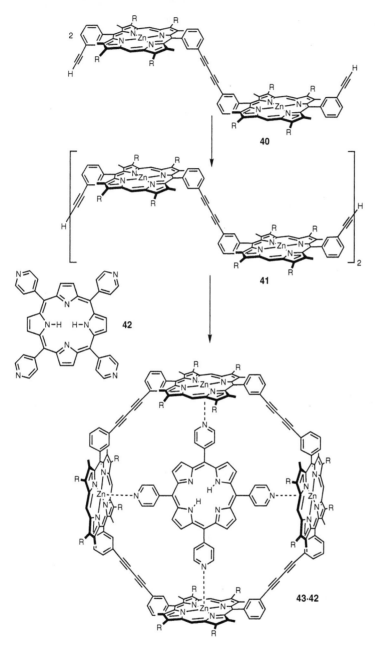

Scheme 1-11 Tetrapyridylporphyrin **42** templated formation of cyclic porphyrin tetramer **43**.

The first step in both templated reactions is the combination of two porphyrin units **36** to yield a linear porphyrin dimer **40**. This linear dimer binds the 4,4′-bipyridine **37** template and adopts a conformation in which intramolecular cyclization is strongly favored. Conversely, when the tridentate ligand **22** binds the linear dimer, it forces it into a conformation in which cyclization is disfavored; the linear dimer then undergoes intermolecular reaction to yield linear trimer, which cyclizes to cyclic trimer **39**. The tridentate ligand **22** performs a largely preventative, negative, role in this trimer synthesis, by inhibiting linear dimer from cyclizing – it seems to be slightly too small to enhance cyclization of the linear trimer. The oligopyridine templates can be easily removed from these cyclic porphyrin oligomers by treatment with acid.

Tetrapyridylporphyrin **42** templates the cyclization of linear porphyrin tetramer **41** and also the conversion of linear dimer **40** to cyclic tetramer **43** (Scheme 1-11). In this latter case, the template binds the linear tetramer formed in the reaction, inducing a conformation favoring cyclization. If the porphyrins are converted to dioxoporphyrins [41], or ruthenium(II)carbonyl porphyrins [42], the specificity of the templates is increased because the binding becomes stronger: the monomeric dioxoporphyrins preassemble around the templates before coupling. In the case of the ruthenium porphyrins it is less clear whether the mechanism is one of preassembly followed by coupling or coupling followed by strong binding of linear intermediates in favorable conformations for intramolecular cyclization. The preassembly route is not significant with zinc porphyrins because the stability constant for the porphyrin monomer–pyridine complex is too low. The drawback of strong binding in the case of dioxo and ruthenium porphyrins is that the template is so strongly bound by the product that the free macrocycle cannot readily be released.

1.3.3.2 Linear templates

Linear templates enhance reaction between two bound substrates rather than between two ends of the same substrate. The classical examples of this type of templating are replication of DNA and synthesis of RNA. In both cases, a single strand of DNA acts as a template, the base sequence on the template strand being complementary to the sequence produced on the new strand. The use of an oligonucleotide to template the formation of an imine, by Lynn and co-workers [30], has already been mentioned (Section 1.3.1, Scheme 1-5). One of the first examples of a non-nucleotide based organic linear template was developed by Kelly et al. [43]. The template **44** binds the two substrate molecules in a transient ternary complex **45·44·46**, enhancing the bimolecular reaction (Scheme 1-12). Following on from this system, Kelly developed a template with two different binding sites so that inactive ternary complexes, where two of the same substrates bind to the template, are disfavored. Thus the effectiveness of the template was enhanced.

Mock et al. [44] have shown that cucurbituril **48** catalyzes the 1,3-dipolar cycloaddition shown in Scheme 1-13. Here, cucurbituril acts as a linear template; note that "linear" refers to the reaction topology and not to the shape of the template. **49** and **50** bind inside the cavity of **48** by hydrogen bonding between the ammonium groups and the rims of the cucurbituril. The reaction between **49** and **50** is made faster and more regiospecific; clean formation of **51** occurs.

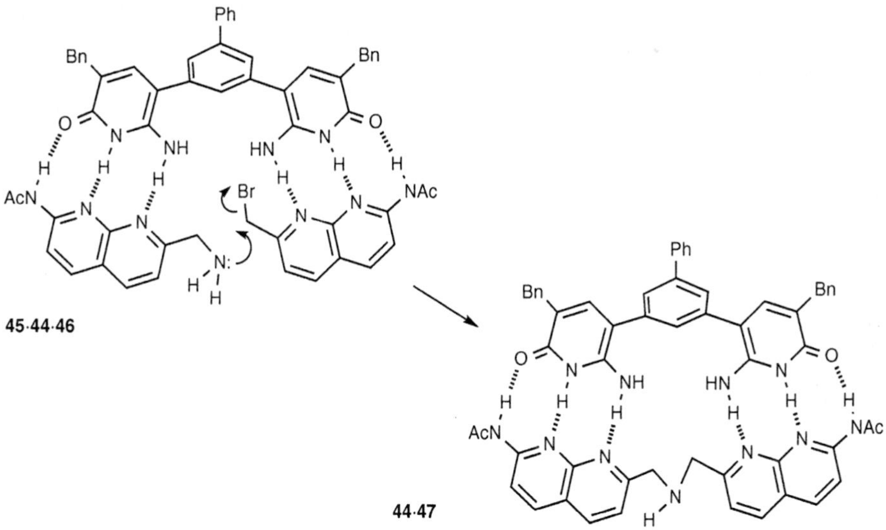

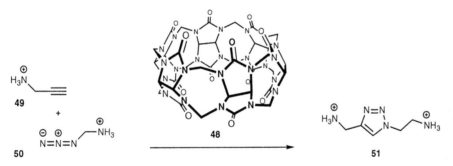

Scheme 1-12   The template **44** simultaneously binds two substrates (**45** and **46**) in a ternary complex, accelerating reaction between them.

Scheme 1-13   Cucurbituril catalyzed regiospecific formation of **51** in aqueous formic acid.

Diels–Alder reactions have also been accelerated inside the cavities of cyclic porphyrin trimers [45]. For example, diene **52** and dienophile **53** react selectively to give the *exo* adduct **54** (Scheme 1-14) inside the cavity of **39** [45a,b]. Cyclic porphyrin trimer **39** is an effective kinetic and thermodynamic template for this reversible Diels–Alder reaction, of which the stereochemical outcome may be reversed to give mainly the *endo* adduct if a smaller porphyrin trimer is used [45c].

Linear templates have potential for self-replication, since in principle the product can be identical to the template. Non-enzymic replication has been achieved with both nucleotide [46] and non-nucleotide [47] model systems. Scheme 1-15 outlines one of the first

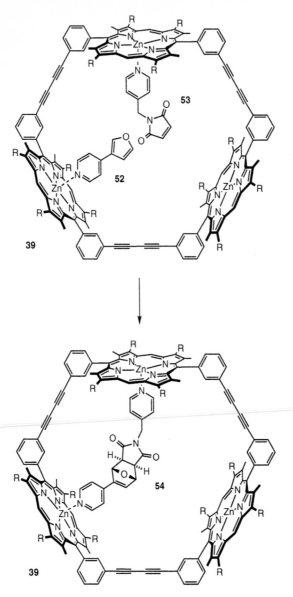

Scheme 1-14 Inside the porphyrin trimer **39** cavity, diene **52** and dienophile **53** react selectively to give the *exo* adduct **54**. The smaller version of trimer **39**, with two butadiyne links replaced with ethyne units, templates the reaction with the opposite stereochemistry, to give the *endo* isomer of **54**.

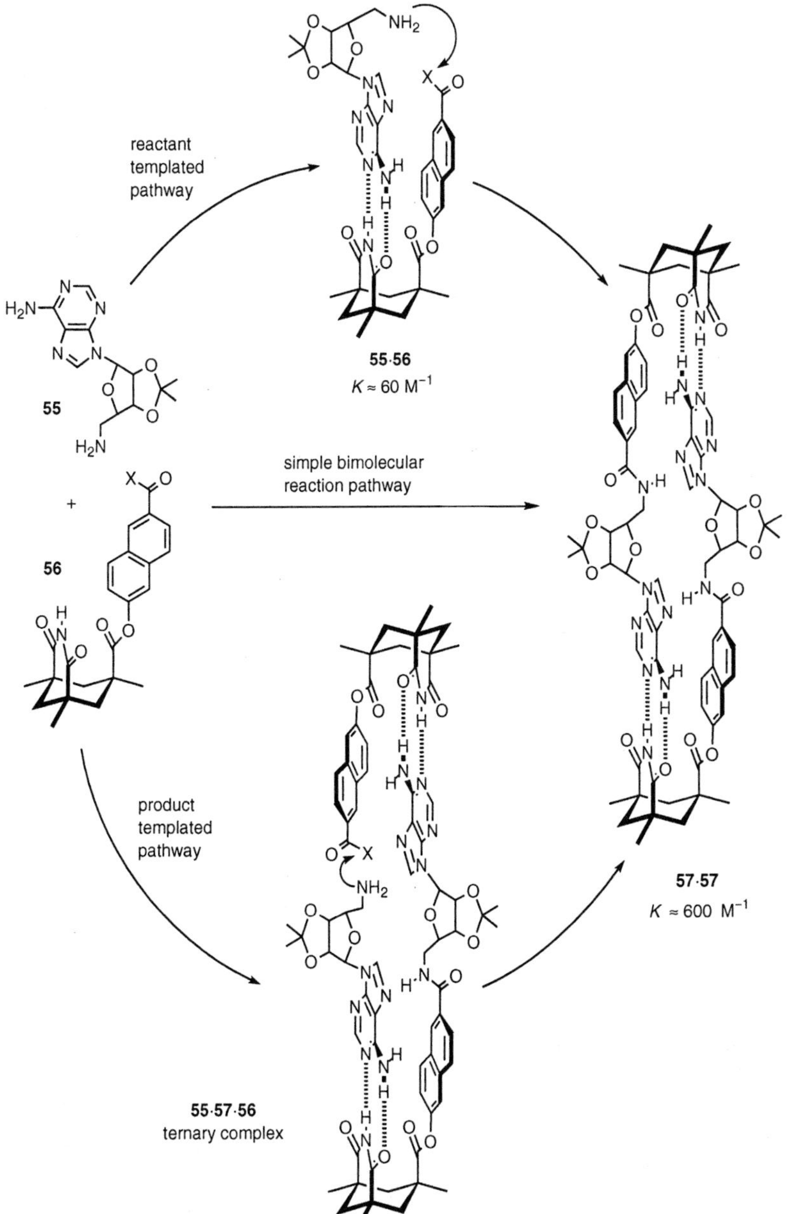

Scheme 1-15 Self-replicating system designed by Rebek and co-workers. The template **57** (X = OC$_6$F$_5$) binds to the two substrates **55** and **56**, generating a ternary complex **55·57·56**. Then **55** and **56** react to generate another molecule of template **57**. Stability constants are shown for the **55·56** and **57·57** complexes.

self-replicating system designed by Rebek, based on Kemp's triacid. The template **57** binds to the two substrates **55** and **56** generating a ternary complex **55·57·56**. Reaction between these template bound substrates is favored over reaction in free solution, to yield the dimerized template **57·57**. When the product dimer dissociates, two template molecules are released for further reaction. The reaction is autocatalytic; seeding the reaction mixture with its product leads to the expected rate enhancement. The initial rates of reaction are proportional to the square root of the product concentration. This "square root law", identified by von Kiedrowski [48] to characterize nucleic acid replicators, applies because the template exists largely as its dimer and only a very small amount is bound to two different substrates as the ternary complex ready for reaction. By changing the naphthalene unit for a biphenyl, the rate of product formation via the ternary complex may be increased. In Section 1.4.2.1, the efficiency of the system illustrated in Scheme 1-15 is quantified in terms of effective molarity. Template **57** may also be formed by a competing route: substrate **55** may bind to substrate **56** to form a bimolecular complex **55·56** which also brings the reactive groups into close proximity.

1.3.3.3 Interweaving templates

An interweaving template forces the substrates to form an interwoven topology; covalent modification locks this information permanently within the molecule – two or more rings being interlocked. Sauvage and co-workers' contributions to the templated synthesis of catenanes and knots have been described in excellent reviews [49]. His group and others have used Cu^I as a template to place phenanthroline ligands in an orthogonal arrangement around the metal. On the addition of a linking agent, the catenate is formed in good yield. The strategy can be extended to the synthesis of a trefoil knot **60·60** (Scheme 1-16). A thread containing two phenanthroline units can wrap around two Cu^I centers to yield a double helix. If the linking group can be persuaded to react with the correct ends, then a trefoil knot will form. This is not the only product, but the fact that even 3% of the trefoil knot forms is amazing. The copper ion can be removed with potassium cyanide to yield the metal-free product. Here the template provides topological information but it does not accelerate covalent bond formation; an interweaving template does not necessarily act as a cyclization template.

The templates used by Stoddart [39a, b, c] in the syntheses of catenanes and rotaxanes use $\pi-\pi$ interactions to enhance both cyclization and interweaving. The template directed cyclization (Scheme 1-9) results in rotaxane formation if the ends of the template **33** are made so bulky that it cannot escape from the macrocycle, and in catenane formation if the two ends of the template are joined to form a macrocycle interlocked with the new macrocycle. The template is an integral part of the final product [39a, d]. Cucurbituril **48** can act as both a linear template and a topological template in the formation of rotaxane, as discussed later (Scheme 1-20).

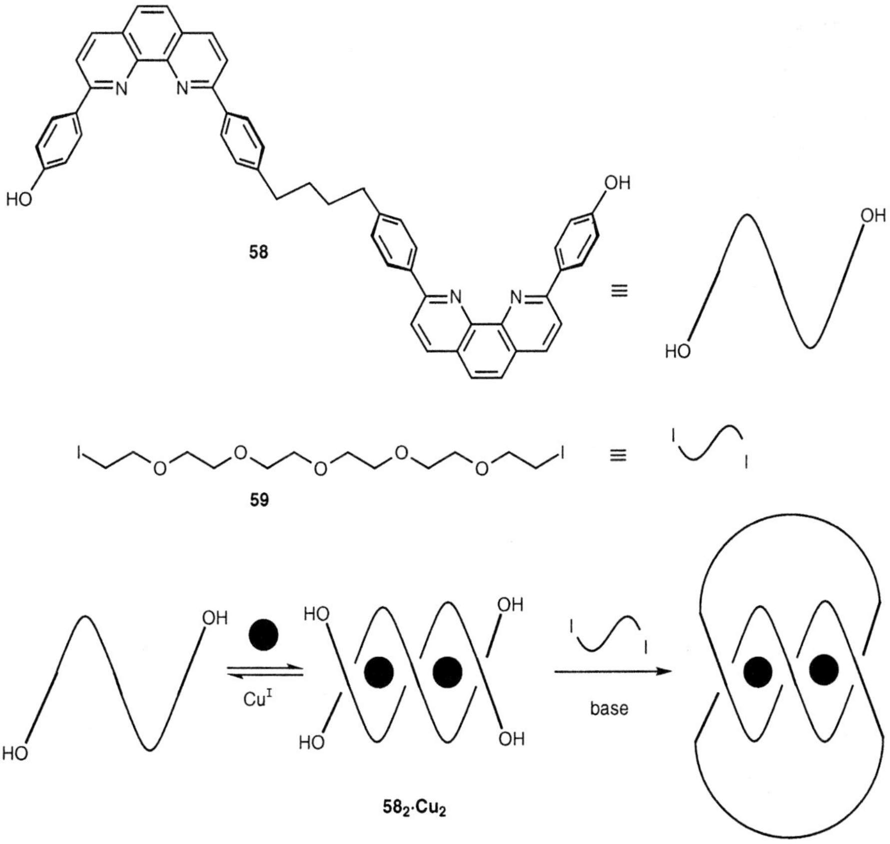

Scheme 1-16   Copper(I) as an interweaving template in the synthesis of a trefoil knot.

1.3.4   Scavenger Templates

Sanders and co-workers have shown that cyclization templates can be used to scavenge cyclizable material in a reaction mixture and thus facilitate the formation of linear oligomers [16, 50]. Consider the general Scheme 1-17. The aim of this reaction is to prepare product **D**. Dimerization of the partially deprotected intermediate **B** can only be carried out efficiently in the absence of fully deprotected molecules **C**, because the latter can couple with mono-deprotected material **B** to generate a new reactive oligomer and thus give a complex mixture. Conventionally, this problem is avoided by isolation and purification of **B** before coupling, but separation becomes increasingly difficult with longer chains. Scavenger templates overcome this problem by enforcing intramolecular reaction of doubly reactive molecules **C**; mono-reactive molecules **B** have no choice but to couple to each other to give **D**. The separation may be postponed until after coupling, when it becomes easier because the desired linear compound **D** is twice as massive as either the starting material **A** or the cyclic by-product **E**. The template does not promote the formation of the longer oligomer directly. This method has facilitated the preparation of a linear porphyrin octamer [50].

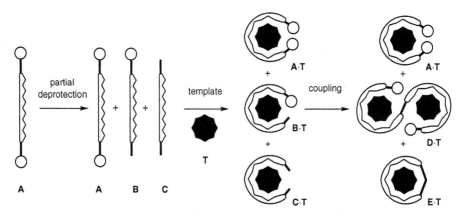

Scheme 1-17   A scavenger template induces molecules with two reactive ends to cyclize, so allowing efficient coupling of mono-protected material. The open circles represent protecting groups.

1.3.5   Negative Templates

The templates discussed so far favor reaction between bound substrates; they are positive templates. When a template disfavors reaction between bound substrates, it is called a "negative template" [16, 40, 50] (Scheme 1-18). A negative template disfavors the formation of a given product not by accelerating a competitive reaction but by specifically disfavoring the particular spatial ordering of atoms which would give that product. It holds the substrate in the wrong geometry for reaction; in Scheme 1-18 the negative template acts as a straitjacket, preventing the substrate from cyclizing. In ideal cases, the negative template may also have a dual role as a positive template in a subsequent step, or in an alternative reaction.

   In principle, any system which can be induced to adopt a particular conformation by a positive template can be prevented from doing so by a negative template. Negative templating of undesirable reactions is beneficial in systems which are under kinetic control, since there are many pathways down which such a reaction can proceed, only one of which leads to the desired product. A good example of negative templating is described by Sanders and co-workers [40, 50] (Scheme 1-11). The linear porphyrin dimer **40** can be prevented from cyclizing to cyclic dimer **38** by tetrapyridylporphyrin **42**. In the sec-

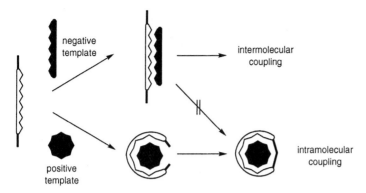

Scheme 1-18   Schematic illustration of positive and negative templating.

ond step, the template acts positively and cyclizes linear porphyrin tetramer **41** to cyclic tetramer **43**.

1.4   Measuring Template Effects

1.4.1   Qualitative Detection of Template Effects

Kinetic templates can be evaluated by measuring rates of reaction, but it is easier to analyze their effect on the yield of product. The practical test for a "template effect" is to see whether the template increases the yield. If the yield is unaffected by the template, this implies that the template is the wrong shape, or that substrate–template binding is too weak under the chosen conditions, or that the reaction conditions are too concentrated so that the substrate proximity induced by the template is insignificant, or that there are so few competing reactions that the template has no effect on the product distribution. If the yield of product is enhanced by the template, it is still necessary to demonstrate that the template is binding to, and controlling the spatial arrangement of, the starting materials or intermediates, rather than acting as a conventional catalyst or reagent.

In cyclizations, the main competing reaction which is reduced by the template is intermolecular coupling, giving linear oligomers, large rings, and polymers. In linear coupling reactions there is no built-in competition of this type, so it can be difficult to determine whether a linear template is operational without measuring rates: the product may be the same whether it is formed on the template or not. Sanders and co-workers have used a competition experiment to overcome this problem [40], enabling quantitative and qualitative information about linear template efficiency to be obtained, as outlined in Scheme 1-19. There are two different types of substrate: those which can bind to the template (shown in black) and those that cannot (shown in white). With an ideal template, all the substrate capable of binding to the template would bind and react quickly,

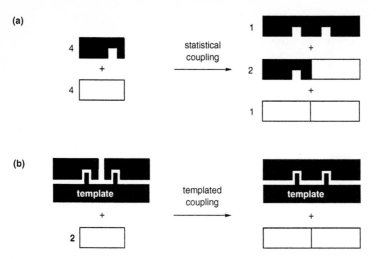

Scheme 1-19   (a) Statistical reaction and (b) coupling in the presence of a linear template.

leaving the unbound substrate in free solution to react more slowly (Scheme 1-19b). Thus, by comparing the product distributions with and without the template, it is possible to determine whether there is a template effect.

In general, product distribution analysis is only used to provide qualitative evidence for template effects. In the next section, we will discuss the parameters which are important in the quantification of template effects and show how choice of concentration regime can be used to optimize the effectiveness of a template.

1.4.2   Quantification of Kinetic Template Effects in Terms of Effective Molarity, Substrate Affinity, and Maximum Rate Enhancement

Kinetic templates accelerate reaction of bound substrates, which makes it tempting to quantify template effects in terms of "rate enhancement". In this section, we will show how this can be misleading because such rate enhancements are concentration dependent. We will elucidate the parameters which determine the rate enhancement achieved with a kinetic template, by analyzing the thermodynamic and kinetic behavior of simple theoretical models, and applying these models to published template systems. Our theoretical models are similar to the Michaelis–Menten analysis of enzyme catalyzed reactions [51], except that we assume there is no catalytic turnover. First, we consider linear templates, then cyclization templates. In general, the rate of reaction varies as the reaction proceeds; whenever we refer to rates in the following discussion, we mean initial rates.

1.4.2.1 Linear templates

Consider the situation where two molecules of a substrate **S**, of initial concentration $[S]_0$, react together to yield a product **P**:

$$S + S \xrightarrow{k_2} P \qquad (1\text{-}1)$$

The bimolecular rate constant for the untemplated reaction is k_2, so the initial rate of the reaction in the absence of a template is:

$$Rate_{\text{Untemp}} = k_2[S]_0^2 \qquad (1\text{-}2)$$

The reaction is facilitated by a linear template **T**. We will assume that this template has two independent, identical, binding sites; both bind substrates with the same microscopic binding constant K, to give a binary complex **S·T** and a ternary complex **S·T·S**.

$$S + T \xrightleftharpoons{2K} S \cdot T \qquad (1\text{-}3)$$

$$S + S + T \xrightleftharpoons{K^2} S \cdot T \cdot S \qquad (1\text{-}4)$$

The concentrations of unbound substrate $[S]$ and unbound template $[T]$ are related to the concentration of these complexes by the equilibrium constant.

$$2K = [S \cdot T]/([S][T]) \qquad (1\text{-}5)$$

$$K^2 = [S \cdot T \cdot S]/([S]^2[T]) \qquad (1\text{-}6)$$

If the total concentration of template is half the total concentration of substrate (i.e., $[S]_0 = 2[T]_0$), then the concentrations of **S**, **S·T**, and **S·T·S** vary with $[S]_0$, as illustrated by curves (a), (b), and (c), respectively, in Figure 1-5 (calculated for the arbitrary value $K = 10^3 \ \text{M}^{-1}$). The termolecular complex **S·T·S** becomes the main species in solution when $[S]_0 \gg 1/K$.

The unimolecular rate constant for conversion of the ternary complex **S·T·S** to the template product **T·P** is k_1.

$$S \cdot T \cdot S \xrightarrow{k_1} T \cdot P \qquad (1\text{-}7)$$

So the initial rate of intramolecular reaction within the termolecular **S·T·S** complex is:

$$Rate_{\text{Intra}} = k_1[S \cdot T \cdot S] = K^2 k_1[S]^2[T] \qquad (1\text{-}8)$$

Let us assume that when a substrate molecule binds a template, this has no effect on the rate of intermolecular reaction of that substrate with other free substrates, or with sub-

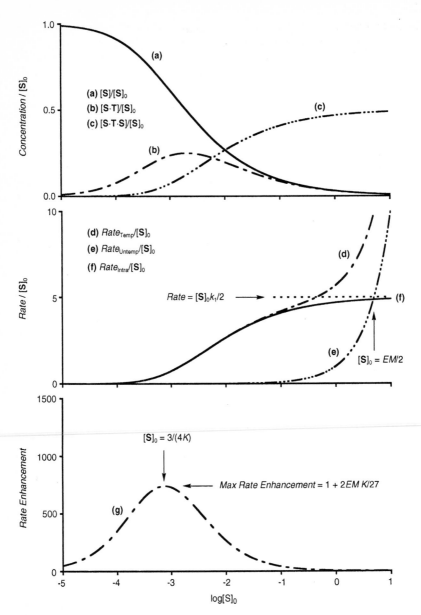

Figure 1-5 Curves (a)–(g) show how **[S]**, **[S·T]**, **[S·T·S]**, $Rate_{\text{Temp}}$, $Rate_{\text{Untemp}}$, $Rate_{\text{Intra}}$, and *Rate Enhancement* vary with $[\mathbf{S}]_0$. Note that concentrations and rates have been normalized by dividing by $[\mathbf{S}]_0$, and that rates are initial rates. (The following are arbitrarily assumed: $[\mathbf{S}]_0 = 2\,[\mathbf{T}]_0$, $K = 10^3\ \text{M}^{-1}$, $k_1 = 10\ \text{s}^{-1}$, and $k_2 = 1\ \text{s}^{-1}\ \text{M}^{-1}$; see Appendix 1a for details of these curves.)

strates bound to other templates. So the template has no effect on the rate of intermolecular reaction, and the rate of reaction in the presence of the template is:

$$Rate_{Temp} = Rate_{Untemp} + Rate_{Intra} \qquad (1\text{-}9)$$

$Rate_{Temp}$ (d), $Rate_{Untemp}$ (e) and $Rate_{Intra}$ (f) are plotted as functions of $[S]_0$ in Figure 1-5 (assuming $k_1 = 10 \text{ s}^{-1}$ and $k_2 = 1 \text{ s}^{-1} \text{ M}^{-1}$). The rate of the reaction in the absence of template (e) increases very steeply with concentration, because it is bimolecular, whereas the rate of reaction via $S \cdot T \cdot S$ (f) levels off where $[S]_0 \gg 1/K$, when all of the substrate is bound, and is asymptotic with a limiting value of $[S]_0 k_1/2$. There is a certain value of $[S]_0$ at which the reaction in the absence of template (e) becomes greater than the maximum possible rate of reaction in the ternary complex (i.e., when $k_2[S]_0 > k_1/2$). When $[S]_0 = k_1/(2k_2)$, the presence of the template simply doubles the overall rate of reaction (d) and as $[S]_0$ increases above this value, the effect of the template becomes insignificant. The critical concentration k_1/k_2 is a key quantity for defining the effectiveness of a template and it is known as the "effective molarity" of the system, *EM* [52].

$$EM = k_1/k_2 \qquad (1\text{-}10)$$

When $[S]_0 = EM/2$, the termolecular complex $S \cdot T \cdot S$ undergoes intermolecular and intramolecular reaction at equal rates. The factor of 2 results from the fact that we have a reaction between two identical substrate molecules.

Now let us see how the rate enhancement varies with $[S]_0$. The rate enhancement can be defined as the ratio of the initial rates of reaction in the presence and absence of the template, which gives:

$$Rate\ Enhancement = \frac{Rate_{Temp}}{Rate_{Untemp}}$$

$$= 1 + \frac{EM\ K^2 [S]^4}{2[S]_0^3} \qquad (1\text{-}11)$$

The rate enhancement, plotted as a function of $[S]_0$ in Figure 1-5 (g), reaches a maximum value of $1 + 2EM \cdot K/27$ when $[S]_0 = 3/(4K)$ (these are exact algebraic results, independent of the values of k_1, k_2, and K). Curve (g) shows that it is misleading to state a rate enhancement for a linear template at some arbitrary concentration $[S]_0$. The maximum rate enhancement occurs when the rates of the templated and untemplated reactions are both low; the intermolecular reaction is barely perceptible whereas the reaction in the ternary complex is proceeding several orders of magnitude faster; the observed rate enhancement is therefore very high.

From inspection of the curves in Figure 1-5, it is clear that there are two important parameters which need to be determined in order to evaluate a linear template system: the binding constant K and the effective molarity *EM*. At concentrations significantly lower than $3/(4K)$, there will be very little of the ternary complex $S \cdot T \cdot S$ present and so the template directed reaction will be insignificant. At concentrations higher than $3/(4K)$, intermolecular reaction starts to compete, even if the substrates are bound on a template in close proximity. It is interesting to note that the optimum concentration depends only on the binding constant K, whereas the magnitude of the peak rate enhancement depends on the product of the binding constant and the effective molarity.

Values of *EM* and *K* for five real linear template systems are compared in Figures 1-6–1-9 and Scheme 1-20. Effective molarities usually combine the uncertainties from a number of experimental binding constants and rate constants, so the values quoted here should be regarded as order of magnitude estimates. The situation which most resembles our theoretical model is that shown in Figure 1-6 [40], because the tetrapyridylporphyrin template **42** has two identical, essentially independent, binding sites for the linear dimer substrate **61**. Although the effective molarity of this system is low ($EM = 5 \times 10^{-4}$ M), this corresponds to a respectable calculated maximum rate enhancement because the binding constant K is so high. Note that if this reaction had been studied at concentrations higher than 10^{-3} M, no template effect would have been observed. The cyclic porphyrin trimer templated Diels–Alder reaction [45] (Figure 1-7) also fits reasonably well to our simple model because the binding sites are equivalent and independent; although

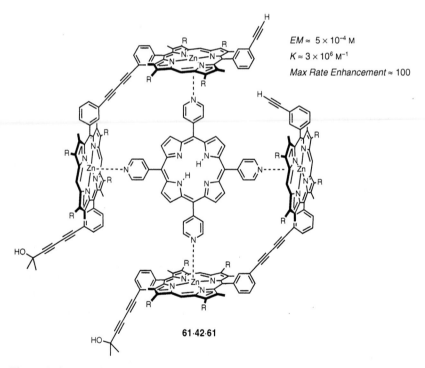

$EM \approx 5 \times 10^{-4}$ M

$K \approx 3 \times 10^6$ M^{-1}

Max Rate Enhancement $\approx$ 100

61·42·61

Figure 1-6   Template efficiency parameters for the tetrapyridylporphyrin templated Glaser coupling of two zinc porphyrin dimers, in dichloromethane.

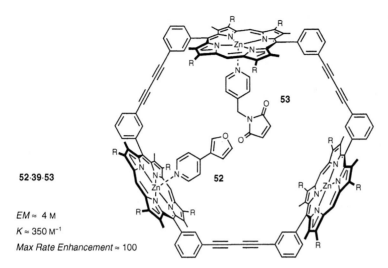

52·39·53

EM ≈ 4 M

K ≈ 350 M⁻¹

Max Rate Enhancement ≈ 100

Figure 1-7 Parameters for the porphyrin trimer templated *exo* Diels–Alder reaction, in 1,1,2,2-tetrachloroethane at 60 °C (cf. Scheme 1-14).

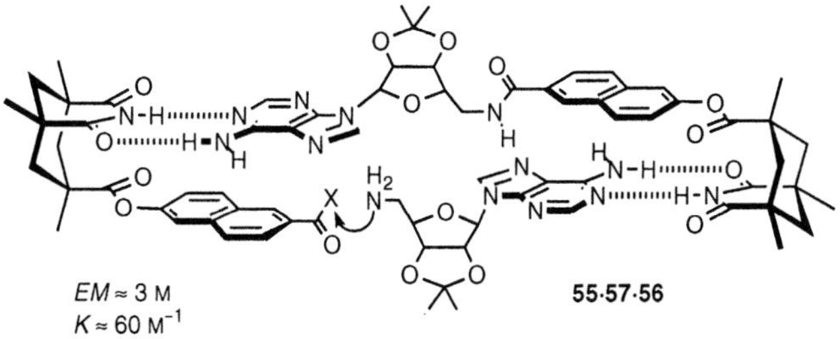

EM ≈ 3 M

K ≈ 60 M⁻¹

55·57·56

Figure 1-8 Effective molarity and stability constant for the ternary complex in Rebek's self-replicator (cf. Scheme 1-15).

there are two substrates (**52** and **53**), they have similar binding constants (500 M^{-1} and 250 M^{-1} respectively at 60 °C). The effective molarity is about 4 M, corresponding to a rate enhancement similar to that in the tetrapyridylporphyrin templated reaction. The rate enhancement becomes higher at lower temperatures (30 °C), due to increased binding, as expected from Equation (1-11). The cucurbituril templated cycloaddition (Scheme 1-20) has a much higher effective molarity (*EM* = 4×10^4 M). This pair of substrates was chosen for detailed kinetic analysis, rather than **49** and **50** (Scheme 1-13) because in this case catalytic turnover is prevented, because the product **48·64** is a rotaxane [44]. The substrates **62** and **63** have fairly similar affinities for the template (2.3×10^3 M^{-1} and ca. 7.9×10^2 M^{-1} respectively), but binding at the two sites shows a strong negative cooperativity. The stability constant of the ternary complex **62·48·63** is only 2.4×10^4 M^{-2}; for

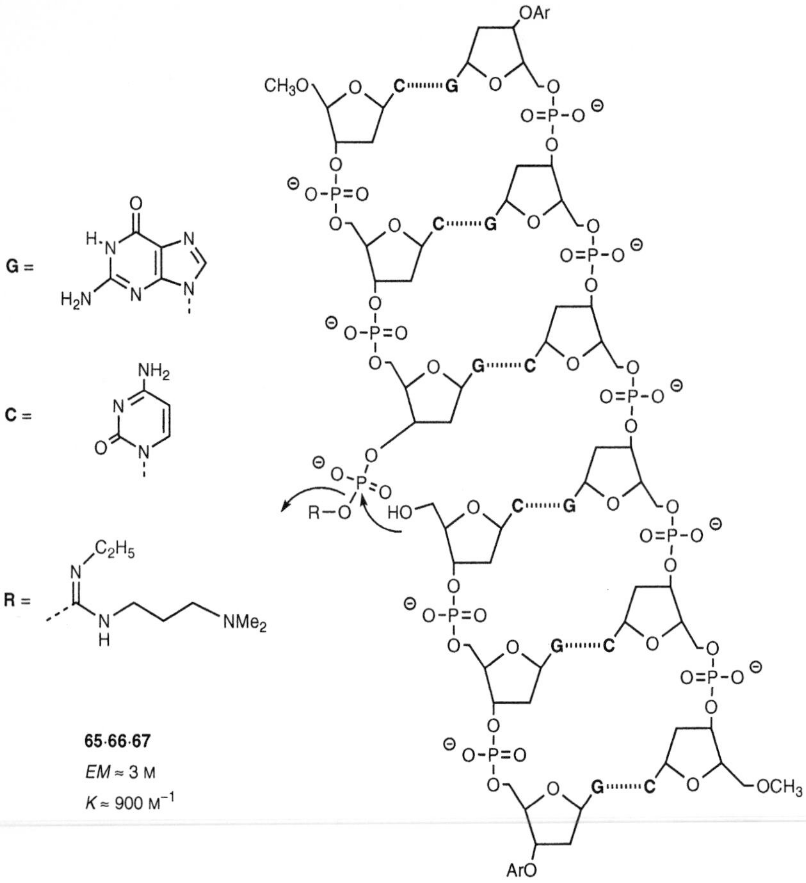

Figure 1-9   Effective molarity and stability constant for a hexadeoxynucleotide template at 0 °C.

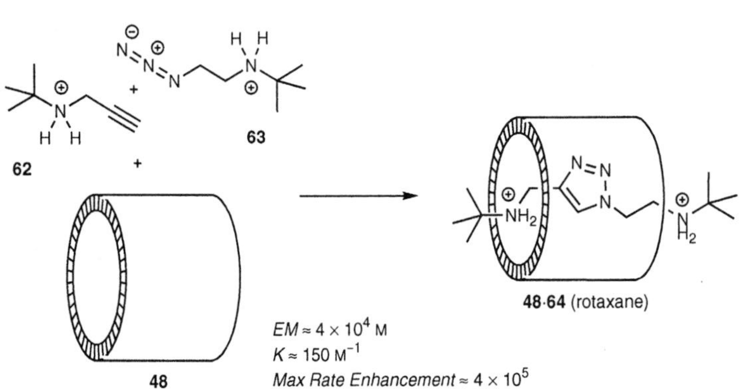

$EM \approx 4 \times 10^4$ M
$K \approx 150$ M^{-1}
Max Rate Enhancement $\approx 4 \times 10^5$

Scheme 1-20   Parameters for cucurbituril **48** acting as linear and topological templates.

our analysis, we need to take an effective binding constant, K, which is the square root of this, giving a maximum rate enhancement of 4×10^5.

The kinetic analysis of a self-replicating system is complicated because of template dimerization, so it is difficult to estimate maximum rate enhancements for these systems. However, we can still compare the effective molarities and binding constants of self-replicating linear templates, as illustrated by Figures 1-8 [47d] and 1-9. The substrate binding constant of von Kiedrowski's hexadeoxynucleotide template [46b] (Figure 1-9) has not been measured but can be estimated to be ca. $900 \, \text{M}^{-1}$ under the reaction conditions using the nearest-neighbor parameters for predicting DNA duplex stability [53].

1.4.2.2   Quantitative analysis of template effects in tethered reactions

Tethered reactions can be regarded as linear templated systems in which the template–substrate binding is infinitely strong. The "template efficiency" is therefore defined simply by the effective molarity for the intramolecular reaction; the rate enhancement becomes greater than 2 when $[S]_0 < EM$, and becomes infinite at infinite dilution.

1.4.2.3   Cyclization templates

The kinetics of templated cyclization reactions can be considered in the same way as for the linear coupling described in Section 1.4.2.1. There is one significant difference: cyclization reactions are intramolecular, so they have characteristic effective molarities even in the absence of a template.

Consider a substrate **S** which can cyclize intramolecularly to a give a product **P** or react intermolecularly to give a linear dimer **D** (which can go on to give longer linear and cyclic oligomers).

$$\mathbf{S} \xrightarrow{k_1} \mathbf{P} \tag{1-12}$$

$$\mathbf{S} + \mathbf{S} \xrightarrow{k_2} \mathbf{D} \tag{1-13}$$

In the absence of a template, the initial rates of untemplated cyclization ($Rate_{\text{UntempCyc}}$) and intermolecular coupling ($Rate_{\text{D}}$) are:

$$Rate_{\text{UntempCyc}} = k_1 [\mathbf{S}]_0 \tag{1-14}$$

$$Rate_{\text{D}} = k_2 [\mathbf{S}]_0^2 \tag{1-15}$$

The substrate can bind a template **T**, with binding constant K, to form a complex **S·T** which cyclizes more rapidly.

$$\mathbf{S} + \mathbf{T} \xrightleftharpoons{K} \mathbf{S \cdot T} \tag{1-16}$$

$$K = [\mathbf{S \cdot T}]/([\mathbf{S}][\mathbf{T}]) \tag{1-17}$$

The rate of cyclization of the substrate, when bound to the template, $Rate_{TempCyc}$, is given by Equation (1-19), where k_3 is the unimolecular rate constant for substrate cyclization in the presence of the template.

$$\mathbf{S \cdot T} \xrightarrow{k_3} \mathbf{P \cdot T} \tag{1-18}$$

$$Rate_{TempCyc} = k_3[\mathbf{S \cdot T}] \tag{1-19}$$

EM_{Temp} is the effective molarity for the cyclization of the substrate bound to the template and EM_{Untemp} is the effective molarity for the cyclization of the free substrate. EM_{Untemp} indicates the inherent tendency of the substrate to cyclize.

$$EM_{Temp} = k_3/k_2 \tag{1-20}$$

$$EM_{Untemp} = k_1/k_2 \tag{1-21}$$

If we assume that template binding has no effect on the rate of intermolecular coupling (as we assumed for linear templates), then the overall initial rate of product formation in the presence of the template is:

$$Rate_{Temp} = Rate_{UntempCyc} + Rate_{TempCyc} \tag{1-22}$$

This allows the following expression to be derived for the *Rate Enhancement*:

$$Rate\ Enhancement = \frac{Rate_{Temp}}{Rate_{UntempCyc}} \tag{1-23}$$

$$= \frac{[\mathbf{S}]}{[\mathbf{S}]_0} + \frac{EM_{Temp}\,K[\mathbf{S}]^2}{EM_{Untemp}[\mathbf{S}]_0} \tag{1-24}$$

In Figure 1-10, curve (h) shows how the rate enhancement varies as a function of $[\mathbf{S}]_0$ (for $[\mathbf{S}]_0 = [\mathbf{T}]_0$, $K = 10^3$ M^{-1}, $EM_{Temp} = 0.1$ M and $EM_{Untemp} = 10^{-4}$ M). *Rate Enhancement* is concentration dependent and is therefore a misleading quantity unless the concentration at which it is measured is quoted together with the binding constant K. At high concentrations, when the substrate is all bound to the template, the rate enhancement tends towards the ratio of the effective molarities EM_{Temp}/EM_{Untemp}; when $[\mathbf{S}]_0 \ll 1/K$, the *Rate Enhancement* tends towards unity (no effect) because there is no binding. Curve (i) shows the fraction of $\mathbf{S}$ which cyclizes in the absence of template and curve (j) the fraction of $\mathbf{S}$ which cyclizes in the presence of the template; at low concentrations the inherent tendency of the substrate to cyclize means that the fraction cyclizing approaches unity, whereas at high concentrations, intermolecular reaction competes with cyclization. Subtraction of curve (i) from (j) results in curve (k) which illustrates the concentration regime over which the template is effective, which is $EM_{Untemp} < [\mathbf{S}]_0 < EM_{Temp}$.

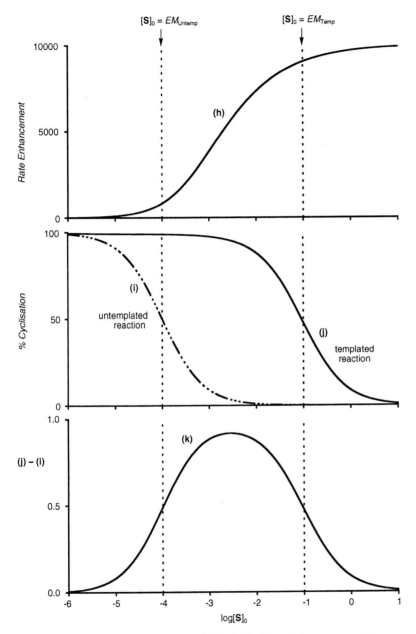

Figure 1-10   Curve (h) shows the variation in the *Rate Enhancement* with respect to $\log[S]_0$ (for $K = 10^3$ M^{-1}, $EM_{\text{Temp}} = 0.1$ M and $EM_{\text{Untemp}} = 10^{-4}$ M). Curve (i) shows the fraction of **S** which cyclizes in the absence of template and curve (j) the fraction of **S** which cyclizes in the presence of the template. Subtraction of curve (i) from (j) results in curve (k). (See Appendix 1b for details.)

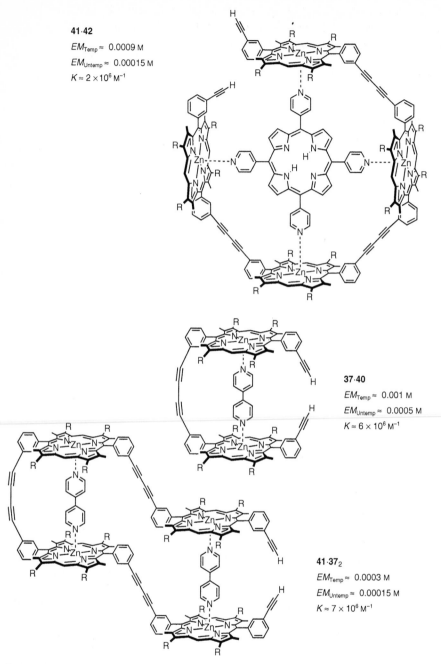

41·42

$EM_{Temp} \approx 0.0009$ M

$EM_{Untemp} \approx 0.00015$ M

$K \approx 2 \times 10^6$ M^{-1}

37·40

$EM_{Temp} \approx 0.001$ M

$EM_{Untemp} \approx 0.0005$ M

$K \approx 6 \times 10^6$ M^{-1}

41·37$_2$

$EM_{Temp} \approx 0.0003$ M

$EM_{Untemp} \approx 0.00015$ M

$K \approx 7 \times 10^6$ M^{-1}

Figure 1-11 Effective molarities and binding constants for cyclization of linear porphyrin oligomers, by Glaser coupling in dichloromethane.

At concentrations higher than EM_{Temp}, intermolecular reaction is favored even in the presence of template, and at concentrations lower than EM_{Untemp} the substrate cyclizes so efficiently that the template has little effect.

Table 1-1 provides some examples of cyclization effective molarities for the formation of benzocrown ethers in the presence of a variety of group I metal cation templates [37a,b]. The highest effective molarity in this series is achieved by the potassium cation when it templates the formation of benzo-18-crown-6 ($y = 5$, Scheme 1-8); here the EM_{Temp} of 123 M is 1500 times higher than EM_{Untemp}. Figure 1-11 gives effective molarities for the cyclization of some linear porphyrin oligomers by Glaser coupling [40]. These effective molarities are all very low but cyclization is significantly favored by the template and the strong binding constants enable coupling to be carried out efficiently at very low concentration.

Table 1-1 Effective molarities (M) for various metal cations used as templates for the cyclization of **68** ($y = 3, 4, 5, 6, 9,$ and 15) in Me_2SO at 25 °C.

68 (OCH₂CH₂)yBr	EM_{Temp}				EM_{Untemp}
y	Na⁺	K⁺	Rb⁺	Cs⁺	
3	1.8	1	0.7	0.6	0.08
4	26	10	6	4	0.09
5	14	123	48	22	0.08
6	0.5	8	20	25	0.07
9	0.1	0.9	2.1	1.2	0.04
15	0.04	0.1	0.2	0.2	0.02

1.5 Conclusion

This chapter illustrates how templates have been used to aid organic synthesis and attempts to refine and expand the definition of a molecular template. The later chapters in this book demonstrate that templates are emerging as powerful tools in many areas of molecular construction. We have highlighted the key parameters required to evaluate template effects quantitatively. We hope this will encourage pioneers in the field to calibrate new systems, so that we can gauge their progress as they escalate to higher levels of efficiency and sophistication.

Appendix 1a: Equations for Figure 1-5

$$[S]_0 = [S] + [S \cdot T] + 2[S \cdot T \cdot S] \tag{1-25}$$

$$[T]_0 = [T] + [S \cdot T] + [S \cdot T \cdot S] \tag{1-26}$$

Combination of mass balance Equations (1-25) and (1-26) with the thermodynamic Equations (1-5) and (1-6) gives Equations (1-27), (1-28), and (1-29). These functions are illustrated in Figure 1-5 (a) to (c).

$$\text{(a)} \quad [S] \quad = \frac{-1 + \sqrt{1 + 4K[S]_0}}{2K} \tag{1-27}$$

$$\text{(b)} \quad [S \cdot T] \quad = (2K[S]^3)/[S]_0^2 \tag{1-28}$$

$$\text{(c)} \quad [S \cdot T \cdot S] = (K^2[S]^4)/[S]_0^2 \tag{1-29}$$

$$\quad [T] \quad = [S]^2/(2[S]_0) \tag{1-30}$$

From Equations (1-9) and (1-30), Equations (1-31), (1-2), and (1-32) for $Rate_{Temp}$, $Rate_{Untemp}$, and $Rate_{Intra}$ can be obtained. These functions are illustrated in Figure 1-5 (d) to (g).

$$\text{(d)} \quad Rate_{Temp} \quad = k_2[S]_0^2 + \frac{k_1 K^2[S]^4}{2[S]_0} \tag{1-31}$$

$$\text{(e)} \quad Rate_{Untemp} = k_2[S]_0^2 \tag{1-2}$$

$$\text{(f)} \quad Rate_{Intra} \quad = \frac{k_1 K^2[S]^4}{2[S]_0} \tag{1-32}$$

Figure 1-5 (g) shows the concentration dependence of the *Rate Enhancement*.

$$\text{(g)} \quad Rate\ Enhancement = 1 + \frac{EM\ K^2[S]^4}{2[S]_0^3} \tag{1-11}$$

Appendix 1b: Equations for Figure 1-10

$$[T]_0 = [T] + [S \cdot T] \tag{1-33}$$

$$[S]_0 = [S] + [S \cdot T] \tag{1-34}$$

Combination of mass balance Equations (1-33) and (1-34) with the thermodynamic Equation (1-17) gives:

$$[S] = \frac{-1 + \sqrt{1 + 4K[S]_0}}{2K} \qquad (1\text{-}35)$$

The functions (1-24), (1-36), and (1-37) are illustrated in Figure 1-10 (h) to (j). They describe the concentration dependence of the effectiveness of a cyclization template.

(h) *Rate Enhancement*
$$= \frac{Rate_{Temp}}{Rate_{UntempCyc}} \qquad (1\text{-}23)$$

$$= \frac{[S]}{[S]_0} + \frac{EM_{Temp} K[S]^2}{EM_{Untemp}[S]_0} \qquad (1\text{-}24)$$

(i) *Fraction Cyclization*$_{Untemp}$
$$= \frac{Rate_{UntempCyc}}{Rate_{UntempCyc} + Rate_D}$$

$$= \frac{EM_{Untemp}}{EM_{Untemp} + [S]_0} \qquad (1\text{-}36)$$

(j) *Fraction Cyclization*$_{Temp}$
$$= \frac{Rate_{TempCyc} + Rate_{UntempCyc}}{Rate_{TempCyc} + Rate_{UntempCyc} + Rate_D}$$

$$= \frac{EM_{Untemp}[S] + K \cdot EM_{Temp}[S]^2}{[S]_0^2 + EM_{Untemp}[S] + K \cdot EM_{Temp}[S]^2} \qquad (1\text{-}37)$$

References

[1] (a) J. Turner (Ed.), *The Dictionary of Art*, Macmillan, London, **1996**, *Vol. 30*; (b) F. S. Merritt (Ed.), *Building Design and Construction Handbook,* 4th edn., McGraw-Hill, New York, **1982**.

[2] A. Didron, *Ann. Archéol.* **1845**, *2*, 150.

[3] (a) F. B. Andrews, *The Mediaeval Builder and His Methods,* Barnes and Noble, New York, **1993**; (b) L. F. Salzman, *Building in England Down to 1540: a Documentary History*, Oxford University Press, Oxford, **1952**.

[4] J. D. Watson, F. H. C. Crick, *Nature (London)* **1953**, *171*, 737–738.

[5] A. R. Todd (Ed.), *Perspectives in Organic Chemistry*; Interscience, London, **1956**, p. 263.

[6] F. Seidel, *Chem. Ber.* **1926**, *59B*, 1894–1908. This reaction had been previously investigated by T. Posner, *Berichte* **1898**, 656–660, but from the limited characterization data it seems that the macrocycle 2 was not formed, maybe because the reaction mixture contained water.

[7]   (a) G. A. Melson, D. H. Busch, *Proc. Chem. Soc., London*, **1963**, 223–224; (b) D. H. Busch, *Rec. Chem. Prog.* **1964**, *25*, 106–126.

[8]   (a) R. P. Linstead, *J. Chem. Soc.* **1934**, 1016–1017; (b) G. T. Byrne, R. P. Linstead, A. R. Lowe, *J. Chem. Soc.* **1934**, 1017–1022.

[9]   N. F. Curtis, *Coord. Chem. Rev.* **1968**, *3*, 3–47.

[10] (a) N. F. Curtis, R. W. Hay, *J. Chem. Soc., Chem. Commun.* **1966**, 524–525; (b) N. F. Curtis, D. A. House, *Chem. Ind.* **1961**, 1708–1709.

[11] C. J. Pedersen, *Angew. Chem., Int. Ed. Engl.* **1988**, *27*, 1021–1027.

[12] (a) D. H. Busch, J. A. Burke Jr., D. C. Jicha, M. C. Thompson, M. L. Morris, *Adv. Chem. Ser.* **1962**, *37*, 125–142; (b) *Chem. Eng. News* **1962**, *40*, 57; (c) D. H. Busch, *Adv. Chem. Ser.* **1963**, *37*, 1–18.

[13] (a) M. C. Thompson, D. H. Busch, *J. Am. Chem. Soc.* **1962**, *84*, 1762–1763; (b) M. C. Thompson, D. H. Busch, *J. Am. Chem. Soc.* **1964**, *86*, 3651–3656.

[14] D. H. Busch, A. L. Vance, A. G. Kolchinksii, in *Comprehensive Supramolecular Chemistry*, *Vol. 9*, Eds.: J.-P. Sauvage, M. W. Hosseini, Elsevier, Oxford, **1996**, pp. 1–42.

[15] R. Hoss, F. Vögtle, *Angew. Chem., Int. Ed. Engl.* **1994**, *33*, 375–384.

[16] S. Anderson, H. L. Anderson, J. K. M. Sanders, *Acc. Chem. Res.* **1993**, *26*, 469–475.

[17] J. F. Stoddart, in *Frontiers in Supramolecular Organic Chemistry and Photochemistry*, Eds.: H.-J. Schneider, H. Dürr, VCH, Weinheim, **1991**, pp. 251–263.

[18] B. Dietrich, P. Viout, J.-M. Lehn, *Macrocyclic Chemistry*, VCH, Weinheim, **1993**.

[19] L. F. Lindoy, *The Chemistry of Macrocyclic Ligand Complexes*, Cambridge University Press, Cambridge, **1989**.

[20] E. C. Constable, *Metals and Ligand Reactivity*, Ellis Horwood, Chichester, **1990**.

[21] (a) D. H. Busch, *J. Inclusion Phenom.* **1992**, *12*, 389–395; (b) D. H. Busch, N. A. Stephenson, *Coord. Chem. Rev.* **1990**, *100*, 119–154.

[22] (a) H. Gies, B. Marler, *Zeolites* **1992**, *12*, 42–49; (b) T. Loiseau, G. Férey, *J. Chem. Soc., Chem. Commun.* **1992**, 1197–1198.

[23] J. S. Beck, J. C. Vartuli, W. J. Roth, M. E. Leonowicz, C. T. Kresge, K. D. Schmitt, C. T.-W. Chu, D. H. Olson, E. W. Sheppard, S. B. McCullen, J. B. Higgins, J. L. Schlenker, *J. Am. Chem. Soc.* **1992**, *114*, 10834–10843.

[24] (a) C. T. Kresge, M. E. Leonowicz, W. J. Roth, J. C. Vartuli, J. S. Beck, *Nature (London)* **1992**, *359*, 710–712; (b) C. R. Martin, *Acc. Chem. Res.* **1995**, *28*, 61–68.

[25] Y. A. Vlasov, N. Yao, D. J. Norris, *Adv. Mater.* **1999**, *11*, 165–169.

[26] M. C. Thompson, D. H. Busch, *J. Am. Chem. Soc.* **1964**, *86*, 213–217.

[27] (a) J. D. Curry, D. H. Busch, *J. Am. Chem. Soc.* **1964**, *86*, 592–594; (b) G. A. Melson, D. H. Busch, *J. Am. Chem. Soc.* **1965**, *87*, 1706–1710.

[28] E. L. Blinn, D. H. Busch, *Inorg. Chem.* **1968**, *7*, 820–824.

[29] (a) A. Meller, A. Ossko, *Monatsh. Chem.* **1972**, *103*, 150–155; (b) T. J. Marks, D. R. Stojakovic, *J. Am. Chem. Soc.* **1978**, *100*, 1695–1705; (c) M. Geyer, F. Plenzig, J. Rauschnabel, M. Hanack, B. del Ray, A. Sastre, T. Torres, *Synthesis* **1996**, 1139–1151; (d) N. Kabayashi, R. Kondo, S. Nakajima, T. Osa, *J. Am. Chem. Soc.* **1990**, *112*, 9640–9641; (e) A. Weitemeyer, H. Kliesch, D. Wöhrle, *J. Org. Chem.* **1995**, *60*, 4900–4904.

[30] (a) Z.-Y. J. Zhan, D. G. Lynn, *J. Am. Chem. Soc.* **1997**, *119*, 12420–12421; (b) J. T. Goodwin, D. G. Lynn, *J. Am. Chem. Soc.* **1992**, *114*, 9197–9198; (c) P. Luo, J. C. Leitzel, Z.-Y. J. Zhan, D. G. Lynn, *J. Am. Chem. Soc.* **1998**, *120*, 3019–3031; (d) J. T. Goodwin, P. Luo, J. C. Leitzel, D. G. Lynn, in *Self-Production of Supramolecular Structures: From Synthetic Structures to Models of Minimal Living Systems*, Eds.: G. R. Fleischaker, S. Colonna, P. L. Luisi, Kluwer Academic Publishers, Dordrecht, The Netherlands, **1994**, pp. 99–104.

[31] (a) F. Ibukuro, T. Kusukawa, M. Fujita, *J. Am. Chem. Soc.* **1998**, *120*, 8561–8562; (b) M. Fujita, *Chem. Soc. Rev.* **1998**, 417–425; (c) M. Fujita, *Acc. Chem. Res.* **1999**, *32*, 53–61.

[32] (a) K. S. Feldman, Y. B. Lee, *J. Am. Chem. Soc.* **1987**, *109*, 5850–5851; (b) K. S. Feldman, J. S. Bobo, G. L. Tewalt, *J. Org. Chem.* **1992**, *57*, 4573–4574.

[33] A. Pelter, M. G. Hutchings, K. Rowe, K. Smith, *J. Chem. Soc., Perkin 1* **1975**, 138–143.

[34] G. Tuchscherer, L. Scheibler, P. Dumy, M. Mutter, *Biopolymers* (*Peptide Science*) **1998**, *47*, 63–73.

[35] C. D. Bain, G. M. Whitesides, *Angew. Chem., Int. Ed. Engl.* **1989**, *28*, 506–512.

[36] (a) J. S. Lindsey, *New J. Chem.* **1991**, *15*, 153–180; (b) G. M. Whitesides, J. P. Mathais, C. T. Seto, *Science* (Washington, DC) **1991**, *254*, 1312–1319; (c) D. Philp, J. F. Stoddart, *Angew. Chem., Int. Ed. Engl.* **1996**, *35*, 1154–1196.

[37] (a) G. Illuminati, L. Mandolini, B. Masci, *J. Am. Chem. Soc.* **1983**, *105*, 555–563; (b) L. Mandolini, B. Masci, *J. Am. Chem. Soc.* **1984**, *106*, 168–174; (c) R. Cacciapaglia, L. Mandolini, *Chem. Soc. Rev.* **1993**, 221–231.

[38] (a) D. J. Cram, J. M. Cram, *Container Molecules and Their Guests*, Royal Society of Chemistry, Cambridge, **1994**; (b) J. A. Bryant, M. T. Blanda, M. Vincenti, D. J. Cram, *J. Am. Chem. Soc.* **1991**, *113*, 2167–2172.

[39] (a) D. B. Amabilino, J. F. Stoddart, *Chem. Rev.* **1995**, *95*, 2725–2828; (b) S. Capobianchi, G. Doddi, G. Ercolani, J. W. Keyes, P. Mencarelli, *J. Org. Chem.* **1997**, *62*, 7015–7017; (c) D. B. Amabilino, P.-L. Anelli, P. R. Ashton, G. R. Brown, E. Córdova, L. A. Godínez, W. Hayes, A. E. Kaifer, D. Philp, A. M. Z. Slawin, N. Spencer, J. F. Stoddart, M. S. Tolley, D. J. Williams, *J. Am. Chem. Soc.* **1995**, *117*, 11142–11170; (d) D. G. Hamilton, J. E. Davies, L. Prodi, J. K. M. Sanders, *Chem. Eur. J.* **1998**, *4*, 608–620.

[40] S. Anderson, H. L. Anderson, J. K. M. Sanders, *J. Chem. Soc., Perkin 1* **1995**, 2255–2267.

[41] D. W. J. McCallien, J. K. M. Sanders, *J. Am. Chem. Soc.* **1995**, *117*, 6611–6612.

[42] V. Marvaud, A. Vidal-Ferran, S. J. Webb, J. K. M. Sanders, *J. Chem. Soc., Dalton Trans.* **1997**, 985–990.

[43] (a) T. R. Kelly, C. Zhao, G. J. Bridger, *J. Am. Chem. Soc.* **1989**, *111*, 3744–3745; (b) T. R. Kelly, G. J. Bridger, C. Zhao, *J. Am. Chem. Soc.* **1990**, *112*, 8024–8034.

[44] W. L. Mock, T. A. Irra, J. P. Wepsiec, M. Adhya, *J. Org. Chem.* **1989**, *54*, 5302–5308.

[45] (a) C. J. Walter, H. L. Anderson, J. K. M. Sanders, *J. Chem. Soc., Chem. Commun.* **1993**, 458–460; (b) C. J. Walter, J. K. M. Sanders, *Angew. Chem., Int. Ed. Engl.* **1995**, *34*, 217–219; (c) Z. Clyde-Watson, A. Vidal-Ferran, L. J. Twyman, C. J. Walter, D. W. J. McCallien, S. Fanni, N. Bampos, R. S. Wylie, J. K. M. Sanders, *New. J. Chem.* **1998**, 493–502; (d) M. Marty, Z. Clyde-Watson, L. J. Twyman, M. Nakash, J. K. M. Sanders, *Chem. Commun.* **1998**, 2265–2266.

[46] (a) L. E. Orgel, *Acc. Chem. Res.* **1995**, *28*, 109–118; (b) G. von Kiedrowski, *Angew. Chem., Int. Ed. Engl.* **1986**, *25*, 932–935; (c) D. Sievers, G. von Kiedrowski, *Chem. Eur. J.* **1998**, *4*, 629–641; (d) A. Luther, R. Brandsch, G. von Kiedrowski, *Nature (London)* **1998**, *396*, 245–248; (e) T. Li, K. C. Nicolaou, *Nature (London)* **1994**, *369*, 218–221.

[47] (a) D. H. Lee, J. R. Granja, J. A. Martinez, K. Severin, M. R. Ghadiri, *Nature (London)* **1996**, *382*, 525–528; (b) K. Severin, D. H. Lee, J. A. Martinez, M. R. Ghadiri, *Chem. Eur. J.* **1997**, *3*, 1017–1024; (c) T. Tjivikua, P. Ballester, J. Rebek, Jr., *J. Am. Chem. Soc.*, **1990**, *112*, 1249–1250; (d) J. S. Nowick, Q. Feng, T. Tjivikua, P. Ballester, J. Rebek, Jr., *J. Am. Chem. Soc.* **1991**, *113*, 8831–8839; (e) E. A. Wintner, M. M. Conn, J. Rebek, Jr., *Acc. Chem. Res.* **1994**, *27*, 198–203; (f) E. A. Wintner, M. M. Conn, J. Rebek, Jr., *J. Am. Chem. Soc.* **1994**, *116*, 8877–8884; (g) E. A. Wintner, J. Rebek, Jr., *Acta Chem. Scand.* **1996**, *50*, 467–485;

(h) A. Terfort, G. von Kiedrowski, *Angew. Chem., Int. Ed. Engl.* **1992**, *31*, 654–656; (i) S. Yao, I. Ghosh, R. Zutshi, J. Chmielewski, *Angew. Chem., Int. Ed. Engl.* **1998**, *37*, 478–481.

[48] G. von Kiedrowski, *Bioorg. Chem. Front.* **1993**, *3*, 113–146.

[49] (a) J.-P. Sauvage, *Acc. Chem. Res.* **1990**, *23*, 319–327; (b) J.-C. Chambron, C. O. Dietrich-Buchecker, V. Heitz, J.-F. Nierengarten, J.-P. Sauvage, C. Pascard, J. Guilhem, *Pure Appl. Chem.* **1995**, *67*, 233–240.

[50] (a) S. Anderson, H. L. Anderson, J. K. M. Sanders, *Angew. Chem., Int. Ed. Engl.* **1992**, *31*, 907–910; (b) S. Anderson, H. L. Anderson, J. K. M. Sanders, *J. Chem. Soc., Perkin Trans. 1* **1995**, 2247–2254.

[51] A. R. Schulz, *Enzyme Kinetics*, Cambridge University Press, Cambridge, **1994**.

[52] (a) A. J. Kirby, *Adv. Phys. Org. Chem.* **1980**, *17*, 183–278; (b) L. Mandolini, *Adv. Phys. Org. Chem.* **1986**, *22*, 1–111; (c) M. I. Page, *Chem. Soc. Rev.* **1973**, 295–323; (d) K. A. Connors, *Chemical Kinetics*, VCH, New York, **1990**, 363–368.

[53] J. SantaLucia, Jr., H. T. Allawi, P. A. Seneviratne, *Biochemistry* **1996**, *35*, 3555–3562.

2 Templated Synthesis of Polymers – Molecularly Imprinted Materials for Recognition and Catalysis

Günter Wulff

2.1 Introduction

Templates that are non-covalently or covalently attached to reacting substrates can control the reaction pathway with regard to the chemical composition and the stereochemistry of the products formed. Thus, in many cases it is possible to transfer properties of the template to the reaction product. Looking at this concept from the standpoint of a polymer chemist, at least four different cases can be distinguished:

1) A low-molecular-weight template controls the synthesis of low-molecular-weight compounds. This principle is often used for the preparation of cyclic compounds, but many asymmetric syntheses also fall into this scheme.
2) A polymeric template controls the outcome of a reaction of low-molecular-weight compounds. Enzyme-catalyzed reactions can be looked at in this manner.
3) High-molecular-weight templates control the formation of macromolecules. Apart from the frequent examples in biochemistry (e.g., biosynthesis of nucleic acids, proteins, etc.), this is the case in the so-called "template polymerization" in which monomers are bound to a polymer that acts as a template to give, after polymerization and removal of the template, defined linear polymers [1, 2].
4) In this review article, templated syntheses of macromolecules and polymeric materials will be discussed in which a low-molecular-weight template controls the structure of macromolecules. The composition (e.g., the proportion of co-monomers), the sequence of co-monomers as well as the stereochemistry (including the chirality) of the newly formed stereogenic centers is controlled during the polymerization of double bonds.

The general idea of the research work presented here focuses on attempts to control the constitution and the configuration of complex supramolecular arrangements by specific interactions of the constituents with a suitable template molecule [3]. If in this case the constituents bear polymerizable groups, the whole constituent–template complex can be polymerized and, thus, the original structure will be stabilized and frozen. After removal of the template molecules, polymers with defined structures are obtained by transfer of structural information from template to polymer.

Mechanistic studies revealed that this polymerization proceeds through an asymmetric cyclocopolymerization involving an unusually large 19-membered ring [18] (see Scheme 2-1). The optical activity stems from the formation of chiral distyryl dyads in the polymer main chain, which are induced by the chiral template molecule. The dyads are separated from one another by one or more co-monomeric units. The chirality of the polymers is, however, independent of the configuration at these co-monomeric centers. The absolute configuration of the prevailing asymmetric dyad building block has been established to be (*S,S*), as evidenced by the synthesis of oligomeric model compounds. Conformational analysis shows that both so-called "racemic" as well as "*meso*" dyads must be expected. The ratio of *meso* and racemic dyads could be determined by radical cyclization of **1** in the presence of an excess of azobis(isobutyronitrile) (AIBN) to yield a monomeric cyclic compound [3, 23] (Scheme 2-2). Hydrolysis of this cyclic compound and deboronation yielded four reaction products. Investigation of the reaction products showed that the two diastereomeric dyads are produced in a comparable ratio; in both cases the enantiomeric excess is very high, resulting in structures that correspond in the polymer to a mixture of (*S,S*) as well as (*R,S*) dyads. Only the (*S,S*) dyads ("racemic dyads") will contribute to the chiroptical properties of the copolymers but the *meso* dyads (*R,S*) have nearly no influence.

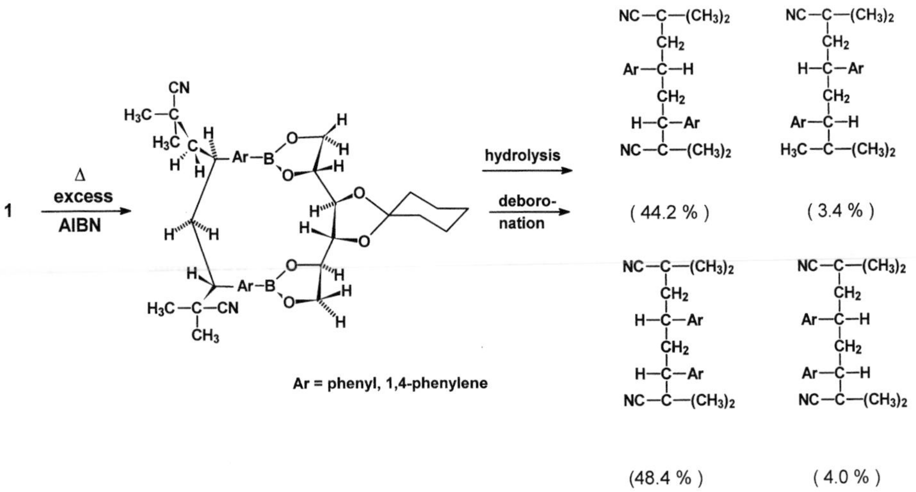

Scheme 2-2   Schematic representation of the radical cyclization of monomer **1** with AIBN and the preparation of the four stereoisomeric dimers. Only one stereoisomer is represented in the cyclization product [23].

2.3   Exact Placement of Functional Groups on the Surfaces of Rigid Polymeric Materials Using Template Molecules

The two-dimensional information transfer from low-molecular-weight templates to polymers can be achieved on the surfaces, e.g., of wide-pore silica. We have attempted to introduce two amino groups onto a more or less planar surface of silica, arranged at a specific distance apart with the aid of a template (Scheme 2-3a). With this method, it was possible to locate two functional groups on a silica surface via siloxane bond formation (see Scheme 2-3b), thus enabling an investigation of the selectivity due to accuracy in distance alone [8, 9].

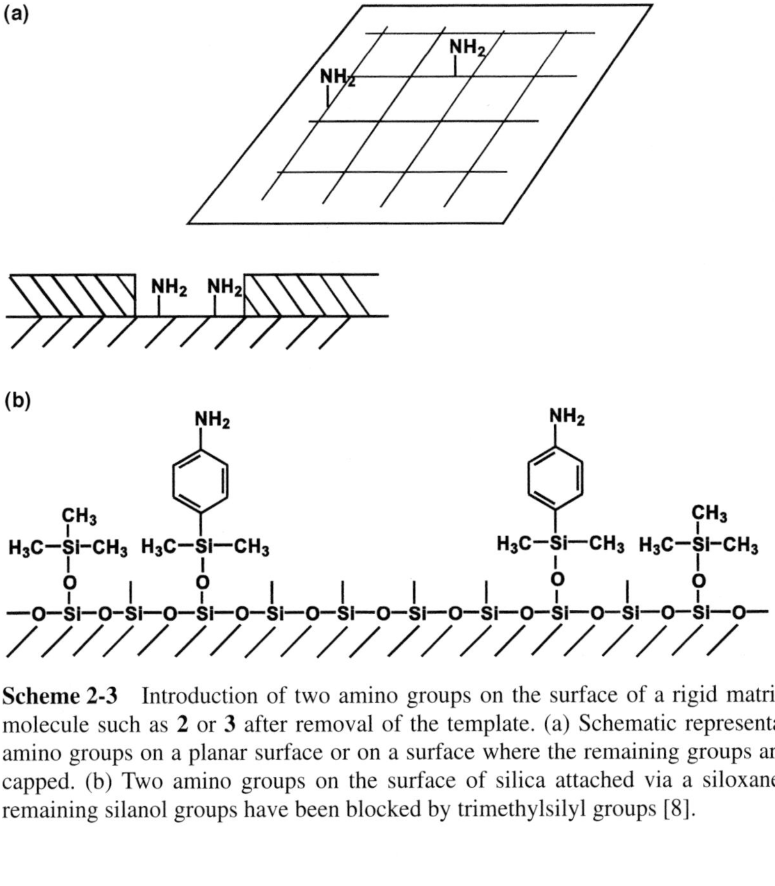

(a)

(b)

Scheme 2-3 Introduction of two amino groups on the surface of a rigid matrix by a template molecule such as **2** or **3** after removal of the template. (a) Schematic representation of the two amino groups on a planar surface or on a surface where the remaining groups around have been capped. (b) Two amino groups on the surface of silica attached via a siloxane grouping. The remaining silanol groups have been blocked by trimethylsilyl groups [8].

Monomer 2

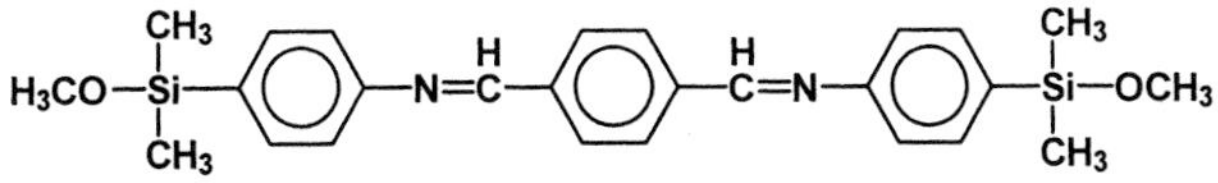

Monomer 3

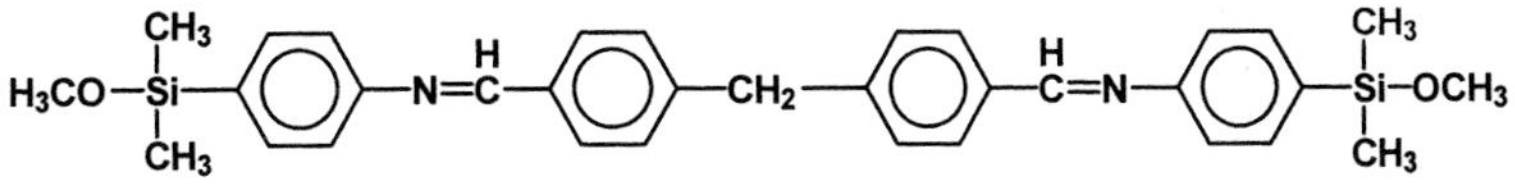

Monomer 4 $H_2N-(CH_2)_3-Si(OCH_3)_3$

Two amino groups were attached to the surface of silica at a distance of 0.72 and 1.05 nm from one another using the template monomers **2** and **3**, respectively. The attachment to the surface is achieved through the formation of siloxane bonds by condensation between the methoxysilane group of **2** and **3** and the silanol groups on the surface of the silica. Most of the remaining silanol groups were afterwards capped by reaction with hexamethyldisilazane to avoid non-specific adsorption (Scheme 2-3). Over 95% of the templates could be split off. Unlike the situation with polymers, in this case the position of the two amino groups should not be changed as a result of chain mobility, swelling, or shrinking. The distance can only be altered by conformational changes within the functional group part. For comparison purposes, a silica with randomly distributed amino groups was prepared from **4**. In order to elucidate the role of accuracy in distance, the selectivity α was determined by equilibration with an equimolar mixture of the two template dialdehydes **5** and **6** (see Table 2-2). Both the silicas showed a significant difference in binding, preferring their own templates, with α values of 1.74 and 1.67. This clearly suggests that by using distance selectivity alone and with differences of only 0.33 nm (between **5** and **6**), substrate selectivity can be observed.

Table 2-2   Selectivity of modified silicas with each of the two amino groups at a defined distance [8].

	Split-ting per-cen-tage	Dis-tance r of groups (nm)	Apparent binding constant		Selec-tivity α'
			OHC—⬡—CHO **5**	OHC—⬡—CH₂—⬡—CHO **6**	
Silica modified with **2**	<95%	0.72	4.91	2.58	1.74
Silica modified with **3**	>95%	1.05	9.07	13.77	1.67
Silica modified with **4** (at random)	–	–	2.26	2.05	–

Chromatography with silicas of this type could also be used to separate dicarboxylic acids that differ in the distance between their carboxyl groups. Good separations were observed compared with those obtained under identical conditions on silica with randomly distributed amino groups [10].

2.4 Molecular Imprinting in Polymeric Materials Using Template Molecules

2.4.1 The Principle

A three-dimensional transfer of information from a template to a polymer consists in the preparation of crosslinked polymers by molecular imprinting with templates [24–26] (for reviews, see [10–16, 27, 28]). For this, polymerizable vinyl monomers containing functional groups were attached to suitable template molecules by covalent or non-covalent interactions. Subsequent copolymerization in the presence of solvents and relatively large concentrations of crosslinking agents produced rigid macroporous polymers. Removal of the template molecules (Scheme 2-4) left behind chiral cavities in the polymer whose shape and arrangement of functional groups were determined by the structure of the template molecules.

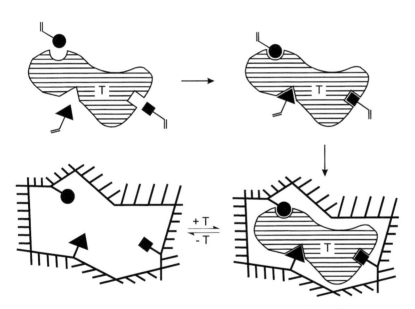

Scheme 2-4   Schematic representation of the imprinting of specific cavities in a crosslinked polymer by a template (T) with three different binding groups [10].

This technique resembles the formation of antibodies from haptens, and actually a similar mechanism to the imprinting was formerly thought to be the mechanism of formation of antibodies [29]. The functional groups in these cavities are located at various points in the polymer chain, and are held in a definite mutual orientation simply by the crosslinking. In this case, the stereochemical information is not carried by a low-molecular-weight part of the molecule. Instead, the entire arrangement of the polymer chains (the topochemistry) is responsible for the stereochemical structure. This is reminiscent of the structure of the active centers of enzymes; we have therefore termed these polymers "enzyme-analogous". The relationship of the template to the imprinted cavity corresponds to the key/lock principle proposed by Emil Fischer for enzyme catalysis more than 100 years ago [30]. The imprinting process has a certain analogy to F. H. Dickey's attempts to imprint silica gel by precipitation in the presence of template molecules [31].

An example of the imprinting procedure is the polymerization of the template monomer **7**, which was used for the optimization of the method [32–35]. In this case, phenyl-α-D-mannopyranoside **7a** acts as template. Two molecules of 4-vinylphenylboronic acid are bound by diester linkages to this template. The binding sites (the boronic acids) are bound by a covalent interaction. The monomer was copolymerized by free radical initiation in the presence of an inert solvent with a large amount of bifunctional crosslinking agent. Polymers thus obtained are macroporous and have a permanent pore structure and a high inner surface area. Polymers of this type possess good accessibility on the surface of the pores and a rather rigid structure with low mobility of the polymer chains.

7   7 a

The template **7a** can be split off by water or methanol to an extent of up to 95% (Scheme 2-5). The accuracy of the steric arrangement of the binding sites in the cavity can be tested by the ability of the polymer to resolve the racemate of the template, namely of phenyl-α-D,L-mannopyranoside. Therefore the polymer is equilibrated in a batch procedure with a solution of the racemate under conditions under which rebinding in equilibrium is possible. The enrichment of the antipodes on the polymer and in solution is determined by measuring the specific optical rotation and the separation factor α, i.e., the ratio of the distribution coefficients of the D and L compounds between polymer and solution, is calculated. After extensive optimization of the procedure, α values between 3.5 and 6.0 were obtained [10]. This is an extremely high selectivity for racemic resolution that cannot be reached by most other methods.

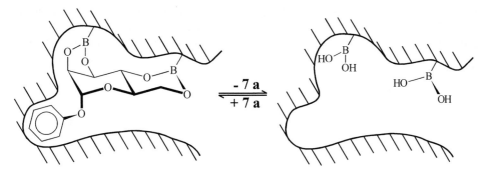

Scheme 2-5 Schematic representation of the polymerization of **7** to obtain a specific cavity. The template **7a** can be removed with water or methanol to give the free cavity [10].

Polymers obtained by this procedure can be used for the chromatographic separation of the racemates of the template molecules [10, 32, 35]. The selectivity of the separation process is fairly high (separation factors up to $\alpha = 4.56$), and, at higher temperatures with gradient elution, resolution values of $R_s = 4.2$ with baseline separation have been obtained (Figure 2-2). These sorbents can be prepared conveniently and possess excellent thermomechanical stability. Even when used at 80 °C under high pressure for a long time, no leakage of the stationary phase or decrease in selectivity during chromatography was observed.

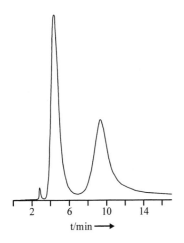

t/min ⟶

Figure 2-2 Chromatographic resolution of D,L-**7a** on a polymer imprinted with **7** (elution with a solvent gradient at 90 °C) [35].

In the meanwhile, a large number of different templates have been used by us and many other groups in the world. An interesting extension of the concept of molecular imprinting was introduced by Mosbach and co-workers (for reviews see [12, 36]), who used only non-covalent interactions during imprinting and the resulting equilibration studies. This approach offers new possibilities for binding because only a limited number of practical linkages are available for fast and reversible covalent binding, and because it has the advantage of easy preparation of the template assemblies. Mostly racemic amino acids were separated in this case, and high selectivity could be demonstrated (Scheme 2-6).

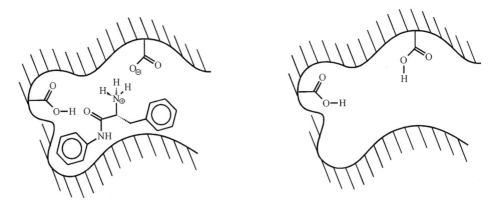

Scheme 2-6 Schematic representation of a cavity produced in the presence of L-phenylalanine anilide. The polymerization takes place in the presence of acrylic acid. Non-covalent electrostatic interactions and hydrogen bond formation occur [12].

Table 2-3 lists a wide range of examples for different chemical classes of templates that have been used for molecular imprinting in crosslinked polymers. The type of binding during the imprinting and the polymer structure will be discussed in other sections.

Table 2-3 Examples of templates used for the preparation of molecularly imprinted polymers.

Class of substance	Compound	Binding type[a]	Application	Reference
Sugars	Mannose deriv.	A	Enant. resolution	[32–35]
	Fructose, galactose	A	Enant. resolution	[37, 38]
	Sucrose	A	Microreactor	[39]
	Glucose	B	Separation	[40]
	Glucose	C	Sensor	[41]
	Gluconamide	B	Specific membranes	[42]
	Sialic acid	A+B	Sensor	[43]
Diols, polyols	Propanediol	A	Enant. resolution	[44]
	Mannitol	A	Enant. resolution	[25, 26, 44]
Hydroxycarboxylic acids	Glyceric acid	A+D	Enant. resolution	[24–26, 45–47]
	Mandelic acid	A	Enant. resolution	[48]

Table 2-3   (Continued).

Class of substance	Compound	Binding type[a]	Application	Reference
Amino acid derivatives	Anilides	A	Enant. resolution	[49]
	Anilides	B	Enant. resolution	[12, 14, 36, 50]
	Dansyl deriv.	B	Sensor	[51]
	DOPA	D	Microreactor	[52, 53]
	Amino acids	B	Enant. resolution	[54–56]
	Amino acids	B	Sensor	[57]
	Amino acids	C	Enant. resolution	[58, 59]
Peptides	*N*-Ac-L-Phe-L-Trp-OMe	B	Enant. resolution	[60, 61]
	Leu-5-enkephalin	B	Separation	[62]
	(Z)-L-Ala-L-Ala-OMe	B	Enant. resolution	[63]
Proteins	Urease	B	Separation	[64]
	Ribonuclease A	B	Separation	[65]
	Myoglobin	B	Separation	[66]
Bacteria	*Listeria* spp.	B	Sensor	[67, 68]
Nucleosides and nucleotides	Adenosine	B	Sensor	[69]
	AMP[b]	B	Separation	[70, 71]
	Nucleic acid	B	Separation	[72]
Purine derivatives	9-Ethyladenine	B	Separation	[73]
	Theophylline	B	Immunoassay	[74]
	Theopyhlline, caffeine	B	Membranes	[75]
Steroids	Cholesterol	B	Sensors	[71]
	Cholesterol	D	Separation	[76]
	Cholesterol	A	Separation	[68]
	Androstane deriv.	D	Microreactor	[77]
	Testosterone	B	Separation	[78]
Drugs	Pentamidine	B	Separation	[79]
	Nicotine	B	Separation	[80]
	Diazepam	B	Immunoassay	[74]
	Cinchonine	B	Separation	[81]
	Morphine	B	Immunoassay	[82]
	Propranolol	B	Separation, enant. resolution	[83–85]
	Propranolol	B	Immunoassay	[86]
	Chloramphenicol	B	Sensor	[87]
Dicarbonyl compounds	Dialdehydes	A	Separation	[8, 9]
	Dialdehydes	A[c]	Separation	[8, 9]
	Dialdehydes	A	Catalyst	[88]
	Diketones	A	Separation	[89, 90]
	Diketones	C	Separation	[91]
	Dicarboxylic acids	D	Microreactor	[92, 93]
	Dicarboxylic acids	A	Catalyst	[94]
Disulfides	Benzyl disulfides	A	Microreactor	[95, 96]
Imidazole derivatives	Bisimidazoles	C	Separation	[13, 97, 98]

Table 2-3   (Continued).

Class of substance	Compound	Binding type[a]	Application	Reference
Dyes	Methyl orange	B[c]	Separation	[31, 99]
	Methyl orange	B	Separation	[100]
	Rhodanile Blue	B	Separation	[101]
Pesticides	Artrazine	B	Separation	[102–105]
	Artrazine	B	Sensor	[106]
	Simazine	B	Separation	[107]
Phosphonate esters	Transition state	B	Catalyst	[108–112]
	analogues	B[c]	Catalyst	[113]

[a] A, covalent binding during imprinting and equilibration; B, noncovalent binding during imprinting and equilibration; C, metal coordination binding during imprinting and equilibration; D, covalent binding during imprinting, noncovalent binding during equilibration.
[b] Adenosine monophosphate.
[c] Imprinted in silica.

    As Table 2-3 shows, imprinted polymers have been mainly used as separation media (mostly in chromatography). Of special interest is the enantiomeric resolution of race-mates. Further applications are as immunosorbents and chemosensors. The cavities in the imprinted polymers have also been used as microreactors for selective reactions and, more interestingly, as the active sites of catalytically active polymers. In 1998 nearly 100 papers appeared in the literature on molecular imprinting, together with one book [114]; another book is imminent [115].

2.4.2   The Optimization of the Structure of the Polymer Network

The optimization of the polymer structure was rather complicated. On one hand, the polymers should be rather rigid to preserve the structure of the cavity after splitting off the template. On the other hand, a high flexibility of the polymers should be present to facilitate a fast equilibrium between release and reuptake of the template in the cavity. These two properties are contradictory to each other and a careful optimization has to be performed in these cases. Furthermore, good accessibility of as many cavities as possible should be possible. Good thermal and mechanical stability is also necessary. Ever since the first experiments, in nearly all cases macroporous polymers have been employed which possess a high inner surface area ($100–600 \ m^2 \ g^{-1}$) and which show, after optimization, good accessibility as well as good thermal and mechanical stability.

    The selectivity is mostly influenced by the kind and amount of crosslinking agent used. Figure 2-3 shows the dependence of the selectivity for racemic resolution of the racemate of **7a** on polymers prepared from **7** [33, 34]. Below a certain amount of cross-linking in the polymer (around 10%) no selectivity can be observed; the cavities are not sufficently stabilized. Above 10% crosslinking a steady increase in selectivity is ob-

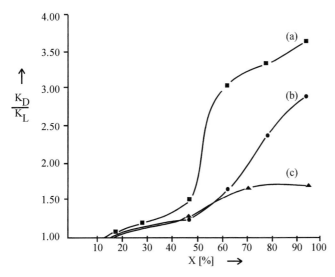

Figure 2-3 Selectivity of polymers as a function of the type and amount (X) of crosslinking agent [33]. The polymers were prepared in the presence of **7** with various proportions of the crosslinking agents: (a) ethylene dimethacrylate; (b) tetramethylene dimethacrylate; (c) divinylbenzene. After removal of the template **7a**, the separation factor $\alpha = K_D/K_L$ was determined for the resolution of D,L-**7a** in a batch process.

served. Between 50 and 66% a surprisingly high increase in selectivity occurred, especially if ethylene dimethacrylate is used as a crosslinker. In this investigation, ethylene dimethacrylate appears to be superior to technical divinylbenzene or butylene dimethacrylate as a crosslinker.

The inert solvent used during polymerization also has a strong influence on the polymer structure [27], but the influence on selectivity is astonishingly low. A stronger influence on the selectivity is observed if the amount of inert solvent during polymerization is raised from 0.29 to 1.76 mL g^{-1} monomeric mixture, the optimum being around 1.0 mL g^{-1} [47].

Aside from a high selectivity, the polymers should be able to undergo a fast and reversible binding of the substrate within the cavities. For this, a certain flexibility is necessary. Cavities of accurate shape but without any flexibility present kinetic hindrance to reversible binding.

Polymers obtained with ethylene dimethacrylate as crosslinker retained their specificity for a long period. Even under high pressure in a high-performance liquid chromatography (HPLC) column, the activity remained for months. This was true even when the column was used at 70–80 °C. On the other hand, polymers crosslinked with divinylbenzene gradually lost their specificity at higher temperatures. Interestingly, at 60 °C the α value for racemic resolution was further increased to 5.11. Higher selectivity at increased temperature had been observed during earlier chromatographic studies [116, 117].

In most cases, the macroporous imprinted polymers are prepared in bulk and are then crushed and sieved. Thus, by a rather tedious procedure, irregularly broken polymers are obtained. Usual suspension polymerization is not possible with most non-covalent and even with some covalent bindings since water interferes with the binding reaction, and hampers an efficient imprinting. By using new types of stoichiometric non-covalent binding, these difficulties can be overcome and polymers can be prepared by standard suspension polymerization methods [118]. Uniformly sized particles are thus easily obtained.

Another possibility is a suspension polymerization in media other than water. For example, a suspension polymerization using a liquid perfluorinated alkane as the dispersing phase was reported [119]. In other cases, a two-step swelling and polymerization method was applied to prepare molecularly imprinted beads [120].

A further possibility is to use silica-based supports, which are available in a broad range of bead sizes and pore diameters. These beads can easily be modified by silanizing reagents carrying polymerizable groups. On this surface, a thin layer of a monomeric mixture containing the template is polymerized. Thus the beads can be coated with imprinted polymers for the separation of dyes [121], enantiomeric sugars [122, 123], or bisimidazoles [124]. These coated silicas show very good properties as chromatographic supports; their drawback is the lower loading capacity compared to macroporous polymers. It is also possible to use trimethylol trimethacrylate-based beads and to cover these as described before [125].

Imprinting in thin layers on the surface of other materials offers the possibility of a surface imprinting procedure. High-molecular-weight templates cannot be imprinted in the usual manner in a bulk polymerization since, after crosslinking, the templates cannot be removed. Therefore enzymes have been used as templates by a surface imprinting method [64, 65]. Especially interesting is the possibility of imprinting even such large objects as bacteria. First reports of work in this direction have appeared [67, 68, 126], and it remains to be seen how selectively these imprints distinguish between different species of bacteria.

It appears advantageous to use for molecular imprinting an in-situ method that has been developed previously for reversed-phase HPLC [127]. In this case, the polymer is directly prepared inside the column. It has been shown that this method is applicable for molecular imprinting [54, 79, 127, 128] but the selectivity for separation in these columns is somewhat reduced.

2.4.3   The Role of the Binding-site Interactions

In the imprinting procedure, the binding groups have several functions [10, 14, 129, 130]. On one hand, the bond between the template and the binding group should be as strong as possible during the polymerization. This enables the binding groups to be fixed by the template in a definite orientation to the polymer chains during the crosslinking polymerization. It should then be possible to split off the templates as completely as possible. A very important function is the interaction of the binding groups with the substrates to be bonded: for example, with the compound that acted as the template. This interaction should take place as rapidly and reversibly as possible, so that the chromatographic process or catalysis will be rapid. Thus, although a high activation energy is desirable for the first function, it should be as low as possible for the other two.

Detailed investigations have shown that the selectivity in enantiomeric resolution is dependent both on the orientation of the functional groups inside the cavities and the shape of the cavities [37, 38, 131, 132]. The dominating factor, though, is the orientation of the functional group inside the cavity [37]. If two binding sites per template have been used, several mono-point bindings can occur but only one two-point binding [10, 123]. It is the two-point binding that provides high selectivity. Therefore, this portion has to be increased, which is possible, e.g., by increasing the temperature [133].

Another problem of binding is the reuptake of template in the cavity. In the case of covalent binding and all other types of stoichiometric binding, binding sites are situated only in the cavity. After removal of the template, this usually leads to a swelling in the cavities which guarantees a high proportion (90–95%) of reuptake after the first removal [10]. At the same time, it facilitates a quick mass transfer during equilibration of the template with the polymer (Scheme 2-7). On reuptake of the template, the cavity shrinks to its original volume. This behaviour is similar to the *induced fit* known from enzyme chemistry.

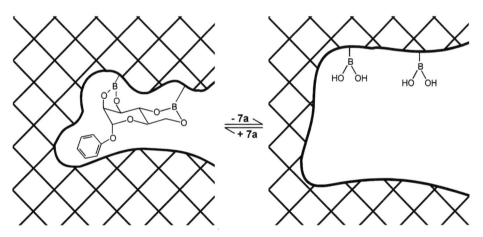

Scheme 2-7 Schematic representation of the removal from a polymer of the template **7a** bound by covalent binding and the swelling of the cavity. Afterwards, 90–95% of the cavities can be reoccupied [10].

On the other hand, during non-covalent imprinting, e.g., with acrylic acid, a fourfold excess of binding sites has to be used in order to obtain good selectivity. Under these conditions the binding sites are distributed all over the polymer (Scheme 2-8). As was found recently [50] under these conditions, only 15% of the cavities can take up a template again, the remaining 85% being lost irreversibly for separation. This might be due to a shrinking of the majority of the cavities. Therefore these imprinted polymers are not well suited for preparative separations and for investigations on catalysis. In consequence, new and better binding sites should be devised.

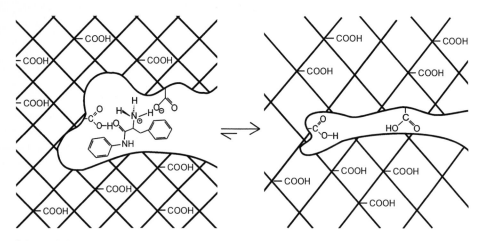

Scheme 2-8   Schematic representation of the removal of the L-phenylalanine-anilide template bound by non-covalent interaction. Owing to the large excess of carboxyl groups, a substantial part of the cavities may shrink. Only around 15% of the cavities can be reoccupied [10].

The boronic acid behaves quite satisfactorily as a binding site since it gives a strong interaction during the imprinting, and the binding during equilibration can be accelerated by addition of suitable bases such as ammonia or piperidine. In this case, the trigonal boron is transformed to the more reactive tetragonal form (Equations (a) and (b)). Unfortunately, there are not many examples of good binding through covalent bonds. For non-covalent binding, methacrylic acid has been replaced by more acidic compounds such as sulfonic acids [134] or fluorinated carboxylic acids [135], but there is no major improvement.

An interesting type of binding during polymerization and later in the final polymer can be achieved by coordinative bonds to metals. This type of bond is analogous to that used in ligand exchange chromatography. The advantage of this kind of bond is that its strength can be controlled by experimental conditions. Definite interactions occur during the polymerization, and an excess of binding groups is not necessary. The subsequent binding of the substrate to the polymer is so rapid that in many cases even rapid chromatography is possible. This method was first used for imprinting by Fujii et al. [58]. They obtained a remarkably high selectivity in the optical resolution of N-benzyl-D,L-valine using a chiral Schiff base ligand of (1R,2R)-1,2-diaminocyclohexane and 4-(4-vinylbenzyloxy)salicylaldehyde to which amino acids could be bound through a Co^{3+} cation. The exceptionally high separation factor α for the racemate resolution of the template molecule was found to be 682. The enantiomeric excess (e.e.) was 95.5% in the batch process. More detailed investigations showed that most of the separating ability was due to the cavity effect. The observed selectivity is greater than that of many enzymes. Unfortunately, the mass transfer rate in the case of the Co^{3+} complex is very slow in the chromatographic separation, so that practical separations are very difficult.

Recently, Arnold, Dhal et al. investigated the bonding of imidazole-containing compounds with Cu^{2+} complexes such as **8** in great detail (Equation (c)) [13, 97, 98]. Model experiments were first carried out in which bisimidazoles with various distances between the imidazole groups were used as templates to position polymerizable iminodiacetate groups in the polymer. These experiments were aimed at developing an effective recognition for proteins that depends on the correct spatial arrangement of a few binding sites on a polymer [13].

Another even more promising approach uses tetraazacyclononane–Cu^{2+} chelates **9** for binding. With this binding site it is possible to bind, e.g., glucose [41] as well as free amino acids [59].

The amidine binding site possesses very promising properties for the binding of carboxylic acids or phosphonic monoesters. Since these complexes tend to be insoluble, a number of derivatives have been prepared [136, 137]. Especially suitable is the 4-vinyl-*N,N'*-diethylphenylamidine **10** [112] (see Equation (d)). Association constants in these equilibria of 10^4–10^6 dm^3 mol^{-1}, depending on the solvent, are observed. Splitting off the templates is easily possible and also the equilibrations are very quick. A reverse interaction of an amidine-containing drug (pentamidine) and acrylic acid has been used previously [79].

10   **a**   R = HC=CH$_2$

     b   R = H$_2$C—CH$_3$

Binding constants in hydrogen bonding increase with multiple interactions during ring formation. Thus an investigation on the binding of amidopyrazoles with dipeptides was performed in our institute [138, 139]. It is expected that under these conditions one amidopyrazole is bound to the top face of the dipeptide via three-point binding whereas the second goes to the bottom face via two-point binding. Complexation should stabilize the β-sheet conformation in the dipeptide (Figure 2-4). NMR titrations of Ac-L-Val-L-Val-OMe with 3-methacryloylaminopyrazole show large but markedly different downfield shifts for both peptide amide protons, one oriented to the upper and the other to the lower site. When one equivalent of 5-amidopyrazole is added, only complexation from the top by three-point binding with an association constant of 80.0 dm^3 mol^{-1} is observed. On further addition of 5-amidopyrazole, complexation from below by two-point binding occurs with an association constant of only 2.0 dm^3 mol^{-1}. In both complexes the dipeptide possesses a β-sheet conformation. By variation of the acid part of the amide the association constant can be further substantially increased ($K = 890$ dm^3 mol^{-1} for CF$_3$–CO–). Thus the binding site monomer **11** has been used successfully for imprinting with dipeptides and for racemic resolution of the racemate of the dipeptide template [139].

    Multiple interactions are also used with the new binding site 2,6-bis(acrylamido)pyridine. In this case, different barbiturate drugs, such as cyclobarbital, act as templates [140]. Retentions vary, depending on the barbiturate.

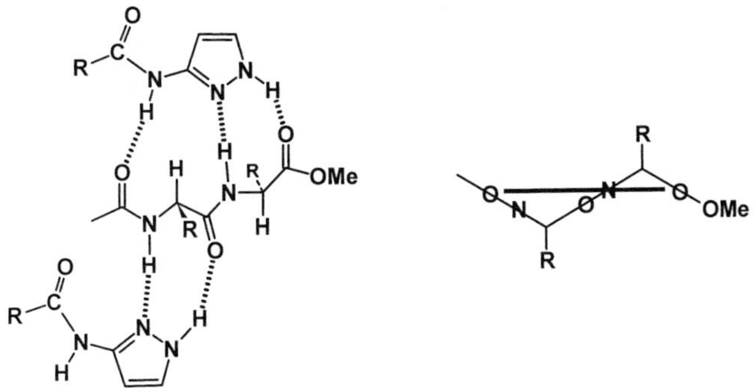

Figure 2-4   Side and top views of the computer-calculated 2:1 dipeptide–amidopyrazole complex [138]. When seen from above, the heterocycle on the top side is symbolized by a horizontal bar (for clarity, the second heterocycle is omitted).

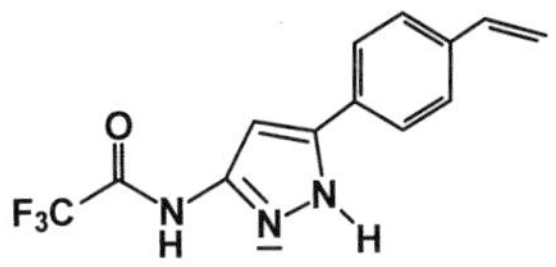

11

2.4.4   Chiroptical Properties of the Crosslinked Polymers

Usually the chiroptical properties of highly crosslinked polymers cannot be measured. The asymmetry of the empty cavities can be analyzed by the excellent racemate resolution ability. By a new method it can now also be directly detected by measuring the optical activity [133]. This is measured by suspending the polymer in a solvent which has the same refractive index as the polymer, a technique which was developed for other types of insoluble polymers [141]. The molar optical rotation values thus measured are shown in Table 2-4.

Table 2-4   Molar optical rotation values for polymers with chiral cavities [133].

	Template monomer 7	Polymer **P E′** with template **7a**	Polymer **P E** template **7a** split off[a]
$[M]_{546}^{20}$	$-448.9°$	$-61.7°$	$+110.0°$

[a]   As the ethylene glycol ester.

If we compare the value $-61.7°$ for a polymer prepared from **7** (calculated for the molar content of **7**), still containing the template, with the value $-448.9°$ for the template monomer **7**, it becomes apparent that the molar optical rotation has decreased considerably as a result of polymerization. This could have several causes, one being the influence of the polymer matrix. Its effect can be determined by splitting off the optically active template **7a**. If the boronic acids are converted with an achiral diol to the corresponding ethylene glycol ester, the polymer gives a positive molar rotation ($[M]_{546}^{20} = +110.0°$). This shows that in **P E** (Table 4), the imprints generated in the polymer make a positive contribution to the optical rotation value. Measuring the optical rotation in the solid phase thus allows the properties of chiral cavities in the polymer, such as the binding situation of a bound substrate, different swelling situations, etc., to be determined directly.

If, for example, a polymer with empty cavities is recharged with the template, the resulting polymer, in contrast to **P E′** (with template), has a very large positive rotation value ($[M]_{546}^{20} = +323°$). If this loaded polymer is heated in acetonitrile in the presence of a 3 Å molecular sieve, the molar rotation $[M]_{546}^{20}$ changes to $-68°$, i.e,. to about the same value as for the original polymer with template. Evidently most of the template molecules are first bound only by a single point of attachment (formed by esterification of only one boronic acid group in the cavity), and this then changes to a double attachment. At higher temperature most of the template is quickly bound by a two-point binding.

The optical rotation of these polymers without templates is not caused by individual chiral centers, as is usual, but by the boundaries of the empty imprints as a whole. Their chiral construction and conformation are stabilized by means of the crosslinking of the polymer chains.

It is unlikely that the chiral configurations of the linear portions of the chains contribute to the asymmetry of the cavity since no asymmetric cyclocopolymerization is possible for the template monomer **7**. With other types of template monomers, though, such contributions of backbone chiral portions of the polymer might be expected (see Section 2.2).

2.4.5   Chromatography Using Molecularly Imprinted Polymers

Molecularly imprinted polymer networks have been used as stationary phases in chromatography, especially for the resolution of racemates. Although the selectivity was good, strong peak broadening in the chromatographic process at first made it impossible to obtain complete separation, for example, of the racemate of template **7a**. When significantly more selective adsorbents and improved chromatographic processes coupled with higher temperatures became available, in 1986 we were able, for the first time, to perform a complete racemate resolution with $R_s = 2.1$ [117]. Later, we achieved resolutions of $R_s = 4.3$ by using gradient elution [35] (see Figure 2-2).

With non-covalent interactions, for example using the polymer described in Scheme 2-6, it was thought that a separation would have been much easier to achieve. However, these systems were found to be very complicated in chromatographic practice. Strong interactions with the medium occurred, and the mass transfer was also very slow. In 1988 Mosbach et al. [36] obtained a complete resolution with $R_s = 1.2$, and Sellergren and Shea [142] achieved considerable improvement five years later after very careful optimization work. Protonation of the substrate could be obtained in an aqueous buffer system, pH = 4.0, in which the carboxyl groups inside the cavities were partially deprotonated and those outside the cavities were not deprotonated at all. This gave selective binding by an ion-exchange mechanism only inside the cavities. Kinetics were good, and could be further improved by use of a heat-treated polymer to give a resolution of $R_s = 2.2$.

Remarkably, in these separations the more strongly retarded enantiomer (corresponding to the template molecule) gives much more pronounced peak broadening if a solvent gradient is not used during chromatography. The number of theoretical plates for this compound is at most half that of the less strongly retarded compound. The reason must be essentially the two-point binding of the template molecule and the one-point binding of the other enantiomer. The difference also shows itself in the temperature dependence of the number of theoretical plates for the two enantiomers. Whereas the template molecule shows hardly any temperature dependence because of the increased two-point binding with slow binding rates at higher temperatures, the number of theoretical plates for the second enantiomer increases rapidly, as expected (Figure 2-5a) [117]. The effect of concentration on retention is also very different for the two enantiomers (Figure 2-5b). The "wrong" enantiomer does not show much effect, whereas the retention in the case of the template molecule is greatly increased at lower concentrations at which only the most selective cavities are occupied by the template molecule. The selectivity therefore increases sharply [50].

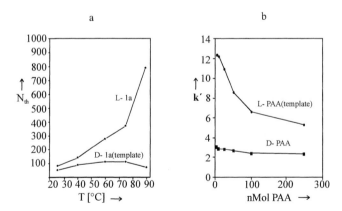

Figure 2-5 Differences in the effect of temperature on the number of theoretical plates, and in the effect of the amount chromatographed on the retention for template molecules and their enantiomers. (a) Temperature dependence of the number of theoretical plates (N_{th}) in the resolution of D-**7a** and L-**7a** on a polymer imprinted with **7** [117]. (b) Dependence of the capacity factor k' on amount treated in the resolution of L- and of D-phenylalanine-anilide on a polymer imprinted with L-phenylalanine-anilide [50].

Preparative separations are also possible. For example, 30 mg of the racemate of **7b** were effectively resolved by 20 g of polymer[143].

2.4.6   Catalysis With Molecularly Imprinted Polymers

One reason for producing cavities with a definite shape and a predetermined arrangement of functional groups is the desire to perform stereoselective and regioselective reactions in these cavities, which are then termed "microreactors". This is an interesting example of the "template" terminology. A so-called template is used to prepare an imprint of a specific shape containing functional groups in a certain stereochemistry. This imprint is then used as a real template to control reactions performed inside the templated cavity. This might be one of the rare examples of using real templates.

The cavity is first imprinted with the end product of the reaction, and a precursor is then embedded into the cavity so that a reaction can convert it into the product. The first experiments were carried out by the research groups of Shea [92] and Neckers [93], who performed cycloadditions to obtain cyclopropanedicarboxylic and cyclobutanedicarboxylic acids. The latter compounds were obtained with remarkable regio- and diastereoselectivity.

The first asymmetric syntheses in the chiral cavity were achieved in our research group [27, 52, 53]. A cavity was made with an L-DOPA methyl ester. After removal of the template, glycine was embedded in the cavity, deprotonated, and alkylated. So far, the highest enantiomeric excesses (36% e.e.) on using imprinted polymers have been with the amino acids formed in this way. This excess is purely a result of the shape of the asymmetric cavity.

In some very remarkable experiments, Byström et al. [77] were recently able to demonstrate high regio- and stereoselectivity in reactions inside the imprinted cavity. The steroid **12** was copolymerized as the template monomer, and removed by reduction. The hydroxyl group in the polymer newly formed from the carboxyl group was converted into an active hydride by $LiAlH_4$. With the help of this polymer, androstan-3,17-dione was reduced to the alcohol exclusively in position 17, whereas in solution or with a polymer with statistically distributed hydride groups it is reduced exclusively in position 3.

12

Imprinting should also be an excellent method to prepare active sites of enzyme analogues. It has already been reported that antibodies prepared against the transition state of a reaction show considerable catalytic activity (for review, see [144]). Thus, antibodies prepared against a phosphonic ester (as a transition state analogue for the alkaline ester hydrolysis) enhanced the rate of ester hydrolysis by 10^3-10^4 fold. In this case, it was also possible to obtain an asymmetric catalysis since the antibodies provide an asymmetric active site. The same should be possible with imprinted polymers (for reviews see [145, 146]). Initial attempts of different groups [108–111] to use this concept for the preparation of catalytically active imprinted polymers for ester hydrolysis had been rather disappointing. Enhancements of 1.6- and 2-fold, in one case of 6.7-fold, are quite low and should be improved.

Better results were obtained for the catalysis of the dehydrofluorination of 4-fluoro-4-(4-nitrophenyl)butanone by Shea et al. [94] and Mosbach et al [147]. Shea used benzylmalonic acid as the template to position two polymerizable amines in a definite arrangement in the cavity. After removal of the template, dehydrofluorination was enhanced by this catalyst by a factor of 8.6.

Even stronger accelerations (25-fold) were obtained during decarboxylations of 3-carboxybenzisoxazoles [88]. In this case two amino groups were placed a suitable distance apart in the cavity by means of a polymerizable di-Schiff base.

Morihara's research group worked very intensively on preparing catalysts on the surface of silica gel (see, e.g., [113, 148]). In a process described as "footprint catalysis", commercial silica gel is treated with Al^{3+} ions and imprinted with a transition-state analogue. Similar accelerations to those of the foregoing examples were obtained.

It appears that the shape of the transition state alone does not lead to sufficient catalysis; in addition, catalytically active groups have to be introduced. This is also true for catalytic antibodies, since Benkovic et al. [149] showed that, e.g., a guanidinium group (of the amino acid L-arginine) plays an important role in the catalysis of the basic hydrolysis of esters by a catalytic antibody.

We have therefore applied amidine groups both for binding and catalysis, investigating the alkaline hydrolysis of ester **13** [112]. Phosphonic monoester **14** was used as a transition state analogue for templating. Addition of two equivalents of the new binding site monomer **10** furnished the bisamidinium salt. By the usual polymerization, workup, and removal of the template, active sites were obtained with two amidine groups each. Owing to the stoichiometric interaction of the binding sites, the amidine groups are only located in the active sites (Scheme 2-9).

At pH 7.6, the imprinted polymer accelerated the rate of hydrolysis of ester **13** by more than 100-fold compared to the reaction in solution at the same pH (Equation (e)). Addition of an equivalent amount of monomeric amidine to the solution only slightly increased the rate. Polymerizing the amidinium benzoate gave a somewhat stronger enhancement in rate.

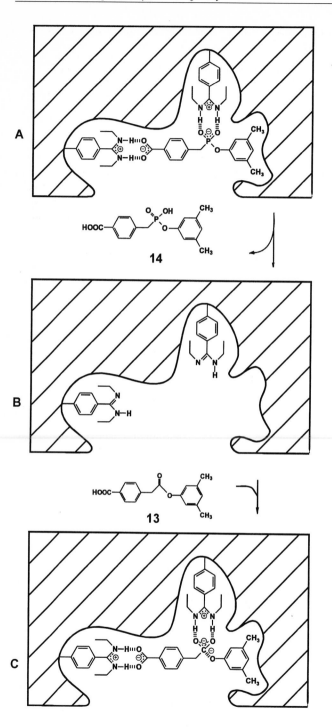

Scheme 2-9 Schematic representation of the polymerization of monomer **14** in the presence of two equivalents of **10a** (A), splitting off **14** (B), and catalysis causing alkaline hydrolysis of **13** through a tetrahedral transition state (C) [112].

Table 2-5   Relative rates of hydrolysis of **13** with different catalysts in buffer solution at pH = 7.6 (Equation (e)) [112].

Blank	With **10b**	With a polymer imprinted with **10a** benzoate	With a polymer imprinted with **14** and **10a**
1.0	2.4	20.5	102.2

These examples showed the strongest catalytic effects obtained until that time by the imprinting method. Even when compared to antibodies, only one to two orders of magnitude are lacking. This is especially remarkable since we used "polyclonal" active sites and rigid, insoluble polymers. It should also be mentioned that these hydrolyses occur with non-activated phenol esters and not, as in nearly all other cases, with activated 4-nitrophenyl esters.

In order to see whether or not these polymers show typical enzyme analogue properties, we investigated the kinetics of the catalyzed reaction in the presence of various excesses of substrate with respect to the catalyst. Figure 2-6 shows the typical Michaelis–Menten kinetics observed. Saturation phenomena occur at higher concentrations.

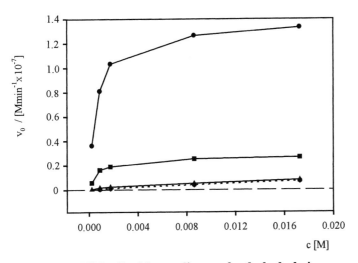

Michaelis - Menten diagram for the hydrolysis of the model ester in presence of :

— ● — transition state **14** imprinted polymer
— ■ — polymer imprinted with amidinium-benzoate
— ▲ — hydrolysis in solution of pH = 7.6 in presence of amidine **10b**
· · ● · hydrolysis in solution of pH = 7.6

Figure 2-6   Michaelis–Menten kinetics of the hydrolysis of **13** according to Equation (e) in the presence of different catalysts. The initial rates are plotted versus substrate concentration [112].

This shows that all active sites are then occupied and that the reaction becomes independent of substrate concentration. In contrast, in solution, and in solution with the addition of amidine, a much slower linear relationship is observed. The amidinium benzoate shows also some type of Michaelis–Menten kinetics. The benzoate therefore acts as a less effective template.

The Michaelis constant K_m was determined to be 0.60 mM. Turnover is relatively low ($k_{cat} = 0.4 \times 10^{-2}$ min^{-1}) but is definitely present. Furthermore, we found that the template molecule itself is a powerful competitive inhibitor with $K_i = 0.025$ mM, i.e., it is bound more strongly than the substrate by a factor of 20. It is remarkable that such a strong binding of substrate and template occurs in water–acetonitrile (1:1). Binding in aqueous solution by usual electrostatic or hydrogen bonding is much weaker.

We observed some product inhibition similar to that by catalytic antibodies. In the case mentioned, the rate of reaction was calculated from the amount of released acid. If the calculation is based on phenol release, the rate of enhancement is nearly doubled. Hydrolysis of carbonates should avoid this difficulty. Therefore, diphenyl phosphate was used as a template, and the hydrolysis of diphenyl carbonate was then investigated [150]. Compared to solution, an enhancement of 982-fold was obtained and typical Michaelis–Menten kinetics was observed ($v_{max} = 0.023$ mM min^{-1}, $K_m = 5.01$ mM, $k_{cat} = 0.0115$ min^{-1}, $k_{cat}/K_m = 2.306$ min^{-1} M^{-1}). In very recent experiments [151] we could show that by optimizing the polymer structure the enhancement of reaction rate of the imprinted polymer compared to the polymer with statistically distributed amidine groups could be further improved.

In another approach, an imprinting using labile covalent interactions was used very recently [152]. A template monomer was used to introduce a transition state analogous structure and at the same time a dicarboxylate moiety. In this case an enhancement of 120-fold compared to the solution and 55-fold compared to a control polymer containing statistically distributed dicarboxylates was obtained.

2.4.7 Outlook

The preparation of polymers and other materials with molecularly imprinted cavities has now reached a high degree of sophistication. The application of these materials is becoming more and more interesting. First industrial applications of imprinted materials are envisaged, especially as stationary phases in chromatography, e.g., for the resolution of racemates. Other interesting applications can be seen in membranes and in sensors.

At present, several research groups are engaged in preparing suitable layers or membranes for this purpose. Compared to biosensors, these layers are far more stable and they can be prepared for a large variety of compounds. Compared to standard chemosensors they are far more selective, so that there is a good chance of a broad application. It is still necessary to develop extremely sensitive methods for detecting substances bound to the imprinted membrane. At present, conductometry [57, 70, 71, 106, 153, 154], capacitance [155], pH-potentiometry [41], voltammetry [69], optical detection [87, 156], fluorescence [43, 51, 67, 68], and polarized UV [157] have been used for detection (for review see [158, 159]).

The use of imprinted polymers for radioimmunoassays has also been described [74, 82, 160]. In this case, imprinted polymers were used instead of antibodies. It is clear that these compounds cannot compete with monoclonal antibodies with regard to selectivity, but they are much easier to prepare and might therefore find application in several cases.

Reactions inside imprinted cavities are another interesting area. Of great importance to the field is catalysis with imprinted polymers and imprinted silicas. For a broader application of molecularly imprinted polymers further improvement of the method will be necessary. The following problems are in the forefront of investigation today:

a) Direct preparation of microparticles by suspension or emulsion polymerization. This problem has already been discussed.
b) Imprinting procedures in aqueous solutions. As mentioned previously, water interferes with binding since hydrogen bonding is drastically reduced in water and, e.g., boronic esters are hydrolyzed. However, it will still be necessary for many templates to work in water. Furthermore, if the imprinting is performed in an organic solvent and the equilibrations are done in water, the binding mechanism and the selectivity might be completely different.
c) Imprinting with high-molecular-weight biopolymers or even with bacteria. This is difficult: high-molecular-weight compounds cannot be extracted from a highly crosslinked bulk polymer, and therefore a type of surface imprinting has to be applied.
d) Development of new and better binding sites in imprinting, as has been indicated in Section 2.4.3.
e) Improvement of the mass transfer in imprinted polymers. This is necessary for rapid chromatographic separations, especially on a preparative scale.
f) Reduction of "polyclonality" of the cavities, which is still a severe problem.
g) Increase in capacity of chromatographic columns, especially with non-stoichiometric non-covalent interactions, as pointed out before.
h) Development of extremely sensitive detection methods for use in chemosensors.
i) Development of suitable groupings for catalysis.

2.5 Experimental Procedures

2.5.1 Polymer from Scheme 2-5

2.5.1.1 Preparation of template monomer 7 [34]

Phenyl-α-D-mannopyranoside (**7a**) (6 g; 23.4 mmol) and tris(4-vinylphenyl)boroxin (6.075 g; 15.6 mmol) were heated in benzene with removal of the water by azeotropic distillation. After complete removal of water, the solvent was evaporated, and the residue was crystallized from diethyl ether. Yield: 9.7 g (86%), m.p. 139 °C.

2.5.1.2   Preparation of the polymer [35]

Template monomer **7** (0.75 g), AIBN initiator (120 mg), crosslinker ethylene dimeth-acrylate (15.0 g) in tetrahydrofuran (15.0 g) were filled into a tube, carefully degassed by three freeze–thaw cycles, sealed under argon, and polymerized for four days at 65 °C. The tube was then cooled and broken, and the polymer was milled with an Alpine Con-traplex 63 C, and sieved to a grain size of 125–163 μm. Alternatively, the polymer could be milled to a finer powder and separated to a particle diameter fraction of 8–15 μm by a wind-sieving machine (Alpine Multiplex 100 MRZ). This material was first extracted with dry diethyl ether before being dried in vacuum at 40 °C. The template was removed from the polymer by a continuous extraction with methanol–water (250 mL per g of polymer). After the solvent was evaporated, the residue was dissolved in a defined vol-ume of methanol and the content of **7a** was determined polarimetrically.

This polymer possessed an inner surface area of 322 $m^2\ g^{-1}$, a splitting percentage of the template of 82%, a swelling ability in methanol of 1.20, and an α value of 4.52.

2.5.2   Polymer from Scheme 2-6

2.5.2.1   Thermally initiated polymerization [36]

Ethylene dimethacrylate (5.64 g), methacrylic acid (0.52 g; 6.0 mmol), L-phenylalanine anilide (0.25 g; 1.5 mmol), and of AIBN (60 mg) in acetonitrile 98.2 mL) were mixed in a glass tube. After degassing, the tube was sealed under nitrogen and consecutively heat-ed for 24 h each at 60, 90, and 120 °C. Subsequently, the polymer was ground and sub-jected to continuous extraction in acetonitrile for 24 h. To determine the recovery of imprint molecules, the extracted anilide was quantitatively determined and the polymer was investigated by nitrogen elemental analysis before and after extraction of the imprint molecule. According to these methods about 90% of the imprint molecules had been removed from the polymers.

For chromatographic purposes the polymers were first milled, then sieved in a wind-sieving machine (Alpine Multiplex 100 MZR). Separation factor α for the separation of D,L-phenylalanine anilide was 3.5. The chromatographic separation showed a resolution R_s for the racemate of 1.2.

2.5.2.2   Photochemically initiated polymerization [50]

To ethylene dimethacrylate (3.8 mL; 20 mmol), methacrylic acid (0.34 mL; 4 mmol), and L-phenylalanine anilide (240 mg; 1 mmol) in CH_2Cl_2 (porogen) (5.6 mL) were add-ed AIBN (40 mg; 0.25 mmol) as initiator. The mixture was transferred to a 50 mL thick-walled glass tube, degassed during three freeze–thaw cycles, and sealed under vacuum. The polymerization was performed at 15 °C in a thermostated water bath. The tube was placed at approx. 10 cm distance from the UV light source (medium-pressure mercury vapor lamp (Conrad–Hanovia) of 550 W with 33 W in the 320–400 nm interval) and turned at regular intervals for a symmetric exposure. After 24 h, the tubes were crushed, and the polymer was then ground in a mortar, followed by Soxhlet extraction with methanol for 12 h. Template recovery was determined by HPLC on a reversed-phase C$_{18}$ column using *p*-aminophenyl acetate as internal standard and MeOH/3% HOAc (1:1) as eluent. By this method 60–70% of template was directly recovered. The polymers were then dried overnight under vacuum at 50 °C and sieved to a 150–250 μm and a 25–38 μm particle size fraction.

The polymer thus obtained was gel-like, with an inner surface area of 3.8 m^2 g^{-1}. Swelling in acetonitrile was 2.01. For chromatographic separations the performance was improved by dry heating of the sorbent for 17 h at 120 °C. The α value for the separation of D,L-phenylalanine anilide was 2.3; chromatographic separation showed a resolution for the racemate of $R_s = 1.9$.

2.5.3 Polymer from Scheme 2-9 [136, 137]

2.5.3.1 *N*-Ethyl-4-vinylbenzamide

4-Vinylbenzoic acid chloride [161, 162] (81.60 g; 0.49 mol) in dry CH$_2$Cl$_2$ (100 mL) was dropped into a solution of ethylamine (45.09 g; 1.00 mol) in dry CH$_2$Cl$_2$ (250 mL) at −20 °C. After it had been warmed to ambient temperature, phenothiazine (0.20 g) was added. The solution was stirred for 15 h. The precipitated ethylammonium chloride was filtered off and washed with a small amount of dry CH$_2$Cl$_2$. The solvent was removed in vacuo and the residue recrystallized from EtOAc to give colorless crystals (172.4 g, 98%).

2.5.3.2 *N*-Ethyl-4-vinylbenzocarboximide acid ethyl ester

N-Ethyl-4-vinylbenzamide (39.95 g; 0.228 mol) was added to a solution of triethyloxonium tetrafluoroborate (58.5 g; 0.308 mol) in dry CH$_2$Cl$_2$ (150 mL) under argon.The solution was stirred for 2 h at ambient temperature, phenothiazine (0.20 g) was added, the stirring was continued for 36 h, and the solvent removed in vacuo. Remaining oil was treated with 3 M NaOH (80 mL)and directly extracted with ice-cooled Et$_2$O (250 mL). The aqueous layer was extracted another four times with Et$_2$O, the combined organic layers were dried over Na$_2$SO$_4$, filtered, concentrated in vacuo, and distilled to give a colorless liquid (46.19 g, approx. 100%, b.p. 53 °C, 10^{-3} mbar).

2.5.3.3 *N,N′*-Diethyl-4-vinylbenzamidine (10a)

Dry ethylammonium chloride (63.61 g; 0.78 mol) was added to a solution of *N*-ethyl-4-vinylbenzocarboximide acid ethyl ester (121.97 g; 0.60 mol) in dry EtOH (270 mL) under argon. After the mixture had been stirred at 15 °C for 5 h, 4-*tert*-butylbrenzcatechol (0.5 g) was added, the stirring was continued for five days at ambient temperature, and the solvent was removed in vacuo. The residue was treated with ice-cooled 6 M NaOH (500 mL) and directly extracted with an ice-cooled mixture of EtOAc/Et$_2$O (1:1). The aqueous layer was extracted another four times, and the combined organic layers were dried with Na_2SO_4. The solvent was removed in vacuo, and the residue was sublimed twice at 0.01 mbar to give white crystals (105.24 g, 87%, m.p. 76 °C).

2.5.4 Preparation of the polymer [112]

The polymer was prepared similarly to the first example from a mixture of ethylene di-methacrylate (4.67 g), methyl methacrylate (0.22 g), *N,N′*-diethyl-4-vinylbenzamidine (**10a**) (0.442 g; 2.188 mmol) template **14** (0.350 g; 1.094 mmol), AIBN (56.8 mg), and tetrahydrofuran (5.64 mL) as the porogen. Polymerization in a sealed ampoule was carried out for 70 h at 60 °C. After the usual workup procedure, the template was removed from the particles by extraction with methanol, followed by washing twice with 0.1 M NaOH/acetonitrile (1:1), water, and acetonitrile. The recovery of the template from the washings was determined by HPLC [RP8-column, eluent 0.2% trifluoroacetic acid in water/acetonitrile (55:45)]. Recovery of template was 85%. The inner surface area of the polymer amounted to 228 m^2 g^{-1}, and the swelling ratio in methanol was 1.45.

This polymer was used for the experiments described in Table 2-5 and Figure 2-6.

Acknowledgment

These investigations were supported by financial grants from the Deutsche Forschungs-gemeinschaft, Minister für Wissenschaft und Forschung des Landes Nordrhein-West-falen, and Fonds der Chemischen Industrie.

References

[1] C. H. Bamford, in *Development in Polymerisation – 2* (Ed.: R. N. Haward), Applied Science, London, **1979**, pp. 215–277.
[2] Y. Y. Tan, G. Challa, *Encycl. Polym. Sci. Eng.* **1985**, *16*, 554–569.
[3] G. Wulff, in *Supramolecular Stereochemistry* (Ed.: J. S. Siegel), Kluwer, Dordrecht, **1995**, pp. 13–19.
[4] G. Wulff, *Angew. Chem.* **1989**, *101*, 22–36; *Angew. Chem., Int. Ed. Engl.* **1989**, *28*, 21–37.
[5] G. Wulff, *CHEMTECH* **1991**, 364–370.
[6] G. Wulff, *Polymer News* **1991**, *16*, 167–173.
[7] G. Wulff, in *Synthesis of Polymers* (Ed.: A. D. Schlüter), Wiley–VCH Verlag, Weinheim, **1998**, pp. 375–401.
[8] G. Wulff, B. Heide, G. Helfmeier, *J. Am. Chem. Soc.* **1986**, *108*, 1089–1091.
[9] G. Wulff, B. Heide, G. Helfmeier, *React. Polym. Ion Exch. Sorbents* **1987**, *6*, 299–310.
[10] G. Wulff, *Angew. Chem.* **1995**, *107*, 1959–1979; *Angew. Chem., Int. Ed. Engl.* **1995**, *34*, 1812–1832.
[11] K. J. Shea, *Trends Polym. Sci.* **1994**, *2*, 166–173.
[12] K. Mosbach, O. Ramström, *Biotechnology* **1996**, *14*, 163–170.

[13] S. Mallik, S. D. Plunkett, P. K. Dhal, R. D. Johnson, D. Pack, D. Shnek, F. H. Arnold, *New J. Chem.* **1994**, *18*, 299–304.

[14] B. Sellergren, in *Practical Approach to Chiral Separations by Liquid Chromatography* (Ed.: G. Subramanian), VCH, Weinheim, **1994**, pp. 69–93.

[15] J. H. G. Steinke, I. R. Dunkin, D. C. Sherrington, *Adv. Polym. Sci.* **1995**, *123*, 81–126.

[16] T. Takeuchi, J. Matsui, *Acta Polym.* **1996**, *47*, 471–480.

[17] G. Wulff, K. Zabrocki, J. Hohn, *Angew. Chem.* **1978**, *90*, 567–568; *Angew. Chem., Int. Ed. Engl.* **1978**, *17*, 535–537.

[18] G. Wulff, R. Kemmerer, B. Vogt, *J. Am. Chem. Soc.* **1987**, *109*, 7449–7457.

[19] G. Wulff, P. K. Dhal, *Angew. Chem.* **1989**, *101*, 198–200; *Angew. Chem., Int. Ed. Engl.* **1989**, *28*, 196–198.

[20] K. Yokota, O. Haba, T. Satoh, T. Kakuchi, *Macromol. Chem. Phys.* **1995**, *196*, 2383–2416.

[21] G. Wulff, P. K. Dhal, *Macromolecules* **1988**, *21*, 571–578.

[22] G. Wulff, P. K. Dhal, *Macromolecules* **1990**, *23*, 100–111.

[23] G. Wulff, B. Kühneweg, *J. Org. Chem.* **1997**, *62*, 5785–5792.

[24] G. Wulff, A. Sarhan, *Angew. Chem.* **1972**, *84*, 364; *Angew. Chem., Int. Ed. Engl.* **1972**, *11*, 341.

[25] G. Wulff, A. Sarhan, *German Patent, Offenlegungsschrift* DE-A 2242796; *Chem. Abstr.* **1974**, *83*, 60300w.

[26] G. Wulff, A. Sarhan, K. Zabrocki, *Tetrahedron Lett.* **1973**, *44*, 4329–4332.

[27] G. Wulff, in *Polymeric Reagents and Catalysts* (Ed.: W. T. Ford), ACS Symposium Series, Vol. 308, Washington, **1986**, pp. 186–230.

[28] G. Wulff, in *Bioorganic Chemistry in Healthcare and Technology* (Eds.: U. K. Pandit, F. C. Alderweireldt), Plenum Press, New York, **1991**, pp. 55–69.

[29] L. Pauling, *J. Am. Chem. Soc.* **1940**, *62*, 2643–2657.

[30] E. Fischer, *Ber. Dtsch. Chem. Ges.* **1894**, *27*, 2985–2993.

[31] F. H. Dickey, *Proc. Natl. Acad. Sci. U. S. A.* **1949**, *35*, 227–229.

[32] G. Wulff, W. Vesper, R. Grobe-Einsler, A. Sarhan, *Makromol. Chem.* **1977**, *178*, 2799–2819.

[33] G. Wulff, R. Kemmerer, J. Vietmeier, H.-G. Poll, *Nouv. J. Chim.* **1982**, *6*, 681–687.

[34] G. Wulff, J. Vietmeier, H.-G. Poll, *Makromol. Chem.* **1987**, *188*, 731–740.

[35] G. Wulff, M. Minárik, *J. Liq. Chromatogr.* **1990**, *13*, 2987–3000.

[36] B. Sellergren, M. Lepistö, K. Mosbach, *J. Am. Chem. Soc.* **1988**, *110*, 5853–5860.

[37] G. Wulff, J. Schauhoff, *J. Org. Chem.* **1991**, *56*, 395–400.

[38] G. Wulff, J. Haarer, *Makromol. Chem.* **1991**, *192*, 1329–1338.

[39] W. M. Macindoe, M. Jenner, A. Williams, *Carbohydr. Res.* **1996**, *284*, 151–161.

[40] A. G. Mayes, L. I. Andersson, K. Mosbach, *Anal. Biochem.* **1994**, *222*, 483–488.

[41] G. Chen, Z. Guan, C.–T. Chen, S. Fu, V. Sundaresah, F. H. Arnold, *Nature Biotechnology* **1997**, *15*, 354–357.

[42] R. J. H. Hafkamp, B. P. A. Kokke, I. M. Danke, H. P. M. Geurts, A. E. Rowan, M. C. Feiters, R. J. M. Nolte, *Chem. Commun.* **1997**, 54––547.

[43] S. A. Piletsky, E. V. Piletskaya, K. Yano, A. Kugimiya, A. V. Elgersma, R. Levi, U. Kahlow, T. Takeuchi, I. Karube, *Analyt. Lett.* **1996**, *29*, 157–170.

[44] G. Wulff, I. Schulze, K. Zabrocki, W. Vesper, *Makromol. Chem.* **1980**, *181*, 531–544.

[45] A. Sarhan, G. Wulff, *Makromol. Chem.* **1982**, *183*, 1603–1614.

[46] G. Wulff, A. Sarhan, in *Chemical Approaches to Understanding Enzyme Catalysis: Biomimetic Chemistry and Transition-state Aalogs* (Eds.: B. S. Green, Y. Ashani, D. Chipman), Elsevier, Amsterdam, **1982**, pp. 106–118.

[47] A. Sarhan, G. Wulff, *Makromol. Chem.* **1982**, *183*, 85–92.
[48] A. Sarhan, *Makromol. Chem. Rapid Commun.* **1982**, *3*, 489–493.
[49] G. Wulff, W. Best, A. Akelah, *Reactive Polym.* **1984**, *2*, 167–174.
[50] B. Sellergren, K. J. Shea, *J. Chromatogr.* **1993**, *635*, 31–49.
[51] D. Kriz, O. Ramström, A. Svensson, K. Mosbach, *Anal. Chem.* **1995**, *67*, 2142–2144.
[52] G. Wulff, J. Vietmeier, *Makromol. Chem.* **1989**, *190*, 1727–1735.
[53] G. Wulff, J. Vietmeier, *Makromol. Chem.* **1989**, *190*, 1717–1726.
[54] J. Matsui, T. Kato, T. Takeuchi, M. Suzuki, K. Yokoyama, E. Tamiya, I. Karube, *Anal. Chem.* **1993**, *65*, 2223–2224.
[55] J.-M. Lin, T. Nakagama, X.-Z. Wu, K. Uchiyama, T. Hobo, *Fresenius J. Anal. Chem.* **1997**, *357*, 130–132.
[56] M. Yoshikawa, J.-I. Izumi, T. Kitao, S. Sakamoto, *Macromolecules* **1996**, *29*, 8197–8203.
[57] S. A. Piletsky, I. A. Butovich, V. P. Kukhar, *Zh. Anal. Khim.* **1992**, *47*, 1681–1684.
[58] Y. Fujii, K. Kikuchi, K. Matsutani, K. Ota, M. Adachi, M. Syoji, I. Haneishi, Y. Kuwana, *Chem. Lett.* **1984**, 1487–1490.
[59] S. Vidyasankar, M. Ru, F. H. Arnold, *J. Chromatogr. A* **1997**, *775*, 51–63.
[60] O. Ramström, I. A. Nicholls, K. Mosbach, *Tetrahedron: Asymmetry* **1994**, *5*, 649–656.
[61] I. A. Nicholls, O. Ramström, K. Mosbach, *J. Chromatogr. Sect. A* **1995**, *691*, 349–353.
[62] L. I. Andersson, R. Müller, K. Mosbach, *Macromol. Chem., Rapid Commun.* **1996**, *17*, 65–71.
[63] M. Kempe, K. Mosbach, *J. Chromatogr. Sect. A* **1995**, *691*, 317–323.
[64] D. L. Venton, E. Gudipati, *Biochim. Biophys. Acta* **1995**, *1250*, 126–136.
[65] M. Kempe, M. Glad, K. Mosbach, *J. Mol. Recognit.* **1995**, *8*, 35–39.
[66] S. Hjertén, J.-L. Liao, K. Nakazoto, Y. Wang, G. Zamaratskaia, H.-X. Zhang, *Chromatographia* **1997**, *44*, 227–234.
[67] A. Atherne, C. Alexander, M. J. Payne, N. Perez, E. N. Vulfson, *J. Am. Chem. Soc.* **1996**, *118*, 8771–8772.
[68] M. J. Whitcombe, C. Alexander, E. N. Vulfson, *Trends Food Sci. Technol.* **1997**, *8*, 140–145.
[69] L. D. Spurlock, A. Jaramillo, A. Praserthdam, J. Lewis, A. Brajter-Toth, *Anal. Chim. Acta* **1996**, *336*, 37–46.
[70] S. A. Piletsky, D. M. Fedoryak, V. P. Kukhar, *Dokl. Akad. Nauk. Ukr. SSR Ser. B (4)* **1990**, 53–54.
[71] S. A. Piletsky, Y. P. Parhometz, N. V. Lauryk, T. L. Panasyuk, A. V. El'skaya, *Sens. Actuators B* **1994**, *18–19*, 629–631.
[72] H. Bühnemann, N. Dattagupta, H. J. Schuetz, W. Müller, *Biochemistry* **1981**, *20*, 2864–2874.
[73] D. Spivak, M. A. Gilmore, K. J. Shea, *J. Am. Chem. Soc.* **1997**, *119*, 4388–4393.
[74] G. Vlatakis, L. I. Andersson, R. Müller, K. Mosbach, *Nature (London)* **1993**, *361*, 645–647.
[75] T. Kobayashi, H. Y. Wang, N. Fujii, *Chem. Lett.* **1995**, *10*, 927–928.
[76] M. J. Whitcombe, M. E. Rodriguez, P. Villar, E. N. Vulfson, *J. Am. Chem. Soc.* **1995**, *117*, 7105–7111.
[77] S. E. Byström, A. Boerje, B. Akermark, *J. Am. Chem. Soc.* **1993**, *115*, 2081–2083.
[78] S. H. Cheong, S. McNiven, A. Rachkov, R. Levi, K. Yono, I. Karube, *Macromolecules* **1997**, *30*, 1317–1322.
[79] B. Sellergren, *Anal. Chem.* **1994**, *66*, 1578–1582.
[80] J. Matsui, A. Kaneko, Y. Miyoshi, K. Yokoyama, E. Tamiya, T. Takeuchi, *Anal. Lett.* **1996**, *29*, 2071–2078.

[81] J. Matsui, I. A. Nicholls, T. Takeuchi, *Tetrahedron: Asymmetry* **1996**, *7*, 1357–1361.

[82] L. I. Andersson, R. Müller, G. Vlatakis, K. Mosbach, *Proc. Natl. Acad. Sci. U. S. A.* **1995**, *92*, 4788–4792.

[83] L. Schweitz, L. J. Andersson, S. Nilsson, *Anal. Chem.* **1997**, *69*, 1179–1183.

[84] P. Martin, I. D. Wilson, D. E. Morgan, G. R. Jones, K. Jones, *Anal. Commun.* **1997**, *34*, 45–47.

[85] M. Walshe, E. Garcia, J. Howarth, M. R. Smyth, M. T. Kelly, *Anal. Commun.* **1997**, *34*, 119–122.

[86] L. I. Andersson, *Anal. Chem.* **1996**, *68*, 111–117.

[87] R. Levi, S. McNiven, S. A. Piletsky, S.-H. Cheong, K. Yano, I. Karube, *Anal. Chem.* **1997**, *69*, 2017–2021.

[88] S. Kato, K. J. Shea, unpublished results presented at the ACS Spring Meeting in San Francisco, **1997**.

[89] K. J. Shea, T. K. Dougherty, *J. Am. Chem. Soc.* **1986**, *108*, 1091–1093.

[90] K. J. Shea, D. Y. Sasaki, *J. Am. Chem. Soc.* **1991**, *113*, 4109–4120.

[91] J. Matsui, I. A. Nicholls, T. Takeuchi, K. Mosbach, I. Karube, *Anal. Chim. Acta* **1996**, *335*, 71–77.

[92] K. J. Shea, E. A. Thompson, S. D. Pandey, P. S. Beauchamp, *J. Am. Chem. Soc.* **1980**, *102*, 3149–3151.

[93] J. Damen, D. C. Neckers, *J. Am. Chem. Soc.* **1980**, *102*, 3265–3267.

[94] J. V. Beach, K. J. Shea, *J. Am. Chem. Soc.* **1994**, *116*, 379–380.

[95] G. Wulff, I. Schulze, *Angew. Chem.* **1978**, *90*, 568–570; *Angew. Chem., Int. Ed. Engl.* **1978**, *17*, 537–538.

[96] G. Wulff, I. Schulze, *Isr. J. Chem.* **1978**, *17*, 291–297.

[97] P. K. Dhal, F. H. Arnold, *Macromolecules* **1992**, *25*, 7051–7059.

[98] S. Mallik, R. D. Johnson, F. H. Arnold, *J. Am. Chem. Soc.* **1994**, *116*, 8902–8911.

[99] F. H. Dickey, *J. Phys. Chem.* **1955**, *59*, 695–707.

[100] T. Takagishi, I. M. Klotz, *Biopolymers* **1972**, *11*, 483–491.

[101] R. Arshady, K. Mosbach, *Makromol. Chem.* **1981**, *182*, 687–692.

[102] J. Matsui, Y. Miyoshi, O. Doblhoff-Dier, T. Takeuchi, *Anal. Chem.* **1995**, *67*, 4404–4408.

[103] M. Siemann, L. I. Andersson, K. Mosbach, *J. Agric. Food Chem.* **1996**, *44*, 141–145.

[104] M. T. Muldoon, L. H. Stanker, *Anal. Chem.* **1997**, *69*, 803–808.

[105] B. Sellergren, C. Dauwe, T. Schneider, *Macromolecules* **1997**, *30*, 2454–2459.

[106] S. A. Piletsky, E. V. Piletskaya, A. V. Elgersma, K. Yano, I. Karube, Y. P. Parhometz, *Biosens. Biolelectron.* **1995**, *10*, 959–964.

[107] J. Matsui, M. Okada, M. Tsuruoka, T. Takeuchi, *Anal. Commun.* **1997**, *34*, 85–87.

[108] D. K. Robinson, K. Mosbach, *J. Chem. Soc., Chem. Commun.* **1989**, 969–970.

[109] B. Sellergren, K. J. Shea, *Tetrahedron Asymmetry* **1994**, *5*, 1403–1406.

[110] K. Ohkubo, Y. Urata, S. Hirota, Y. Honda, T. Sagawa, *J. Mol. Catal.* **1994**, *87*, L21–L24.

[111] K. Ohkubo, Y. Funakoshi, Y. Urata, S. Hirota, S. Usui, T. Sagawa, *J. Chem. Soc., Chem. Commun.* **1995**, 2143–2144.

[112] G. Wulff, T. Gross, R. Schönfeld, *Angew. Chem.* **1997**, *109*, 2050–2052; *Angew. Chem., Int. Ed. Engl.* **1997**, *36*, 1961–1964.

[113] T. Shimada, K. Nakanishi, K. Morihara, *Bull. Chem. Soc. Jpn.* **1992**, *65*, 954–958.

[114] R. A. Bartsch, M. Maeda (eds.), *Molecular and Ionic Recognition with Imprinted Polymers*, ACS Symposium Series, Vol. 703, Washington, **1998**.

[115] B. Sellergren (ed.), *Molecularly Imprinted Polymers. Man Made Mimics of Antibodies and their Application in Analytical Chemistry*, Elsevier, Amsterdam, in press.

[116] G. Wulff, W. Vesper, *J. Chromatogr.* **1978**, *167*, 171–186.
[117] G. Wulff, M. Minárik, *J. High Resolut. Chromatogr. Commun.* **1986**, *9*, 607–608.
[118] A. Strikovsky, J. Hradil, G. Wulff, unpublished results.
[119] A. G. Mayes, K. Mosbach, *Anal. Chem.* **1996**, *68*, 3769–3774.
[120] K. Hosoya, K. Yoshizako, N. Tanaka, K. Kimata, T. Araki, J. Haginaka, *Chem. Lett.* **1994**, 1437–1438.
[121] O. Norrlöw, M. Glad, K. Mosbach, *J. Chromatogr.* **1984**, *299*, 29–41.
[122] G. Wulff, D. Oberkobusch, M. Minárik, *React. Polym. Ion Exch. Sorbents* **1985**, *3*, 261–275.
[123] G. Wulff, D. Oberkobusch, M. Minárik, in *Proceedings of the XVIIIth Solvay Conference on Chemistry, Brussels 1983* (Ed.: G. v. Binst), Springer Verlag, Berlin, **1986**, pp. 229–233.
[124] S. D. Plunkett, F. H. Arnold, *J. Chromatogr. Sect. A* **1995**, *708*, 19–29.
[125] F. H. Arnold, S. Plunkett, P. K. Dhal, S. Vidyasankar, *Polym. Prepr. (Am. Chem. Soc. Div. Polym. Chem.)* **1995**, *36*, 97–98.
[126] E. Vulfson, C. Alexander, M. Whitcombe, *Chem. Br.* **1997**, 23–26.
[127] J. Matsui, Y. Miyoshi, R. Matsui, T. Takeuchi, *Anal. Sci.* **1995**, *11*, 1017–1019.
[128] B. Sellergren, *J. Chromatogr. Sect. A* **1994**, *673*, 133–141.
[129] G. Wulff, *TIBTECH* **1993**, *11*, 85–87.
[130] G. Wulff, *Mol. Cryst. Liq. Cryst.* **1996**, *276*, 1–6.
[131] D. J. O'Shannessy, L. I. Andersson, K. Mosbach, *J. Mol. Recognit.* **1989**, *2*, 1–5.
[132] K. J. Shea, D. Y. Sasaki, *J. Am. Chem. Soc.* **1989**, *111*, 3442–3444.
[133] G. Wulff, G. Kirstein, *Angew. Chem.* **1990**, *102*, 706–708; *Angew. Chem., Int. Ed. Engl.* **1990**, *29*, 684–686.
[134] I. R. Dunkin, J. Lenfeld, D. C. Sherrington, *Polymer* **1993**, *34*, 77–84.
[135] J. Matsui, Y. Miyoshi, T. Takeuchi, *Chem. Lett.* **1995**, 1007–1008.
[136] G. Wulff, R. Schönfeld, M. Grün, R. Baumstark, G. Wildburg, L. Häußling (BASF AG), *German Patent, Offenlegungsschrift* DE A 197 20 345 A 1, **1998**.
[137] G. Wulff, R. Schönfeld, unpublished work; R. Schönfeld, *PhD Thesis*, University of Düsseldorf, **1998**.
[138] T. Schrader, C. Kirsten, *J. Chem. Soc., Chem. Commun.* **1996**, 2089–2090.
[139] C. Kirsten, T. Schrader, *J. Am. Chem. Soc.* **1997**, *119*, 12061–12068; C. Kirsten, *Doctoral Dissertation*, University of Düsseldorf, **1997**.
[140] K. Tanabe, T. Takeuchi, J. Matsui, K. Yano, K. Ikebukuro, I. Karube, *J. Chem. Soc., Chem. Commun.* **1995**, 2303–2304.
[141] P. Pino, I. Bartus, O. Vogl, *Polym. Prepr. (Am. Chem. Soc. Div. Polym. Chem.)* **1988**, *29*, 254–255.
[142] B. Sellergren, K. J. Shea, *J. Chromatogr. Sect. A* **1993**, *654*, 17–28.
[143] G. Wulff, M. Minárik, in *Chromatographic Chiral Separations* (Eds.: M. Zief, L. J. Crane), Marcel Dekker, New York, **1988**, pp. 15–52.
[144] R. A. Lerner, S. J. Benkovic, P. G. Schultz, *Science* **1991**, *252*, 659–667.
[145] F. Locatelli, P. Gamez, M. Lemaire, *Stud. Surface Sci. Catal.* **1997**, *108*, 517–522.
[146] M. E. Davis, A. Katz, W. R. Ahmad, *Chem. Mater.* **1996**, *8*, 1820–1839.
[147] R. Müller, L. I. Andersson, K. Mosbach, *Makromol. Chem., Rapid Commun.* **1993**, *14*, 637–641.
[148] K. Morihara, M. Kurokawa, Y. Kamata, T. Shimada, *J. Chem. Soc., Chem. Commun.* **1992**, 358–360.
[149] J. D. Stewart, L. J. Liotta, S. J. Benkovic, *Acc. Chem. Res.* **1993**, *26*, 396–404.
[150] G. Wulff, D. Kasper, unpublished results; D. Kasper, *PhD Thesis*, University of Düsseldorf, **1999**.

[151] K. Shiraishi, G. Wulff, unpublished results.

[152] J.-M. Kim, K.-D. Ahn, G. Wulff, *J. Am. Chem. Soc.*, submitted.

[153] D. Kriz, M. Kempe, K. Mosbach, *Sens. Actuators B* **1996**, *33*, 178–181.

[154] S. A. Piletsky, E. V. Piletskaya, T. I. Pansyuk, A. V. El'skaya, A. E. Rachhov, R. Levi, I. Karube, G. Wulff, *Macromolecules* **1998**, *31*, 2137–2140.

[155] E. Hedborg, F. Winquist, I. Lundström, L. Andersson, K. Mosbach, *Sens. Actuators A* **1993**, *37–38*, 796–799.

[156] S. Marx-Tibbon, I. Willner, *Chem. Commun.* **1994**, 1261–1262.

[157] J. H. G. Steinke, I. R. Dunkin, D. C. Sherrington, *Macromolecules* **1996**, *29*, 407–415.

[158] G. Wulff, in *Frontiers in Biosensorics I—Fundamental Aspects* (Eds.: F. W. Scheller, F. Schubert, J. Fedrowitz), Birkhäuser, Basel, **1997**, pp. 13–26.

[159] D. Kriz, O. Ramström, K. Mosbach, *Anal. Chem.* **1997**, *69*, A 345–349.

[160] H. Bengtsson, U. Roos, L. I. Andersson, *Anal. Commun.* **1997**, *34*, 233–235.

[161] R. Broos, D. Tavernier, M. Anteunis, *J. Chem. Ed.* **1978**, *55*, 813.

[162] T. Ishizone, A. Hirao, S. Nakahama, *Macromolecules* **1989**, *22*, 2895–2901.

3 Templated Synthesis of Catenanes and Rotaxanes

Françisco M. Raymo, J. Fraser Stoddart

3.1 Introduction

Catenanes [1] are molecules incorporating two or more component macrocycles, mechanically interlocked like links in a chain. Rotaxanes [1] are molecules which incorporate one or more macrocyclic components encircling the linear portion of a dumbbell-shaped component in a manner reminiscent of an abacus. In both instances, no covalent bonds hold the components together. Rather, mechanical bonds – in most situations reinforced by noncovalent bonds – are responsible for linking the components. Dismembering a catenane or a rotaxane into its individual components can occur only if one or more covalent bonds are cleaved in the mechanically interlocked molecule. Thus, catenanes and rotaxanes (Figure 3-1) behave as well-defined molecular compounds with properties significantly different from those of their separate components when they are isolated as individual molecular compounds in their own right. These unique features of catenanes and rotaxanes offer the possibility of designing the molecular-sized counterparts of bearings, chains, joints, machines, motors, and many other macroscopic objects composed of mechanically interlocked parts.

The first catenanes and rotaxanes were constructed by either statistical [2] or directed [3] synthetic approaches. The statistical method relies [2] on the formation of small quantities of a species in which a cyclic molecule is threaded by an acyclic molecule. After experiencing appropriate covalent bond formation, these threaded species are con-

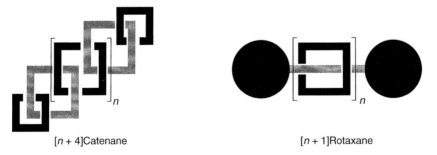

[n + 4]Catenane   [n + 1]Rotaxane

Figure 3-1   Schematic representations of [n + 4]catenanes and [n + 1]rotaxanes.

verted into mechanically interlocked molecular compounds. However, threading of the cyclic onto the acyclic molecule is disfavored entropically and very small amounts of the threaded species are formed in the absence of any significant noncovalent bonding interactions between their components. As a result, the yields of the resulting catenanes and rotaxanes are extremely low. The directed syntheses of catenanes and rotaxanes involve [3] the multistep construction of so-called precatenanes and prerotaxanes. These precursors incorporate covalently bonded macrocyclic and acyclic components. After the cleavage of the covalent bonds holding the components together, mechanically interlocked molecules result. These multistep syntheses, however, are laborious, time-consuming, and, without exception, low-yielding overall. Fortunately, however, efficient methods for the production of complexes incorporating macrocyclic hosts threaded onto acyclic guests have become available with the advent of supramolecular chemistry [4]. Indeed, template-directed [5] synthetic approaches to mechanically interlocked molecular compounds have so far been developed that rely upon transition metal coordination [6, 7], hydrogen bonding [8, 9], hydrophobic interactions [7, 10], and donor/acceptor interactions [11, 12]. In most of the transition metal-based-synthetic strategies, metal templates assist in the formation of catenanes or of rotaxanes and then the templates are removed without destroying the mechanically interlocked nature of the molecules. By contrast, when organic templates are used, they become an integral part of the mechanically interlocked structures and cannot be removed without extensive modification of the catenane's or rotaxane's constitution or redox state.

3.2   Metal-Templated Syntheses

The first efficient synthesis of a [2]catenane, which was realized (Figure 3-2) in 1983 by Sauvage et al. [6, 13], relied upon the assistance of a metal template. The Strasbourg group made use of the fact that Cu^+ ions coordinate tetrahedrally with phenanthroline ligands. Thus, when equimolar amounts of the phenanthroline-based diol **1**, the preformed macrocycle **2**, and $Cu(MeCN)_4BF_4$ are mixed together in solution, the threaded species **3** is obtained. Reaction of this intermediate with $I[(CH_2)_2O]_4(CH_2)_2I$ in the presence of Cs_2CO_3 also affords the [2]catenate **4**, but this time only in a yield of 42%. Similarly, when $Cu(MeCN)_4BF_4$ is mixed with two molar equivalents of the phenanthroline-based diol **1** in solution, the complex **5** is obtained. Its reaction with $I[(CH_2)_2O]_4(CH_2)_2I$ in the presence of Cs_2CO_3 affords the [2]catenate **4** in a yield of 27%. The interlocked structure of **4** was confirmed unequivocally by X-ray crystallography which revealed the tetrahedral coordination of the Cu^+ ion by the two phenanthroline ligands. Demetallation of the [2]catenate **4** was achieved quantitatively by treating it with a large excess of KCN in aqueous acetonitrile. The solid state structure of the demetallated form of **4** revealed that the two macrocyclic components are still interlocked. However, in the absence of the Cu^+ ions, the molecules adopt a completely different co-conformation [14] wherein the two phenanthroline ligands become positioned well away from each other in the component macrocycles.

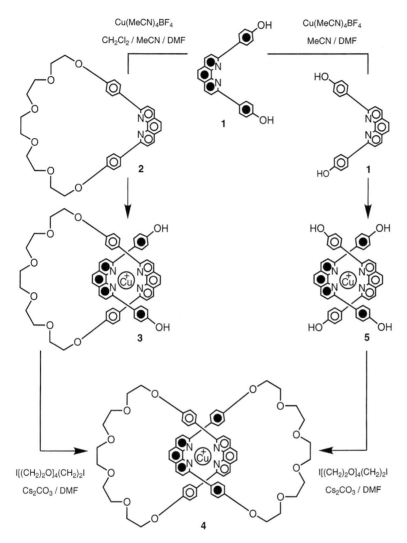

Figure 3-2 Metal-templated synthetic approaches to the [2]catenate **4**.

Metal templates have also been employed [6, 15] to synthesize (Figure 3-3) phenanthroline-containing rotaxanes. The threaded species **6** is prepared by combining equimolar amounts of its cyclic and acyclic components and Cu(MeCN)$_4$BF$_4$ in solution. Reaction of **6** with 3,5-di-*t*-butylbenzaldehyde and 3,3′-diethyl-4,4′-dimethyl-2,2′-dipyrrylmethane in the presence of CF$_3$CO$_2$H, followed by treatment with chloranil and then with Zn(OAc)$_2$ · H$_2$O, affords the [2]rotaxane **7**. Removal of the Cu$^+$ ions is achieved by treating a CH$_2$Cl$_2$/MeCN/H$_2$O solution of the [2]rotaxane **7** with KCN. As in the case of the [2]catenate **4**, an ^{1}H-NMR spectroscopic investigation revealed that demetallation is accompanied by a co-conformational change which involves the circumrotation of the macrocyclic component through 180°. Thus, in the demetallated [2]rotaxane **8**, the two phenanthroline ligands are also remote from each other in the dumbbell and macrocycle components. However, complete reorganization to regenerate a co-conformation similar to that adopted by the metallated [2]rotaxane **7** occurs when the demetallated [2]rotaxane **8** is mixed with Zn(OAc)$_2$ · H$_2$O in a CH$_2$Cl$_2$/MeOH solution.

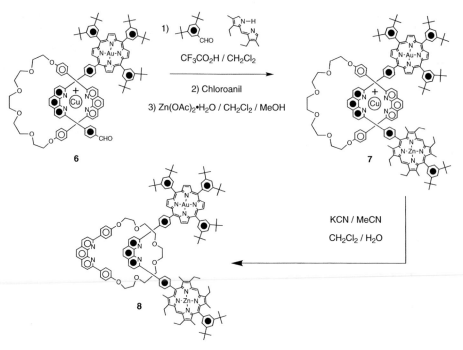

Figure 3-3   Metal-templated synthesis of the [2]rotaxane **8** under kinetic control.

3.3   Hydrogen Bonding-assisted Syntheses

Hydrogen bonding interactions between amide groups can be employed [8] (Figure 3-4) to template the synthesis of [2]catenanes composed of mechanically interlocked macrocyclic lactams. This discovery was made independently by Hunter and by Vögtle. In Hunter's experiments, the diamine **9** was reacted [16] in CH$_2$Cl$_2$ with the bis(acid chloride) **10** in the presence of Et$_3$N to afford the diamine **11**. In an attempt to synthesize a macrocyclic lactam which was expected to bind *p*-benzoquinone, diamine **11** was reacted with bis(acid chloride) **10** under high-dilution conditions. However, in addition to the expected macrocycle, the [2]catenane **12** was also isolated, in a yield of 34%. The interlocked nature of the structure of the [2]catenane **12** was deduced initially from a detailed high-field NMR spectroscopic investigation and was confirmed, subsequently, by a single-crystal X-ray analysis, which revealed a network of hydrogen bonding interactions between the amide groups in the component macrocycles. A similar synthetic outcome was attained by Vögtle et al. [17, 18] when the diamine **9** was reacted with the bis(acid chloride) **13** under high-dilution conditions. In this instance, the [2]catenane **14** was obtained directly in a yield of 18%. The mechanism of these template-directed syntheses was unraveled [19] by analyzing the distribution of products for one- and two-step syntheses performed using starting components having various R groups on their 5-R-*m*-phenylene rings (Figure 3-4). It is believed that the formation of a macrocyclic lactam is followed by the binding of an acyclic precursor inside its cavity as a result of hydrogen bond formation. The macrocyclization of the bound acyclic species between its termini interlocks mechanically the two macrocyclic lactams.

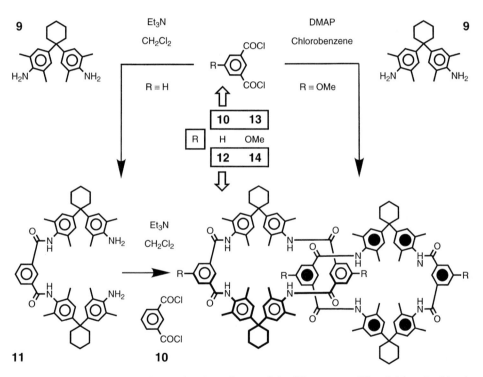

Figure 3-4 Hydrogen bonding-assisted syntheses of the [2]catenanes **12** and **14** under kinetic control.

A similar recognition motif has been employed by Vögtle et al. [8, 20] and by Leigh et al. [21, 22] in the template-directed synthesis (Figure 3-5) of [2]rotaxanes. The glycine-based compounds **15** and **16**, incorporating terminal diphenylmethane stoppers, are able to template the macrocyclization to a macrocyclic lactam of an appropriate size, insofar as the Et_3N-promoted reaction of isophthaloyl chloride **10** with *p*-xylylenediamine **17** in the presence of **15** or **16** affords [21] the [2]rotaxane **18** or **19** in yields of 30 or 28%, respectively. As a result of the presence of dual amide recognition sites at each end of the linear portion in the dumbbell-shaped component, the macrocyclic lactam can shuttle degenerately back and forth along the acyclic backbone in $CDCl_3$. Variable-temperature ^{1}H-NMR spectroscopic investigations revealed that the rate of shuttling is affected by the number (*n*) of methylene groups linking the peptide units. At 298 K, the rates of shuttling are ca. 37 000 and 5 200 s^{-1} for **18** and **19**, respectively. In $(CD_3)_2SO$, the amide recognition sites become highly solvated, the macrocyclic lactam resides preferentially on the central polymethylene chain of the dumbbell component, and no shuttling of the macrocycle is observed. The dependence of the shuttling motion on the polarity of the medium can be exploited to tune the rates of this dynamic process by controlling the degree of solvation of the recognition sites using appropriate mixtures of solvents.

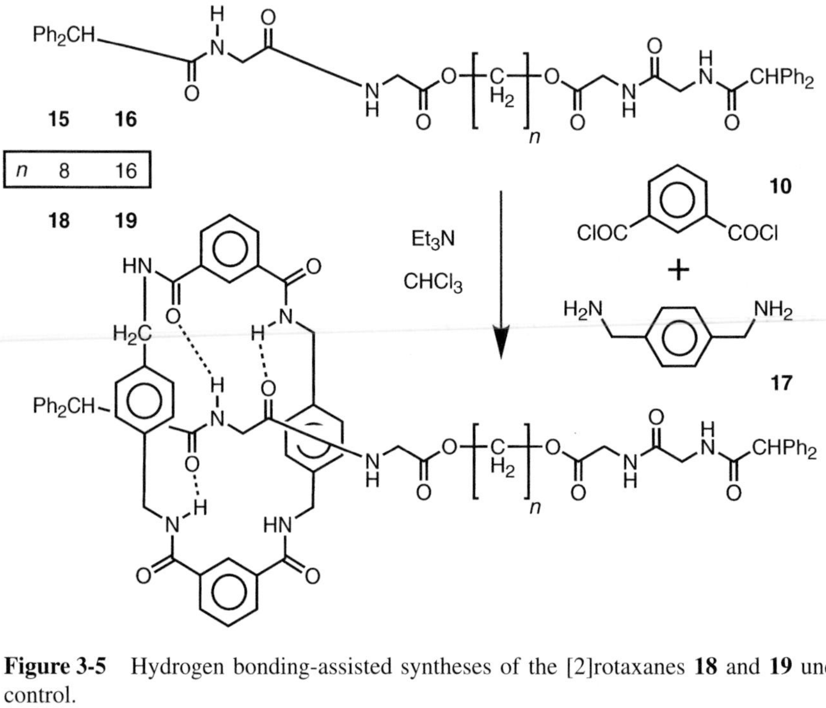

Figure 3-5 Hydrogen bonding-assisted syntheses of the [2]rotaxanes **18** and **19** under kinetic control.

The hydrogen bond-directed assembly of a [2]catenane under thermodynamic control has been realized (Figure 3-6) by Leigh et al. [23] using a reversible ring-opening and ring-closing metathesis (RORCM) reaction. An isomeric mixture of the macrocyclic lactam **20** (0.2 M in CH_2Cl_2) and the Grubbs ruthenium carbene catalyst was stirred at ambient temperature. The progress of the reaction was monitored by high-performance liquid chromatography (HPLC), which showed the gradual formation of the [2]catenane **21**. When the equilibrium between the free macrocycle **20** and the interlocked species **21** was reached, the percentage of [2]catenane in solution was higher than 95%. However, diluting the solution shifted the equilibrium toward the free macrocycle, demonstrating that this reaction is one which is truly occurring under thermodynamic control.

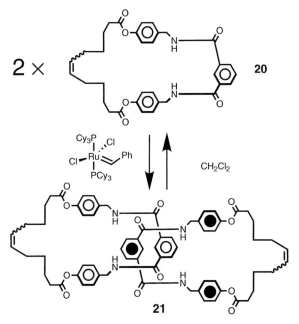

Figure 3-6 Hydrogen bonding-assisted synthesis of the [2]catenane **21** under thermodynamic control.

3.4 Hydrophobically Driven Syntheses

In aqueous solution, cyclodextrins bind [10] organic guests inside their cavities with pseudorotaxane geometries. Thus, if the binding event is followed by the intramolecular linking of the termini of the threaded guest, mechanical interlocking of the two components occurs, yielding a [2]catenane. As a result of reasoning along these lines, Lüttringhaus et al. [24] attempted unsuccessfully, as early as 1958, to synthesize a cyclodextrin-containing [2]catenane. By combining α-cyclodextrin with a 1,4-dioxybenzene-containing dithiol in H_2O, a 1:1 complex was indeed obtained. However, the oxidation of the terminal thiol groups of the guest to afford a macrocyclic disulfide did not occur. It was some 35 years later that the first cyclodextrin-containing catenane (Figure 3-7) was synthesized [25] using a similar design strategy. The biphenylene-containing bisa-mine **22** is threaded through the cavity of the methylated β-cyclodextrin derivative **24** in aqueous solution. Thus, when the diamine **22** is reacted with terephthaloyl chloride **23** in the presence of **24** and sodium hydroxide, the [2]catenanes **25** and **26**, as well as the [3]catenanes **27** and **28**, are obtained, but in very low yields. It is interesting to note that the [3]catenanes **27** and **28** are topological stereoisomers and differ only in the relative orientation of the two toroidal cyclodextrin components. They were separated by reverse-phase high-performance liquid chromatography and their topologies were deduced by careful analyses of their ^{13}C-NMR spectra.

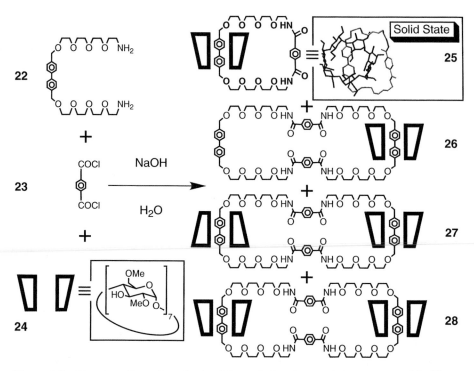

Figure 3-7   Template-directed synthesis of the cyclodextrin-containing catenanes **25–28**.

The formation of pseudorotaxane complexes in aqueous solutions has also been exploited [10] to template the synthesis of cyclodextrin-containing rotaxanes. In the example illustrated in Figure 3-8, the diamine **29** is mixed [26, 27] with a methylated α-cyclodextrin derivative – either **30** or **31** – in aqueous solution. In both instances, a pseu-dorotaxane complex self-assembles spontaneously. The subsequent covalent attachment of bulky stoppers is achieved by reacting these complexes with sodium 2,4,6-trinitroben-zenesulfonate **32**. The resulting [2]rotaxanes **33** and **34** can be isolated in yields of 42 and

48%, respectively. The incorporation of the 2,4,6-trinitrophenyl stoppers was confirmed by the characteristic absorption band in the UV-Vis spectra of the [2]rotaxanes. Furthermore, fast atom bombardment mass spectrometry, ^{1}H-NMR and ^{13}C-NMR spectroscopies, elemental analysis, and the marked differences in solubilities between the [2]rotaxanes and their parent components confirmed their interlocked structures.

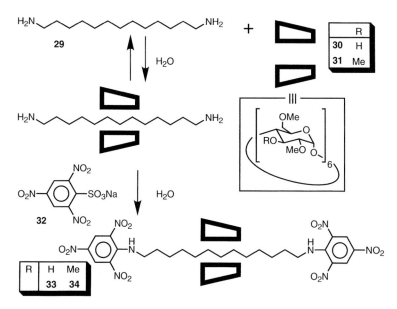

Figure 3-8 Template-directed synthesis of cyclodextrin-containing [2]rotaxanes **33** and **34**.

Fujita et al. [28] have achieved the quantitative hydrophobically driven syntheses of [2]catenanes by employing pyridine-based ligands and transition metals possessing square planar geometries. In the example illustrated in Figure 3-9, the [2]catenane **36** self-assembles quantitatively from the preformed dinuclear macrocycle **35** when heated at 100 °C in a concentrated aqueous solution of NaNO$_3$. Under these forcing conditions, the ligand-platinum bonds are constantly opening and closing. Sequential ligand exchange between the two individual dinuclear macrocycles produces an intermediate Möbius strip compound which – once again as a result of ligand exchange – is transformed into the [2]catenane **36**. Presumably, strong face-to-face and edge-to-face intercomponent interactions between aromatic units within the [2]catenane drive the process thermodynamically toward the mechanically interlocked compound.

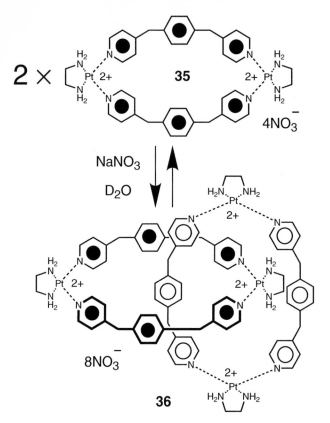

Figure 3-9   Metal-templated synthesis of the [2]catenane **36** under thermodynamic control.

3.5   Aromatic Templates

The 1,4-dioxybenzene-based macrocyclic polyether **37** binds the bipyridinium-based compound **38** as a result of [C−H···O] hydrogen bonding [29] between some of the polyether oxygen atoms and the α-bipyridinium protons and [π···π] stacking between the complementary aromatic units [30]. The same noncovalent bonds, supplemented by [C−H···π] interactions between the 1,4-dioxybenzene protons and the *p*-phenylene spacers, assist [31] in the complexation of the 1,4-dioxybenzene derivative **39** by the bipyridinium-based cyclophane **40**. In both complexes – i.e., **37·38** and **39·40** shown in Figure 3-10 – the acyclic guest is inserted through the macrocyclic host with its ends protruding outside the cavity from opposite sides. This recognition motif paved [11] the way for the design of template-directed approaches to catenanes and rotaxanes.

The first [2]catenane incorporating π-electron-rich and π-electron-deficient recognition sites was synthesized [31, 32] as outlined in Figure 3-11. The reaction of the

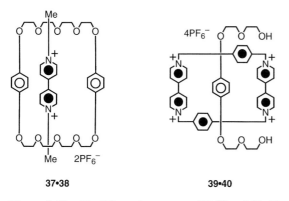

37•38   **39•40**

Figure 3-10   The [2]pseudorotaxanes **37·38** and **39·40**.

bis(pyridylpyridinium)-containing dicationic salt **41** with the dibromide **42** produced a tricationic intermediate incorporating one bipyridinium and one pyridylpyridinium unit. Once formed, this intermediate is bound immediately by the preformed macrocyclic polyether **37**, which enhances greatly the rate of the subsequent ring-closure reaction under kinetic control. After counterion exchange, the [2]catenane **43** can be isolated in a yield of 70%. A kinetic investigation of this template-directed synthesis has demonstrated [33] that the macrocyclic template accelerates the ring-closure reaction ca. 232-fold at a concentration of ca. 0.1 M. The interlocked structure of this [2]catenane was confirmed (Figure 3-11) by single-crystal X-ray analysis which also revealed a combination of intracatenane [C−H···O] hydrogen bonds, [π···π] stacking, and [C−H···π] interactions. In solution, the two mechanically interlocked rings undergo a number of co-conformational changes with respect to each other. The macrocyclic polyether circumrotates through the cavity of the tetracationic cyclophane, exchanging the "inside" and "alongside" 1,4-dioxybenzene rings. Similarly, the tetracationic cyclophane circumrotates through the cavity of the macrocyclic polyether, exchanging the "inside" and "'alongside" bipyridinium units. Variable-temperature ^{1}H-NMR spectroscopy revealed that the free-energy barrier associated with the circumrotation of the π-electron-rich macrocyclic polyether through the cavity of the π-electron-deficient cyclophane is ca. 4 kcal mol^{-1} higher than that of the other circumrotation process. Interestingly, when the π-donor character of the 1,4-dioxybenzene rings is reduced by introducing fluorine substituents, the intracatenane noncovalent bonding interactions are weakened [34] significantly and, as a result, the free-energy barriers associated with the circumrotation processes decrease substantially, i.e., by ca. 3 kcal mol^{-1}. Consistently, when better π-donors (e.g., 1,5-dioxynaphthalene ring systems) are used in place of the 1,4-dioxybenzene rings, the intracatenane noncovalent bonding interactions are reinforced [35] and the free-energy barriers associated with the circumrotation processes increase significantly, i.e., by ca. 2 kcal mol^{-1}. It is interesting to note that the π-donor character of the π-electron-rich recognition sites also affects [34, 35] the efficiency of the template-directed syntheses. Thus, when the π-donor character is diminished by introducing electron-withdrawing

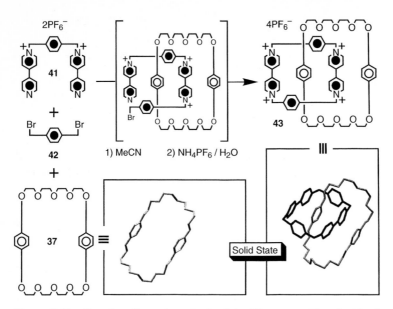

Figure 3-11   Template-directed synthesis of the [2]catenane **43** under kinetic control.

substituents (e.g., fluorine atoms), the yields of the resulting [2]catenanes drop signifi-cantly to a point where it becomes impossible to obtain the catenane. By contrast, when the π-donor character is enhanced (e.g., using 1,5-dioxynaphthalene ring systems), the yields increase to more than 80% in the best cases.

This synthetic strategy has been extended [11, 36] to the template-directed synthesis of catenanes incorporating more than two mechanically interlocked components. Indeed, by enlarging the cavities of the π-electron-rich and of the π-electron-deficient ring com-ponents, as many as seven macrocycles have been interlocked. Reaction of the bis(pyri-dylpyridinium)-containing dicationic salt **44** with the dibromide **45** in the presence of the macrocyclic polyether **46** affords [37] (Figure 3-12) the [3]catenane **47** in a yield of 10%. This [3]catenane incorporates two macrocyclic polyethers which are large enough to accommodate one more bipyridinium unit in each of their cavities. Thus, when the bis(pyridylpyridinium)-containing dicationic salt **41** is reacted with the dibromide **42** at ambient temperature and pressure in the presence of the [3]catenane **47**, the [5]catenane (Olympiadane) **48** is obtained, along with a [4]catenane (not shown in Figure 3-12) in yields of 5 and 31%, respectively. When the same reaction is performed under high pres-sure (12 kbar) conditions, the [7]catenane **49**, a [6]catenane (not shown in Figure 3-12), and Olympiadane (**48**) are obtained in yields of 26, 28, and 30%, respectively. These [n]catenanes were characterized by liquid secondary ion mass spectrometry which revealed peaks for $[M-nPF_6]^+$ corresponding to the loss of hexafluorophosphate coun-terions. Their interlocked structures were demonstrated unequivocally by single-crystal X-ray analyses. These studies revealed that the $[\pi\cdots\pi]$ stacking interactions between the

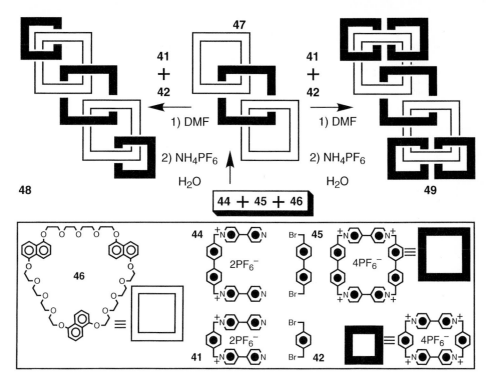

Figure 3-12 Conversion of the [3]catenane **47** into the [5]catenane **48** and into the [7]catenane **49**.

π-donors and π-acceptors are accompanied by [C–H$\cdots$O] hydrogen bonds between the α-bipyridinium hydrogen atoms and some of the polyether oxygen atoms, as well as by [C–H$\cdots\pi$] interactions between some of the 1,5-dioxynaphthalene hydrogen atoms and the phenylene rings in the tetracationic cyclophane components.

The combination of noncovalent bonding interactions with the binding of bipyridinium recognition sites by π-electron-rich macrocyclic polyethers was also employed [11] to synthesize rotaxanes under template control. In the example illustrated in Figure 3-13, the reaction of the bis(pyridylpyridinium)-containing dicationic salt **41** with the chloride **50** affords [38, 39] a tricationic intermediate (not shown in Figure 3-13) incorporating one bipyridinium and one pyridylpyridinium unit. This intermediate is bound immediately by the preformed macrocyclic polyether **37** and, after the covalent attachment of a second tetraarylmethane-based stopper at the other end of the acyclic components, the [2]rotaxane **51** is obtained in a yield of 18%. In solution, the macrocyclic component "shuttles" from one bipyridinium recognition site to the other. Variable-temperature ^{1}H-NMR spectroscopy showed that this degenerate site exchange process occurs at a rate of ca. 300 000 s^{-1} in (CD$_3$)$_2$CO at 25 °C.

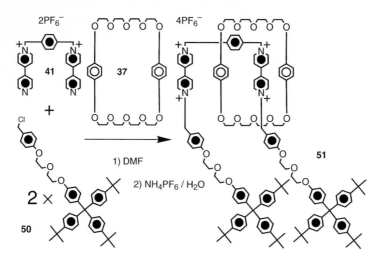

Figure 3-13   Template-directed synthesis of the [2]rotaxane **51** under kinetic control.

An alternative approach that is entirely thermodynamically controlled has also been developed [38b, 40] for the template-directed synthesis of [2]rotaxanes incorporating bipyridinium-based dumbbell-shaped components encircled by π-electron-rich macrocyclic polyethers. In this instance, a preformed macrocyclic polyether is combined with a preformed dumbbell-shaped compound. If size complementarity between the macrocycle's cavity and the dumbbell's stoppers is achieved, slippage of the macrocyclic component over the stoppers of the dumbbell-shaped compound occurs upon heating. Thus, when an MeCN solution of the macrocyclic polyether **37** and any one of the dumbbell-shaped compounds **52**–**54** is heated [38b, 40a] (Figure 3-14) at 50 °C for 10 days, the corresponding [2]rotaxanes **56**–**58** are obtained in yields of 52, 45, and 47%, respectively. However, when the dumbbell-shaped compound **55** incorporating bis(4-*t*-butylphenyl)-4-isopropylphenylmethyl-based stoppers is employed, under otherwise identical conditions, no rotaxane corresponding to structure **59** can be isolated. The replacement of ethyl groups with isopropyl groups is sufficient to prevent the slippage of the macrocycle over the tetraarylmethane-based stoppers. The mechanism associated with the passage of the macrocyclic component over the tetraarylmethane-based stoppers has been investigated [41] computationally in order to gain further insight into such an "all-or-nothing" substituent effect. Model stoppers incorporating a methoxy group instead of the acyclic portion of the dumbbell-shaped compound were oriented relative to the macrocycle as illustrated in Figure 3-15. The distances D were varied stepwise from 58.0 to 35.0 Å, while the coordinates of the oxygen atom of the stopper, and those of a reference point located at a distance of 60.0 Å, were fixed. At each step, the superstructure was subjected to molecular dynamics at a simulated temperature of 500 K and then the energies of 200 randomly selected co-conformers were minimized. The energies calculated for the lowest energy point obtained for each step were plotted (Figure 3-16) against D.

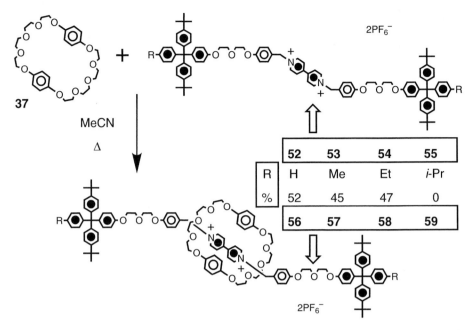

Figure 3-14   Syntheses of the [2]rotaxanes **56–58** by slippage under thermodynamic control.

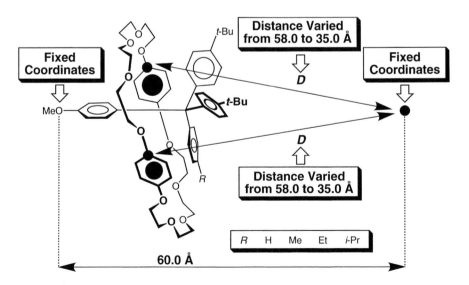

Figure 3-15   Initial geometries and constraints used for simulating the slippage processes.

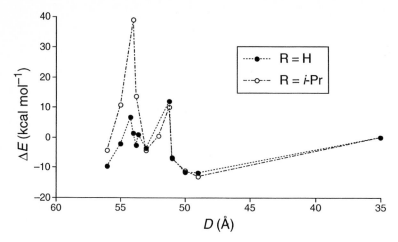

Figure 3-16 Energy profiles associated with the slippage of the macrocycle **37** over the stoppers of **52** and **55**.

This simulation protocol was employed to study the slippage processes associated with R equal to H, Me, Et, and *i*-Pr. In all cases, the resulting energy profiles displayed two energy barriers. The first of these is associated with the slippage of one of the polyether chains over the R group, while the second one corresponds to the slippage of the other polyether chain over one of the two *t*-Bu groups. When the size of the R group is increased, the first energy barrier rises, whereas the second one remains constant. When R is equal to either H or Me, the second energy barrier is rate determining. On the other hand, when R is equivalent to either Et or *i*-Pr, the first energy barrier is rate-determining. However, the energy barrier observed for slippage of the macrocycle over the *i*-Pr-bearing stopper is more than 20 kcal mol^{-1} higher than for any of its smaller congeners. Figure 3-16 shows the plots of the energy profiles associated with R equal to H and *i*-Pr. It is interesting to note that the first energy barrier increases significantly and becomes rate-determining when R is *i*-Pr.

Aromatic templates, in conjunction with coordinative bonds, have been employed by Sanders et al. [42] to self-assemble a [2]catenane incorporating a chiral metallomacrocycle. The 1,5-dioxynaphthalene-based macrocyclic polyether **60** threads onto the π-electron-deficient compound **61** in MeCN. Thus, when both compounds and $Zn(OSO_2CF_3)_2$ are mixed in this solvent, threading of **60** onto **61** is followed by the [2+2] assembly of a helical metallomacrocycle as a result of the tetrahedral coordination of two Zn^{2+} centers by the bipyridine ligands appended to the π-electron-deficient recognition sites. The resulting [2]catenane **62** was characterized by a combination of ^{1}H-NMR spectroscopy and electrospray mass spectrometry.

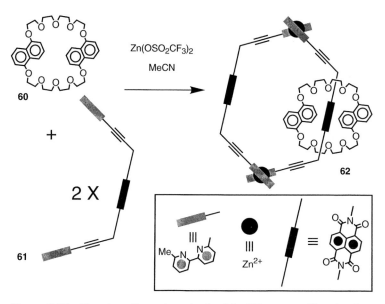

Figure 3-17 Template-directed synthesis of the [2]catenane **62** under thermodynamic control.

3.6 Dialkylammonium-containing Rotaxanes

The macrocyclic polyether **64** binds (Figure 3-18) the dialkylammonium salt **63** with pseudorotaxane geometry in solution and in the solid state [43]. The formation of the complex **63·64** is determined by [N$^+$–H···O] and [C–H···O] hydrogen bonds between the hydrogen atoms of the ammonium center and of the two adjacent methylene groups, respectively, and the polyether oxygen atoms, as well as by [π···π] stacking between the aromatic units of host and guest. By enlarging the size of the host, two, three, and four dialkylammonium guests can be complexed inside the macrocyclic cavity. Thus, the [3]pseudorotaxane **(63)$_2$·37**, the [4]pseudorotaxane **(63)$_3$·65**, and the [5]pseudorotaxane **(63)$_4$·66** self-assemble [44] spontaneously when the dialkylammonium salt **63** is combined with the corresponding macrocyclic host.

The threading of a dialkylammonium salt through the cavity of a crown ether was exploited [45] to synthesize (Figure 3-19) a [2]rotaxane after the covalent attachment under kinetic control of bulky groups at both ends of the guest. Upon mixing the crown ether **64** with the dialkylammonium salt **67** in CH$_2$Cl$_2$, a [2]pseudorotaxane forms spontaneously. The azide groups at the termini of its guest component are capable of undergoing cycloadditions under relatively mild conditions. Thus, reaction with di-*t*-butyl acetylenedicarboxylate converts the [2]pseudorotaxane intermediate into the [2]rotaxane **68** in a yield of 31% as a result of the formation of trisubstituted triazole stoppers.

This strategy was extended [46] (Figure 3-20) to the template-directed synthesis of a [2]rotaxane incorporating a dumbbell-shaped component possessing one ammonium and

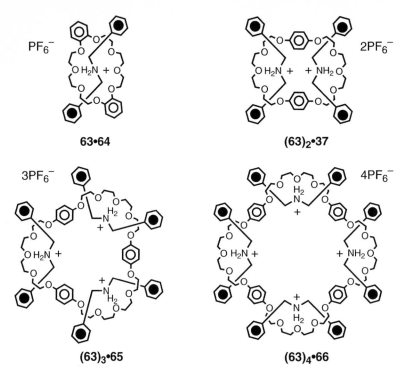

63•64 (63)$_2$•37

(63)$_3$•65 **(63)$_4$•66**

Figure 3-18   The [*n*]pseudorotaxanes **63·64**, **(63)$_2$·37**, **(63)$_3$·65**, and **(63)$_4$·66**.

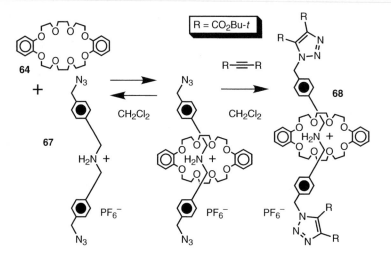

Figure 3-19   Template-directed synthesis of the [2]rotaxane **68** under kinetic control.

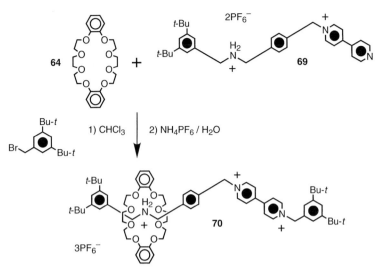

Figure 3-20   Template-directed synthesis of the [2]rotaxane **70** under kinetic control.

one bipyridinium recognition site. Threading of the macrocyclic polyether **64** onto the dicationic salt **69**, which has one stopper already attached to one of its two ends, occurs spontaneously in $CHCl_3$ at ambient temperature. Subsequent addition of 3,5-di-*t*-butyl-benzyl bromide, and heating the solution under reflux for four days, afforded the [2]rotaxane **70** in a yield of 38%. The ^{1}H-NMR spectrum of the [2]rotaxane **70** in $(CD_3)_2CO$ at room temperature showed the selective binding of the ammonium recognition site by the macrocyclic polyether. However, addition of *i*-Pr_2NEt deprotonates the ammonium group, "switching off" its recognition properties. As a result, the macrocyclic polyether component moves and occupies the bipyridinium recognition site. The outcome is the appearance of a red color associated with the charge-transfer interactions between the catechol rings and the bipyridinium unit. Addition of CF_3CO_2H regenerates the ammonium recognition site, resulting in the disappearance of the red color, since the macrocyclic polyether component moves back onto the ammonium recognition site. The [2]rotaxane **70** is a remarkable example of a "clean" on/off molecular-sized switch [47] which can be operated reversibly by external stimuli.

The guest component of the *supramolecular* complex **63·64** (Figure 3-18) incorporates phenyl rings at its two ends. These groups are small enough to penetrate the macrocyclic cavity of the host but they are not large enough to prevent dethreading. Thus, a thermodynamic equilibrium between complexed and uncomplexed species is established in solution. By contrast, the dialkylammonium salt **71** incorporates terminal groups which are slightly bulkier. Indeed, heating for several days is required [48] (Figure 3-21) to permit the threading of the host **62** onto the guest **71**. After cooling the solution to ambient temperature, the threaded species **72** becomes kinetically stable and can be isolated after chromatography in a yield of >90%. Furthermore, no dissociation of **72** into

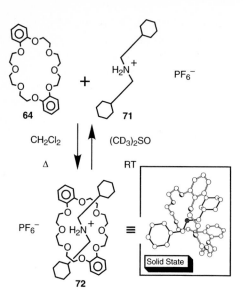

Figure 3-21   Reversible complexation of the salt **71** by the macrocyclic polyether **64**.

its separate components was observed at ambient temperature in a CDCl$_3$/CD$_3$CN (3:1) solution, even after several weeks. Interestingly, however, when (CD$_3$)$_2$SO is employed instead as the solvent, the quantitative dissociation of **72** into **64** and **71** occurs at ambient temperature in only 18 h, presumably since the intercomponent hydrogen bonds are destroyed in this highly solvating medium.

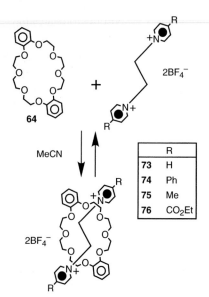

	R
73	H
74	Ph
75	Me
76	CO$_2$Et

Figure 3-22   Complexation of the salts **73–76** by the macrocyclic polyether **64**.

A similar approach has been employed (Figure 3-22) by Loeb and Wisner [49] to self-assemble [2]pseudorotaxanes composed of 1,2-bis(pyridinium)ethane axles and crown ether wheels. Indeed, by mixing any of the bis(pyridinium)-based dicationic guests **73–76** with the macrocyclic polyether **64**, pseudorotaxane complexes self-assemble in MeCN. Threading of the macrocyclic components onto the dicationic guests was confirmed by the X-ray crystallographic analyses of some of these complexes which revealed the presence of [C–H···O] hydrogen bonds between the pyridinium hydrogen atoms in the α-position with respect to the nitrogen atoms and the polyether oxygen atoms.

3.7 Conclusions

Mechanically interlocked molecular compounds – namely, catenanes and rotaxanes – can now be synthesized efficiently by relying on the assistance of inorganic or organic templates. These templates can guide the covalent synthesis of catenanes and rotaxanes under either kinetic or thermodynamic control. In the first instance, the irreversible formation of one or more covalent bonds determines the outcome of the overall process, i.e., it depends upon the relative stabilities of transition states leading to products. In the second case, reversible bonding interactions guide the outcome of the template-directed synthesis according to the relative stabilities of the products. When organic templates are employed, the formation of intermediate complexes is stabilized by noncovalent bonding interactions. These supramolecular intermediates are converted into mechanically interlocked molecules after the formation of either "dynamic" or "static" covalent/coordinative bonds. In both instances, the template becomes an integral part of the final product and cannot be removed without destroying its mechanically interlocked structure. When metal templates are used, intermediate complexes are formed as a result of metal coordination. Once again, the conversion of such intermediates into mechanically interlocked products requires the formation of either "dynamic" or "static" covalent/coordinative bonds. However, in some instances, the metal template can be easily removed once the mechanically interlocked structure has been assembled. These efficient template-directed syntheses not only offer the opportunity of reproducing the features of mechanical interlocking at the molecular level, but they also pave the way for the construction of molecular-level switches, motors, and machines [50]. The reason is that the information used to template their formation lives on in the molecules thereafter.

3.8 Experimental Procedures

3.8.1 [2]Catenane 4 [13]

A solution of Cu(MeCN)$_4$BF$_4$ (1.8 mmol) in degassed MeCN (30 mL) was added to a solution of **1** (3.0 mmol) in degassed DMF (20 mL). The mixture was stirred for 1 h at ambient temperature and under an atmosphere of Ar. The solvent was evaporated under reduced pressure to afford **5** quantitatively. A solution of **5** (1.5 mmol) and I[(CH$_2$)O]$_4$(CH$_2$)$_2$I (3.3 mmol) in DMF (200 mL) was added dropwise over 20 h to a suspension of Cs$_2$CO$_3$ (9.20 mmol) in DMF (400 mL) maintained at 60 °C. Another portion of I[(CH$_2$)O]$_4$(CH$_2$)$_2$I (1.2 mmol) in DMF (50 mL) was added over 20 h and the resulting mixture was stirred for a further 24 h at 60 °C. After cooling down to ambient temperature, the solvent was removed under reduced pressure and the residue was partitioned between H$_2$O and CH$_2$Cl$_2$. The aqueous layer was extracted with CH$_2$Cl$_2$ (3×100 mL). The organic layers were combined, and HBF$_4$ (34%, 100 mL) was added. The mixture was stirred for 12 h at ambient temperature. After decantation, the solution was washed with H$_2$O (2×100 mL) and a large excess of NaBF$_4$, dissolved in the minimum amount of H$_2$O, was added. After 3 h, the organic layer was washed with H$_2$O (2×100 mL) and dried (MgSO$_4$). The solvent was removed under reduced pressure and the residue was purified by column chromatography (SiO$_2$, CH$_2$Cl$_2$) to afford the [2]catenane **4** (42%).

3.8.2   [2]Catenane 12 [16]

A solution of diamine **9** (1.3 mmol) and Et$_3$N (0.4 mL) in dry CH$_2$Cl$_2$ (250 mL) and a solution of isophthaloyl chloride **10** (1.8 g) in dry CH$_2$Cl$_2$ (250 mL) were added simultaneously over 4 h to dry CH$_2$Cl$_2$ (1200 mL) from two separate dropping funnels. The mixture was stirred for 12 h at ambient temperature. The solvent was evaporated under reduced pressure and the residue was purified by column chromatography (SiO$_2$, CHCl$_3$). After crystallization from CHCl$_3$–pentane, the [2]catenane **12** (34%) was obtained.

3.8.3   [2]Catenane 43 [31]

A solution of the macrocyclic polyether **37** (0.18 mmol), the salt **41** (0.07 mmol), and the dibromide **42** (0.07 mmol) in dry MeCN (3 mL) was stirred for 14 days at ambient temperature. The solvent was removed under reduced pressure and the resulting solid residue was purified by column chromatography [SiO$_2$, MeOH–2 M NH$_4$Cl–MeNO$_2$ (7:2:1)] to afford a red solid which was dissolved in H$_2$O (60 mL). After the addition of a saturated aqueous solution of NH$_4$PF$_6$, the [2]catenane **43** (70%) precipitated out of solution.

3.8.4   [2]Rotaxane 51 [38]

A solution of the macrocyclic polyether **37** (0.22 mmol), the salt **41** (0.20 mmol), and the chloride **50** (0.43 mmol) in dry DMF (8 mL) was subjected to a pressure of 12 kbar for two days at 30 °C. The solvent was evaporated under reduced pressure, and the residue was purified by column chromatography [SiO$_2$, MeOH–CH$_2$Cl$_2$–CH$_3$NO$_2$–2 M NH$_4$Cl (70:16:9:5)] to give a red solid which was dissolved in H$_2$O (60 mL). After the addition of a saturated aqueous solution of NH$_4$PF$_6$, the [2]rotaxane **51** (18%) precipitated out of solution.

3.8.5   [2]Rotaxane 56 [38]

A solution of the macrocyclic polyether **37** (0.28 mmol) and the salt **52** (0.07 mmol) in MeCN (7 mL) was stirred for 10 days at 50 °C. The solvent was evaporated under reduced pressure, and the residue was purified by column chromatography [SiO$_2$, MeOH–MeNO$_2$–CH$_2$Cl$_2$ (6:1:1)] to give **56** (52%).

3.8.6   [2]Rotaxane 68 [45]

A solution of the macrocyclic polyether **64** (0.662 mmol), the salt **67** (0.221 mmol), and di-*t*-butyl acetylenedicarboxylate (186 µmol) in CH$_2$Cl$_2$ (10 mL) was heated under reflux for nine days. After cooling to ambient temperature, the solvent was removed under reduced pressure and the residue was purified by column chromatography [SiO$_2$, CH$_2$Cl$_2$–MeOH (from 100:0 to 90:10)] to afford the [2]rotaxane **68** (31%).

References

[1] For books and reviews on catenanes and rotaxanes, see: (a) G. Schill, *Catenanes, Rotaxanes and Knots*, Academic Press, New York, **1971**. (b) D. M. Walba, *Tetrahedron* **1985**, *41*, 3161–3212. (c) C. O. Dietrich-Buchecker, J.-P. Sauvage, *Chem. Rev.* **1987**, *87*, 795–810. (d) C. O. Dietrich-Buchecker, J.-P. Sauvage, *Bioorg. Chem. Front.* **1991**, *2*, 195–248. (e) J.-C. Chambron, C. O. Dietrich-Buchecker, J.-P. Sauvage, *Top. Curr. Chem.* **1993**, *165*, 131–162. (f) H. W. Gibson, H. Marand, *Adv. Mater.* **1993**, *5*, 11–21. (g) H. W. Gibson, M. C. Bheda, P. T. Engen, *Prog. Polym. Sci.* **1994**, *19*, 843–945. (h) D. B. Amabilino, I. W. Parsons, J. F. Stoddart, *Trends Polym. Sci.* **1994**, *2*, 146–152. (i) D. B. Amabilino, J. F. Stoddart, *Chem. Rev.* **1995**, *95*, 2725–2828. (j) H. W. Gibson, *Large Ring Molecules*, Ed. J. A. Semlyen, Wiley, New York, **1996**, 191–202. (k) M. Belohradsky, F. M. Raymo, J. F. Stoddart, *Collect. Czech. Chem. Commun.* **1996**, *61*, 1–43. (l) F. M. Raymo, J. F. Stoddart, *Trends Polym. Sci.* **1996**, *4*, 208–211. (m) M. Belohradsky, F. M. Raymo, J. F. Stoddart, *Collect. Czech. Chem. Commun.* **1997**, *62*, 527–557. (n) R. Jäger, F. Vögtle, *Angew. Chem., Int. Ed. Engl.* **1997**, *36*, 930–944. (o) Eds. C. O. Dietrich-Buckecher, J.-P. Sauvage, *Molecular Catenanes, Rotaxanes and Knots*, VCH-Wiley, Weinheim, **1999**. (p) F. M. Raymo, J. F. Stoddart, *Chem. Rev.* **1999**, *99*, 1643–1664. (q) D. A. Leigh, A. Murphy, *Chem. Ind.* **1999**, 178–183. (r) G. A. Breault, C. A. Hunter, P. C. Mayers, *Tetrahedron* **1999**, *55*, 5265–5293.

[2] (a) E. Wasserman, *J. Am. Chem. Soc.* **1960**, *82*, 4433–4434. (b) H. L. Frisch, E. Wasserman, *J. Am. Chem. Soc.* **1961**, *83*, 3789–3795. (c) E. Wasserman, *Sci. Am.* **1962**, *207*(5), 94–102. (d) I. T. Harrison, S. Harrison, *J. Am. Chem. Soc.* **1967**, *89*, 5723–5724. (e) I. T. Harrison, *J. Chem. Soc., Chem. Commun.* **1972**, 231–232. (f) I. T. Harrison, *J. Chem. Soc., Perkin Trans. 1* **1974**, 301–304. (g) G. Agam, D. Graiver, A. Zilkha, *J. Am. Chem. Soc.* **1976**, *98*, 5206–5214. (h) G. Agam, A. Zilkha, *J. Am. Chem. Soc.* **1976**, *98*, 5214–5216.

[3] (a) G. Schill, A. Luttringhaus, *Angew. Chem., Int. Ed. Engl.* **1964**, *3*, 546–547. (b) G. Schill, *Chem. Ber.* **1965**, *98*, 3439–3445. (c) G. Schill, *Chem. Ber.* **1967**, *100*, 2021–2037. (d) G. Schill, C. Zurcher, *Angew. Chem., Int. Ed. Engl.* **1969**, *8*, 988. (e) G. Schill, H. Zollenkopf, *Liebigs Ann. Chem.* **1969**, *721*, 53–74. (f) G. Schill, E. Logemann, W. Vetter, *Angew. Chem., Int. Ed. Engl.* **1972**, *11*, 1089–1090. (g) G. Schill, W. Beckmann, W. Vetter, *Angew. Chem., Int. Ed. Engl.* **1973**, *12*, 665–666. (h) G. Schill, C. Zürcher, W. Vetter, *Chem. Ber.* **1973**, *106*, 228–235. (i) G. Schill, C. Zürcher, *Chem. Ber.* **1977**, *110*, 2046–2066. (j) E. Logemann, G. Schill, *Chem. Ber.* **1978**, *111*, 2615–2629. (k) G. Schill, G. Doerjer, E. Logemann, W. Vetter, *Chem. Ber.* **1980**, *112*, 3697–3705. (l) G. Schill, K. Rissler, H. Fritz, W. Vetter, *Angew. Chem., Int. Ed. Engl.* **1981**, *20*, 187–189. (m) G. Schill, N. Schweickert, H. Fritz, W. Vetter, *Angew. Chem., Int. Ed. Engl.* **1983**, *22*, 889–891. (n) K. Rißler, G. Schill, H. Fritz, W. Vetter, *Chem. Ber.* **1986**, *119*, 1374–1399. (o) G. Schill, W. Beckmann, N. Schweickert, H. Fritz, *Chem. Ber.* **1986**, *119*, 2647–2655. (p) G. Schill, N. Schweickert, H. Fritz, W. Vetter, *Chem. Ber.* **1988**, *121*, 961–970.

[4] (a) F. Vögtle, *Supramolecular Chemistry*, Wiley, New York, **1991**. (b) J.-M. Lehn, *Supramolecular Chemistry*, VCH, Weinheim, **1995**. (c) Eds. J.-M. Lehn, J. L. Atwood, J. E. D. Davies, D. D. Mac Nicol, F. Vögtle, *Comprehensive Supramolecular Chemistry*, Pergamon, Oxford, **1996**.

[5] For accounts and reviews on template-directed syntheses, see: (a) D. H. Busch, N. A. Stephenson, *Coord. Chem. Rev.* **1990**, *100*, 119–154. (b) J. S. Lindsey, *New J. Chem.* **1991**, *15*, 153–180. (c) G. M. Whitesides, J. P. Mathias, C. T. Seto, *Science* **1991**, *254*, 1312–1319. (d) D. Philp, J. F. Stoddart, *Synlett* **1991**, 445–458. (e) D. H. Busch, *J. Inclusion Phenom.* **1992**, *12*, 389–395. (f) S. Anderson, H. L. Anderson, J. K. M. Sanders, *Acc. Chem. Res.* **1993**, *26*, 469–475. (g) R. Cacciapaglia, L. Mandolini, *Chem. Soc. Rev.* **1993**, *22*, 221–231. (h) R. Hoss, F. Vögtle, *Angew. Chem., Int. Ed. Engl.* **1994**, *33*, 375–384. (i) J. P. Schneider, J. W. Kelly, *Chem. Rev.* **1995**, *95*, 2169–2187. (j) D. Philp, J. F. Stoddart, *Angew. Chem., Int. Ed. Engl.* **1996**, *35*, 1155–1196. (k) F. M. Raymo, J. F. Stoddart, *Pure Appl. Chem.* **1996**, *68*, 313–322. (l) M. C. T. Fyfe, J. F. Stoddart, *Acc. Chem. Res.* **1997**, *30*, 393–401. (m) T. J. Hubin, A. G. Kolchinski, A. L. Vance, D. L. Busch, *Adv. Supramol. Chem.* **1999**, *5*, 237–357.

[6] (a) J.-P. Sauvage, *Acc. Chem. Res.* **1990**, *23*, 319–327. (b) J.-C. Chambron, C. O. Dietrich-Buchecker, C. Hemmert, A. K. Khemiss, D. Mitchell, J.-P. Sauvage, J. Weiss, *Pure Appl. Chem.* **1990**, *62*, 1027–1034. (c) J.-C. Chambron, S. Chardon-Noblat, A. Harriman, V. Heitz, J.-P. Sauvage, *Pure Appl. Chem.* **1993**, *65*, 2343–2349. (d) J.-C. Chambron, C. O. Dietrich-Buchecker, J.-F. Nierengarten, J.-P. Sauvage, *Pure Appl. Chem.* **1994**, *66*, 1543–1550. (e) J.-C. Chambron, C. O. Dietrich-Buchecker, V. Heitz, J.-F. Nierengarten, J.-P. Sauvage, C. Pascard, J. Guilhem, *Pure Appl. Chem.* **1995**, *67*, 233–240. (f) J.-C. Chambron, C. O. Dietrich-Buchecker, J.-P. Sauvage, *Comprehensive Supramolecular Chemistry*, Vol. 9, Eds. M. W. Hosseini, J.-P. Sauvage, Pergamon, Oxford, **1996**, 43–83.

[7] (a) F. Bickelhaupt, *J. Organomet. Chem.* **1994**, *475*, 1–14. (b) M. Fujita, K. Ogura, *Coord. Chem. Rev.* **1996**, *148*, 249–264. (c) M. Fujita, *Comprehensive Supramolecular Chemistry*, Vol. 9, Eds. M. W. Hosseini, J.-P. Sauvage, Pergamon, Oxford, **1996**, 253–282. (d) Y. M. Jeon, D. Whang, J. Kim, K. Kim, *Chem. Lett.* **1996**, 503–504. (e) D. Whang, Y. M. Jeon, J. Heo, K. Kim, *J. Am. Chem. Soc.* **1996**, *118*, 11333–11334. (f) D. Whang, J. Heo, C. A. Kim, K. Kim, *Chem. Commun.* **1997**, 2361–2362. (g) D. Whang, K. Kim, *J. Am. Chem. Soc.* **1997**, *119*, 451–452. (h) D. Whang, K. M. Park, J. Heo, P. R. Ashton, K. Kim, *J. Am. Chem. Soc.* **1998**, *120*, 4899–4900. (i) S. G. Roh, K. M. Park, G. J. Park, S. Sakamoto, K. Yamaguchi, K. Kim, *Angew. Chem. Int. Ed.* **1999**, *38*, 638–641.

[8] (a) C. A. Hunter, *Chem. Soc. Rev.* **1994**, *23*, 101–109. (b) G. Brodesser, R. Güther, R. Hoss, S. Meier, S. Ottens-Hildebrandt, J. Schmitz, F. Vögtle, *Pure Appl. Chem.* **1993**, *65*, 2325–

2328. (c) F. Vögtle, T. Dünnwald, T. Schmidt, *Acc. Chem. Res.* **1996**, *29*, 451–460. (d) F. Vögtle, R. Jäger, M. Händel, S. Ottens-Hildebrandt, *Pure. Appl. Chem.* **1996**, *68*, 225–232. (e) A. G. Johnston, D. A. Leigh, R. J. Pritchard, M. D. Deegan, *Angew. Chem., Int. Ed. Engl.* **1995**, *34*, 1209–1212. (f) A. G. Johnston, D. A. Leigh, L. Nezhat, J. P. Smart, M. D. Deegan, *Angew. Chem., Int. Ed. Engl.* **1995**, *34*, 1212–1216. (g) D. A. Leigh, K. Moody, J. P. Smart, K. J. Watson, A. M. Z. Slawin, *Angew. Chem., Int. Ed. Engl.* **1996**, *35*, 306–310. (h) D. A. Leigh, A. Murphy, J. P. Smart, M. S. Deleuze, F. Zerbetto, *J. Am. Chem. Soc.* **1998**, *120*, 6458–6467. (i) M. Fanti, C. A. Fustin, D. A. Leigh, A. Murphy, P. Rudolf, R. Caudano, R. Zamboni, F. Zerbetto, *J. Phys. Chem. A* **1998**, *102*, 5782–5788. (j) C. A. Fustin, P. Rudolf, A. F. Taminiaux, F. Zerbetto, D. A. Leigh, R. Caudano, *Thin Solid Films* **1998**, *327–328*, 321–325. (k) T. J. Kidd, D. A. Leigh, A. J. Wilson, *J. Am. Chem. Soc.* **1999**, *121*, 1599–1600. (l) M. S. Deleuze, D. A. Leigh, F. Zerbetto, *J. Am. Chem. Soc.* **1999**, *121*, 2364–2379. (m) F. Biscarini, W. Gebauer, D. Di Domenico, R. Zamboni, J. I. Pascual, D. A. Leigh, A. Murphy, D. Tetard, *Synth. Met.* **1999**, *102*, 1466–1467. (n) R. Zamboni, M. Muccini, W. Gebauer, F. Biscarini, M. Murgia, G. Ruani, D. A. Leigh, A. Murphy, D. Tetard, *Synth. Met.* **1999**, *102*, 1556–1557.

[9] (a) A. G. Kolchinski, D. H. Busch, N. W. Alcock, *J. Chem. Soc., Chem. Commun.* **1995**, 1289–1291. (b) A. G. Kolchinski, N. W. Alcock, R. A. Roesner, D. H. Busch, *Chem. Commun.* **1998**, 1437–1438. (c) P. T. Glink, C. Schiavo, J. F. Stoddart, D. J. Williams, *Chem. Commun.* **1996**, 1483–1490. (d) P. T. Glink, J. F. Stoddart, *Pure Appl. Chem.* **1998**, *70*, 419–424. (e) M. C. T. Fyfe, J. F. Stoddart, *Adv. Supramol. Chem.* **1999**, *5*, 1–53. (f) M. C. T. Fyfe, J. F. Stoddart, *Coord. Chem. Rev.* **1999**, *183*, 139–155.

[10] (a) J. F. Stoddart, *Angew. Chem., Int. Ed. Engl.* **1992**, *31*, 846–848. (b) H. Ogino, *New J. Chem.* **1993**, *17*, 683–688. (c) G. Wenz, F. Wolf, M. Wagner, S. Kubik, *New J. Chem.* **1993**, *17*, 729–738. (d) R. Isnin, A. E. Kaifer, *Pure Appl. Chem.* **1993**, *65*, 495–498. (e) A. Harada, *Polym. News* **1993**, *18*, 358–363. (f) A. Harada, J. Li, M. Kamachi, *Proc. Jpn. Acad.* **1993**, *69*, 39–44. (g) G. Wenz, *Angew. Chem., Int. Ed. Engl.* **1994**, *33*, 802–822. (h) A. Harada, *Coord. Chem. Rev.* **1996**, *148*, 115–133. (i) A. Harada, *Large Ring Molecules*, Ed. J. A. Semlyen, Wiley, New York, **1996**, 406–432. (j) A. Harada, *Supramol. Sci.* **1996**, *3*, 19–23. (k) A. Harada, *Adv. Polym.* **1997**, *133*, 142–191. (l) A. Harada, *Carbohydr. Polym.* **1997**, *34*, 183–188. (m) A. Harada, *Acta Polym.* **1998**, *49*, 3–17. (n) S. A. Nepogodiev, J. F. Stoddart, *Chem. Rev.* **1998**, *98*, 1959–1976.

[11] (a) D. B. Amabilino, J. F. Stoddart, *Pure Appl. Chem.* **1993**, *65*, 2351–2359. (b) D. Pasini, F. M. Raymo, J. F. Stoddart, *Gazz. Chim. Ital.* **1995**, *125*, 431–435. (c) S. J. Langford, J. F. Stoddart, *Pure Appl. Chem.* **1996**, *68*, 1255–1260. (d) D. B. Amabilino, F. M. Raymo, J. F. Stoddart, *Comprehensive Supramolecular Chemistry*, Vol. 9, Eds. M. W. Hosseini, J.-P. Sauvage, Pergamon, Oxford, **1996**, 85–130. (e) F. M. Raymo, J. F. Stoddart, *Pure Appl. Chem.* **1997**, *69*, 1987–1997. (f) R. E. Gillard, F. M. Raymo, J. F. Stoddart, *Chem. Eur. J.* **1997**, *3*, 1933–1940. (g) F. M. Raymo, J. F. Stoddart, *Chemtracts* **1998**, *11*, 491–511.

[12] (a) D. G. Hamilton, J. K. M. Sanders, J. E. Davies, W. Clegg, S. J. Teat, *Chem. Commun.*, **1997**, 897–898. (b) A. C. Try, M. M. Harding, D. G. Hamilton, J. K. M. Sanders, *Chem. Commun.* **1998**, 723–724. (c) D. G. Hamilton, J. E. Davies, L. Prodi, J. K. M. Sanders, *Chem. Eur. J.* **1998**, *4*, 608–620. (d) D. G. Hamilton, N. Feeder, L. Prodi, S. J. Teat, W. Clegg, J. K. M. Sanders, *J. Am. Chem. Soc.* **1998**, *120*, 1096–1097. (e) D. G. Hamilton, N. Feeder, S. J. Teat, J. K. M. Sanders, *New J. Chem.* **1999**, 493–502. (f) D. G. Hamilton, L. Prodi, N. Feeder, J. K. M. Sanders, *J. Chem. Soc., Perkin Trans. 1* **1999**, 1057–1065.

[13] (a) C. O. Dietrich-Buchecker, J.-P. Sauvage, J.-P. Kintzinger, *Tetrahedron Lett.* **1983**, *24*, 5095–5098. (b) C. O. Dietrich-Buchecker, J.-P. Sauvage, J.-M. Kern, *J. Am. Chem. Soc.*

1984, *106*, 3043–3045. (c) C. O. Dietrich-Buchecker, J.-P. Sauvage, *Tetrahedron* **1990**, *46*, 503–512.

[14] For a definition of the term "co-conformation", see: M. C. T. Fyfe, P. T. Glink, S. Menzer, J. F. Stoddart, A. J. P. White, D. J. Williams, *Angew. Chem., Int. Ed. Engl.* **1997**, *36*, 2068–2070.

[15] J.-C. Chambron, V. Heitz, J.-P. Sauvage, *J. Am. Chem. Soc.* **1993**, *115*, 12378–12384.

[16] (a) C. A. Hunter, D. H. Purvis, *Angew. Chem., Int. Ed. Engl.* **1992**, *31*, 792–795. (b) C. A. Hunter, *J. Am. Chem. Soc.* **1992**, *114*, 5303–5311. (c) H. Adams, F. J. Carver, C. A. Hunter, *J. Chem. Soc., Chem. Commun.* **1995**, 809–810.

[17] F. Vögtle, S. Meier, R. Hoss, *Angew. Chem., Int. Ed. Engl.* **1992**, *31*, 1619–1622.

[18] For the template-directed syntheses of related catenanes, see: (a) S. Ottens-Hildebrandt, M. Nieger, K. Rissanen, J. Rouvinen, S. Meier, G. Harder, F. Vögtle, *J. Chem. Soc., Chem. Commun.* **1995**, 777–778. (b) S. Ottens-Hildebrandt, T. Schmidt, F. Vögtle, *Liebigs Ann.* **1995**, 1855–1860. (c) R. Jäger, T. Schmidt, D. Karbach, F. Vögtle, *Synlett* **1996**, 723–725. (d) C. Yamamoto, Y. Okamoto, T. Schmidt, R. Jäger, F. Vögtle, *J. Am. Chem. Soc.* **1997**, *119*, 10547–10548. (e) S. Baumann, R. Jäger, F. Ahuis, B. Kray, F. Vögtle, *Liebigs Ann./Recueil* **1997**, 761–766. (f) A. Andrievsky, F. Ahuis, J. L. Sessler, F. Vögtle, D. Gudat, M. Moini, *J. Am. Chem. Soc.* **1998**, *120*, 9712–9713.

[19] (a) S. Ottens-Hildebrandt, S. Meier, W. Schmidt, F. Vögtle, *Angew. Chem., Int. Ed. Engl.* **1994**, *33*, 1767–1770. (b) F. J. Carver, C. A. Hunter, R. J. Shannon, *J. Chem. Soc., Chem. Commun.* **1994**, 1277–1280.

[20] (a) F. Vögtle, M. Händel, S. Meier, S. Ottens-Hildebrandt, F. Ott, T. Schmidt, *Liebigs Ann.* **1995**, 739–743. (b) F. Vögtle, T. Dünnwald, M. Händel, R. Jäger, S. Meier, *Chem. Eur. J.* **1996**, *2*, 640–643. (c) F. Vögtle, F. Ahuis, S. Baumann, J. L. Sessler, *Liebigs Ann.* **1996**, 921–926. (d) R. Jäger, M. Händel, J. Harren, K. Rissanen, F. Vögtle, *Liebigs Ann.* **1996**, 1201–1207. (e) F. Vögtle, R. Jäger, M. Händel, S. Ottens-Hildebrandt, W. Schmidt, *Synthesis* **1996**, 353–356. (f) M. Händel, M. Plevoets, S. Gestermann, F. Vögtle, *Angew. Chem., Int. Ed. Engl.* **1997**, *36*, 1199–1201. (g) T. Dünnwald, R. Jäger, F. Vögtle, *Chem. Eur. J.* **1997**, *3*, 2043–2051. (h) R. Jäger, S. Baumann, M. Fischer, O. Safarowsky, M. Nieger, F. Vögtle, *Liebigs Ann./Recueil* **1997**, 2269–2273. (i) O. Braun, F. Vögtle, *Synlett* **1997**, 1184–1185. (j) C. Fischer, M. Nieger, O. Mogck, V. Böhmer, R. Ungaro, F. Vögtle, *Eur. J. Org. Chem.* **1998**, 155–161. (k) T. Dünnwald, A. Hassain, F. Vögtle, *Synthesis* **1998**, 339–348. (l) T. Schmidt, R. Schmieder, W. M. Müller, B. Kiupel, F. Vögtle, *Eur. J. Org. Chem.* **1998**, 2003–2007. (m) G. M. Hübner, J. Gläser, C. Seel, *Angew. Chem. Int. Ed.* **1999**, *38*, 383–386. (n) A. Hossain Parham, B. Windisch, F. Vögtle, *Eur. J. Org. Chem.* **1999**, 1233–1238. (o) C. Heim, A. Affeld, M. Nieger, F. Vögtle, *Helv. Chim. Acta* **1999**, *82*, 746–759. (p) C. Kauffmann, W. M. Müller, F. Vögtle, S. Weinman, S. Abramson, B. Fuchs, *Synthesis* **1999**, 849–853.

[21] A. S. Lane, D. A. Leigh, A. Murphy, *J. Am. Chem. Soc.* **1997**, *119*, 11092–11093.

[22] For the template-directed syntheses of related rotaxanes, see: (a) A. G. Johnston, D. A. Leigh, A. Murphy, J. P. Smart, M. D. Deegan, *J. Am. Chem. Soc.* **1996**, *118*, 10662–10663. (b) D. A. Leigh, A. Murphy, J. P. Smart, A. M. Z. Slawin, *Angew. Chem., Int. Ed. Engl.* **1997**, *36*, 728–732. (c) W. Clegg, C. Gimenez-Saiz, D. A. Leigh, A. Murphy, A. M. Z. Slawin, S. J. Teat, *J. Am. Chem. Soc.* **1999**, *121*, 4124–4129.

[23] T. J. Kidd, D. A. Leigh, A. J. Wilson, *J. Am. Chem. Soc.* **1999**, *121*, 1399–1400.

[24] A. Lüttringhaus, F. Cramer, H. Prinzbach, F. M. Henglein, *Justus Liebigs Ann. Chem.* **1958**, *613*, 185–198.

[25] (a) D. Armspach, P. R. Ashton, C. P. Moore, N. Spencer, J. F. Stoddart, T. J. Wear, D. J. Williams, *Angew. Chem., Int. Ed. Engl.* **1993**, *32*, 854–858. (b) D. Armspach, P. R. Ashton, R.

Ballardini, V. Balzani, A. Godi, C. P. Moore, L. Prodi, N. Spencer, J. F. Stoddart, M. S. Tolley, T. J. Wear, D. J. Williams, *Chem. Eur. J.* **1995**, *1*, 33–55.

[26] A. Harada, J. Li, M. Kamachi, *Chem. Commun.* **1997**, 1413–1414.

[27] For related examples of cyclodextrin-containing rotaxanes, see: (a) H. Ogino, *J. Am. Chem. Soc.* **1981**, *103*, 1303–1304. (b) K. Yamanari, Y. Shimura, *Bull. Chem. Soc. Jpn.* **1983**, *56*, 2283–2289. (c) K. Yamanari, Y. Shimura, *Bull. Chem. Soc. Jpn.* **1984**, *57*, 1596–1603. (d) H. Ogino, K. Ohata, *Inorg. Chem.* **1984**, *23*, 3312–3316. (e) J. S. Manka, D. S. Lawrence, *J. Am. Chem. Soc.* **1990**, *112*, 2440–2442. (f) T. Venkata, S. Rao, D. S. Lawrence, *J. Am. Chem. Soc.* **1990**, *112*, 3614–3615. (g) R. Isnin, A. E. Kaifer, *J. Am. Chem. Soc.* **1991**, *113*, 8188–8190. (h) G. Wenz, E. von der Bey, L. Schmidt, *Angew. Chem., Int. Ed. Engl.* **1992**, *31*, 783–785. (i) R. S. Wylie, D. H. Macartney, *J. Am. Chem. Soc.* **1992**, *114*, 3136–3138. (f) M. Kunitake, K. Kotoo, O. Manabe, T. Muramatsu, N. Nakashima, *Chem. Lett.* **1993**, 1033–1036. (g) R. S. Wylie, D. H. Macartney, *Supramol. Chem.* **1993**, *3*, 29–35. (h) S. Anderson, T. D. W. Claridge, H. L. Anderson, *Angew. Chem., Int. Ed. Engl.* **1997**, *36*, 1310–1313. (i) A. P. Lyon, D. H. Macartney, *Inorg. Chem.* **1997**, *36*, 729–736. (j) R. B. Hannak, G. Färber, R. Konrat, B. Kräuter, *J. Am. Chem. Soc.* **1997**, *119*, 2313–2314. (k) H. Murakami, A. Kawabuchi, K. Kotoo, M. Kunitake, N. Nakashima, *J. Am. Chem. Soc.* **1997**, *119*, 7605–7606. (l) M. Tamura, A. Ueno, *Chem. Lett.* **1998**, 369–370. (m) S. Anderson, R. T. Aplin, T. D. W. Claridge, T. Goodson III, A. C. Maciel, G. Rumbles, J. F. Ryan, H. L. Anderson, *J. Chem. Soc.,Perkin Trans. 1* **1998**, 2383–2397.

[28] (a) M. Fujita, F. Ibukuro, H. Hagihara, K. Ogura, *Nature (London)* **1994**, *367*, 720–723. (b) M. Fujita, F. Ibukuro, K. Yamaguchi, K. Ogura, *J. Am. Chem. Soc.* **1995**, *117*, 4175–4176. (c) M. Fujita, F. Ibukuro, H. Seki, O. Kamo, M. Imanari, K. Ogura, *J. Am. Chem. Soc.* **1996**, *118*, 899–900. (d) M. Fujita, *Acc. Chem. Res.* **1998**, *32*, 53–61. (e) M. Fujita, *Chem. Soc. Rev.* **1998**, *27*, 417–425. (f) M. Fujita, M. Aoyagi, F. Ibukuro, K. Ogura, K. Yamaguchi, *J. Am. Chem. Soc.* **1998**, *120*, 611–612. (g) M. Fujita, N. Fujita, K. Ogura, K. Yamaguchi, *Nature* **1999**, *400*, 52–55.

[29] K. N. Houk, S. Menzer, S. P. Newton, F. M. Raymo, J. F. Stoddart, D. J. Williams, *J. Am. Chem. Soc.* **1999**, *121*, 1479–1487.

[30] B. L. Allwood, N. Spencer, H. Shahriari-Zavareh, J. F. Stoddart, D. J. Williams, *J. Chem. Soc., Chem. Commun.* **1987**, 1064–1066.

[31] P.-L. Anelli, P. R. Ashton, R. Ballardini, V. Balzani, M. Delgado, M. T. Gandolfi, T. T. Goodnow, A. E. Kaifer, D. Philp, M. Pietraszkiewicz, L. Prodi, M. V. Reddington, A. M. Z. Slawin, N. Spencer, J. F. Stoddart, C. Vicent, D. J. Williams, *J. Am. Chem. Soc.* **1992**, *114*, 193–218.

[32] P. R. Ashton, T. T. Goodnow, A. E. Kaifer, M. V. Reddington, A. M. Z. Slawin, N. Spencer, J. F. Stoddart, C. Vicent, D. J. Williams, *Angew. Chem., Int. Ed. Engl.* **1989**, *28*, 1396–1399.

[33] S. Capobianchi, G. Doddi, G. Ercolani, P. Mencarelli, *J. Org. Chem.* **1998**, *63*, 8088–8089.

[34] (a) R. E. Gillard, J. F. Stoddart, A. J. P. White, B. J. Williams, D. J. Williams, *J. Org. Chem.* **1996**, *61*, 4504–4505. (b) R. Ballardini, V. Balzani, C. L. Brown, A. Credi, R. E. Gillard, M. Montalti, D. Philp, J. F. Stoddart, M. Venturi, A. J. P. White, B. J. Williams, D. J. Williams, *J. Am. Chem. Soc.* **1997**, *119*, 12503–12513.

[35] (a) P. R. Ashton, C. L. Brown, E. J. T. Chrystal, T. T. Goodnow, A. E. Kaifer, K. P. Parry, D. Philp, A. M. Z. Slawin, N. Spencer, J. F. Stoddart, D. J. Williams, *J. Chem. Soc., Chem. Commun.* **1991**, 634–639. (b) M. Asakawa, P. R. Ashton, S. E. Boyd, C. L. Brown, R. E. Gillard, O. Kocian, F. M. Raymo, J. F. Stoddart, M. S. Tolley, A. J. P. White, D. J. Williams, *J. Org. Chem.* **1997**, *62*, 26–37.

[36] For examples of catenanes incorporating more than two π-electron-rich and π-electron defi-
cient macrocyclic components, see: (a) P. R. Ashton, C. L. Brown, E. J. T. Chrystal, T. T:
Goodnow, A. E. Kaifer, K. P. Parry, A. M. Z. Slawin, N. Spencer, J. F. Stoddart, D. J. Williams, *Angew. Chem., Int. Ed. Engl.* **1991**, *30*, 1039–1042. (b) P. R. Ashton, C. L. Brown,
E. J. T. Chrystal, K. P. Parry, M. Pietraszkiewicz, N. Spencer, J. F. Stoddart, *Angew. Chem.,
Int. Ed. Engl.* **1991**, *30*, 1042–1045. (c) P. R. Ashton, C. L. Brown, J. R. Chapman, R. T. Gal-
lagher, J. F. Stoddart, *Tetrahedron Lett.* **1992**, *33*, 7771–7774. (d) D. B. Amabilino, P. R.
Ashton, J. F. Stoddart, S. Menzer, D. J. Williams, *J. Chem. Soc., Chem. Commun.* **1994**,
2475–2478. (e) D. B. Amabilino, P. R. Ashton, C. L. Brown, E. Córdova, L. A. Godínez, T.
T. Goodnow, A. E. Kaifer, S. P. Newton, M. Pietraszkiewicz, D. Philp, F. M. Raymo, A. S.
Reder, M. T. Rutland, A. M. Z. Slawin, N. Spencer, J. F. Stoddart, D. J. Williams, *J. Am.
Chem. Soc.* **1995**, *117*, 1271–1293. (f) M. Asakawa, P. R. Ashton, S. Menzer, F. M. Raymo,
J. F. Stoddart, A. J. P. White, D. J. Williams, *Chem. Eur. J.* **1996**, *2*, 877–893. (g) P. R.
Ashton, S. E. Boyd, C. G. Claessens, R. E. Gillard, S. Menzer, J. F. Stoddart, M. S. Tolley, A.
J. P. White, D. J. Williams, *Chem. Eur. J.* **1997**, *3*, 788–798. (h) M. Asakawa, P. R. Ashton,
C. L. Brown, M. C. T. Fyfe, S. Menzer, D. Pasini, C. Scheuer, N. Spencer, J. F. Stoddart, A.
J. P. White, D. J. Williams, *Chem. Eur. J.* **1997**, *3*, 1136–1150. (i) D. B. Amabilino, P. R.
Ashton, J. F. Stoddart, A. J. P. White, D. J. Williams, *Chem. Eur. J.* **1998**, *4*, 460–468. (j) P.
R. Ashton, V. Balzani, A. Credi, O. Kocian, D. Pasini, L. Prodi, N. Spencer, J. F. Stoddart, M.
S. Tolley, A. J. P. White, D. J. Williams, *Chem. Eur. J.* **1998**, *4*, 590–607.

[37] (a) D. B. Amabilino, P. R. Ashton, A. S. Reder, N. Spencer, J. F. Stoddart, *Angew. Chem., Int.
Ed. Engl.* **1994**, *33*, 1286–1290. (b) D. B. Amabilino, P. R. Ashton, S. E. Boyd, J. Y. Lee, S.
Menzer, J. F. Stoddart, D. J. Williams, *Angew. Chem., Int. Ed. Engl.* **1997**, *36*, 2070–2072.
(c) D. B. Amabilino, P. R. Ashton, S. E. Boyd, J. Y. Lee, S. Menzer, J. F. Stoddart, D. J. Williams, *J. Am. Chem. Soc.* **1998**, *120*, 4295–4307.

[38] (a) P. R. Ashton, D. Philp, N. Spencer, J. F. Stoddart, *J. Chem. Soc., Chem. Commun.* **1992**,
1124–1128. (b) P. R. Ashton, R. Ballardini, V. Balzani, M. Belohradsky, M. T. Gandolfi, D.
Philp, L. Prodi, F. M. Raymo, M. V. Reddington, N. Spencer, J. F. Stoddart, M. Venturi, D. J.
Williams, *J. Am. Chem. Soc.* **1996**, *118*, 4931–4951.

[39] A similar template-directed approach has been employed to self-assemble [*n*]rotaxanes
incorporating up to four mechanically interlocked components and large dendritic stoppers:
D. B. Amabilino, P. R. Ashton, V. Balzani, C. L. Brown, A. Credi, J. M. J. Fréchet, J. W.
Leon, F. M. Raymo, N. Spencer, J. F. Stoddart, M. Venturi, *J. Am. Chem. Soc.* **1996**, *118*,
12012–12020.

[40] (a) P. R. Ashton, M. Belohradsky, D. Philp, J. F. Stoddart, *J. Chem. Soc., Chem. Commun.*
1993, 1269–1274. (b) P. R. Ashton, M. Belohradsky, D. Philp, N. Spencer, J. F. Stoddart,
J. Chem. Soc., Chem. Commun. **1993**, 1274–1277. (b) D. B. Amabilino, P. R. Ashton, M.
Belohradsky, F. M. Raymo, J. F. Stoddart, *J. Chem. Soc., Chem. Commun.* **1995**, 747–750.
(c) D. B. Amabilino, P. R. Ashton, M. Belohradsky, F. M. Raymo, J. F. Stoddart, *J. Chem.
Soc., Chem. Commun.* **1995**, 751–753. (d) M. Asakawa, P. R. Ashton, R. Ballardini, V.
Balzani, M. Belohradsky, M. T. Gandolfi, O. Kocian, L. Prodi, F. M. Raymo, J. F. Stoddart,
M. Venturi, *J. Am. Chem. Soc.* **1997**, *119*, 302–310. (e) D. B. Amabilino, M. Asakawa, P. R.
Ashton, R. Ballardini, V. Balzani, M. Belohradsky, A. Credi, M. Higuchi, F. M. Raymo, T.
Shimizu, J. F. Stoddart, M. Venturi, K. Yase, *New J. Chem.* **1998**, 959–972.

[41] F. M. Raymo, K. N. Houk, J. F. Stoddart, *J. Am. Chem. Soc.* **1998**, 120, 9318–9322.

[42] A. C. Try, M. M. Harding, D. G. Hamilton, J. K. M. Sanders, *Chem. Commun.* **1998**, 723–724.

[43] P. R. Ashton, P. J. Campbell, E. J. T. Chrystal, P. T. Glink, S. Menzer, D. Philp, N. Spencer, J.
F. Stoddart, P. A. Tasker, D. J. Williams, *Angew. Chem., Int. Ed. Engl.* **1995**, *34*, 1865–1869.

[44] (a) P. R. Ashton, E. J. T. Chrystal, P. T. Glink, S. Menzer, C. Schiavo, J. F. Stoddart, P. A. Tasker, D. J. Williams, *Angew. Chem., Int. Ed. Engl.* **1995**, *34*, 1869–1871. (b) M. C. T. Fyfe, P. T. Glink, S. Menzer, J. F. Stoddart, A. J. P. White, D. J. Williams, *Angew. Chem., Int. Ed. Engl.* **1997**, *36*, 2068–2070. (c) P. R. Ashton, M. C. T. Fyfe, P. T. Glink, S. Menzer, J. F. Stoddart, A. J. P. White, D. J. Williams, *J. Am. Chem. Soc.* **1997**, *119*, 12514–12524.

[45] P. R. Ashton, P. T. Glink, J. F. Stoddart, P. A. Tasker, A. J. P. White, D. J. Williams, *Chem. Eur. J.* **1996**, *2*, 729–736.

[46] M.-V. Martínez-Díaz, N. Spencer, J. F. Stoddart, *Angew. Chem., Int. Ed. Engl.* **1997**, *36*, 1904–1907.

[47] For examples of mechanically interlocked molecules behaving as molecular switches, see: (a) A. Livoreil, C. O. Dietrich-Buchecker, J.-P. Sauvage, *J. Am. Chem. Soc.* **1994**, *116*, 9399–9400. (b) D. B. Amabilino, C. O. Dietrich-Buchecker, A. Livoreil, L. Pérez-García, J.-P. Sauvage, J. F. Stoddart, *J. Am. Chem. Soc.* **1996**, *118*, 3905–3913. (c) D. J. Cárdenas, A. Livoreil, J.-P. Sauvage, *J. Am. Chem. Soc.* **1996**, *118*, 11980–11981. (d) F. Baumann, A. Livoreil, W. Kaim, J.-P. Sauvage, *Chem. Commun.* **1997**, 35–36. (e) A. Livoreil, J.-P. Sauvage, N. Armaroli, V. Balzani, L. Flamigni, B. Ventura, *J. Am. Chem. Soc.* **1997**, *119*, 12114–12124. (f) A. C. Benniston, A. Harriman, *Angew. Chem., Int. Ed. Engl.* **1993**, *32*, 1459–1461. (g) A. C. Benniston, A. Harriman, D. Philp, J. F. Stoddart, *J. Am. Chem. Soc.* **1993**, *115*, 5298–5299. (h) A. C. Benniston, A. Harriman, V. M. Lynch, *Tetrahedron Lett.* **1994**, *35*, 1473–1476. (i) A. C. Benniston, A. Harriman, V. M. Lynch, *J. Am. Chem. Soc.* **1995**, *117*, 5275–5291. (j) A. C. Benniston, *Chem. Soc. Rev.* **1996**, *25*, 427–435. (k) A. C. Benniston, A. Harriman, D. S. Yufit, *Angew. Chem., Int. Ed. Engl.* **1997**, *36*, 2356–2358. (l) R. Ballardini, V. Balzani, M. T. Gandolfi, L. Prodi, M. Venturi, D. Philp, H. G. Ricketts, J. F. Stoddart, *Angew. Chem., Int. Ed. Engl.* **1993**, *32*, 1301–1303. (m) P. R. Ashton, R. Ballardini, V. Balzani, A. Credi, M. T. Gandolfi, S. Menzer, L. Pérez-García, L. Prodi, J. F. Stoddart, M. Venturi, A. J. P. White, D. J. Williams, *J. Am. Chem. Soc.* **1995**, *117*, 11171–11197. (n) R. Ballardini, V. Balzani, A. Credi, M. T. Gandolfi, S. J. Langford, S. Menzer, L. Prodi, J. F. Stoddart, M. Venturi, D. J. Williams, *Angew. Chem., Int. Ed. Engl.* **1996**, *35*, 978–981. (o) P. R. Ashton, R. Ballardini, V. Balzani, S. E. Boyd, A. Credi, M. T. Gandolfi, M. Gómez-López, S. Iqbal, D. Philp, J. A. Preece, L. Prodi, H. G. Ricketts, J. F. Stoddart, M. S. Tolley, M. Venturi, A. J. P. White, D. J. Williams, *Chem. Eur. J.* **1997**, *3*, 152–170. (p) P. L. Anelli, M. Asakawa, P. R. Ashton, R. A. Bissell, G. Clavier, R. Górski, A. E. Kaifer, S. J. Langford, G. Mattersteig, S. Menzer, D. Philp, A. M. Z. Slawin, N. Spencer, J. F. Stoddart, M. S. Tolley, D. J. Williams, *Chem. Eur. J.* **1997**, *3*, 1113–1135. (q) M. Asakawa, P. R. Ashton, V. Balzani, A. Credi, G. Mattersteig, O. A. Matthews, M. Montalti, N. Spencer, J. F. Stoddart, M. Venturi, *Chem. Eur. J.* **1997**, *3*, 1992–1996. (r) A. Credi, V. Balzani, S. J. Langford, J. F. Stoddart, *J. Am. Chem. Soc.* **1997**, *119*, 2679–2681. (s) P. R. Ashton, R. Ballardini, V. Balzani, M. Gómez-López, S. E. Lawrence, M. V. Martínez-Díaz, M. Montalti, A. Piersanti, L. Prodi, J. F. Stoddart, D. J. Williams, *J. Am. Chem. Soc.* **1997**, *119*, 10641–10651. (t) M. Asakawa, P. R. Ashton, V. Balzani, A. Credi, C. Hamers, G. Mattersteig, M. Montalti, A. N. Shipway, N. Spencer, J. F. Stoddart, M. S. Tolley, M. Venturi, A. J. P. White, D. J. Williams, *Angew. Chem., Int. Ed. Engl.* **1998**, *37*, 333–337. (u) P. R. Ashton, V. Balzani, O. Kocian, L. Prodi, N. Spencer, J. F. Stoddart, *J. Am. Chem. Soc.* **1998**, *120*, 11190–11191. (v) P. R. Ashton, R. Ballardini, V. Balzani, I. Baxter, A. Credi, M. C. T. Fyfe, M. T. Gandolfi, M. Gómez-López, M.-V. Martínez-Diaz, A. Piersanti, N. Spencer, J. F. Stoddart, M. Venturi, A. J. P. White, D. J. Williams, *J. Am. Chem. Soc.* **1998**, *120*, 11932–11942. (w) M. Asakawa, P. R. Ashton, V. Balzani, C. L. Brown, A. Credi, O. A. Matthews, S. P. Newton, F. M. Raymo, A. N. Shipway, N. Spencer, A. Quick, J. F. Stoddart, A. J. P. White, D. J. Williams, *Chem. Eur.*

4.2   Carceplexes

4.2.1   The First Soluble Carceplex

Cram reported the preparation of the first soluble carceplex, carceplex **2**·guest, from the shell closure reaction between two bowl-shaped tetrol molecules **1a**, base, and four molecules of bromochloromethane under conditions of high dilution in dipolar, aprotic solvents (Scheme 4-1) [12, 13].

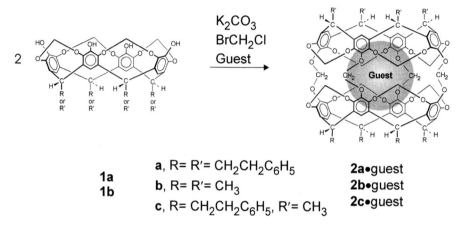

K_2CO_3
$BrCH_2Cl$
Guest

1a
1b

a, R= R′= $CH_2CH_2C_6H_5$
b, R= R′= CH_3
c, R= $CH_2CH_2C_6H_5$, R′= CH_3

2a•guest
2b•guest
2c•guest

Scheme 4-1   Synthesis of the first soluble carceplex **2**·guest.

Product yields as high as 87% have been achieved for formation of carceplex **2a**·guest [14]. These yields are remarkably high for a reaction that joins seven molecules and makes eight new covalent bonds. Regarding templating, no carceplex or carcerand products were observed in the reaction mixture when it was conducted using the solvent *N*-formylpiperidine (NFP), a molecule too big for the interior of carcerand **2a** [13]. However, when the reaction was repeated doping the NFP solvent with 5% *N,N*-dimethylacetamide (DMA), carceplex **2a**·DMA was obtained in 10% yield, which suggests that the formation of carceplex **2a**·guest requires a template such as DMA [13].

4.2.1.1   Template ratios in the formation of an acetal-bridged carceplex

Further investigation in our laboratories led to the discovery of a one million-fold range in selectivity for various small molecules found to be suitable guests (Table 4-1) [14]. Template ratios were calculated from direct competition experiments between pairs of guests through measurement of the carceplex product ratios obtained from integration of bound guest ^{1}H-NMR signals in the product mixtures. Just as product ratios reflect relative rates of rate-determining steps in irreversible reactions, template ratios reflect the

Table 4-1   Template ratios for the formation of acetal-bridged carceplex **2**·guest

Guest	Template ratio	Guest	Template ratio
Pyrazine	1 000 000	Thiophene	5 800
Methyl acetate	470 000	1,3-Dithiolane	4 400
1,4-Dioxane	290 000	(±) 2-Butanol	2 800
Dimethyl sulfide	180 000	Benzene	2 400
Ethyl methyl sulfide	130 000	2-Propanol	1 500
Dimethyl carbonate	73 000	Pyrrole	1 000
DMSO	70 000	Tetrahydrothiophene	410
1,3-Dioxolane	38 000	1,3-Dioxane	200
2-Butanone	37 000	Acetamide	160
Pyridine	34 000	Trioxane	100
Dimethyl sulfone	19 000	Acetonitrile	73
1,4-Thioxane	14 000	Ethanol	61
2,3-Dihydrofuran	13 000	Ethyl acetate	45
Furan	12 000	Diethyl ether	21
Tetrahydrofuran	12 000	DMA	20
Pyridazine	8 600	DMF	7
Acetone	6 700	NMP	1

relative rates of the guest-determining step (GDS) for each of the suitable templates. The GDS refers to the step occurring along the reaction pathway during which the guest becomes permanently entrapped within the forming host [14]. Therefore, the template effect observed is kinetic in origin. Unlike most template effects, the carceplex product is "tagged" with the template, and therefore large template ratios can be measured with good precision without the need for the determination of the rate of reaction. Yields are often compared in template studies, but these do not provide quantitative measurement of template ratios, and the range in template effects measured by yields is small.

Table 4-1 shows that pyrazine is the best template and is a million times better than the poorest measurable template, *N*-methyl-2-pyrrolidinone (NMP) [14]. Also, considering the series of suitable guest molecules, the selectivity is unusually high, where small perturbations in guest structure can lead to large differences in templating abilities [14]. For example, methyl acetate is 10 000 times better than ethyl acetate [14]. A similar template effect has also been observed in the formation of carceplex **2c**·guest [15].

4.2.1.2   Formation of a charged hydrogen bonded complex

Further investigation into the driving forces responsible for the formation of carceplex **2b**·guest led to the discovery of a complex (**3b**·guest, Scheme 4-2) that forms in solution from two tetrol molecules (**1b**) in the presence of base, where a guest is reversibly encapsulated between two tetrol bowls, held together by four charged hydrogen bonds (CHBs) [16]. Measurement of relative stabilities of complexes formed with selected guests from the series in Table 4-1 showed that complex **3b**·guest expresses the same guest selectivity as carceplex **2a**·guest. Thus, complex **3b**·guest serves as a good transition-state model for the GDS in the formation of carceplex **2**·guest [16]. Although complex **3b**·guest is not at all a carceplex (it binds reversibly), it is a highly relevant model for the formation of carceplex **2a**·guest. Thus, discussion of **3b**·guest is presented here, rather than in Section 4.4, which deals with this type of tilled hydrogen bonded capsule.

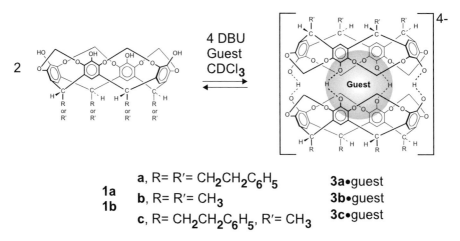

a, R= R′= CH$_2$CH$_2$C$_6$H$_5$ **3a**•guest

1a
1b   **b**, R= R′= CH$_3$ **3b**•guest

c, R= CH$_2$CH$_2$C$_6$H$_5$, R′= CH$_3$ **3c**•guest

Scheme 4-2   Formation of hydrogen bonded complex **3**· guest.

Striking similarities are manifested in the ^{1}H NMR spectra of **2a–c**·pyrazine [14, 15] and **3b**·pyrazine [16, 17], and X-ray crystal structures of carceplex **2a**·pyrazine [15] and complex **3b**·pyrazine [17]. These similarities, along with the guest selectivities observed for each, indicate that the same interactions are at play in driving the thermodynamic formation of **3**·guest and the kinetic formation of **2**·guest [17]. Favorable interactions include CHBs between the bowls, favorable van der Waals contacts, CH$\cdots\pi$ interactions, CH$\cdots$X (X=O) hydrogen bonding, conjugation of O–H$\cdots$O$^-$ and OCH$_2$O bonds to their respective aromatic rings, and $\pi\cdots\pi$ interactions [12–17]. The general trend observed experimentally was reproduced by theoretical analyses [18].

4.2.1.3   Mechanism of formation for an acetal-bridged carceplex

To delineate further the forces at play in forming carceplex **2a**·guest, the isolation of reaction intermediates [14b,19], as well as the effects of base and solvent [14b], were probed, and the following conclusions were made. The first step in the formation of carceplex **2**·guest is the reversible formation of complex **3**·guest [14b]. Installation of the first acetal bridge to form a monobridged intermediate then occurs, with guests still in rapid exchange [14b]. The GDS follows during the formation of a second bridge at any position to form doubly bridged intermediates, where guest exchange ceases under the reaction conditions [14b]. The ability of a particular guest to align the phenoxides of each hemisphere of the mono-bridged intermediate ultimately determines the rate at which the second bridge is installed [14b]. Hence, the most stable complexes (i.e., those containing the best template molecule) form the second bridge the fastest, thereby ensnaring more of the preferred guest under the given reaction conditions. The relative rates of the GDSs are largely dictated by ground-state effects, where the more stable complexes are formed in higher concentration. In other words, the rate constants for forming the second bridge are likely to be very similar for the 34 guests studied. Installations of the third and fourth bridges then ensue, providing the completed carceplex product [14b]. (Incidentally, the formation of the fourth bridge is the *rate*-determining step, but not the GDS.)

4.2.2 Large Carceplexes from Cyclic Arrays of Cavitands

Knowledge gleaned in understanding the driving forces for formation of carceplex **2a**·guest has benefited our research group in the desire to construct molecular hosts with larger cavities, or with multiple cavities. We have reported the synthesis of cyclic trimer **5** and tetramer **6** through reaction of A,C-bisbenzyl ether **4** with base and BrCH$_2$Cl linker (Scheme 4-3) in dimethylformamide (DMF) to render 5% and 15% yields of each,

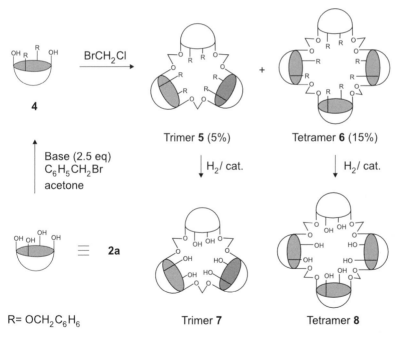

Scheme 4-3 Synthesis of cyclic trimeric and tetrameric assemblies from A,C bisbenzyl ether **4**. The four OH-groups in the cavitand are labelled A, B, C, D.

respectively [20]. The free hydroxyl groups of trimer **7** and tetramer **8** were obtained by hydrogenolysis [20]. Cyclic trimer **7** possesses a fairly large cavity for a synthetic host, and is conformationally very rigid. Whereas cyclic tetramer **8** has an even larger interior, its structure is far more flexible at room temperature. Sealing the upper and lower rims of these assemblies with suitable capping molecules could potentially lead to the formation of carceplexes with cavities larger than any known at present!

4.2.2.1   Synthesis of a trimer carceplex containing three DMF molecules

A cap molecule of suitable size and symmetry could effectively close off the interior of trimer **7** to create a very large molecular container. Such a feat has recently been accomplished in our laboratories, where trimer carceplex **9·3** DMF was synthesized in 37% yield, by reaction of trimer **7** with 2,4,6-tris(bromomethyl)mesitylene under basic conditions (Scheme 4-4) [21]. No empty trimer carcerand has yet been isolated, which implies that a template (or templates) is required for formation. Can a large single template molecule form trimer carceplex **9·**guest? We are currently investigating this possibility. Such an investigation begs the question: when does a template effect become a solvent effect?

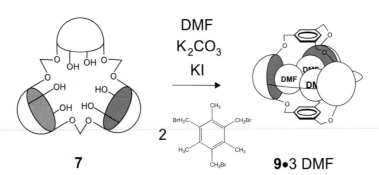

Scheme 4-4   Synthesis of trimer carceplex **9·** 3 DMF.

4.2.2.2   Synthesis of a bis(carceplex)

As in the case for cyclic trimer **7**, capping of the upper and lower rims of cyclic tetramer **8** with an appropriate cap molecule could potentially render a carceplex with an unprecedentedly huge cavity. However, due to its flexible structure, tetramer **8** can fold in upon itself, forming a carceplex with two separate chambers (Scheme 4-5). Subjecting tetramer **8** to optimal carceplex-forming conditions found for **2·**pyrazine effectively gave the bis(carceplex) **10·2** pyrazines in 74% yield [20]. MM2 calculations performed on **10·2** pyrazine suggest that the two neighboring capsules are twisted 90° with respect to each other [20]. The conformation of these two capsules and the relative positions of each guest inside could possibly lead to some kind of intercapsular communication process, which could be important in the development of molecular switching devices [20].

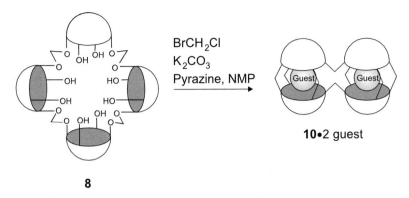

8

Scheme 4-5 Synthesis of bis-carceplex **10**·2 pyrazines.

4.2.3 Benzylthia-bridged Carceplex

Carceplexes have also been synthesized using templates from cavitands with functionalities other than phenols. Cram synthesized benzylthia-bridged carceplex **13**·guest through the shell-closure reaction between tetra(benzyl chloride) cavitand **11** and tetrabenzylthiol cavitand **12** in the solvents 2-butanone, 3-pentanone, ethanol/benzene (1:2), dimethylformamide (DMF), methanol/benzene (2:1), and acetonitrile/benzene (2:1), yielding carceplexes **13**·2-butanone, **13**·3-pentanone, **13**·ethanol, **13**·DMF, **13**·2 methanol, and **13**·2 MeCN respectively (Scheme 4-6) [22]. Technically, **13**·2 MeCN is a hemicarceplex, as one acetonitrile molecule was found to escape the interior of the host via a "billiard-ball" effect, leaving **13**·MeCN when heated in toluene [22]. The forma-

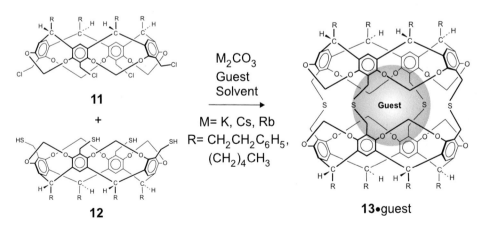

Scheme 4-6 Synthesis of benzylthia-bridged carceplex **13**·guest.

tion of carceplex **13** · guest was found to be a templated process, as indicated by the absence of any empty carcerand (**13**) in any of the product mixtures [22]. Moreover, when the reaction was conducted in an apolar solvent such as benzene, in the absence of any other potential guests, no carceplex **13** · guest was found to form [22]. However, when benzene was used as a co-solvent to help dissolve cavitands **11** and **12** in the polar solvents ethanol, methanol, or acetonitrile, only carceplex **13** · guest containing a polar solvent molecule as guest was isolated.

Preliminary experiments in our laboratories suggest that carceplex **13** · guest forms selectively in the presence of more than one suitable guest [23]. The range in selectivity observed was calculated to span two million-fold.

4.2.4   Calix[4]arene–Cavitand Hybrid Carceplexes

All carceplexes mentioned so far have used only cavitand building blocks as their precursors. In fact, there are very few examples in the literature of carceplexes that significantly deviate from the two cavitand topology of **2** · guest or **13** · guest. Two examples already mentioned are **9** · 3 DMF and **10** · 2 pyrazine. Reinhoudt and co-workers have prepared asymmetric carceplexes composed of both calixarenes and cavitands [24]. The synthesis of **15** · guest involved intramolecular cyclization of *endo*-coupled compound **14**, where the *t*-butyldimethylsilyl (TBDMS) groups are removed in situ with CsF, followed by displacement of the chlorides with the phenoxides of the resorcinarene (Scheme 4-7) [24]. Amide and sulfoxide guests/templates were incarcerated by solvent inclusion, while other potential guests incapable of acting as the solvent were encapsulated through doped inclusion [25]. Suitable guests/templates reported are DMF, DMA, NMP, 1,5-dimethyl-2-pyrrolidinone, dimethyl sulfoscide (DMSO), ethyl methyl sulfoxide, thiolane-1-oxide, 3-sulfolene, and 2-butanone [25]. The relative templating abilities of several guests were

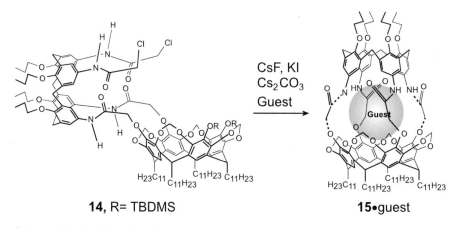

14, R= TBDMS   **15•guest**

Scheme 4-7   Reinhoudt's calix[4]arene/resorc[4]arene carceplex **15** · guest.

investigated through competition reactions by adding 5 vol% of each competing guest to the reaction in 1,5-dimethyl-2-pyrrolidinone solvent [25]. The results are displayed in Table 4-2, where DMA is the best template for the synthesis of carceplex **15**·guest. The relative templating abilities of these guests were described as comparable to the association strengths between the calix[4]- and resorcin[4]arene cavity of the host and the guest [25]. DMA is a superior template as it provides the best solvation of the transition state during shell closure. Stabilization of this transition state, thus promoting formation of the resulting product, is believed to arise partly from hydrogen bonding between the N*H* protons of the host and the incarcerated guests [25]. Slow association between the host and guests that are poorer templates would lead to the formation of intermolecularly coupled products, and/or decomposition of the chloroacetamido groups of precursor **14** [25]. Carceplexes were not formed using the potential guest molecules N,N-dimethylthioformamide, N,N-dimethylthioacetamide, N,N-dimethylmethanesulfonamide, cyclopentanone, N-ethyl-N'-methylacetamide, and biacetal [25]. Attempts at including six-membered ring molecules, and the best template in the formation of **2**·guest, pyrazine, failed as well [25].

Reinhoudt's preparation of diastereomeric complexes **15**·guest, each differing in the orientation of the contained guest, provided means for the discovery of a new form of stereoisomerism, carceroisomerism [26]. Application of these novel host–guest structures as molecular switches has been suggested [25, 26].

Table 4-2 Template ratios and yields for the formation of Reinhoudt's carceplex **15**·guest

Guest	Templating ability[a]	Yield (%)[b]
DMA	100	27
DMSO	63	16[c]
DMF	27	13[c]
2-Butanone	27	16

[a] DMA is set at 100. [b] Isolated carceplex when only one guest is used during doped inclusion. [c] Yield of deuterated guests.

4.2.5 Metal-bridged Carceplexes

Carceplexes traditionally consist of *neutral* organic molecules imprisoned by the *covalently-linked, neutral* host carcerands. Dalcanale and co-workers at the University of Parma have expanded the scope of carceplexes by characterizing carceplex **17**·guest, synthesized through the metal-induced self-assembly of two molecules of tetracyanocavitand **16** connected with four Pd[II] or Pt[II] square-planar linkages, where the incarcerated guest is a triflate anion, $CF_3SO_3^-$ (Scheme 4-8) [27].

The structure of **17**·guest was confirmed through extensive characterization involving NMR (^{1}H, ^{13}C, ^{31}P, ^{19}F) and FT-IR spectroscopies, electrospray-MS, and vapor-phase osmometry [27]. Carceplex **17b**·guest was also discovered to disassemble by ligand exchange upon addition of eight equivalents of triethylamine to give two molecules of precursor bowl **16** and four $PtL(NEt_3)_2(OTf)_2$ (L=1,3-bis(diphenylphosphino)propane) [27]. Subsequently, **17** · guest reforms upon addition of eight equivalents of triflic acid (CF_3SO_3H) [27]. Ligand exchange controls the reversible cage formation [27]. As no empty cage **17** was ever observed as a product, $CF_3SO_3^-$ is required as a template [27].

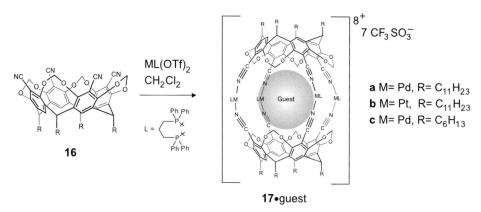

Scheme 4-8 Dalcanale's metal-bridged cage complex **17** · guest. L=1,3-bis(diphenylphosphino)propane.

4.3   Hemicarceplexes

Closely related to carceplexes are hemicarceplexes, for which some template studies have been undertaken. A wide range of hemicarceplexes have been synthesized to date, most of which involve the linkage of two cavitands, where a wide variety of differently sized spacers have been incorporated. Thus, fairly large-ring portals have been created in the hemicarceplex skin, which allow entrapped guest molecules to escape into the external medium.

4.3.1   Template Effects in the Formation of a Trismethylene-Bridged Hemicarceplex

Template effects in the formation of hemicarceplex **19** · guest (Scheme 4-9) have been investigated [28]. Very similar in structure to carceplex **2** · guest, tris-bridged **19** · guest is formed by reaction of two triol bowls (**8**) with three molecules of bromochloromethane. Notice that the shell of hemicarcerand **19** has a single modest-sized hole, through which, with sufficient time and heat, the trapped guest can escape. High yields of **19** · guest were obtained using templates such as pyrazine, 1,4-dioxane, DMSO, THF, acetone, pyrrole, 1,3,5-trioxane, DMA, and NMP [28]. The yields were found to surpass theoretical yields expected from statistical analyses [28, 29]. The template ratios determined for these guests in hemicarceplex **19** mirrored those for carceplex **2**·guest [28]. Also, a complex similar to **3**·guest (**34**·guest, Figure 4-4) has been discovered to form, where two triol bowls (**18**) are prealigned about a guest so that the maximum number of inter-bowl hydrogen bonds (three) form [28]. Non-covalent interactions between the forming host and suitable templates promote the most stable alignment of the triol bowl precursors, and thus the templates facilitate the formation of hemicarceplex **19**·guest [28].

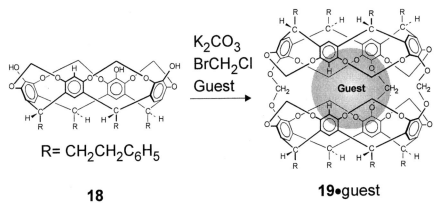

18   **19•guest**

Scheme 4-9   Synthesis of tris(acetal)-bridged hemicarceplex **19**·guest.

4.3.2   Hemicarceplexes Containing Four Slotted Portals

Using the same shell-closure approach as for the synthesis of carceplex **2**·guest, a large number of hemicarceplexes with differently sized portals have been formed [10, 30]. Few template studies exist on these hemicarceplexes, as most have been prepared by Cram to investigate complexation/decomplexation behaviors and novel reactivity of the entrapped species [30]. Our research group is currently probing the template effect in the formation of hemicarceplex **20a**·guest. Preliminary results feature *p*-xylene as the best template, which is over three thousand times better than the poorest measurable template, *N*-formylpiperidine [31]. Template effects have also been observed in separate shell-closure reactions forming **20b**·guest and **20c**·guest (Figure 4-1). Yields of **20b**·guest and **20c**·guest were increased from 20 and 18% to 51 and 27%, respectively, upon addition of 1,2-dimethoxybenzene to the reactants [30g]. This is the first of two reported examples where the formation of a hemicarceplex is templated, but the product is not "tagged" with the template, as it escapes after the host is formed.

Some intriguing results involving template effects in the formation of *m*-xylyl-bridged hemicarceplexes (Figure 4-2) have been observed [32]. The robustness of **21**·guest prompted Cram and co-workers to investigate how binding ability is changed upon altering the cavity size of the host. This was accomplished through various shell-closure reac-

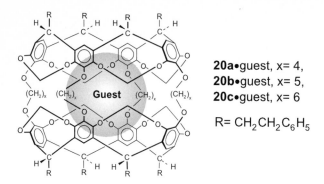

20a•guest, x= 4,
20b•guest, x= 5,
20c•guest, x= 6

R= CH$_2$CH$_2$C$_6$H$_5$

Figure 4-1   Tris-bridged hemicarcerand **20**.

tions using cavitands that differ in their interhemispheric "spanners": methylene (**27a, b**), ethylene (**28a, b**), and propylene (**29a, b**) (Figure 4-3) were explored [32]. Hemicarceplexes having like (**21–23**) and unlike (**24–26**) northern and southern hemispheres were prepared by two different strategies, each revealing striking differences in yields [32]. These methods were analogous to the methods used to form carceplex **2** · guest ("2 + 4" addition of two tetrol bowls and four linker molecules), and carceplex **13** · guest ("1 + 1" addition of one tetrol with one tetrachloride bowl).

Applying the "2 + 4" conditions, **21** · NMP was synthesized in 50% yield from tetrol **27a** and linker in NMP [32]. In contrast, the "1 + 1" conditions produced only a 2.2% yield of **21** · NMP from tetrol **27a** plus tetrachloride **27b** [32]. These results suggest that NMP acts as a template, enabling two tetrol bowls to preassociate to form complex **3** · NMP (R = R′ = C$_5$H$_{11}$). In the first reaction, NMP can be considered as a "positive" template, as the two preorganized tetrols react with the linker to form bridges leading to the desired product in high yield. However, NMP can be considered a "negative" template in the second reaction since the formation of the desired product is inhibited as the effective concentration of free tetrol **27a** available for reaction with the tetrachloride bowl **27b** is reduced in forming a dimer of **27a**. In addition, installation of the first bridge would lead to a poorly aligned intermediate that would be unable to form intercavitand hydrogen bonds, and thus it would be poised for oligomerization/polymerization. NMP was also found to act as a "negative" template for the syntheses of hemicarceplexes **24** · guest and **25** · guest. Reaction of **28a** and **27b** gave a much higher product yield of **24** · guest (21%), than the corresponding reverse reaction involving **27a** and **28b** (2%) [32]. Similarly, **25** · guest is formed in 1.8% yield from **29a** and **27b**, and ~0% from **27a** and **29b** [32]. These results are consistent with **27a** promoting a complex **3**·NMP whereas **28a** and **29a** are precluded from such. As no complex akin to **3** · guest forms, the tetrols are available for the "1 + 1" reaction, but are not preorganized to facilitate the "2+4" reaction. Lack of complex formation for **28a** and **29a** also explains the differences in the yields of **22** · guest and **23** · guest, between reactions involving tetrol bowl plus tetrachloride bowl (43% and 6%, respectively) and those involving tetrol bowl plus linker (8% and 0%, respectively).

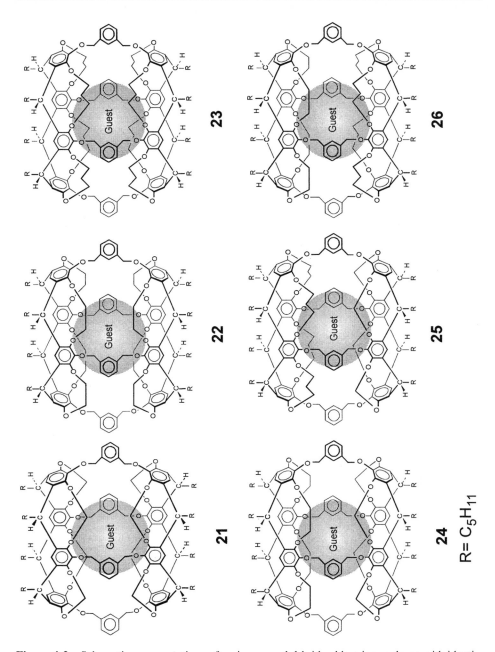

Figure 4-2 Schematic representations of various *m*-xylyl-bridged hemicarceplexes with identical and different polar hemispheres.

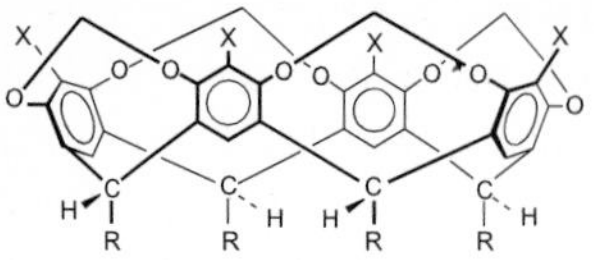

Figure 4-3   Cavitand building blocks used in the synthesis of hemicarceplexes **21–26**.

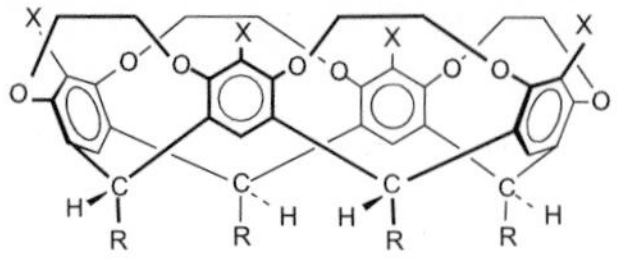

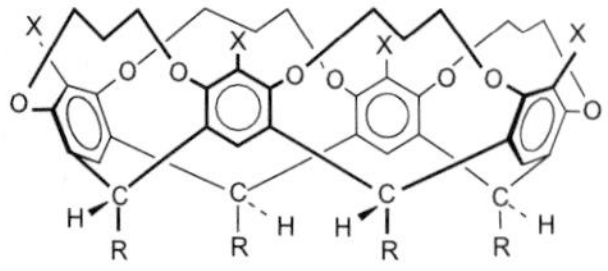

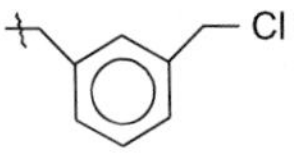

Template effects appear to be absent in three out of four of these reactions. However, addition of **29a** to **29b** only gives the product **23** · guest when 1,2,3-trimethoxybenzene is present in the reaction at 5% (w/w) of the solvent [32]. It was suggested that 1,2,3-trimethoxybenzene templates the formation **23** · guest, which escapes during the work-up, as only the free host **23** was isolated [32]. This is the second example where such a template/escape process occurs in the formation of hemicarceplexes (vide infra). Support for this notion is gleaned from the isolation of **25** · 1,2,3-trimethoxybenzene from the reaction of **29a** and **27b**. **25** · guest did not form in the absence of 1,2,3-trimethoxybenzene [32].

4.4   Capsules

Rebek has defined self-assembling capsules as "receptors with enclosed cavities, formed by the reversible noncovalent interaction of two or more, not necessarily identical, subunits"[11]. Consequently, the resulting capsule has a well-defined structure in solution, and shows binding capabilities that are absent for the individual components alone [11]. Examples provided in this section are restricted to capsules that form only in the presence of a template. As a result, unique sets of spectroscopic data are acquired for solutions of both the individual subunits in the presence and absence of the required templates.

4.4.1   Capsules Composed of Cavitands Linked via Covalent Bonds or Charged Hydrogen Bonds

Referring back to Section 4.2.2.2, we saw that two tetrol bowls (**1**) can wrap themselves around a suitable template in the presence of base to form complex **3** · guest. The relative stabilities of complexes formed with several guests were found to mirror the template effect observed in the formation of carceplex **2** · guest. Further investigation into the significance of complex **3** · guest in the mechanism of carceplex formation has also led to the discovery of eight new complexes, **30–37** · guest, all of which reversibly encapsulate small molecules within their cavities (Figure 4-4), and all of which manifest similar guest selectivity to **3** · guest [19]. Thus, each of these complexes, along with complex **3** · guest, are valid transition state models for the formation of carceplex **2** · guest. Note that **30** · guest, **34** · guest, and **36** · guest do not have CHBs (for definition, see section 4.2.1.2); thus these complexes are neutral. Moreover, **30**, **34**, and **36** do not require a template for their formation, whereas **31**, **32**, **33**, **35**, and **37** do. Enthalpy–entropy compensation appears to be at play in the formation of capsules **3** · guest and **30–37** · guest; exchange rates were calculated to range from microseconds to days, depending strongly on temperature, host, solvent, and guest [19].

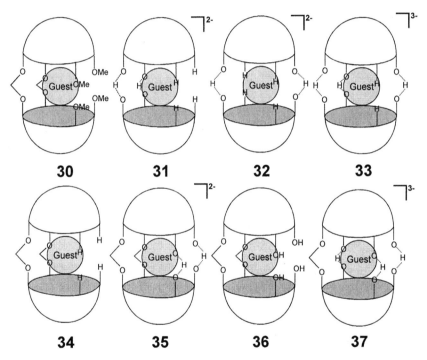

Figure 4-4 Schematic representations of various self-assembling reversible capsules derived from cavitand precursors.

4.4.2 Rebek's "Tennis Balls", "Softballs", and "Wiffle Balls"

Rebek and co-workers have successfully prepared many fascinating self-assembling capsules. These systems are all related, as they involve the oligomerization of two or more concave monomeric subunits [11]. Many of these subunits incorporate glycoluril functionalities at each terminus of the molecule, and contain hydrogen bond donor and acceptor groups, which are separated by a rigid spacer. These concave subunits are self-complementary and can dimerize to form hollow structures seamed together by numerous hydrogen bonds. In the process, cavities large enough for small molecules are created. The first report in the literature of one of these hydrogen bonded dimers, was Rebek's "tennis ball" **38** · **38** (Figure 4-5), which forms in solvents that do not compete for hydrogen bonds [33]. Apparently, the "tennis ball" does not require a template for its formation, as the empty dimer has been observed by ^{1}H-NMR [33]. However, the mystery of whether or not the cavity of this species is truly empty, or contains dissolved gases, is still unanswered [33, 34].

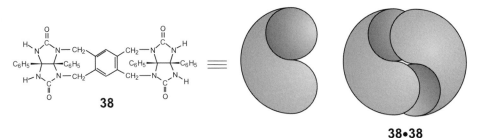

38·38

Figure 4-5   Rebek's "tennis ball".

Modified versions of Rebek's "tennis ball" have been prepared by extending the structure of the spacer between the two glycoluril termini in the monomeric subunit [35]. Figure 4-6 shows the various monomeric subunits (**39–41**) that Rebek and coworkers have used to form dimeric hydrogen-bonded capsules: **39** · guest · **39**, **40** · guest · **40**, and **41** · guest · **41**. These spherically shaped, dimeric capsules are typically identified by the characteristic concentration independent downfield-shifted resonances assigned to the urea N–H protons in the ^{1}H-NMR spectra. In addition host–guest signal integrations are in accord with 2 : 1 complexes being formed. Monomers **39a** and **39b** are both insoluble in NMR solvents such CHCl$_3$ and *p*-xylene-d_{10}, where they form gel-like phases of low-order hydrogen-bonded aggregates.[35] Compound **39** rapidly dissolves upon addition of various adamantyl and ferrocene derivatives to form reversible capsules **39** · guest · **39**, which can dissociate and recombine slowly on the ^{1}H-NMR timescale [35]. Typical ^{1}H-NMR spectra also show upfield signals for encapsulated guests as well as signals for excess free guest [35]. Hence, binding affinities have been quantified through measurement of equilibrium association constants for 1,3,5,7-tetramethyladamantane (6.7), 1-adamantaneamine (190), 1-adamantanecarboxamide (310), 1,3-adamantanedicarboxylic acid, and 1-ferrocene-carboxylic acid (280) [35]. (Note that the values in parentheses are denoted as apparent association constants, K_a(app) (in M^{-1}), because of the unknown aggregation state of the "monomer".) These guests template the formation of dimeric capsules **39a** · guest · **39a** and **39b** · guest · **39b**.

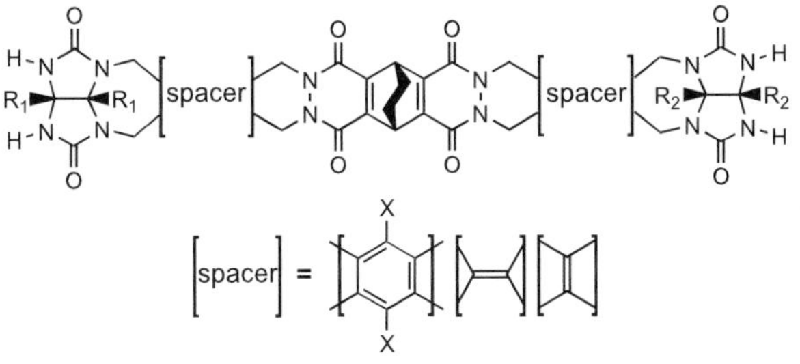

39 **40** **41**

a $R_1 = R_2 = C_6H_5$, X= H
b $R_1 = R_2 = CO_2$-*i*-C_5H_{11}, X= H
c $R_1 = R_2$ = 4-n-heptylphenyl, X= OH
d $R_1 = CO_2$-*i*-C_5H_{11}, R_2 = *p*-$C_6H_4OC_6H_{13}$

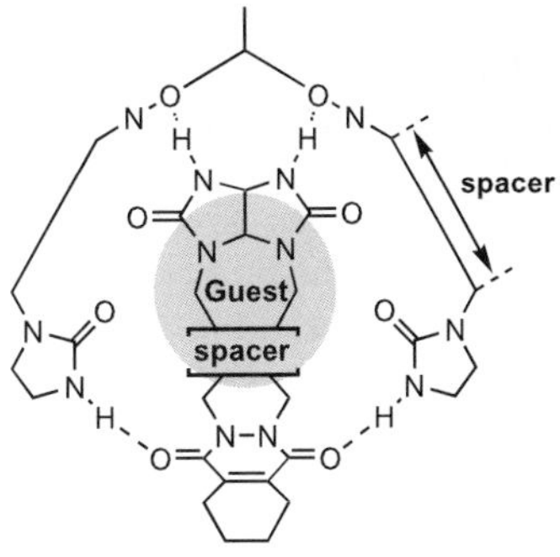

39•guest•39
40•guest•40
41•guest•41

Figure 4-6 Rebek's dimeric "softball" and "wiffle-ball" capsules.

Incorporating four additional hydroxyl groups in the aryl spacer gives a new monomer (**39c**), with improved solubility in common NMR solvents [36]. "Softball" **39c** · guest · **39c** is tied together by up to 16 hydrogen bonds, twice the number in **39a** · guest · **39a** and **39b** · guest · **39b** [36]. These additional hydrogen bonds are derived from participation by the hydroxyl groups, whose proton signals are shifted downfield in the ^{1}H-NMR spectra [36]. Equilibrium association constants were determined for the formation of 2 : 1 host : guest complexes with adamantane (440), 1-adamantanecarboxamide (2 800), 1-ferrocenemethanol (3 800), 1-adamantanecarboxylic acid (5600), 1-ferrocenecarboxylic acid (17 000), and 1-adamantanemethanol (27 000) as guests [36].

Guests that are able to form a maximum number of favorable van der Waals interactions with the interior of "softball" dimers form the most stable complexes [35, 36]. Stabilities of these dimeric complexes also improve when the encapsulated guest possesses functionalities capable of participating in hydrogen bonding with the seam of the host [35, 36]. Enthalpies and entropies of capsule formation were determined in $CDCl_3$, *p*-xylene-d_{10}, and benzene-d_6 solvents [36, 37]. Formation of capsules **39a** · guest · **39a**, **39b** · guest · **39b**, and **39c** · guest · **39c** is entropically driven, and is accompanied by a more modest gain in enthalpy [36, 37]. This is consistent with the displacement of two solvent molecules by a single guest from the interior of these "softball" capsules [36, 37]. As corroboration, three unique capsules have been observed to form when monomers **39a–c** are equilibrated in an equimolar solution of benzene-d_6 and fluorobenzene-d_5 [36, 37]. Dimers containing two benzenes, two fluorobenzenes, and one of each guest were identified [36, 37]. This ability to bind multiple guests in a single cavity has been exploited recently by Rebek and co-workers in the acceleration of Diels–Alder reactions within "softball" dimers [38].

The structural diversity of "softball" capsules has also been analyzed by replacing the aryl spacer in the monomer with isomeric ethylene groups (shown in Figure 4-6) to form the congeners **40** · guest · **40** and **41** · guest · **41**, referred to as "wiffle" balls [39]. Various molecules were encapsulated in **40** · guest · **40**, but no template effects were reported [39]. Out of the five solvents tested (CD_2Cl_2, $CDCl_3$, C_6D_6, toluene-d_8, and p-xylene-d_{10}), heterodimer **40** · guest · **41** was only observed to form in benzene-d_6, which could implicate benzene as a template for **40** · guest · **41** [39].

Chiral "softballs" have also been reported from symmetric (**39a** and **39b**) and asymmetric (**39d**) monomers, when optically active camphor derivatives were encapsulated [40]. For example, dimerization of **39a** in the presence of camphanic acid in p-xylene-d_{10} gave two sets of host resonances in the ^{1}H-NMR spectra [40]. Doubling of these signals is a result of the formation of two isomeric "softball" dimers in which the guest lies in two different orientations [40]. Reorientation of the guest within the intact capsule is energetically unfavorable and can only occur through a stepwise process involving dissociation, then rotation of the guest, followed by recombination [40]. Therefore, chirality is induced from *within* the capsule [40]. Chirality can also be introduced from an asymmetric surface (**39d**) provided by the host in binding camphor as in "softball" **39d** · **39d** [40]. This is evident in the doubling observed of the guest resonances in the ^{1}H-NMR spectrum [40].

4.4.2.1 Template effects of solvent in the synthesis of "softball" dimers

Softball derivatives have been prepared by coupling tetraester **42** with two equivalents of diamine hydrochloride **43** (Scheme 4-10) in a variety of solvents in the presence of Et_3N [41, 42]. Product mixtures were found to contain three stereoisomers: C-shaped **44**, S-shaped **45**, and W-shaped **46** [42]. C-shaped isomer formed in higher yields in aromatic solvents such as benzene-d_6, toluene-d_8, and p-xylene-d_{12} (dimerization of **44** has been previously observed in p-xylene-d_{12}) [42]. Only statistical ratios of the three isomers were obtained in CH_2Cl_2 and $CHCl_3$, which are solvents that are less suitable for dimerization [42]. The overall reaction sequence giving the C, S, and W-shaped isomers proceeds through two steps. The first involves the formation of **47** (*E*) and **48** (*Z*) in each of the above mentioned solvents (Figure 4-7) [42]. Products **44**, **45**, and **46** are then formed in the second step by formation of two more amide bonds. In the second step, suitable solvents (vide supra) template the formation of the desired C-shaped **44** from the (*Z*) intermediate **48**, giving greater than statistical yields [42]. A zwitterionic tetrahedral intermediate resembling the structure of C-shaped **44** was proposed to be involved, whose concave surface provides a complementary fit to aromatic solvents [42]. Further stabilization of the zwitterionic intermediate directly leading to C-shaped **44** was also proposed to be provided through dimerization with another molecule of **44**, with concomitant reversible encapsulation of solvent as guest [42]. However, no experimental evidence supporting this suggestion has yet been put forth.

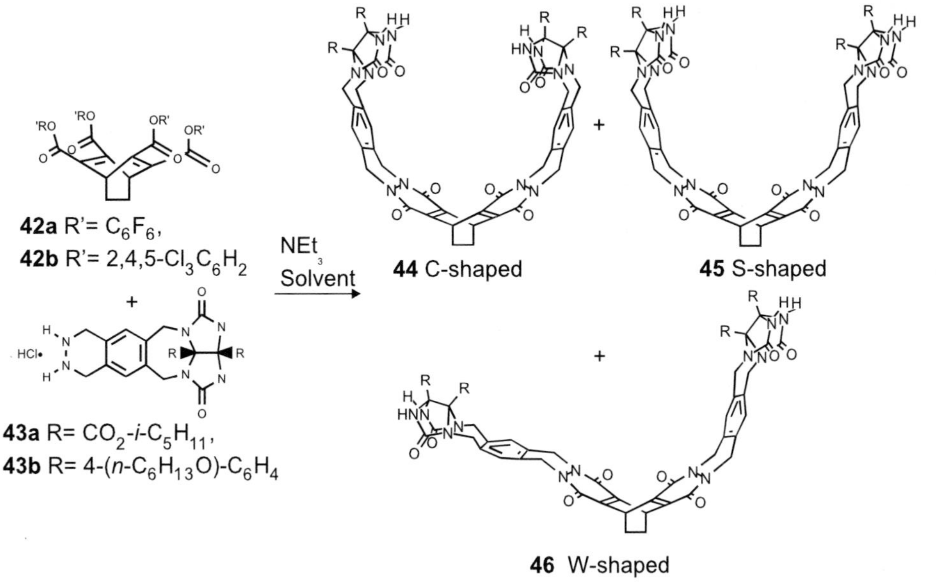
42a R'= C_6F_6,
42b R'= 2,4,5-$Cl_3C_6H_2$
43a R= CO_2-i-C_5H_{11},
43b R= 4-(n-$C_6H_{13}O$)-C_6H_4
NEt$_3$
Solvent
44 C-shaped
45 S-shaped
46 W-shaped

Scheme 4-10 Synthesis of C, S, and W-shaped monomeric diglycoluril isomers **44–46**.

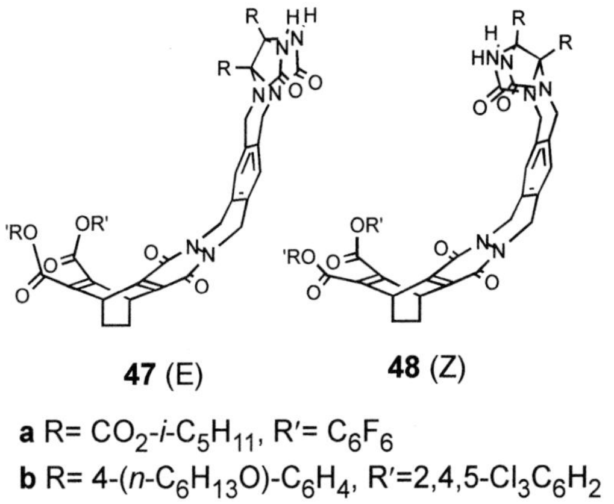
47 (E)
48 (Z)
a R= CO_2-i-C_5H_{11}, R'= C_6F_6
b R= 4-(n-$C_6H_{13}O$)-C_6H_4, R'=2,4,5-$Cl_3C_6H_2$

Figure 4-7 (*E*) and (*Z*) intermediates in the formation of isomeric diglycoluril functionalized monomers **44–46**.

4.4.3 Other Hydrogen Bonded Capsules

Rebek and co-workers have also produced self-assembling capsules where the monomer possesses C_3 symmetry (as opposed to C_2), by incorporating multiple glycoluril moieties into a flattened aromatic surface. For example, "jelly donut" dimer **50** forms in $CDCl_3$ from monomer **49**. This capsule features a flattened spherical interior cavity that can accommodate disk-shaped molecules like benzene and cyclohexane (Scheme 4-11) [43].

= Guest
R=
COO
49
50

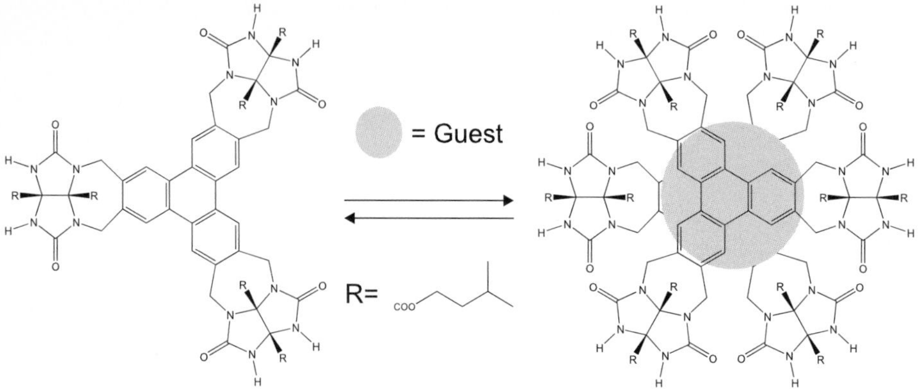

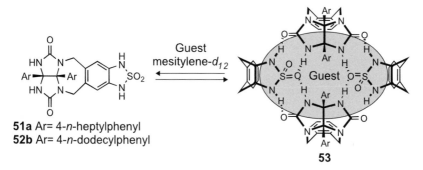

Scheme 4-11 Formation of Rebek's "jelly donut".

Recently, tetrameric assembly **52** (Scheme 4-12) has been reported [44]. Preparation of **52** involves the self-assembly of four discrete concave-shaped, monomeric subunits (**51**). Compound **51** exists in its monomeric state and is soluble in solvents that compete for hydrogen bonds, such as DMSO-d_6 [44]. In solvents such as CD_2Cl_2, $CDCl_3$, C_6D_6, toluene-d_8, and *p*-xylene-d_{10}, monomer **51** remains insoluble, existing as suspensions, which form clear solutions when adamantyl derivatives are added [44]. ^{1}H-NMR spectra of these solutions show the characteristic downfield-shifted glycoluril N–H signals, indicative of hydrogen bonding. A 4:1 ratio for host : guest signals was also measured, thus distinguishing the formation of a tetrameric assembly which truly encapsulates its guests [44]. Apparent association stabilities (in M^{-1}) were reported for complexes formed with the guests adamantane-2,6-dione (3 200), 2-adamantone (160), 1-adamantanol (48), bicyclo[3.3.1]nonane-2,6-dione, and adamantane (19) [44]. These values suggest that hydrogen bonding between the host with guests that contain carbonyl groups

Guest
mesitylene-d_{12}
Guest
51a Ar= 4-n-heptylphenyl
52b Ar= 4-n-dodecylphenyl
53

Scheme 4-12 Formation of tetrameric hydrogen-bonded capsule **52**.

(i.e. glycouril N–H···O=CR$_2$) increases complex stability [44]. The existence of such hydrogen bonding is evident in the ^{1}H-NMR spectra of the corresponding capsules with guests (vide infra) [44].

Cylinder-shaped molecular capsules such as **54** (Scheme 4-13) have also been report-ed. Deep-cavity resorcinarene **53** was found to dimerize through a seam of eight bifur-cated intermolecular hydrogen bonds between the imide and carbonyl functionalities located on the rim of the monomer [45]. Dimerization was observed to be templated in mesitylene-d_{12}, according to the ^{1}H-NMR spectra, which feature downfield shifted N–H host signals upon encapsulation of guests such as bibenzyl, terphenyl, dicyclohexyl car-bodiimide, *trans*-4-stilbene, *N*-phenylbenzylamine, benzyl phenyl ether, *trans*-4-stilbene methanol, and *p*-[*N*-(*p*-tolyl)]toluamide [45]. Although dimer **54** does form in solvents such as CDCl$_3$, benzene-d_6, and toluene-d_8, mesitylene-d_{12} itself appears to be too large for the cavity, as suggested by the presence of multiply hydrogen-bonded N–H signals in the ^{1}H-NMR spectra [45]. Capsules do form in mesitylene-d_{12} upon addition of tem-plating guests such as benzene, *p*-xylene, toluene, *p*-trifluoromethyltoluene, *p*-chloro-toluene, 2,5-lutidine, and *p*-methylbenzylalcohol, where specific pairs are preferably encapsulated over others [45]. For example, in the presence of a 1:1 ratio of benzene:*p*-xylene in mesitylene-d_{12}, the two resorcinarenes (**53**) dimerize, forming capsules con-taining one of each guest, preferably over other identical pairs of guests [45]. It was sug-gested that one benzene together with one *p*-xylene provides the best occupancy factor [46] as two benzenes leave too much and two *p*-xylenes leave too little empty space [45].

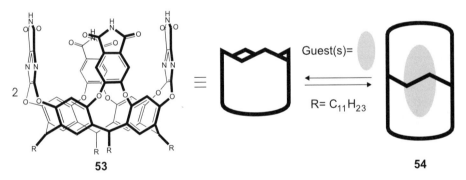

53   **54**

Scheme 4-13   Formation of cylindrical-shaped capsule **54** from deep-cavity resorcinarene **53**.

4.4.4   Capsules that do not Involve Hydrogen Bonding

Using a similar approach to forming synthetic capsules, other structures have been designed that utilize attractive interactions between the host and the guest, other than van der Waals interactions and hydrogen bonding. One example is the formation of a capsule (**56**) from two equivalents of bowl-shaped Lewis acid precursor, aluminum tris(2,6-di-phenylphenoxide) (ATPH) **55**, with a guest such as 1,4-cyclohexanedione (Scheme 4-14)

[47]. Coordinate bonds between the aluminum center of ATPH and the carbonyls of the guest hold the two ATPH subunits together [47]. The resulting dimeric capsule renders the bound guest completely sheltered from the external environment; hence the guest does not dissociate in the presence of other carbonyl substrates [47]. Addition of MeLi (1 equiv) to a solution containing capsule **56** and 4-*tert*-butyl-1-methylcyclohexanone resulted in the recovery of the diketone and 4-*tert*-1-methylcyclohexanol [47]. The interior of capsule **56** was also found to serve as a miniature reaction chamber as Diels–Alder adducts were recovered bound when, at low temperature, two equivalents of ATPH were added to a solution containing a quinone derivative and a suitable diene [47]. Interesting regio- and stereoselective features of some of the Diels–Alder reactions were observed upon analysis of the cyclo-adduct product distributions. Results indicated that preorganization of the diene and quinone (dienophile) with ATPH prior to capsule formation is crucial [47].

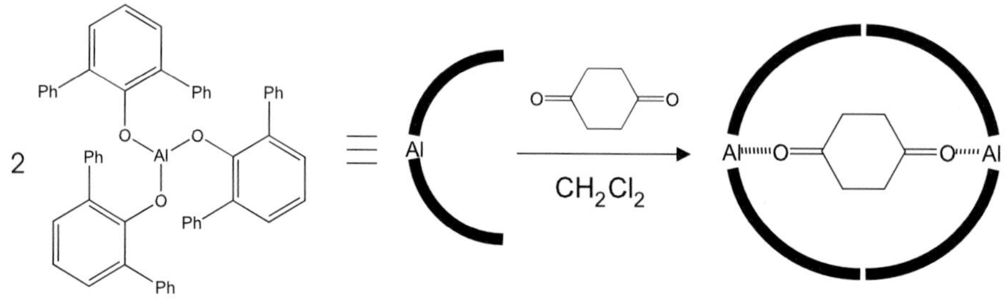

Scheme 4-14   Formation of capsule **56** from Lewis acid receptor ATPH (**55**).

The templating of dimeric capsules incorporating porphyrins has also been reported recently. These reports describe the self-assembly of molecular capsules based on metal-to-ligand interactions of two dizinc(II) bisporphyrins around tetratopic Lewis base template guests [48, 49]. The structure of the monomers used consists of two porphyrin derivatives covalently linked by a concave spacer to give molecules having a "clip-like" topology. Dimerization of two "clips" was found to occur using tetratopic aliphatic or tetrapyridylporphyrin templates, yielding structures that are described as resembling a universal joint.

4.4.5   Other Capsules

Various other examples have been provided in the literature involving the formation of molecular capsules. However, guest binding by these capsules may not have been studied, or evidence of a template requirement for formation may not have been present. These include urea-based calixarene dimers [50], "deep-cavity" resorcinarenes [51], cyclocholate dimers [52], calixarene heterodimers [53], cucurbituril capsules [54], cyclodextrin dimers [55], resorcinarene-based cages [56], self-assembling calix[4]resorcinarene oligomers [57], self-assembled block copolymer aggregates [58], and CTV heterodimers [59].

4.5 Conclusions

This chapter demonstrates the importance of templating in the formation of "molecules within molecules" in the form of carceplexes, hemicarceplexes, and self-assembling capsules. Suitable template molecules can facilitate the formation of large supramolecular structures via favorable non-covalent interactions. Selectivity for particular template molecules can range up to one million-fold. Recognizing and understanding the roles of the molecular forces involved in these processes is important if we are to understand biological assemblies further, including enzymes, cell membranes, and viruses. Information obtained in forming smaller well-defined molecular assemblies will also enable scientists to apply that knowledge in the formation of much larger supramolecular structures, eventually reaching and perhaps even surpassing the size and complexity of those found in nature.

The reversible nature of self-assembling capsules has already been exploited as molecular switching devices and catalysts. Other applications may involve the development of artificial enzymes, and the design of specific drug delivery systems.

Acknowledgements

We thank Christoph Naumann and Ashley Causton for careful reading of this manuscript. We also thank the funding agencies (NIH, PRF, and NSERC) for supporting our work, and we thank the authors of the references discussed for providing a stimulating area to review.

References

[1] (a) C. J. Pederson, *J. Am. Chem. Soc.* **1967**, *89*, 7017–7036; (b) C. J. Pederson, *Aldrichim. Acta* **1971**, *4*, 1; (c) D. J. Busch, A. L. Vance, A. G. Kolchinski, in *Comprehensive Supramolecular Chemistry*; J.-M. Lehn, J. L. Atwood, D. D. MacNicol; F. Vögtle, Series Eds.; Pergamon: New York, **1996**; Vol. 9, pp. 2–42 and references therein.

[2] (a) G. Schill, *Catenanes, Rotaxanes, and Knots*; Academic Press; New York, **1971**; (b) D. B. Amabilino, J. F. Stoddart, *Chem. Rev.* **1995**, *95*, 2725–2628; (c) J.-P. Sauvage, *Acc. Chem. Res.* **1998**, *31*, 611–619.

[3] Some recent papers on catenanes: (a) D. B. Amabilino, P. R. Ashton, V. Balzani, S. E. Boyd, A. Credi, J. Y. Lee, S. Menzer, J. F. Stoddart, M. Venturi, D. J. Williams, *J. Am. Chem. Soc.* **1998**, *120*, 4294–4307; (b) D. G. Hamilton, N. Feeder, L. Prodi, S. J. Teat, W. Clegg, J. K.

M. Sanders, *J. Am. Chem. Soc.* **1998**, *120*, 1096–1097; (c) P. R. Ashton, S. E. Boyd, S. Menzer, D. Pasini, F. M. Raymo, N. Spencer, J. F. Stoddart, A. J. P. White, D. J. Williams, P. G. Wyatt, *Chem. Eur. J.* **1998**, *4* (2), 299–310; (d) S. Capobianchi, G. Doddi, G. Ercolani, P. Mencarelli, *J. Org. Chem.* **1998**, *63*, 8088–8089; For a few recent papers on rotaxanes: (e) C. Fischer, M. Nieger, O. Mogck, V. Böhmer, R. Ungaro, F. Vögtle, *Eur. J. Org. Chem.* **1998**, 155–161; (f) T. Dünnwald, A. H. Parham, F. Vögtle, *Synthesis* **1998**, *3*, 339–348; (g) O. Braun, F. Vögtle, *Synlett* **1997**, *10*, 1184–1186.

[4] (a) G. Wulff, *Template Induced Control of Stereochemistry for the Synthesis of Polymers*; G. Wulff, Ed.; Kluwer Academic Publishers: Netherlands, **1995**; Vol. 473, pp. 13–19; (b) D. Spivak, M. A. Gilmore, K. J. Shea, *J. Am. Chem. Soc.* **1997**, *119*, 4388–4393; (c) G. Wulff, *Angew. Chem. Int. Ed. Engl.* **1995**, *34*, 1812–1832; (d) K. Uezu, H. Nakamura, J.-i. Kanno, T. Sugo, M. Goto, F. Nakashio, *Macromolecules* **1997**, *30*, 3888–3891.

[5] (a) S. I. Zones. Y. Nakagawa, L. T. Yuen, T. V. Harris, *J. Am. Chem. Soc.* **1996**, *118*, 7558–7567; (b) K. Pitchumani, M. Warrier, C. Cui, R. G. Weiss, V. Ramamurthy, *Tetrahedron. Lett.* **1996**, *37*, 6251–6254; (c) T. Xu, E. J. Munson, J. F. Haw, *J. Am. Chem. Soc.* **1994**, *116*, 1962–1972; (d) D. K. Lewis, D. J. Willock, C. R. A. Catlow, J. M. Thomoas, G. J. Hutchings, *Nature (London)* **1996**, *382*, 604–606; (e) C.-H. Tung, L.-Z. Wu, Z. Y. Yuan, N. Su, *J. Am. Chem. Soc.* **1998**, *120*, 11594–11602.

[6] (a) S. Anderson, H. Anderson, J. K. M. Sanders, *Acc. Chem. Res.* **1993**, *26*(9), 469–475; (b) H. L. Anderson, J. K. M. Sanders, *J. Am. Chem. Soc., Perkin Trans. 1* **1995**, *18*, 2223–2229; (c) H. L. Anderson, A. Anderson, J. K. M. Sanders, *J. Chem. Soc., Perkin Trans. 1* **1995**, *18*, 2255–2267; (d) L. G. Mackay. H. L. Anderson, J. K. M. Sanders, *J. Chem. Soc., Perkin Trans. 1* **1995**, *18*, 2269–2273; (e) N. Bampos, V. Marvaud, J. K. M. Sanders, *Chem. Eur. J.* **1998**, *4*(2), 335–343.

[7] (a) T. J. McMurry, K. N. Raymond, P. H. Smith, *Science* 1989, *244*, 938–943; (b) R. Hoss, F. Vögtle, *Angew. Chem. Int. Ed. Engl.* **1994**, *33*, 375–384; (c) B. Hasenknopf, J.-M. Lehn, B. O. Kneisel, G. Baum, D. Fenske, *Angew. Chem., Int. Ed. Engl.* **1996**, *108*, 1838–1840.

[8] R. G. Chapman, J. C. Sherman, *Tetrahedron* **1997**, *53* (47), 15911–15945.

[9] D. Busch, *J. Inclusion Phenom. Mol. Recognit. Chem.* **1992**, *12*, 389–395.

[10] (a) E. Maverick, D. J. Cram, in *Comprehensive Supramolecular Chemistry*; F. Vögtle, Vol. Ed.; J.-M. Lehn, J. L. Atwood, J. E. D. Davies, D. D. MacNicol, F. Vögtle, Series Eds.; Pergamon: New York, **1996**; Vol. 2, pp. 367–418; (b) D. J. Cram, J. M. Cram, *Container Compounds and Their Guests*; The Royal Society of Chemistry: Cambridge, **1994**; (c) J. C. Sherman, in *Large Ring Molecules*; Semlyen, J. A., Ed.; Wiley: Chichester, England, **1996**; pp. 507–524; (d) J. C. Sherman, *Tetrahedron* **1995**, *51*, 3395–3422; (e) A. Jasat, J. C. Sherman, *Chem. Rev.* **1999**, *99*, 931–962.

[11] (a) M. M. Conn, J. Rebek, Jr. *Chem. Rev.* **1997**, *97*, 1647–1668; (b) J. Rebek, Jr. *Chem. Soc. Rev.* **1996**, 255–264.

[12] J. C. Sherman, D. J. Cram, *J. Am. Chem. Soc.* 1989, *111*, 4527–4528.

[13] J. C. Sherman, C. B. Knobler, D. J. Cram, *J. Am. Chem. Soc.* **1991**, *113*, 2194–2204.

[14] (a) R. G. Chapman, N. Chopra, E. D. Cochien, J. C. Sherman, *J. Am. Chem. Soc.* **1994**, *116*, 369–370; (b) R. G. Chapman, J. C. Sherman, *J. Org. Chem.* **1998**, *63* (12), 4103–4110.

[15] J. R. Fraser, B. Borecka, J. Trotter, J. C. Sherman, *J. Org. Chem.* **1995**, *60* (5), 1207–1213.

[16] R. G. Chapman, J. C. Sherman, *J. Am. Chem. Soc.* **1995**, *117*, 9081–9082.

[17] R. G. Chapman, G. Olovsson, J. Trotter, J. C. Sherman, *J. Am. Chem. Soc.* **1998**, *120*, 6252–6260.

[18] K. Nakamura, C. Sheu, A. E. Keating, K. N. Houk, *J. Am. Chem. Soc.* **1997**, *119* (18), 4321–4322.

[19] R. G. Chapman, J. C. Sherman, *J. Am. Chem. Soc.* **1998**, *120*, 9818–9826.

[20] N. Chopra, J. C. Sherman, *Angew. Chem., Int. Ed. Engl.* **1997**, *36*, 16, 1727–1729.

[21] N. Chopra, J. C. Sherman, *Angew. Chem., Int. Ed. Engl.* **1999**, *38*, 1955–1957.

[22] (a) J. A. Bryant, M. T. Blanda, M. Vincenti, D. J. Cram, *J. Chem. Soc., Chem. Commun.* **1990**, 1403–1405; (b) J. A. Bryant, M. T. Blanda, M. Vincenti, D. J. Cram, *J. Am. Chem. Soc.* **1991**, *113*, 6, 2167–2172.

[23] A. Jasat, J. C. Sherman, unpublished results.

[24] A. M. A. van Wageningen, J. P. M. van Duynhoven, W. Verboom, D. N. Reinhoudt, *J. Chem. Soc., Chem. Commun.* **1995**, 1941–1942.

[25] A. M. A. van Wageningen, P. Timmerman, J. P. M. van Duynhoven, W. Verboom, F. C. J. M. van Veggel, D. N. Reinhoudt, *Angew. Chem. Int. Ed. Engl.* **1997**, *3* (4), 639–654.

[26] P. Timmerman, W. Verboom, F. C. J. M. van Veggel, J. P. M. van Duynhoven, D. N. Reinhoudt, *Angew. Chem. Int. Ed. Engl.* **1994**, *33* (22), 2345–2348.

[27] P, Jacopozzi, E. Dalcanale, *Angew. Chem. Int. Ed. Engl.* **1997**, *36* (6), 613–615.

[28] N. Chopra, J. C. Sherman, *Supramol. Chem.* **1995**, *5*, 31–37.

[29] D. J. Cram, M. E. Tanner, C. B. Knobler, *J. Am. Chem. Soc.* **1991**, *113* (20), 7717–7727.

[30] (a) M. L. C. Quan, D. J. Cram, *J. Am. Chem. Soc.* **1991**, *113*, 2754–2755; (b) M. L. C. Quan, C. B. Knobler, D. J. Cram, *J. Chem. Soc., Chem. Commun.* **1991**, 660–662; (c) D. J. Cram, M. T. Blanda, K. Paek, C. B. Knobler, *J. Am. Chem. Soc.* **1992**, *114*, 7765–7773; (d) D. J. Cram, R. Jaeger, K. Deshayes, *J. Am. Chem. Soc.* **1993**, *115*, 10111–10116; (e) Y.-S. Byun, O. Vadhat, M. T. Blanda, C. B. Knobler, D. J. Cram, *J. Chem. Soc., Chem. Commun.* **1995**, 1825–1827; (f) T. A. Robbins, C. B. Knobler, D. R. Bellew, D. J. Cram, *J. Am. Chem. Soc.* **1994**, *116*, 111–122; (g) Y.-S. Byun, T. A. Robbins, C. B. Knobler, D. J. Cram, *J. Chem. Soc., Chem. Commun.* **1995**, 1947–1948; (h) K. J. Judice, D. J. Cram, *J. Am. Chem. Soc.* **1991**, *113*, 2790–2791; (i) J. Yoon, D. J. Cram, *J. Am. Chem. Soc.* **1997**, *119*, 11796–11806; (j) B. S. Park, C. B. Knobler, C. N. Eid, Jr.; R. Warmuth, D. J. Cram, D. J. *J. Chem. Soc., Chem. Commun* **1998**, 55–56; (k) J. Yoon, D. J. Cram, *J. Chem. Soc., Chem. Commun.* **1997**, 497–498; (l) R. Kurdistani, R. G. Helgeson, D. J. Cram, *J. Am. Chem. Soc.* **1995**, *117* (5), 1659–1660; (m) J. Yoon, C. Sheu, K. N. Houk, C. B. Knobler, D. J. Cram, *J. Org. Chem.* **1996**, *61*, 9323–9339; (n) J. Yoon, C. B. Knobler, E. F. Maverick, D. J. Cram, *J. Chem. Soc., Chem. Commun.* **1997**, 1303–1304; (o) J. Yoon, D. J. Cram, *J. Chem. Soc., Chem. Commun.* **1997**, 1505–1506; (p) J. Yoon, D. J. Cram, *J. Chem. Soc., Chem. Commun.* **1997**, 2065–2066.

[31] D. A. Makeiff, D. J. Pope, J. C. Sherman, unpublished results.

[32] R. C. Helgeson, K. Paek, C. B. Knobler, E. F. Maverick, D. J. Cram, *J. Am. Chem. Soc.* **1996**, *118*, 5590–5604.

[33] (a) R. Wyler, J. Mendoza, J. Rebek, Jr. *Angew. Chem. Int. Ed. Engl.* **1993**, *32*(12), 1699–1701; (b) N. Branda, J. Wyler, J. Rebek, Jr. *Science* **1994**, *263*, 1267–1268; (c) C. Valdés, U. P. Spitz, L. M. Toledo, S. W. Kubik, J. Rebek, Jr. *J. Am. Chem. Soc.* **1995**, *117* (51), 12733–12745; (d) N. Branda, R. M. Grotzfeld, C. Valdés, J. Rebek, Jr. *J. Am. Chem. Soc.* **1995**, *117* (1), 85–88; (e) X. Garcías, J. Rebek, Jr. *Angew. Chem. Int. Ed. Engl.* **1996**, *35* (11), 1225–1228.

[34] D. A. Evans, J. S. Evans, *J. Org. Chem.* **1998**, *63*, 8027–8030.

[35] R. S. Meissner, J. Rebek, Jr.; J. de Mendoza, *Science* **1995**, *270*, 1485–1488.

[36] J. Kang, J. Rebek, Jr. *Nature (London)* **1996**, *382*, 239–241.

[37] R. Meissner, X. Garcias, S. Mecozzi, J. Rebek, Jr. *J. Am. Chem. Soc.* **1997**, *119*, 77–85.

[38] (a) J. Kang, J. Rebek, Jr. *Nature (London)* **1997**, *385*, 50–52; (b) J. Kang, G. Hilmersson, J. Santamaría, J. Rebek, Jr. *J. Am. Chem. Soc.* **1998**, *120*, 3650–3656; (c) J. Kang, J. Santamaría, G. Hilmersson, J. Rebek, Jr. *J. Am. Chem. Soc.* **1998**, *120*, 7389–7390.

[39] J. M. Rivera, T. Martín, J. Rebek, Jr. *J. Am. Chem. Soc.* **1998**, *120*, 4, 819–820.

[40] Y. Tokunaga, J. Rebek, Jr. *J. Am. Chem. Soc.* **1998**, *120*, 66–69.

[41] Y. Tokunaga, D. M. Rudkevich, J. Rebek, Jr. *Angew. Chem. Int. Ed. Engl.* **1997**, *36*, 23, 2656–2659.

[42] Y. Tokunaga, D. M. Rudkevich, J. Santamaría, G. Hilmersson, J. Rebek, Jr. *Chem. Eur. J.* **1998**, *4*, 8, 1449–1457.

[43] R. M. Grotzfeld, N. Branda, J. Rebek, Jr. *Science* **1996**, *271*, 487–489.

[44] T. Martin, U. Obst, J. Rebek, Jr. *Science* **1998**, *281*, 1842–1845.

[45] T. Heinz, D. M. Rudkevich, J. Rebek, Jr. *Nature (London)* **1998**, *394*, 764–766.

[46] S. Mecozzi, J. Rebek, Jr. *Chem. Eur. J.* **1998**, *4*(6), 1016–1022.

[47] T. Ooi, Y. Kondo, K. Maruoka, *Angew. Chem. Int. Ed. Engl.* **1998**, *37*(21), 3039–3041.

[48] J. N. H. Reek, A. P. H. J. Schenning, A. W. Bosman, E. W. Meijer, M. J. Crossley, *J. Chem. Soc., Chem. Commun.* **1995**, 11–12.

[49] M. R. Johnston, J. G. Maxwell, R. N. Warrener, *J. Chem. Soc., Chem. Commun.* **1998**, 2739–2740.

[50] (a) K. D. Shimizu, J. Rebek, Jr. *Proc. Nat. Acad. Sci. USA* **1995**, *92*, 12403–12407; (b) B. C. Hamann, K. D. Shimizu, J. Rebek, Jr. *Angew. Chem. Int. Ed. Engl.* **1996**, *35*(12), 1326–1329; (c) R. K. Castellano, D. M. Rudkevich, J. Rebek, Jr. *J. Am. Chem. Soc.* **1996**, *118*, 10002–10003; (d) R. K. Castellano, B. H. Kim, J. Rebek, Jr. *J. Am. Chem. Soc.* **1997**, *119*, 12671–12672; (e) R. K. Castellano, D. M. Rudkevich, J. Rebek, Jr. *Proc. Nat. Acad. Sci. USA* **1997**, *94*, 7132–7137; (f) R. K. Castellano, D. M. Rudkevich, J. Rebek, J. *J. Am. Chem. Soc.* **1998**, *120*, 3657–3663.

[51] (a) S. Ma, D. M. Rudkevich, J. Rebek, Jr. *J. Am. Chem. Soc.* **1998**, *120*, 4977–4981; (b) C. von dem Bussche-Hünnefeld, R. C. Helgeson, D. Bührig, C. B. Knobler, D. J. Cram, *Croat. Chim. Acta* **1996**, *69*, 447–458.

[52] R. P. Boner-Law, J. K. M. Sanders, *Tetrahedron Lett.* **1993**, *34*, 1677–1680; (b) R. P. Boner-Law, J. K. M. Sanders, *J. Am. Chem. Soc.* **1995**, *117*, 259–271.

[53] (a) O. Mogck, V. Böhmer, W. Vogt, *Tetrahedron* **1996**, *52*, 8489–8496; (b) O. Mogck, E. F. Paulus, V. Böhmer, W. Vogt, *J. Chem. Soc., Chem. Commun.* **1996**, 2533–2534; (c) K. Koh, K. Araki, S. Shinkai, *Tetrahedron Lett.* **1994**, *35*, 8255–8258; (d) R. H. Vreekamp, W. Verboom, D. N. Reinhoudt, *J. Org. Chem.* **1996**, *61*, 4282–4288; (e) S. B. Lee, J.-I. Hong, *Tetrahedron Lett.* **1996**, *37*, 8501–8504; (f) J. Costante-Crassous, T. J. Marrone, J. M. Briggs, J. A. McCammon, A. Collet, *J. Am. Chem. Soc.* **1997**, *119*, 3818–3823.

[54] (a) Y. M. Jeon, J. Kim, D. Whang, K. Kim, *J. Am. Chem. Soc.* **1996**, *118*, 9970–9971; (b) P. Cintas, *J. Inclusion Phenom. Mol. Recognit. Chem.* **1994**, *17*, 205–220; (c) D. Whang, J. Heo, J. H. Park, K. Kim, *Angew. Chem. Int. Ed. Engl.* **1998**, *37*(1), 78–80.

[55] (a) P. Klüfers, H. Piotrowski, J. Uhlendorf, *Chem. Eur.* **1997**, *3*, 601–608; (b) R. Fuchs, N. Havermann, P. Klüfers *Angew. Chem. Int. Ed. Engl.* **1993**, *32*, 852–854; (b) F. Hibbert, J. Emsley, *Adv. Phys. Org. Chem.* **1990**, *26*, 255–379; (c) J. Emsley, *Chem. Soc. Rev.* **1980**, *9*, 91–124; (d) A. Novak, *Struct. Bonding* **1974**, *18*, 177–216; (e) P. Klüfers, J. Schuhmacher, *Angew. Chem. Int. Ed. Engl.* **1994**, *33*, 852–854; (f) Z.-I. Yoshida, H. Takekuma, S.-I. Takekuma, Y. Matsubara, *Angew. Chem. Int. Ed. Engl.* **1994**, *33*, 1597–1599; (g) Y. Kikuchi, Y. Tanaka, S. Sutarto, K. Kobayashi, H. Toi, Y. Aoyama, *J. Am. Chem. Soc.* **1992**, *114*, 10302–10306; (h) F. Venema, A. E. Rowan, R. J. M. Nolte, *J. Am. Chem. Soc.* **1996**, *118*, 257–258; (i) A. K. Croft, C. J. Easton, S. F. Lincoln, B. L. May, J. Papageorgiou, *Aust. J. Chem.* **1997**, *50*, 857–859; (j) T. Jiang, D. S. Lawrence, *J. Am. Chem. Soc.* **1995**, *117*, 1857–1858; (k) T. Jiang, D. K. Sukumaran, S.-D. Soni, D. S. Lawrence, *J. Org. Chem.* **1994**, *59*, 5149–5155; B. Hamelin, L. Jullien, C. Derouet, C. H. du Penhoat, P. Berthault, *J. Am. Chem. Soc.* **1998**,

120, 8438–8447; (l) G. Wenz, *Angew. Chem. Int. Ed. Engl.* **1994**, *33*, 803–822; (m) S. Makedonopoulou, I. M. Mavridis, K. Yannakopoulou, J. Papaioannou, *J. Chem. Soc., Chem. Commun.* **1998**, 2133–2134.

[56] O. D. Fox, K. Dalley, R. G. Harrison, *J. Am. Chem. Soc.* **1998**, *120*, 7111–7112.

[57] (a) L. R. MacGillvray, J. L. Atwood, *Nature (London)* **1997**, *389*, 469–472; (b) K. N. Rose, L. J. Barbour, G. W. Orr, J. L. Atwood, *J. Chem. Soc., Chem. Commun.* **1998**, 407–408; (c) K. Murayama, K. Aoki, *J. Chem. Soc., Chem. Commun.* **1998**, 607–608.

[58] S. A. Jenekhe, X. L. Chen, *Science* **1998**, *279*, 1903–1907.

[59] S. B. Lee, J.-I. Hong, *Tetrahedron Lett.* **1996**, *47*, 8501–8504.

5 Template-Directed Ligation: Towards the Synthesis of Sequence Specific Polymers

Yahaloma Gat, David G. Lynn

5.1 Introduction

The structure of the DNA double helix first highlighted the simple elegance of the most sophisticated template reactions known, template directed polymerization. Through these reactions the genomic information is accurately replicated in DNA, transcribed into RNA, and translated into protein. Over 40 years ago, Crick combined these reactions into the "Central Dogma" of biology (Figure 5-1), a formulation suggesting that all living things follow this basic chemical plan [1]. Today, even with the incredible biological diversity known to exist on Earth, no violations of this chemical scheme are known. The nucleic acids store the genetic blueprint, and the co-catalytic proteins function cooperatively with this molecular template to express the encoded information.

Over the last four decades, synthetic non-enzymatic oligonucleotide ligations have been explored as potential models for the evolution of replicating systems on the early Earth [2–11]. These studies have explored ligation reactions that occur on a template without protein/enzymatic catalysis. Such systems have both provided mechanisms for the early evolution of biopolymer catalysts and placed limits on template directed ligation. More recently, the lure of antisense molecules capable of disrupting specific gene expression has resulted in the exploration of backbone-modified nucleic acids [12–33].

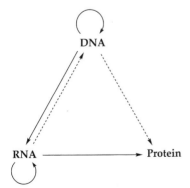

Figure 5-1 The "Central Dogma" as formulated in 1958 with probable (solid arrows) and possible (broken arrows) reactions as indicated (adapted from [1b]).

At the same time, several laboratories have used in-vitro selection strategies to identify RNAs and DNAs that catalyze specific reactions [34–40]. The information from these and related studies now opens the possibility of expanding the molecular skeletons capable of storing sequence information into ones that can be read into complementary materials [41].

Since it has not been possible to cover all the valuable contributions that have been made in these fields of study, we have instead organized the discussion around a minimal scheme for template directed ligation. This minimal scheme will be compared with the autonomous natural genomes that self-replicate via template directed polymerization reactions. The advantages and limitations of each step in the minimal scheme are discussed with respect to the structural diversity in the template that will make it possible to extend these reactions to new materials and new polymerization strategies.

5.2 Template-directed Ligation: Minimalist Scheme

The overall catalytic efficiency of the template, as analyzed by Michaelis–Menten kinetics, is expressed as a function of both binding affinity and rate enhancement. K_m, the substrate concentration necessary for half-maximal rate, can provide an inverse measure of binding affinity, and for native protein enzymes ranges from mM to µM. The catalyst rate enhancement is given by the turnover number, k_{cat}, the number of substrate molecules transformed into product per unit time by a single catalytic site. Values of k_{cat} typically range from 10^3 to 10^7 s^{-1} for enzymes. The ratio of these two constants, k_{cat}/K_m, provides a direct measure of the catalytic efficiency at substrate concentrations that are significantly below template saturation. By considering the oligonucleotide templates as catalysts, a comparison with their enzyme counterparts is inevitable [42]. For example, the RNA component of RNAse P, which catalyzes the hydrolysis of tRNA, has a turnover number of only 1 min^{-1}, a K_m of 0.5 µM, and a catalytic efficiency, k_{cat}/K_m, of 2×10^6 min^{-1} M^{-1} [43]. While RNAse P is a reasonable catalyst, its catalytic efficiency, like all known ribozymes, falls well below the diffusion-limited rates ($>10^8$ s^{-1} M^{-1}) of an ideal catalyst where substrate turnover occurs with every substrate encounter [44].

The minimal scheme for a catalytic template directed ligation is outlined in Figure 5-2. In step **1**, molecular recognition controls preorganization of the substrates along the template. The specificity, or fidelity, of the template directed synthesis is defined by the probability that the complementary substrate, rather than a non-complementary competitor, will be inserted opposite the appropriate site on the template. As can be seen in Figure 5-2, the difference in affinity between two substrates, S1 and S1′, for the template, T, can be expressed as a difference in binding affinity, $\Delta\Delta G = \Delta G_{S1'T} - \Delta G_{S1T}$. The energetics of this pairwise interaction between a substrate and the complementary strand define an effective molarity for the substrates. The association is entropically disfavored, not only due to the three degrees of translational and rotational entropy that are lost, but also because free internal bond rotations can be restricted by complexation [45, 46]. Restraining either the substrate and/or the template to resemble the bound conformation could reduce this entropic cost.

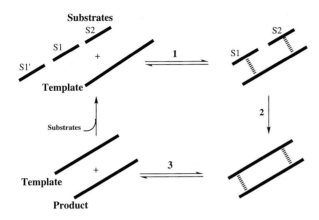

Figure 5-2 Scheme of template-directed ligation with steps **1**, molecular recognition; **2**, ligation; **3**, product dissociation.

The chemical transformation in the process, step **2**, is the ligation. The reaction will be important to both catalytic efficiency and fidelity. k_{cat}/K_m provides a direct measure of catalytic efficiency, and is also sometimes called the specificity constant [47]. On a single catalyst, the relative velocities of the reaction of two substrates S1 and S1′ at equal molar concentrations are proportional to the ratios of their respective k_{cat}/K_m values. Thus, the discrimination between the two competing substrates is determined by this ratio rather than by either quantity alone. In principle, a ligation can exploit any reaction that would couple the substrate chain ends. However, the reaction will be constrained by the geometry required to bind the substrates to the template and little information is currently available to place limits on the selection of appropriate reactions. At least with the nucleic acids, the increasing knowledge and synthetic availability of modified backbones may allow the desired association energies and reaction geometries to be rationally engineered into the templates so that these parameters can be optimized.

Clearly, product dissociation by this scheme, step **3**, will limit turnover, as the product–template duplex will be stabilized over the substrate–template ternary complex. Maximizing $\Delta\Delta G$ in step **1** is required to enhance fidelity, but this increased affinity is generally present in the product duplex. Obviously if the binding affinity could be a tuned variable, high substrate affinity for ligation to maximize accuracy and reduced product affinity to enhance catalytic efficiency, product inhibition could be avoided.

In Nature's genomes, replication fidelity is primarily a kinetic function of the mechanistically complex DNA polymerases [43]. These enzymes employ activated phosphoric acid anhydrides leading to a final product–template duplex. The third step, dissociation of the product strand from the template, requires an additional input of chemical energy and a separate set of protein catalysts. Therefore each step in DNA replication requires a critical and complex set of translation products to ensure the production of a single copy of the template, and for autonomous growth, the template must be of sufficient size to encode all of the required catalysts for the process. The notion that these templates must encode catalysts, or possibly even the templates themselves serve as catalysts for more than one reaction, becomes an important feature for the development of template autonomy.

5.3   Catalytic RNA and DNA

Ribozymes [48–55], discovered in the early 1980s, documented the ability of RNA to fulfill both roles of information storage and catalysis. Prebiotically, in an "RNA-world" scenario which proposes that RNA molecules led to the appearance of early life [56, 57], RNA replication is expected to follow the steps in Figure 5-2. However, known ribozymes catalyze a rather narrow range of reactions at phosphoryl centers [58–60]. In-vitro selection methods have broadened the catalytic repertoire to include also 2′-deoxynucleic acids. Thus, the 2′-hydroxyl group, believed to be critical for the catalytic activity of ribozymes, is unessential in DNAs that cleave RNA [36, 37], promote self-cleavage [38], and catalyze ligation of chemically activated DNAs [35]. For example, E47 is a 47-mer DNA ligase shown to have an initial rate of ligation that is 10^3 greater than the rate of catalysis given by a simple complementary template [35]. The catalyst was however severely limited by product release, and even at saturating substrate concentrations, the turnover number was only 0.66 h^{-1}.

With selection methods, the catalytic activity of RNA molecules now includes reaction types as diverse as ester [61] and amide bond formation [34], bridged biphenyl isomerization [40], and porphyrin metallations [62a]. Selections for amide formation yielded RNAs that were able to catalyze the transfer of an amino acid to their own 5′-hydroxyl group [34]. The selected sequences contained a 13-nucleotide invariant stretch that was hypothesized to serve as an internal template bringing together the aminoacyl group of the substrate and the 5′-hydroxyl of an RNA library. By making a derivative in which the 5′-hydroxyl was replaced with a 5′-amino group, the modified ribozyme was found to transfer the amino acid almost as efficiently as the initially selected ribozyme, but the overall rate enhancement was still low, with a k_{cat} of 0.19 min^{-1}. With porphyrin metallation, selected DNA and RNA catalysts were less efficient than the native ferrochelatase, but the catalytic efficiency of the ribozymes was comparable to that of an antibody selected for the same function [39, 62a]. The size of the nucleic acid, however, was ca. ~8 kDa, whereas the protein was five times larger [62b].

In order to overcome the shortage of catalytically active functional groups in RNA, in-vitro selections have been combined with nucleobase modification [63, 64]. Uridine derivatives modified at C-5 of the nucleobase were first shown to be substrates for RNA polymerase, and then were incorporated into selected RNAs. In one case, an incorporated imidazole-modified uridine was employed to catalyze amide bond formation and a rate enhancement of $10^4 – 10^5$ over the uncatalyzed reaction was observed [63]. RNA sequences using native UTP were completely inactive, establishing that the imidazole-modified uridine was crucial for catalysis. In a second study, uridine bases attached to pyridine were incorporated and found to accelerate the rate of a Diels–Alder cycloaddition [64]. The pyridine was hypothesized to form Lewis acid complexes with transition metals to catalyze the reaction. In contrast to most catalytic RNAs, in these two studies nucleobase templating was not required for catalytic activity. Clearly, like proteins, nucleic acid oligomers are able to catalyze reactions through a variety of mechanisms and their catalytic range can be extended with synthetic modifications.

5.4   Molecular Recognition

The nucleic acids represent the most obvious and best-understood templates in the context of the minimal scheme shown in Figure 5-2. While hydrogen bonding and aromatic stacking of the bases define molecular recognition along the template strand, it is increasingly clear now that the backbone contributes significantly to duplex stability. In this section, the effects of backbone substitution are organized and discussed with respect to duplex stability. The discussion will go beyond Watson–Crick duplexes in order to generalize template directed ligation more broadly to other classes of molecules.

5.4.1   Sugar Substitution

The existing ribose modifications (Figure 5-3) have generally altered backbone flexibility. For example, an open-chain glycerol backbone, effectively arising from the removal of the C-2′ ring carbon of the sugar, has considerably more degrees of freedom than the native backbone (Figure 5-3) and when hybridized with native DNA, the duplexes exhibit dramatically reduced melting temperatures (T_m) [65]. The 9-mer 5′-CTTTTTTTG-3′, with two glycerothymidine substitutions in the center of the strand, gave a T_m of only 11 °C, compared to 40 °C in the native duplex.

In contrast, the hexose-oligonucleotides [66–68] and bicyclo-and-tricyclo-DNAs [69–71] rigidify the backbone and have generally shown greater duplex thermal stability. The bicyclo-DNAs have the C-3′ and C-5′ carbon atoms of the ribose connected via an ethylene bridge, and the related tricyclo-DNAs have an additional annulated cyclopropane. The recently introduced locked-nucleic acids (LNAs), have an oxymethylene linker between C-2′ and C-4′. LNA obeys the Watson–Crick base pairing rules whereas the bicyclo-and-tricyclo-DNAs preferentially accept a complementary strand on the Hoogsteen face. Nevertheless, base mismatch formation is strongly destabilized in these structures, and the more rigid backbones can play a crucial role in duplex stability and base selectivity.

In homo-DNA (h-DNA) and pyranosyl-RNA (p-RNA), the hexitol ring is frozen in a single 3′-*endo*-like conformation, as opposed to the multiple conformations accessible to the furanose of DNA and RNA. The h-DNA duplexes have a quasi-linear orientation as opposed to the superhelical structure of DNA duplexes, and the distance between the base stacking planes is larger relative to that of DNA. The h-DNA duplexes also have different pairing stability, e.g., the pairing priorities of A-A exceeds that of A-T. In contrast to h-DNA, p-RNA, which is also quasi-linear, does not show altered pairing priorities.

The hexose rings in these analogues have a different regiochemistry of attachment to the backbone – C-6′, C-4′ in h-DNA and C-4′, C-2′ in p-RNA – and the base stacking is different. For example, studies of self-complementary sequences, such as 5′-TTTTAAAA-3′ and 5′-AAAATTTT-3′, which show equivalent energies with native duplexes, are quite different with the modified backbones. In p-RNA, the T_m for the duplex of the first sequence is much higher than that of the second. In h-DNA, the situation is reversed and the duplex of the second sequence is more stable than the first. The different backbones have been proposed to alter the inclination between the backbone

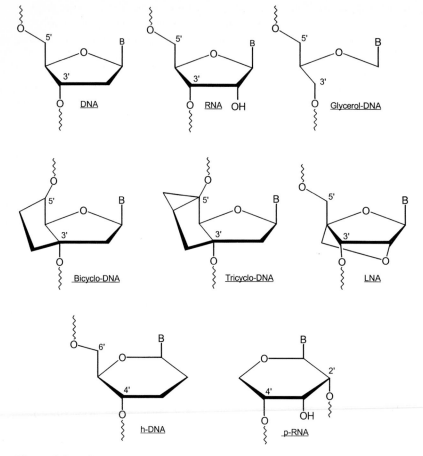

Figure 5-3   Ribose-modified nucleic acids. B designates a nucleobase.

axes and the axes of the Watson–Crick base pairs and to result in an opposite direction-
al orientation of the stacked bases in the two duplexes [68]. It was concluded that an
opposite orientation of the backbone inclination correlates with the differential stability
in the resulting duplexes. These changes in the backbone sugar and the regiochemistry of
attachment clearly have a dramatic effect not only on the association energies but also on
the overall topology of the duplex.

5.4.2   Phosphate Substitution

Modifications at the phosphorous center, including the phosphorothiolate, methylphosphonate, and phosphoramidate derivatives, have generally shown reduced binding affinity for their RNA targets [19]. Replacement of the phosphate generally destabilizes bind-

ing [19, 29, 30], except in a few cases where the linkage restricts conformational flexibility within the complex [19–21, 31, 72]. The repeating charges along the phosphate backbone appear to play a critical role in both strand pre-organization and strand–strand orientations in the duplexes, enforcing an extended conformation in both structures [72].

This conclusion was further explored by the structural analysis of a duplex with a single phosphate replaced with an alkylamine linkage, **1** (Figure 5-4). Solution NMR struc-

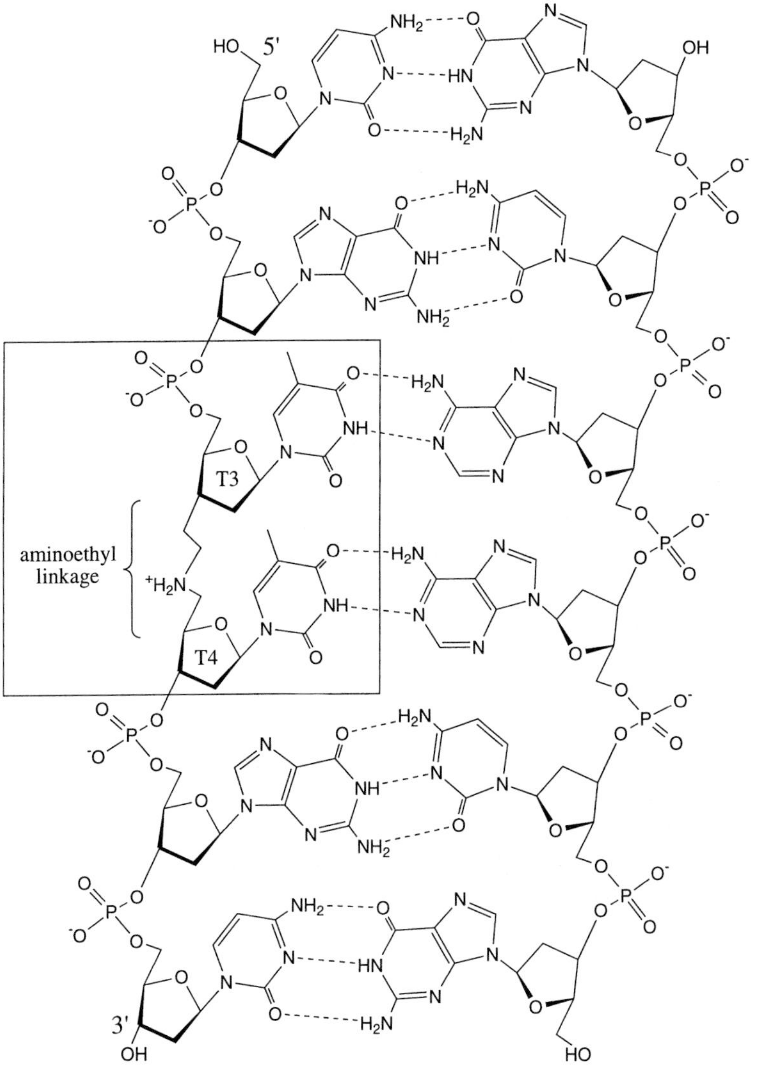

Figure 5-4   The basic structure of duplex **1** with the aminoethyl linkage in the TT central region as indicated with the box.

tures established that both **1** and the native all-phosphate duplexes exist in the canonical B-DNA conformation with Watson–Crick base-pairing preserved. The aminoethyl linkage in **1**, in contrast to the native phosphodiester, permitted detailed NMR analysis of the backbone geometry. Within the minimized structure, Figure 5-5, the CH_2 that replaces the native PO_2^- is flexible and free to collapse into a hydrophobic core formed by the base edges and sugar rings of the flanking TT/AA nucleosides of the duplex (a). This conformation is significantly different from the maximally solvent-exposed orientation of the native phosphate in DNA (b). The stippled rendering of the van der Waals surface highlights the charge distribution.

Strand association energetics of the unmodified duplex were estimated by the nearest-neighbor approximation [73] yielding calculated ΔH values consistent with those determined by both UV hyperchromism and ^{1}H-NMR analyses [72]. The corresponding analyses of the modified duplex **1** showed both that it was destabilized relative to the native duplex and that the overall decrease in stability originated from two opposing factors, a large decrease in enthalpy and a compensating favorable change in entropy. Other structural analogues, prepared to remove the positive charge in the backbone, as well as data from the literature [18, 19] suggested that the positive charge was not the major factor contributing to duplex stability. The entropic difference between the native and the modified duplex **1** was attributed to the ability of the aminoethyl linkage to collapse into the hydrophobic core of **1**.

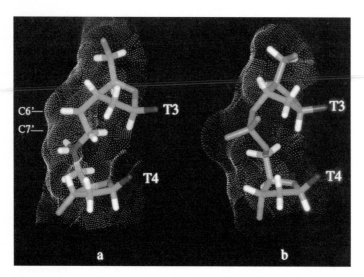

Figure 5-5   Comparison of structures of the TT region of (a) duplex **1**, and (b) native duplex in B-form DNA. Both strands were cut out of the full duplexes, generated (a) from molecular dynamics simulations incorporating NMR constraints, and (b) from crystal structure data [123]. The intruding C7′ methylene with the flat solvent-accessible surface in the modified linkage **1** contrasts markedly with the maximally solvent-accessible phosphate of the phosphodiester linkage. The solvent-accessible surfaces are stippled (adapted from [72]).

A structural model incorporating these data was developed to explain how such a change in the local backbone conformation could disrupt the long-range cooperativity of DNA duplex formation [72]. As outlined in Figure 5-6, the hydrophobic methylene unit can partition from the usual solvent-exposed position of the PO_2^- group (a) into the hydrophobic interior of the double helix (b). This movement accounts for the favorable ΔS, but accentuates the gap between the two flanking phosphates, forcing a movement to shorten their intervening distance. This change destabilizes the optimal topology for duplex stability (c). The overall decrease in the stability of duplex **1** results from a global decrease in base stacking and hydrogen-bonding interactions between the base pairs.

The model in Figure 5-6 suggests that both backbone flexibility and heterogeneity contribute to duplex destabilization. Backbone heterogeneity was investigated in the amide-linked thymidine 8-mer, $(dT^N)_8$. This strand was readily synthesized by solid-phase synthesis (Figure 5-7) and forms a heteroduplex with the native adenine 8-mer, $(dA)_8$, a duplex whose stability and selectively is comparable with that of the all-phosphate duplex [74]. Titration experiments have suggested that these strands form a heteroduplex with a T_m identical to the native duplex. However the heteroduplex stability has a strongly attenuated salt dependence and maintains the same T_m in the absence of salt.

While this backbone homogeneity along the strand is important, it is likely not to be the only factor that is responsible for the stability of the $(dA)_8-(dT^N)_8$ duplex. The dimethylene sulfone RNA, rSNA [28], was studied as a symmetrical octamer and duplex formation was shown to be destabilized as a result of single-strand folding. Therefore, flexibility within the linker is also important, and phosphate charge repulsion and/or conformational rigidity are critical for pre-organizing an extended single strand. The optimal degree of flexibility in the backbone is also likely to be base sequence dependent as the

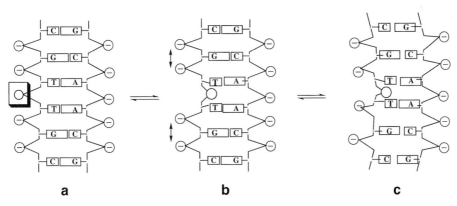

a   b   c

Figure 5-6   (a) Duplex **1** with the negatively charged phosphate removed initiates (b) collapse of the aminoethyl linkage into the hydrophobic interior of the duplex. The remaining phosphates minimize their separation along the strand, causing (c) backbone curvature and loss of the optimal geometry for base-stacking and H-bonding interactions.

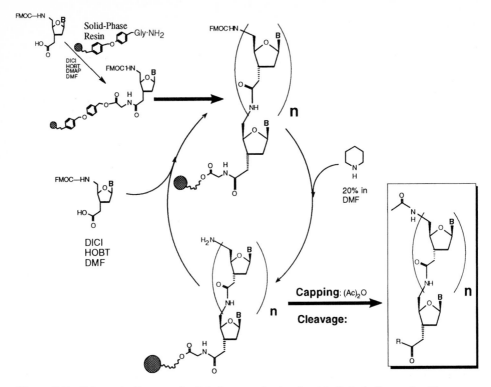

Figure 5-7   Schematic diagram of solid-phase synthesis of amide-linked oligonucleotides.

base affects both the energy of association and the resulting topology within the single strand.

5.4.3   Neutral Acyclic Backbones

The peptide-nucleic acids (PNAs) [22–26], replace both the ribose ring and the phosphate of the backbone with an amide linkage (Figure 5-8). PNAs form strong duplexes and triple-helical structures with DNA, and the binding occurs with a specificity similar to that observed in the formation of the corresponding DNA–DNA structures [24]. The observation that a PNA strand could displace the corresponding DNA strand from a pre-existing duplex highlights both the stability and dynamics of these structures [22]. The three-dimensional structure of a PNA–DNA–PNA triple helix was shown to differ notably from the all-DNA triplexes [25]. It appeared that the flexibility imparted by the replacement of sugar phosphate with the PNA backbone allowed for conformational changes that improved base stacking. However, recent molecular dynamic simulations have challenged PNA flexibility as the reason for triplex stability, and instead have suggested that a conformational preference of the PNA strand itself is responsible [26].

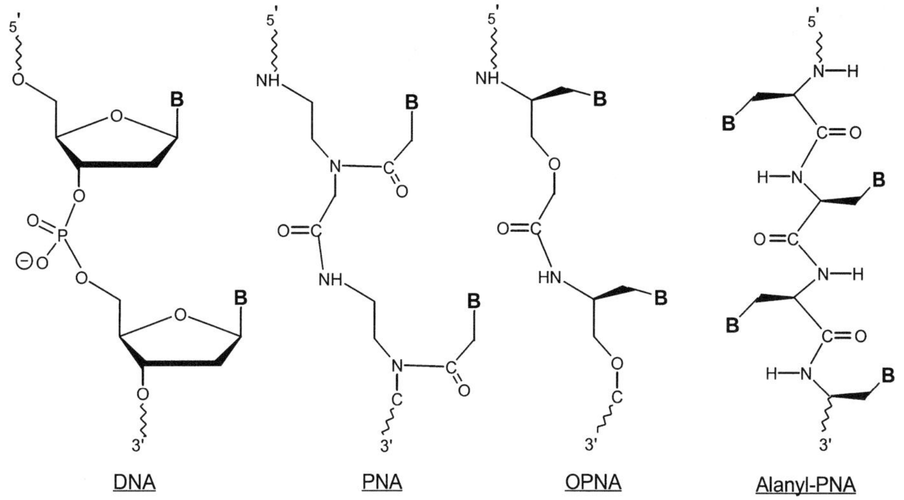

Figure 5-8 Structure of DNA, PNA, OPNA, and alanyl-PNA. In alanyl-PNA, the peptide strand is represented with the absolute chirality alternating at each residue along the strand. B designates a nucleobase in all structures.

Several laboratories have now prepared PNA analogues [27, 75, 76], and among these the oxy-PNA (OPNA) is noteworthy (Figure 5-8) [77]. The ether linkage in OPNA improves water solubility, and enhances sequence specificity in duplex association. The melting temperatures of DNA (thymidine 12-mer) with one equivalent of complementary DNA, OPNA, and PNA (adenine 12-mers) are 30 °C, 43 °C, and 55 °C respectively. The OPNA–DNA duplex has an intermediate T_m and shows a very sharp, cooperative melting transition. The DNA–PNA complex shows mixtures of duplexes and other structures, probably triplexes, which contribute to the high T_m. Circular dichroism (CD) analysis detects no helical structure in the single strand of OPNA, consistently with the flexibility of the backbone. Thermodynamic parameters show that both the enthalpic change (ΔH) and the conformational constraints (ΔS) are largest in the OPNA–DNA complex, suggesting that upon hybridization with native DNA, the randomly coiled OPNA is folded into a regular duplex with optimized stabilization energy. The OPNA strand appears to conform better to the constraints of the more pre-organized DNA complementary strand in duplex formation.

Flexibility within the backbone allows complementary associations to take many forms. Such topological diversity is apparent in the alanyl- and β-homoalanyl-PNA variants [78–82]. Alanyl-PNA (Figure 5-8) contains an α-amino acid backbone carrying the nucleobases as side chains. They differ from the original PNA in two major respects. First, the amide linkage between the nucleobase and the backbone is replaced with a methylene, giving increased conformational flexibility to the base. Secondly, the backbone linkage is shortened, and higher densities of nucleobases exist along the chain.

This combination of both greater nucleobase flexibility and increased base density contributes to alter pairing and geometric arrangements in the duplexes. For example, A–A pairing in alanyl-PNA is more stable than A–T, a structural feature seen in linear stretches of DNA and RNA [83, 84], in h-DNA [66], and in β-homoalanyl PNAs [80–82], which differ from alanyl-PNA by an additional methylene in the backbone. The β-homoalanyl PNAs can exist in higher-ordered structures with linear backbones which self-associate through hydrogen bonding into β-sheet conformations. That these higher-ordered aggregates are driven by strong adenine self-pairing is supported by structures modified with 7-carbaadenine nucleobases. The adenine hexamer has $T_m > 90\,°C$ and more than 70% hyperchromicity, which drops to $T_m = 35\,°C$ with 24% hyperchromicity in the 7-carbaadenine hexamer. Therefore, in the β-homoalanyl PNAs, the nucleobases can be stabilized by hydrogen bonding through both Watson–Crick and Hoogsteen faces. Clearly, as the backbones become increasingly flexible, the nucleobases can dictate self-association in an increasingly diverse range of topologies.

5.4.4   Peptide–Peptide Association

Certain peptide backbones also self-associate, and in some, the association energies can be very large. Collagen fibrils, tropomyosin and actin filaments, viral and other cellular inclusion bodies, hemoglobin S tactoids, prions, and amyloids all represent highly ordered supramolecular aggregates. Like DNA duplexes, these paracrystalline materials attain long-range order, but not in all three dimensions. One of the smallest of the peptides that form amyloids are the Aβ peptides associated with Alzheimer's disease [85]. These proteolytic fragments of the amyloid precursor protein form large fibrillar paracrystalline aggregates (Figure 5-9A).

The Aβ peptide fibers are known to be dominated by β-strand conformations, but further efforts to understand their structures have been limited by the inherent insolubility of the material. Synthetic attachment of a polyethylene block at the hydrophobic C-terminal of a truncated Aβ peptide, residues 10–35, **2**, generated a soluble fibrillous material whose aggregation is controlled by simple changes in concentration, pH, or ionic strength [86]. The increased solubility of **2** appears to be mediated by prevention of fiber–fiber association (Figure 5-9B). β-Strand peptides are notoriously insoluble and the strategy employed in the construction of this block copolymer represents a significant opportunity to investigate the thermodynamics of self-association and the exploitation of these materials as templates.

NH_2-^{10}YEVHHQKLVFFAEDVGSNKGAIIGLM—N

2

Dipolar recoupling solid-state NMR experiments established that the central core region of the peptide Aβ (10–35) was both ordered and homogeneous [87]. The entire length of the peptide has now been shown to exist in an extended β-conformation with each pep-

A

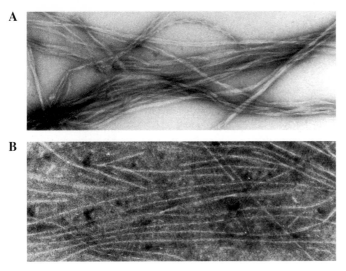

B

Figure 5-9 Electron micrographs of (A) Aβ (10–35), and (B) Aβ (10–35)–PEG fibers (adapted from [86]).

tide arranged parallel to the others and with each amino acid residue in register from peptide to peptide along an extended β-sheet [88]. Similar analyses have now been applied to the protein block of **2**. The peptide domain of the block copolymer is arrayed in the center of the fibril [89] and retains the same parallel, in-register β-structure of Aβ (10–35) [90]. The PEG block is therefore localized to the surface of the fibril, reducing fibril–fibril contact sufficiently to maintain solubility. Preliminary experiments have suggested that the β-helical sheets exist in laminated blocks, again in a parallel, in-register orientation, and held together by side chain–side chain interactions (Figure 5-10) [90]. This unique arrangement is consistent with the fiber dimensions determined by electron microscopy and neutron scattering experiments.

The β-sheet secondary structure is energetically dependent not only on the hydrogen bonding interactions between individual strands, but also on the ability of the resulting sheet to twist, bulge, and fold into multifaceted conformations [91–94]. Consequently, amino acid propensities have proven to be highly dependent on context, greatly complicating the design of β-structure peptides and proteins [95, 96]. Most problematic however are the intermolecular interactions which compromise solubility, the infamous feature of the many amyloid diseases [85, 97]. In the block co-polymer **2**, the energetics of self-association of amyloid peptides dominate the paracrystalline β-sheet structure in the core of the fibril array while the C-terminal polyethylene glycol, localized on the surface of the fibril, ensures solubility. The surface-localized PEG, much like the polar phosphates of the nucleic acid polymers, enforces an overall amphiphilic character of the fiber strand.

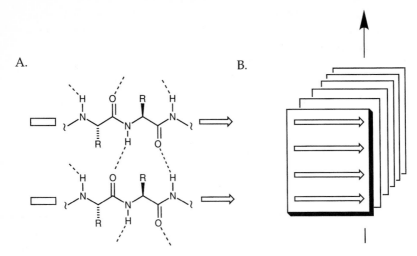

Figure 5-10   (A) The fiber is composed of peptides arrayed in a parallel β-sheet with residues in register. (B) Up to six β-sheets can be laminated together to define the fiber thickness with the peptides oriented perpendicular to the direction of growth of the fiber. Side chain–side chain interactions stabilize the lamination.

The interior of the DNA duplex is organized by hydrogen bonding and aromatic base stacking interactions while the charged backbone ensures a micellar-like orientation to the overall structure. In contrast, inter-strand hydrogen bonding stabilizes the extended backbone β-conformation while side chain interactions, hydrophobic side chain interdigitation, aromatic stacking, and salt bridge formation contribute to tertiary structural sheet and lamination stability (Figure 5-10). The orientation of these interactions are different, with both the peptide backbone and side chains contributing to inter-strand stability, but the end result is remarkably similar. The individual amino acids stack in register along the sheet, suggesting a template capable of dictating sequence-specific ligation. The extent to which these structures can serve in template directed ligation will depend on many factors that have still to be established.

5.5   Ligation

The elegant simplicity of the DNA double helix [98] has provided inspiration to bioorganic chemistry since its discovery in 1953. DNA templates for ligation reactions were suggested [99, 100], but not experimentally explored until 1966 [2]. Deoxyadenosine oligomers were shown to template the ligation of short thymidine oligomers (5- and 6-mers) with carbodiimide activation to the 10- and 12-mer products in 3% and 5% yields, respectively [2]. The first demonstration that a template of defined sequence could catalyze the formation of its complementary strand was shown by Orgel et al. using nucleoside 5′-phosphoimidazolides as activated monomers [3–5]. Under non-enzymatic conditions, dCCGCC templated the synthesis of dGGCGG in 18% yield from a mixture of activated guanosine and cytidine monomers [5].

This oligomerization was shown to be enantioselective by demonstrating that the L-enantiomer terminated the oligomerization of the D-enantiomer. Under these conditions, h-DNA, containing the six-membered hexitol ring frozen in a single 3′-*endo*-like conformation, was more enantioselective than DNA or RNA templates [101, 102]. Since duplex structure does correlate with sugar pucker, 2′-*endo* for B-DNA and 3′-*endo* for A-DNA [103], the effectiveness of the ligation on the hexitol template has led to the suggestion that an A-DNA structure provides a better template. This suggestion is also supported by the observation that addition of Co^{3+} ions, known to induce a B-DNA to A-DNA transition [104], improves the ligation effectiveness.

von Kiedrowski achieved the first successful molecular replication with the self-complementary hexamer dGGCGCC [7]. This template catalyzed the ligation of two complementary trimer substrates using carbodiimide as the dehydrating reagent. The system allowed for each newly assembled strand to be identical to the template, demonstrating autocatalysis, but the hexamer did not allow for exponential amplification. The combined yield after four days was less than 12%, and product duplex dissociation was limiting. In 1994, the system was expanded to include four templates, two complementary and two self-complementary, which through cross-catalytic self-replication, compete for the common trimeric precursors [8]. The replication efficiency was similar with both complementary and self-complementary templates.

In an attempt to exploit the thermodynamics of template association, an imine, formed by reversible condensation (Figure 5-11), was synthetically incorporated in the DNA

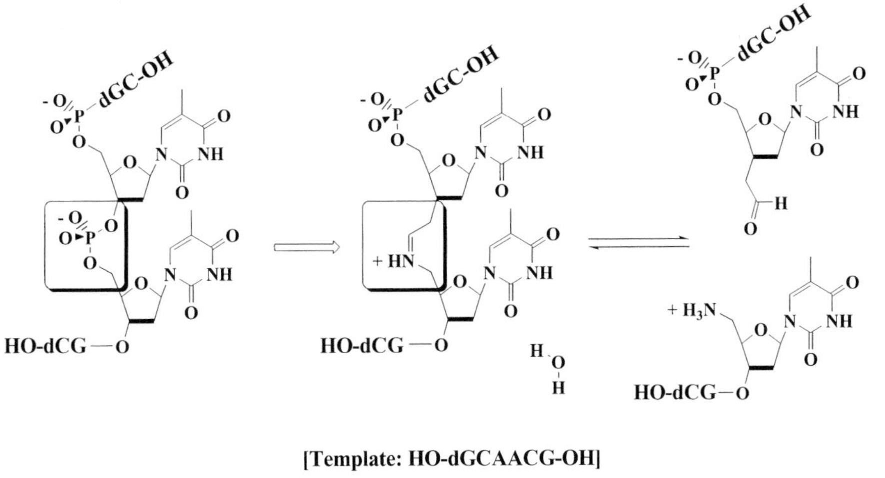

[Template: HO-dGCAACG-OH]

Figure 5-11   The reversible imine-coupling reaction. The modifications were designed to maintain equivalent numbers of atoms in the linking domain.

backbone [105, 106]. The equilibrium constants in Figure 5-12, K_{1-4}, were determined directly from ^{1}H- and ^{15}N-NMR experiments [107]. The relative concentration of the duplex to the free imine single strand, expressed as a function of K_2/K_4, was 10^5 or higher. Under reducing conditions, the rates of reduction were dominated by this large difference in concentration, and significant fidelity was achieved when the complementary amine was allowed to compete with non-complementary substrates [105]. The model shown in Figure 5-6 argued that linker flexibility was critically destabilizing and that the more rigid imine, containing two sp^2 centers and significantly fewer degrees of conformational freedom, should form a more stable duplex [72].

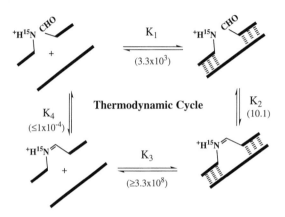

Figure 5-12 The thermodynamic cycle for DNA-directed imine formation and the measured equilibrium constants for each step (adapted from [107]).

This approach of using a reversible initial ligation was extended in the coiled-coil peptide templates [108–112]. The amide synthesis strategy, shown in Figure 5-13 [113], is sufficiently chemoselective to allow utilization of peptide fragments without side chain protection. A thioester at a peptide C-terminus appears to undergo rapid exchange of thiol that is trapped by a spontaneous intramolecular condensation with the amine of the N-terminal cysteine. The generated thioester is not observed as a discrete intermediate because of the rapid S,N-acyl rearrangement through a five-membered-ring transition state to give the peptide bond.

By this strategy, α-helical segments based on the leucine-zipper motif catalyzed template directed ligation of the complementary peptides. The rearrangement step was relatively slow, $t_{1/2} > 1$ min, probably retarded by conformational requirements of the reaction occurring on the template [109]. Nevertheless, the ligation gave rate enhancements of 10^3 over the uncatalyzed reaction. The parabolic growth observed under replicating conditions was attributed to product inhibition [109].

Other catalysts, sensitive to environmental conditions [110–112], were designed by placing ionic residues at the e and g positions of the helical heptad repeats of the coiled coil. With glutamic acids at these positions, the duplex is stable under acidic conditions. Under physiological conditions, the side chain acids are negatively charged and the coiled coil is destabilized. With lysine residues at these positions, either basic conditions,

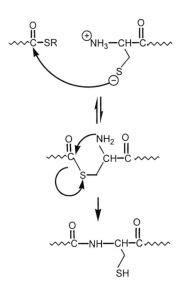

Figure 5-13   A thioester at the C-terminus of one peptide serves as the electrophile for bimolecular condensation with the thiol side chain of an N-terminal cysteine
residue. The *S,N*-acyl intramolecular rearrangement forms
the native amide bond.

or neutral conditions with high salt, allowed for template directed ligation [111]. A four-
fragment auto- and cross-catalytic peptide system under these conditions enhanced replication efficiency [112], but the rate constants derived from this system once again highlighted the inherent contradiction plaguing template directed ligation: a stable template–substrate complex suffers product inhibition, while a weak template-substrate
complex reduces catalytic rate.

5.6   Product Dissociation

Protocols based on cycling the temperature or other environmental parameters capable
of melting the product–template duplex after each cycle shown in Figure 5-2, provide a
general approach to avoid product inhibition [3, 105, 106]. Encoded enzymes provide
this function in natural genomes and physically melting the duplex product at the end of
each cycle controls the polymerase chain reaction [114]. In both cases, exponential
growth of the template is possible. However, several synthetic variations have also been
suggested to overcome product inhibition.

   Luther et al. have circumvented product inhibition by attaching the template to the surface of a solid support [115]. Two complementary DNA 14-mers, immobilized irreversibly on the support, templated the ligation of four 7-mer DNA substrates. After the coupling step, the product strands were released by denaturation, and subsequently immobilized on fresh support to become a template for the next reaction cycle. This procedure
requires several steps, but allows exponential amplification of oligonucleotide templates.

   In a separate approach, the nucleotide template was expanded to a double-helical
structure [116]. The Hoogsteen pairing in the triple helix is relatively weak, thus facili-

tating release of the newly generated strand from the double-helical template. An increase in pH or the addition of excess substrate destabilizes the product complex. The approach requires symmetrical sequences in order to regenerate the template and one of the strands must be all purine for triplex formation. Still, three successive replication cycles were demonstrated giving roughly a five-fold increase in the total amount of the initial 24-mer duplex. Hoogsteen pairing was further employed as a triplex-directing template with a circular DNA [117]. This pyrimidine-rich circular template (44-mer) gave improved thermodynamics of substrate binding (6-mer + 11-mer homopurine); quantitative ligation vs. 51% yield using linear DNA, and a 23-fold rate increase. Thus, this single-turnover ligation on the circular template through triplex formation was shown to be more efficient compared to the linear template. It may be possible with further structural modifications to extend this system to multiple turnovers.

As suggested above, modifications in the linker domain have a dramatic effect on duplex stability and have suggested other strategies to avoid product inhibition. The addition of a reduction step in the imine cycle is shown in Figure 5-14. While the fidelity of the process will be limited by the ratio of the equilibrium constants, K_2/K_4, product inhibition is controlled by the relative duplex stability of the amine product relative to the imine substrate. Based on the model in Figure 5-6, the amine product could be predicted to be less stable than the imine, and indeed, the product duplex **1** was found to be significantly less stable ($>10^6$-fold), hence product inhibition was avoided. With 0.01 molar equivalents of DNA, the turnover number was 13 min^{-1} and there was no effect of

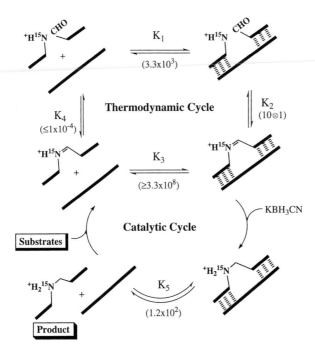

Figure 5-14 Full cycle for the DNA-catalyzed reductive ligation of amine and aldehyde fragment (adapted from [107]).

product concentration, at least through the first half-life where the product/template ratio exceeds 50 : 1. Therefore, the rigidity and resulting stability of the imine duplex, together with the flexibility and instability of the product amine duplex, allow this small nucleic acid template to function as an efficient catalyst [107, 41].

The reaction itself, however, is more similar to a biological translation reaction, converting a phosphate backbone into an amine backbone. The question posed in Figure 5-15 is whether this translation product, which can be prepared with significant amplification, could be used to aid in the replication of the original template. This strategy is different from the natural system in that the translated product serves as the template catalyst for construction of its phospho-linked parent. The aminoethyl linkage is more flexible than that of the phosphate [72], and therefore should less effectively pre-organize the substrates than does DNA. However, catalysis of phosphodiester formation with this template was comparable with that of a DNA template and this effect was attributed to the hydrophobic character of the alkylamine [72]. This assertion was supported by substitution of the nitrogen with alkyl groups, a modification that significantly increased both the hydrophobic character and the catalytic efficiency of the template [118].

Overall, the first step in Figure 5-15 can give greater than 10^6-fold amplification and the product can provide an effective ligation catalyst to reconstruct its parent DNA template. If viewed as a single replication cycle, products far in excess of the exponential growth seen in a single turn of the biological cell cycle are obtained. Extensions of this design for replication systems that achieve greater than exponential growth should be possible.

Another hypothetical extension has emerged in the amyloid area. The simplified scheme shown in Figure 5-2 predicts that any material capable of increasing the effective molarity of properly oriented molecules along its surface could catalyze template directed ligation. While the association energy for fibrillogenesis has not been quantified, it must be large to generate the peptide arrays that exist in the fibers. Both the backbone

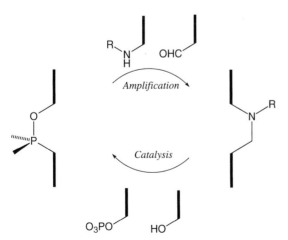

Figure 5-15 Two-stage replication. The initial phospho-linked template is translated into amine product with amplification. The translated product–template catalyzes the synthesis of the original phospho-linked strand.

and the side chains participate in strand–strand association, and, the residues in Aβ (10–35), are aligned in register along the entire length of the peptide. The PEG block copolymer solves the most critical limitation of sheet structures, that of solubility. Therefore, the minimal scheme shown in Figure 5-2 can be modified to that shown in Figure 5-16.

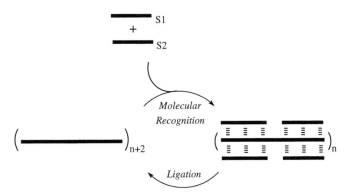

Figure 5-16 Template directed synthesis on a β-sheet template. The dyad template, consisting of n strands, organizes the fragments S1 and S2 through molecular recognition. Each ligation cycle produces a new β-sheet template, consisting of $n+2$ strands.

The product dissociation step is removed from the cycle entirely and the products are recovered after many cycles by disrupting the large soluble fibril arrays. The template is predicted to organize fragments S1 and S2 on both sides of the template such that each turn of the cycle would generate two new strands. For a proposed higher-ordered laminated structures, this stoichiometry should be scale as the number of laminae. The extent to which this strategy can be exploited is unknown and will depend on the mechanism of association of the peptides, the fidelity of the ligation, and the solubility of the fibril. There is clearly now sufficient synthetic flexibility to allow these variables to be optimized and exploited for template directed ligation or polymerization.

5.7 Summary

Intermolecular associations capable of templating specific reactions are central to both biological catalysts and to present-day catalyst design. The nucleic acid polymers have high association energies, well defined topologies, and encode information at an extremely high density. It is equally clear that they are able to catalyze a wide range of chemical reactions. However, oligonucleotide catalysts that use base pairing for molecular recognition are destined to be turnover-limited in the replication cycle of Figure 5-2. Turnover rates are increased by introducing mismatches and shortening recognition sequences [119], or by insertion of mutations that weaken tertiary interactions [120], but such changes lower fidelity and necessitate catalysts of higher molecular weight.

Evolutionarily, in an RNA world scenario where no proteins exist, an autonomous RNA template could encode ribozymes as co-catalysts to provide the origin of self-replicating entities. Product saturation limits the known ribozyme catalysts [121–122], and ribozymes capable of melting duplex structures are certainly required for efficient replication. In addition, oligonucleotide biopolymers appear less adept than proteins in organizing functional groups within active sites for catalysis. This limitation results first from the scarcity of catalytic functional groups in these polymers but, more critically, is a consequence of the phosphate backbone. While this backbone has greater degrees of freedom than the protein backbone, the necessity to shield the anionic phosphate groups enforces an extended conformation and prevents close packing within the structure. This stabilization is readily achieved in the tertiary and quaternary folds of proteins, for example, by creating local environments of low effective dielectric [42]. The solvation equirements in oligonucleotides limit folding into tertiary and quaternary structures that can function cooperatively during catalysis [42].

On the other hand, in a *minimal* catalytic model system, the ribozymes have advantages over protein enzymes. Even short oligonucleotides can effectively use duplex formation to bind substrates and the known ribozymes are generally smaller than their protein counterparts. This substrate recognition through secondary structure is not a common feature of protein binding sites, where tertiary structural folds are necessary to optimize close packing of residues distant in primary sequence [42].

As least conceptually, nucleic acid backbone modifications replacing the phosphate have the potential of both extending the functional groups needed for catalysis and allowing for more densely packed structures with well defined binding pockets. Such hybrid molecules can store information, serve as catalysts for template directed ligation, and be selected for catalytic function. Nature may have evolved the division of labor between the different classes of biopolymers as the genomes became larger and more specialized, but the opportunity for synthetic chemists to develop new materials capable of encoding information that can be translated, replicated, and selected for function is rich indeed. From the information now available, it should be possible to construct new autonomous genomes based on entirely different structural skeletons. We are currently limited only by our structural imagination and synthetic capability.

Acknowledgments

We thank Jay Goodwin, John C. Leitzel, Peizhi Luo, Zheng-Yun J. Zhan, Timothy Burkoth, Tammie Benzinger, David Gregory, Jingdong Ye, and David Morgan for the insightful and creative experiments that made all of this possible and the NIH (Biotechnology Training Grant GM-08369 and R21 RR12723) for support.

References

[1] (a) F. H. C. Crick, *Symp. Soc. Exp. Biol. XII* **1958**, pp. 138–163. (b) D. Thieffry, S. Sarkar, TIBS **1998**, *23*, 312–316.
[2] R. Naylor, P. T. Gilham, *Biochemistry* **1966**, *5*, 2722–2728.

[3] L. E. Orgel, *Nature (London)* **1992**, *358*, 203–209.

[4] L. E. Orgel, R. Lohrmann, *Acc. Chem. Res.* **1974**, *7*, 368–377.

[5] T. Inoue, G. F Joyce, K. Grzeskowiak, L. E. Orgel, J. M. Brown, C. B. Reese, *J. Mol. Biol.* **1984**, *178*, 669–676.

[6] M. Koppitz, P. E. Nielsen, L. E. Orgel, *J. Am. Chem. Soc.* **1998**, *120*, 4563–4569.

[7] G. von Kiedrowski, *Angew. Chem., Int. Ed. Engl.* **1986**, *25*, 932–935.

[8] D. Sievers, G. von Kiedrowski, *Nature (London)* **1994**, *369*, 221–224.

[9] J.-I. Hong, Q. Feng, V. Rotello, J. Rebek Jr., *Science* **1992**, *255*, 848–850.

[10] C. Böhler, P. E. Nielsen, L. E. Orgel, *Nature (London)* **1995**, *376*, 578–581.

[11] R. Stribling, S. L. Miller, *J. Mol. Evol.* **1991**, *32*, 282–288.

[12] P. C. Zamecnik, M. L. Stephenson, *Proc. Natl. Acad. Sci. U. S. A.* **1978**, *75*, 280–284.

[13] M. L. Stephenson, P. C. Zamecnik, *Proc. Natl. Acad. Sci. U. S. A.* **1978**, *75*, 285–288.

[14] A. De Mesmaeker, A. Waldner, J. Lebreton, P. Hoffmann, V. Fritsch, R. Wolf, P. S. Miller, *Oligonucleotides: Antisense Inhibitors of Gene Expression*, J. S. Cohen, ed., CRC Press, Boca Raton, Florida, **1989**, pp. 79–93.

[15] E. Uhlmann, A. Peyman, *Chem. Rev.* **1990**, *90*, 544–584.

[16] M. H. Carruthers, *Acc. Chem. Res.* **1991**, *24*, 278–284.

[17] R. S. Varma, *Synlett* **1993**, 621–637.

[18] A. De Mesmaeker, K.-H. Altmann, A. Waldner, S. Wendeborn, *Curr. Opin. Struc. Biol.* **1995**, *5*, 343–355.

[19] A. De Mesmaeker, R. Häner, P. Martin, H. E. Moser, *Acc. Chem. Res.* **1995**, *28*, 366–374.

[20] A. De Mesmaeker, A. Waldner, J. Lebreton, P. Hoffmann, V. Fritsch, R. M. Wolf, S. M. Freier, *Angew. Chem., Int. Ed. Engl.* **1994**, *33*, 226–229.

[21] A. De Mesmaeker, A. Waldner, Y. S. Sanghvi, J. Lebreton, *Biol. Med. Chem. Lett.* **1994**, *4*, 395–398.

[22] P. E. Nielsen, M. Egholm, R. H. Berg, O. Buchardt, *Science* **1991**, *254*, 1497–1500.

[23] K. L. Dueholm, P. E. Nielsen, *New J. Chem.* **1997**, *21*, 19–31.

[24] M. Egholm, O. Buchardt, L. Christensen, C. Behrens, S. M. Freier, D. A. Driver, R. H. Berg, S. K. Kim, B. Norden, P. E. Nielsen, *Nature (London)* **1993**, *365*, 566–568.

[25] L. Betts, J. A. Josey, J. M. Veal, S. R. Jordan, *Science* **1995**, *270*, 1838–1845.

[26] G. C. Shields, C. A. Laughton, M. Orozco, *J. Am. Chem. Soc.* **1998**, *120*, 5895–5904.

[27] A. Peyman, E. Uhlmann, K. Wagner, S. Augustin, C. Weiser, D. W. Will, G. Breipohl, *Angew. Chem., Int. Ed. Engl.* **1997**, *36*, 2809–2812.

[28] C. Richert, L. A. Roughton, S. A. Benner, *J. Am. Chem. Soc.* **1996**, *118*, 4518–4531.

[29] C. W. Cross, J. S. Rice, X. Gao, *Biochemistry* **1997**, *36*, 4096–4107.

[30] J. S. Rice, X. Gao, *Biochemistry* **1997**, *36*, 399–411.

[31] R. O. Dempcy, K. A. Browne, T. C. Bruice, *J. Am. Chem. Soc.* **1995**, *117*, 6140–6141.

[32] B. Cuenoud, F. Casset, D. Hüsken, F. Natt, R. M. Wolf, K.-H. Altman, P. Martin, H. E. Moser, *Angew. Chem., Int. Ed. Engl.* **1998**, *37*, 1288–1291.

[33] J.-J. Vesseur, F. Debart, Y. S. Sanghvi, P. D. Cook, *J. Am. Chem. Soc.* **1992**, *114*, 4006–4007.

[34] P. A. Lohse, J. W. Szostak, *Nature (London)* **1996**, *381*, 442–444.

[35] B. Cuenoud, J. W. Szostak, *Nature (London)* **1995**, *375*, 611–614.

[36] R. R. Breaker, G. F. Joyce, *Chem. Biol.* **1995**, *2*, 655–660.

[37] R. R. Breaker, G. F. Joyce, *Chem. Biol.* **1994**, *1*, 223–229.

[38] N. Carmi, L. A. Shultz, R. R. Breaker, *Chem. Biol.* **1996**, *3*, 1039–1046.

[39] Y. Li, D. Sen, *Nature Struct. Biol.* **1996**, *3*, 743–747.

[40] J. R. Prudent, T. Uno, P. G. Schultz, *Science* **1994**, *264*, 1924–1927.

[41] Y. Gat, D. G. Lynn, *Biopolymers* **1998**, *48*, 19–28.

[42] G. J. Narlikar, D. Herschlag, *Annu. Rev. Biochem.* **1997**, *66*, 19–59.

[43] B. Lewin, *Genes V*, Oxford University Press, New York, **1994**.

[44] W. J. Albery, J. R. Knowles, *Angew. Chem., Int. Ed. Engl.* **1977**, *16*, 285–293.

[45] W. P. Jencks, *Adv. Enzymol. Relat. Areas Mol. Biol.* **1975**, *43*, 219–410.

[46] E. T. Kool, *Chem. Rev.* **1997**, *97*, 1473–1487.

[47] G. Zubay, *Biochemistry*, Macmillan, New York, **1988**, pp. 266–272.

[48] C. Guerrier-Takada, K. Gardiner, T. Marsh, N. Pace, S. Altman, *Cell* **1983**, *35*, 849–857.

[49] K. Kruger, J. Grabowskip, A. J. Zaung, J. Sands, D. E. Gottschling, T. R. Cech, *Cell* **1982**, *31*, 147–157.

[50] T. R. Cech, *Annu. Rev. Biochem.* **1990**, *59*, 543–568.

[51] J. A. Piccirilli, J. S. Vyle, M. H. Caruthers, T. R. Cech, *Nature (London)* **1993**, *361*, 85–88.

[52] T. R. Cech, *Int. Rev. Cytol.* **1985**, *93*, 3–22.

[53] G. F. Joyce, *Nature (London)* **1989**, *338*, 217–224.

[54] W. Gilbert, *Nature (London)* **1986**, *319*, 618.

[55] R. H. Symons, *Annu. Rev. Biochem.* **1992**, *61*, 641–671.

[56] G. F. Joyce, *Curr. Biol.* **1996**, *6*, 965–967.

[57] K. Bloch, *Chem. Biol.* **1996**, *3*, 405–407.

[58] D. M. Long, O. C. Uhlenbeck, *FASEB J.* **1993**, *7*, 25–30.

[59] T. R. Cech, *Science* **1987**, *236*, 1532–1539.

[60] D. P. Bartel, J. W. Szostak, *Science* **1993**, *261*, 1411–1418.

[61] M. Illangasekare, G. Sanchez, T. Nickles, M. Yarus, *Science* **1995**, *267*, 643–647.

[62] (a) M. M. Conn, J. R. Prudent, P. G. Schultz, *J. Am. Chem. Soc.* **1996**, *118*, 7012–7013. (b) A. G. Cochran, P. G. Schultz, *Science* **1990**, *249*, 781–783.

[63] T. W. Wiegand, R. C. Jenssen, B. E. Eaton, *Chem. Biol.* **1997**, *4*, 675–683.

[64] T. M. Tarasow, S. L. Tarasow, B. E. Eaton, *Nature (London)* **1997**, *389*, 54–57.

[65] K. C. Schneider, S. A. Benner, *J. Am. Chem. Soc.* **1990**, *112*, 453–455.

[66] J. Hunziker, H.-J. Roth, M. Böhringer, A. Giger, U. Diederichsen, M. Göbel, R. Krishnan, B. Jaun, C. Leumann, A. Eschenmoser, *Helv. Chim. Acta* **1993**, *76*, 259–352.

[67] S. Pitsch, R. Krishnamurthy, M. Bolli, S. Wendeborn, A. Holzner, M. Minton, C. Lesueur, I. Schlönvogt, B. Jaun, A. Eschenmoser, *Helv. Chim. Acta* **1995**, *78*, 1621–1635.

[68] R. Micura, R. Kudick, S. Pitsch, A. Eschenmoser, *Angew. Chem., Int. Ed. Engl.* **1999**, *38*, 680–683.

[69] M. Tarköy, M. Bolli, C. Leumann, *Helv. Chim. Acta* **1994**, *77*, 716–744.

[70] R. Steffens, C. Leumann, *J. Am. Chem. Soc.* **1999**, *121*, 3249–3255.

[71] A. A. Koshkin, S. K. Singh, P. Nielsen, V. K. Rajwanshi, R. Kumar, M. Meldgaard, C. E. Olsen, J. Wengel, *Tetrahedron* **1998**, *54*, 3607–3630.

[72] P. Z. Luo, J. C. Leitzel, Z.-Y. J. Zhan, D. G. Lynn, *J. Am. Chem. Soc.* **1998**, *120*, 3019–3031.

[73] (a) K. J. Breslauer, R. Frank, H. Blöcker, L. A. Marky, *Proc. Natl. Acad. Sci. U. S. A.* **1986**, *83*, 3746–3750. (b) G. Vesnaver, K. J. Breslauer, *Proc. Natl. Acad. Sci. U. S. A.* **1991**, *88*, 3569–3573.

[74] J. C. Leitzel, D. G. Lynn, unpublished results.

[75] P. Ciapetti, F. Soccolini, M. Taddei, *Tetrahedron* **1997**, *53*, 1167–1176.

[76] B. Hyrup, M. Egholm, P. E. Nielsen, P. Wittung, B. Nordén, O. Buchardt, *J. Am. Chem. Soc.* **1994**, *116*, 7964–7970.

[77] M. Kuwahara, M. Arimitsu, M. Sisido, *J. Am. Chem. Soc.* **1999**, *121*, 256–257.

[78] U. Diederichsen, *Angew. Chem., Int. Ed. Engl.* **1997**, *36*, 1886–1889.

[79] U. Diederichsen, *Angew. Chem., Int. Ed. Engl.* **1996**, *35*, 445–448.

[80] U. Diederichsen, *Bioorg. Med. Chem. Lett.* **1998**, *8*, 165–168.

[81] U. Diederichsen, H. W. Schmitt, *Eur. J. Org. Chem.* **1998**, 827–835.

[82] U. Diederichsen, H. W. Schmitt, *Angew. Chem. Int. Ed. Engl.* **1998**, *37*, 302–305.

[83] K. J. Baeyens, H. L. De Bondt, A. Pardi, S. R. Holbrook, *Proc. Natl. Acad. Sci. U. S. A.* **1996**, *93*, 12851–12855.

[84] I. Berger, C. Kang, A. Fredian, R. Ratliff, R. Moyzis, A. Rich, *Nature Struct. Biol.* **1995**, *2*, 416–425.

[85] D. B. Teplow, *Int. J. Exp. Clin. Invest.* **1998**, *5*, 121–142.

[86] T. S. Burkoth, T. L . S. Benzinger, D. N. M. Jones, K. Hallenga, S. C. Meredith, D. G. Lynn, *J. Am. Chem. Soc.* **1998**, *120*, 7655–7656.

[87] T. L . S. Benzinger, D. M. Gregory, T. S. Burkoth, H. Miller-Auer, D. G. Lynn, R. E. Botto, S. C. Meredith, *Proc. Natl. Acad. Sci. U. S. A.* **1998**, *95*, 13407–13411.

[88] T. L . S. Benzinger, D. M. Gregory, T. S. Burkoth, H. Miller-Auer, D. G. Lynn, R. E. Botto, S. C. Meredith, submitted for publication.

[89] T. S. Burkoth, T. L. S. Benzinger, V. Urban, D. G. Lynn, S. C. Meredith, P. Thiyagarajan, *J. Am. Chem. Soc.* **1999**, *121*, 7429–7430.

[90] T. S. Burkoth, T. L. S. Benzinger, V. Urban, S. C. Meredith, P. Thiyagarajan, D. G. Lynn, submitted for publication.

[91] F. R. Salemme, *Prog. Biophys. Mol. Biol.* **1983**, *42*, 95–133.

[92] C. Chothia, *Annu. Rev. Biochem.* **1984**, *53*, 537–572.

[93] G. D. Rose, L. M. Gierasch, J. A. Smith, *Adv. Protein Chem.* **1985**, *37*, 1–109.

[94] C. L. Nesloney, J. W. Kelly, *Bioorg. Med. Chem.* **1996**, *4*, 739–766.

[95] D. L. Minor Jr., P. S. Kim, *Nature (London)* **1996**, *380*, 730–734.

[96] C. K. Smith, L. Regan, *Acc. Chem. Res.* **1997**, *30*, 153–161.

[97] J. D. Sipe, *Annu. Rev. Biochem.* **1992**, *61*, 947–975.

[98] J. D. Watson, *The Double Helix*, Weidenfeld and Nicolson, London, **1968**.

[99] *Perspective in Organic Chemistry*, A. R. Todd, ed., Interscience Publishers, London, **1956**, p. 263.

[100] S. Anderson, H. L. Anderson, J. K. M. Sanders, *Acc. Chem. Res.* **1993**, *26*, 469–475.

[101] I. A. Kozlov, P. K. Politis, S. Pitsch, P. Herdewijn, L. E. Orgel, *J. Am. Chem. Soc.* **1999**, *121*, 1108–1109.

[102] I. A. Kozlov, P. K. Politis, A. Van Aerschot, R. Busson, P. Herdewijn, L. E. Orgel, *J. Am. Chem. Soc.* **1999**, *121*, 2653–2656.

[103] *Nucleic Acids in Chemistry and Biology*, G. M. Blackburn, M. J. Gait, eds., Oxford University Press, Oxford, **1996**, pp. 28–29.

[104] H. Robinson, A. H-J. Wang, *Nucl. Acids Res.* **1996**, *24*, 676–682.

[105] J. T. Goodwin, D. G. Lynn, *J. Am. Chem. Soc.* **1992**, *114*, 9197–9198.

[106] J. T. Goodwin, P. Z. Luo, J. C. Leitzel, D. G. Lynn, *Self-Production of Supramolecular Structures: From Synthetic Structures to Models of Minimal Living Systems*, G. R. Fleischaker, S. Colonna, P. L. Luisi, eds., Kluwer Academic Publishers, Dordrecht, **1994**, pp. 99–104.

[107] Z.-Y. J. Zhan, D. G. Lynn, *J. Am. Chem. Soc.* **1997**, *119*, 12420–12421.

[108] D. H. Lee, J. R. Granja, J. A. Martinez, K. Severin, M. R. Ghadiri, *Nature (London)* **1996**, *382*, 525–528.

[109] K. Severin, D. H. Lee, A. J. Kennan, M. R. Ghadiri, *Nature (London)* **1997**, *389*, 706–709.

[110] S. Yao, I. Ghosh, R. Zutshi, J. Chmielewski, *J. Am. Chem. Soc.* **1997**, *119*, 10559–10560.

[111] S. Yao, I. Ghosh, R. Zutshi, J. Chmielewski, *Angew. Chem., Int. Ed. Engl.* **1998**, *37*, 478–481.

[112] S. Yao, I. Ghosh, R. Zutshi, J. Chmielewski, *Nature (London)* **1998**, *396*, 447–450.

[113] P. E. Dawson, T. W. Muir, I. Clark-Lewis, S. B. H. Kent, *Science* **1994**, *266*, 776–779.

[114] K. B. Mullis, F. A. Faloona, *Methods Enzymol.* **1987**, *155*, 335–350.

[115] A. Luther, R. Brandsch, G. von Kiedrowski, *Nature (London)* **1998**, *396*, 245–248.

[116] T. Li, K. C. Nicolaou, *Nature (London)* **1994**, *369*, 218–221.

[117] J. Selvasekaren, K. D. Turnbull, *Nucl. Acids. Res.* **1999**, *27*, 624–627.

[118] J. Ye, D. G. Lynn, submitted for publication.

[119] D. Herschlag, T. R. Cech, *Biochemistry* **1990**, *29*, 10172–10180.

[120] B. Young, D. Herschlag, T. R. Cech, *Cell* **1991**, *67*, 1007–1019.

[121] D. Herschlag, T. R. Cech, *Biochemistry* **1990**, *29*, 10159–10171.

[122] K. J. Hertel, D. Herschlag, O. C. Uhlenbeck, *Biochemistry* **1994**, *33*, 3374–3385.

[123] S. Arnott, D. W. L. Hukins, *J. Mol. Biol.* **1973**, *81*, 93–105.

6 Biomimetic Reactions Directed by Templates and Removable Tethers

Ronald Breslow

6.1 Introduction

Enzymes are remarkable both for their catalytic effectiveness and for their selectivities. Rate accelerations of 10 000 000 000 or more are not unusual, and most enzymes show selectivity toward the substrates they act on, and toward the specific products they form from those substrates. The challenges in the construction of artificial enzymes are to achieve those rate accelerations and those specificities.

In some respects achieving the product specificities is the most important goal. Substrate specificity is critical in the soup of components found in a biological cell, but chemical reactions can usually be performed on pure substrates where substrate specificity of a catalyst is irrelevant. However, the ability of an enzyme to achieve selective reactions on a particular section of the substrate (regioselectivity) and with specific stereochemical consequences (stereoselectivity) are features to be admired and imitated in chemical systems.

Selectivity in simple chemical reactions is normally a reflection of the intrinsic reactivity of the substrate. For example, an oxidation would be performed by a relatively unselective reagent attacking the most easily oxidized part of the substrate. If that is not the desired position, it might be temporarily blocked, or the desired position might be activated in some way. This is not what enzymes do.

Selectivity in enzymatic reactions is normally achieved by imposing the geometric control of an enzyme–substrate complex in which the catalytic group of the enzyme is held near the substrate so as to direct reaction to a specific spot, whether that spot is particularly reactive or not. For example, in the biosynthesis of cholesterol three methyl groups attached to saturated carbons in lanosterol are oxidized and removed even though there are much more reactive double bonds and a hydroxyl group in the molecule. This occurs because the substrate is held next to metalloporphyrin catalytic groups of the enzymes so that only the otherwise unreactive methyl groups are within reach.

Many years ago, we set out to develop such methodology for synthetic chemistry, imposing strong geometric control on reactions that would otherwise show little selectivity, or would show undesired selectivity. We called this approach "biomimetic", non-enzymatic chemistry designed to mimic the biological chemistry performed by enzymes [1]. In some cases, we imposed the desired geometric control by use of a direct tether between the reagent and the substrate, in other cases we used molecular complexing as a closer analogue to the enzyme–substrate complexing that determines biochemical selectivity. In this account of the field, we will describe both approaches. Although space limitations prevent a full account of related work from other laboratories, such related work is described more fully in some of the reviews we have written previously [1–30], particularly in references [4, 10, 15, 16, 19, 24, 26, 30].

6.2 Biomimetic Reactions Using Covalently Linked Tethers and Templates

6.2.1 Photochemical Functionalizations by Tethered Benzophenones

The first example in which a temporarily tethered reagent was directed to attack un-activated C–H bonds remote from the tether point was a study of the products from pho-tolysis of esters of benzophenone-4-carboxylic acid **1** carrying long-chain alkyl groups (Scheme 6-1) [31]. The *n*-tetradecyl ester was attacked over carbons 8–13, with half the

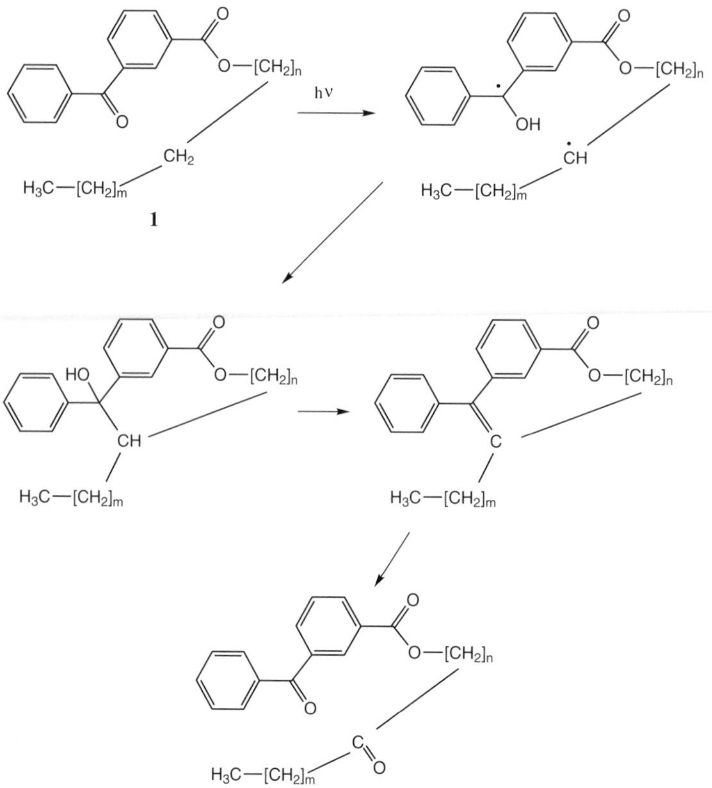

Scheme 6-1   Photochemical remote functionalization of long-chain alkyl esters with a tethered benzophenone.

product from insertion of the benzophenone carbonyl into carbon 12. Products were determined by dehydration of the resulting carbinols and oxidation to produce carbonyl groups at the reaction point, and hydrolysis removed the benzophenone reagent and its tether from the functionalized alkanol. With longer chains, up to 20 carbons, the products indicated attack over greater distances, but again with significant insertion at C-12. The flexibility of the alkyl chains prevented the reactions from being fully selective, but the product distribution furnished information about the conformations of the flexible chains. This was examined further in subsequent studies, to be described.

Steroids are rigid substrates without this flexibility problem, and whose selective functionalization is of considerable practical interest. Thus benzophenone-4-propionic acid was attached to 3α-cholestanol and the ester **2** was photolyzed (Scheme 6-2) [32]. The only methylene group attacked was that at carbon 12 of the steroid, so ester cleavage, dehydration, and olefin oxidation afforded 12-keto-3α-cholestanol **3** as the only ketonic product. However, carbonyl insertion had also occurred at the methine C–H of carbon 14 of the steroid, revealed by lead tetraacetate cleavage and hydrolysis to form Δ^{14}-3α-cholestenol **4** and $\Delta^{8(14)}$-3α-cholestenol **5**. Some of the Δ^{14}-3α-cholestenol **4** was also formed on simple hydrolysis of the tethered reagent, as the result of direct dehydrogenation of the steroid by benzophenone excited-state triplet. Such direct dehydrogenation was later achieved completely selectively with a tethered benzophenone having different geometry, as described below.

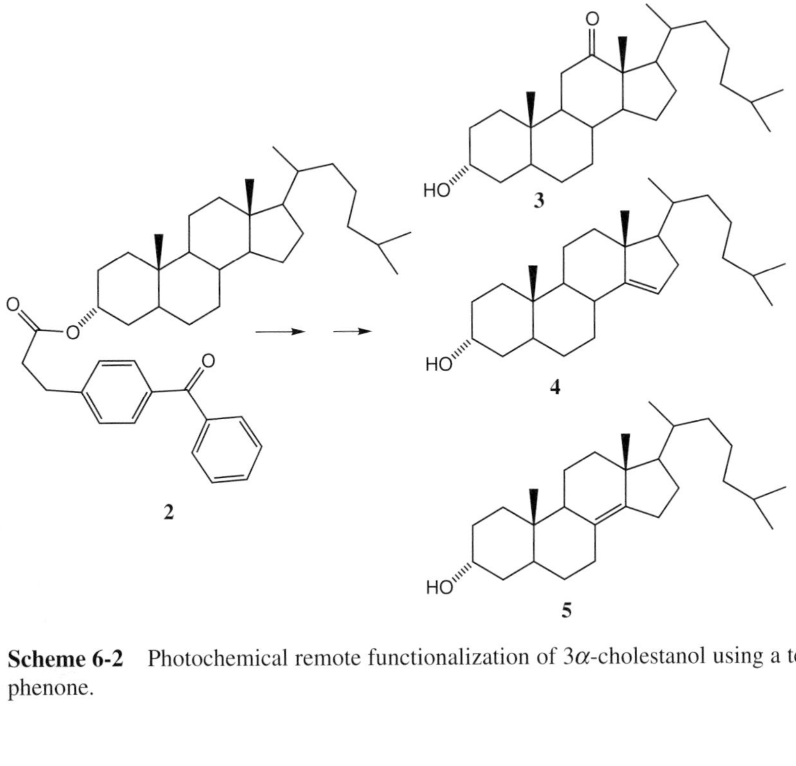

Scheme 6-2 Photochemical remote functionalization of 3α-cholestanol using a tethered benzophenone.

The introduction of a 9(11) double bond into steroids is of practical interest, since such olefins can be converted to the 9-fluoro-11-hydroxy grouping found in most useful corticosteroids. Ester **6** had a benzophenone-4-hexanoic acid tethered to the hydroxyl group of androstan-17β-ol that curled under the steroid ring and photochemically inserted into the C-9 α-hydrogen, forming product **7** with a 9(11) double bond after lead tetraacetate cleavage of the carbinol product and hydrolytic cleavage of the tether (Scheme 6-3) [33]. A product with a 14,15 double bond was also produced.

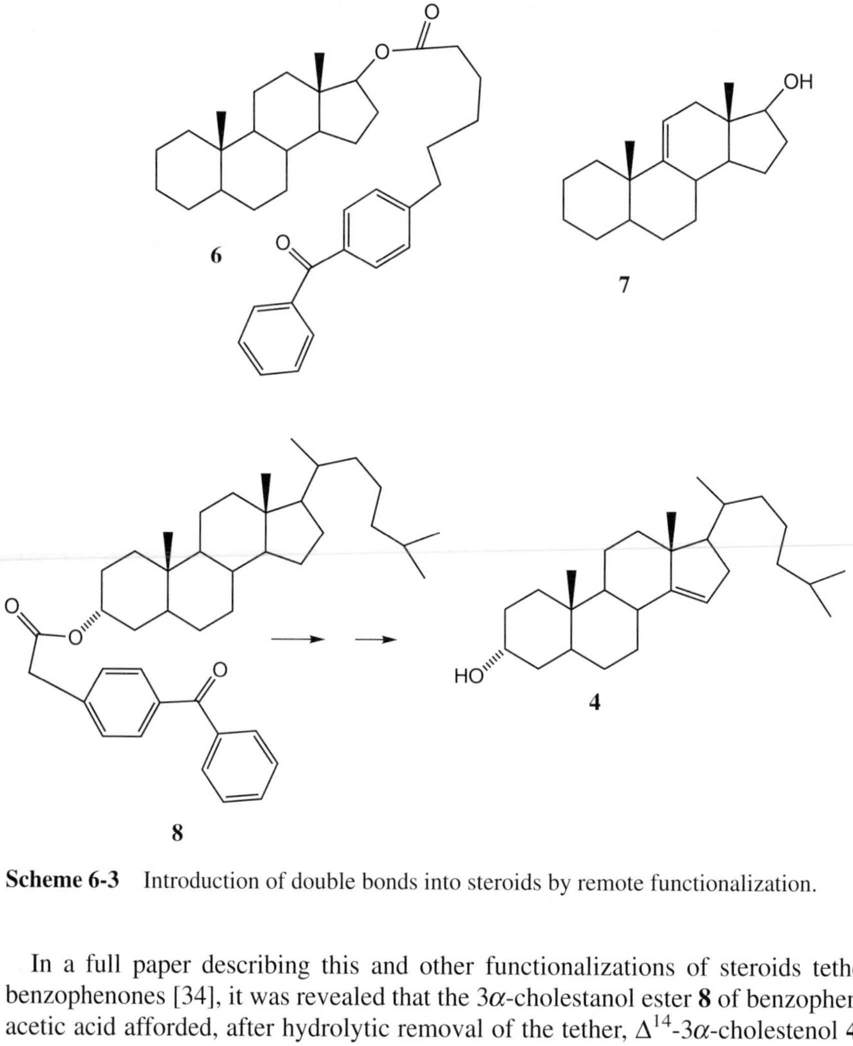

Scheme 6-3   Introduction of double bonds into steroids by remote functionalization.

In a full paper describing this and other functionalizations of steroids tethered to benzophenones [34], it was revealed that the 3α-cholestanol ester **8** of benzophenone-4-acetic acid afforded, after hydrolytic removal of the tether, Δ^{14}-3α-cholestenol **4** as the product along with the diphenylcarbinol from reduction of the benzophenone. The short tether did not permit insertion into the C-14—H bond, so after the oxygen atom of excited benzophenone removed the C-14 hydrogen, the resulting steroid radical inverted to bring the C-15 hydrogen within reach of the benzhydryl radical; it was selectively removed to afford the observed products. In a subsequent study [35], a deuterium was specifically placed in the 15α-position of the steroid, and it was demonstrated that this was indeed the atom that becomes attached to the carbonyl carbon of the benzophenone reagent. It was also found that a small amount of insertion of the benzophenone occurred into C–H bonds at C-12 and C-7 in addition to the previously reported 14–15 dehydrogenation.

The products found in these studies indicated attack on the steroid at a predictable distance from the attachment point, except with highly flexible tethers, but the tether still permitted attack along an arc pivoting around the attachment point. This is a problem that was addressed in later work, to be described.

6.2.2   Free-radical Halogenations by Tethered Reagents

In the course of carrying out the benzophenone photochemistry described in the previous section, we noted that the use of CCl_4 as solvent led to the formation of some chlorinated steroid products. Therefore we examined such halogenations further [36], and found that the simple free-radical chlorination of cholestanyl acetate with phenyliodine dichloride afforded the 9-chloro- and 14-chlorosteroids as the major products. Similar results were seen in bromination by bromotrichloromethane. We extended such selective radical halogenations to suppress reactions other than at C-9 and C-14 by appropriate substituent effects, but more significantly we used tethers to achieve essentially complete selectivity in the halogenations [36].

Attachment of phenyliodine dichloride with a *m*-carboxy group as an ester **9** of 3α-cholestanol led to selective intramolecular chlorination at C-9 (Scheme 6-4). Hydrolysis of the ester led to HCl elimination, affording the $\Delta^{9(11)}$-cholestenol **10** as the main product, along with a little of the 14–15 olefin from intermolecular reaction. By contrast, attaching *p*-iodophenylacetic acid to 3α-cholestanol, then converting it to the dichloride **11**, led to intramolecular chlorination at C-14, with a small amount of more random intermolecular halogenation as well. This demonstrated that free-radical chlorination could be directed geometrically by the use of appropriate tethers. Furthermore, the chemical preference of the intermediate Ar–I–Cl radical for hydrogen abstraction from tertiary positions meant that the position of preferred attack was completely determined by the tether length, and unaffected by the ability of the reagent to swing in an arc about its attachment point. That is because the tertiary carbons with hydrogens on the α-face are all at quite different distances from C-3.

In this work a reagent, an aryl iodide dichloride, was directly attached to the substrate through a tether, but it had disadvantages. With such a scheme only stoichiometric amounts of the reagent could be used, and the reactions led to some recovered starting material. Thus we considered whether a new process was possible in which we would use both a tether and a template to direct halogenations. We decided to try to invent what we called a "radical relay" process.

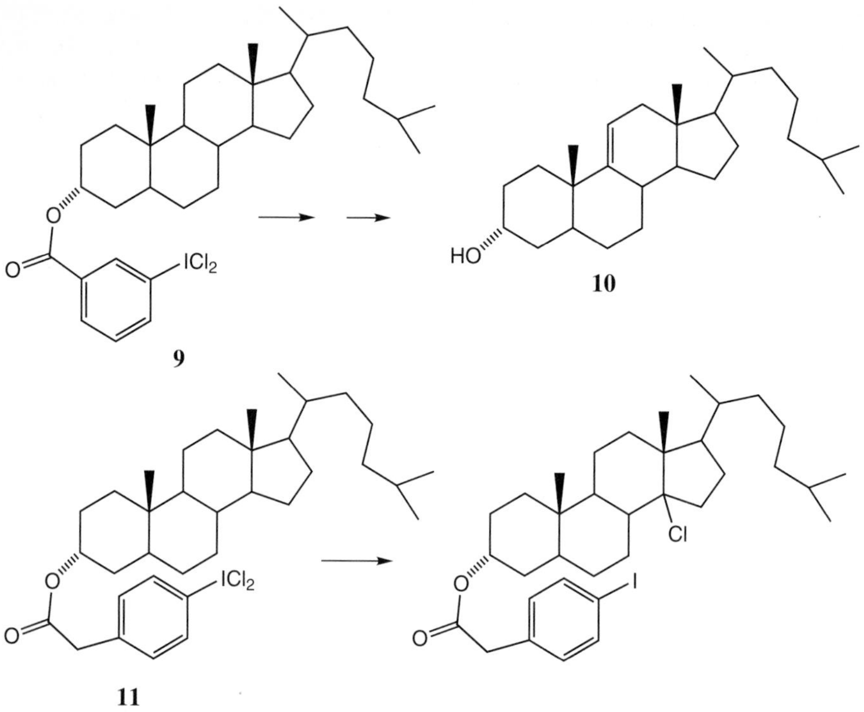

Scheme 6-4 Examples of free-radical halogenations of steroids by tethered reagents.

6.2.3 Free-radical Reactions Directed by Tethered Templates – the Radical Relay Mechanism

In the above halogenations, an iodophenyl group was attached to the steroid substrate, and then converted to the aryl iododichloride with Cl_2 before the free-radical chain process was performed. We wondered whether we could generate the intermediate $Ar-I-Cl$ radical that removes the substrate hydrogen by transferring a chlorine atom to the attached aryl iodide from an external reagent. That is, we wondered whether we could relay a chlorine atom from an external reagent to the iodine atom on the attached aryl iodide and thence to the hydrogen atom within reach. Such a process would be the analogue of throwing a baseball to the catcher who then reached down with it to tag the runner, or of kicking a soccer ball to a player near the goal who then propelled it into the goal itself. The chlorine atom radical relay was similarly successful.

We irradiated a solution of phenyliodine dichloride with the *m*-iodobenzoate ester **12** of 3α-cholestanol and found that C-9 chlorination occurred selectively, while no such selectivity was seen without the iodine atom on the benzoate ester [37]. Various controls established that we were not simply converting the attached iodoaryl template to its dichloride, but were instead performing the desired radical relay process. That is, the free phenyliodine dichloride in solution was converted on photolysis to the Ph–I–Cl radical, and this then transferred its chlorine atom to the template iodobenzoate group to form an attached Ar–I–Cl radical, as shown (Scheme 6-5). As expected from this, when the tethered template was the *p*-iodophenyl acetate, the process afforded the 14-chloro steroid product. Similar results were obtained when the solution chlorinating agent was SO_2Cl_2, and the solution radical was $\cdot SO_2Cl_2$.

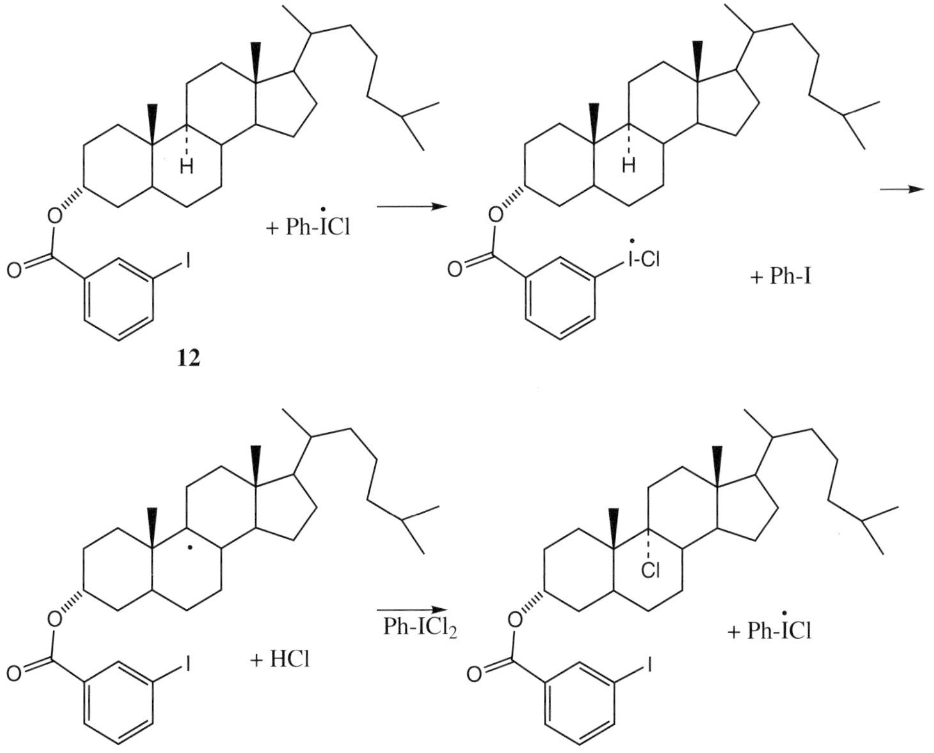

Scheme 6-5 The radical relay mechanism.

Since this study led to a number of extensions, it is desirable to explain why the process works, and what its advantages are. The rate advantage of a radical relay process is that the hydrogen abstraction is intramolecular, rather than the intermolecular abstraction that would occur without the relay by the template. However, this explains it only in part, since the relaying of a chlorine atom from the radical in solution to the iodine of the template is of course an intermolecular process. Why is the two-step sequence – intermolecular chlorine atom transfer, then intramolecular hydrogen abstraction – faster than an intermolecular hydrogen abstraction by the free radical in solution? The answer is related to the reason why many enzymes carry out two-step sequences in hydrolysis reactions. That is, the serine proteases first attack an amide bond with a serine hydroxyl of the enzyme, then bring in a water molecule to hydrolyze the resulting serine ester. Why not simply use the water molecule directly to attack the amide bond?

The advantage of the two-step process with serine proteases has to do with translational entropy [38]. In the first step, attack of an enzyme serine hydroxyl on the amide, the peptide is converted to an ester and the amine is released from the enzyme and recovers some of the translational entropy the substrate lost on binding to the enzyme. Then a water molecule is bound, losing entropy, but that is after the rate-determining step so it is irrelevant. If the second step were rate-determining, as it is when serine proteases hydrolyze esters instead of amides, the entropy lost by the water molecule would still have been compensated by the previous entropy gain from the released leaving-group fragment.

In the radical relay chlorination of steroids, related arguments apply. Chlorine atom transfer to the large reactive iodine atom of the template from a radical in solution is relatively fast and easy, probably not rate-determining, and after the transfer the PhI is freely moving. The otherwise difficult attack at a hydrogen atom in the substrate is fast because it is intramolecular and template directed. The steroid radical then removes a chlorine from the solution reagent, regenerating the solution radical, but this is a fast intermolecular step that is not rate-determining. The result is that the template directed process dominates any random attack by solution radicals, and the reaction is thus positionally selective.

The practical advantages of the radical relay process over the use of a tethered aryliodine dichloride reagent are several. First of all, an excess of the solution chlorinating agent can be used, so complete conversion of the substrate to product is achieved. Secondly, there is no need to premake an attached aryl dichloride by using Cl_2, so sensitive substrates can be used. Finally, other templates can be used that can capture and relay a chlorine atom but cannot themselves be converted to dichloride reagents. These will be described later.

It is important to note that in all of the chlorinations described above a chlorine atom is selective for the removal of a tertiary hydrogen atom, so the templates tethered on the α-face of the steroids can be positionally selective. That is, the tertiary hydrogens at C-5, C-9, C-14, C-17, and in the side chain are all at different distances from C-3, where the tether is attached. Thus with a tether of defined length the intramolecular hydrogen attacks are quite selective. However, it is possible to design non-selectivity into the tethers. Models indicated that a m-iodophenyl acetate tether should be able to direct chlorination to both C-9 and C-14, and this was found experimentally [37]. Both with the premade tethered aryliodine dichloride and with the radical relay process, an essentially equal mixture of C-9 and C-14 halogenation occurred. This is of course not a useful result, but it does indicate the extent to which molecular modeling can be used to predict the products in these reactions.

We examined this further with molecular mechanics computer calculations [39]. The calculations accounted well for our experimental findings, including some we have not yet described here. However, the high selectivities we observed could only be accounted for if we required that the collision of a chlorine atom bound to the iodine of a template had to occur essentially in line with the C–H bond being broken. This is not a general requirement of molecular mechanics calculations, and we suggested that it may reflect a requirement of the dynamics of the process. That is, we argued that collisions are effective only if their energy can be converted into vibrations of the substrate along the reaction direction, and that a more or less linear collision is required for effective energy transfer.

We used our radical relay functionalization of C-9 in steroids to perform a synthesis of cortisone **13** (Scheme 6-6) [40]. The $\Delta^{9(11)}$ double bond in the steroid was hydroborated to allow an oxygen functionality to be placed at C-11, characteristic of corticosteroids. In more practical applications, the double bond can be epoxidized and opened to afford oxygen at C-11 and fluorine at C-9, as in many of the newer corticosteroid pharmaceuticals. We also used other templates to direct chlorination to C-17, where the cholesterol side chain is attached [41]. Further reactions permitted the removal of that side chain; a full paper summarized such aryl iodide template directed reactions in detail [42]. Subsequently we found that template methods could be used both to introduce oxygen at C-11 and to permit very effective removal of the steroid side chain [43]. Related to this, a tethered template with *two* chlorine binding atoms – a bipyridine or an iodophenylpyridine – was able to carry out the simultaneous chlorination of *both* C-9 and C-17, making approaches to corticosteroids even more attractive [44]. Later work [45] also permitted cleavage of the side chain so as to retain an acetyl group at C-17, and with a template tethered to the 6β-position of steroid **14** we were also able to chlorinate directly into the side chain at C-20, as an alternative method to product such 17-acetyl steroids by partial side chain removal.

This work built on our earlier finding [46] that benzophenone photochemical insertion into the side chain could be achieved when the benzophenone was tethered to the β-face of the steroid. The findings, along with X-ray crystal structure data [46], made it clear that the nominally free rotation of the steroid side chain is in fact strongly restricted by steric constraints; the hydrogen on C-20 is pointed to the steroid β-face and requires a β-linked tether to permit its functionalization.

Atoms other than iodine can also serve as chlorine binders in radical relay reactions. The sulfur atom of tethered diaryl sulfides can capture and relay a chlorine atom with predictable geometric control [47]. However, in contrast to the iodine atom templates, the sulfur templates could not capture and relay a chlorine from a free phenyliodochloride radical, only from the more reactive $\cdot SO_2Cl$ species produced when SO_2Cl_2 is the chlorinating reagent. Interestingly, phenyliodine dichloride was able to act as the chlorine atom source in a radical relay mechanism using the sulfur atom of a thioxanthone tethered template in **15** [48].

We also attached *three* steroid substrates to an iodophenyl template in **16** (Scheme 6-7) using a tris-silyl ether link, and saw that the template could direct selective radical-relay chlorination to all three tethered substrate species; in the same work we reported a similar finding when a thiophene ring was the triply tethered template [49]. Thus the sulfur atom of thiophene can perform radical relay. In these triple catalytic functionalizations, the more reactive sulfuryl chloride was the preferred reagent.

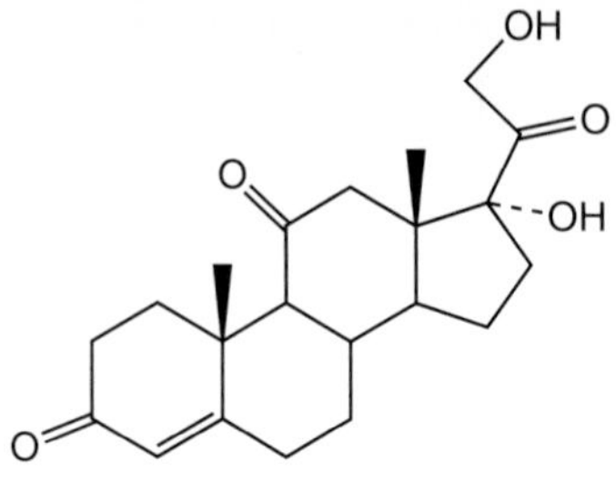

13

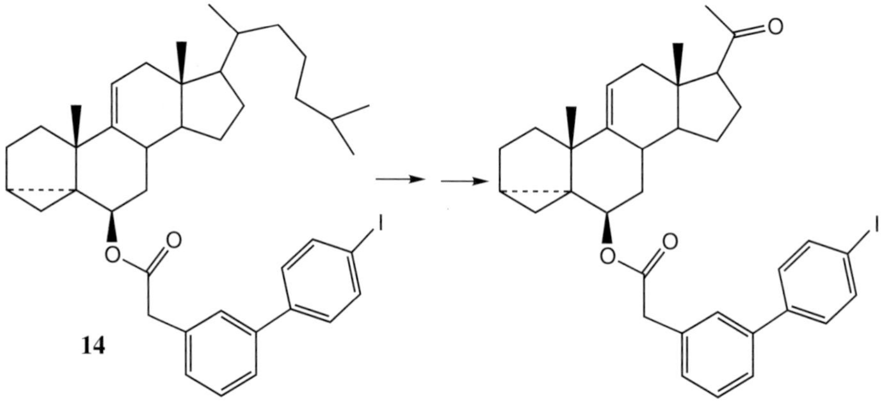

14

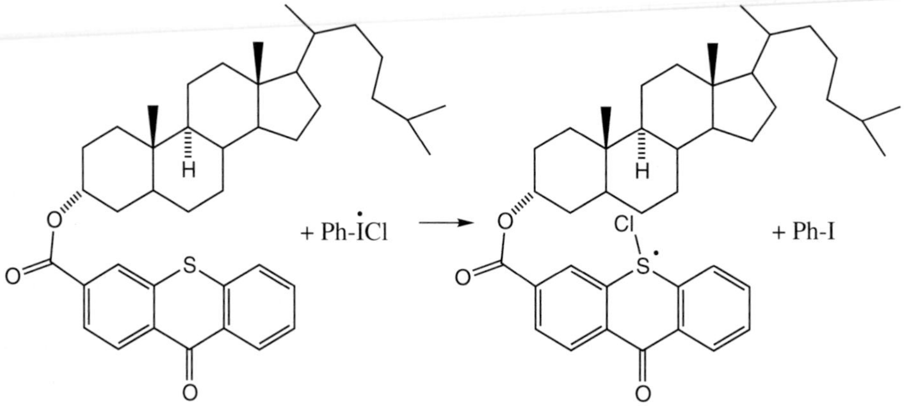

15

Scheme 6-6 Radical relay functionalization in the synthesis of cortisone **13** and with a thioxanthone tethered template.

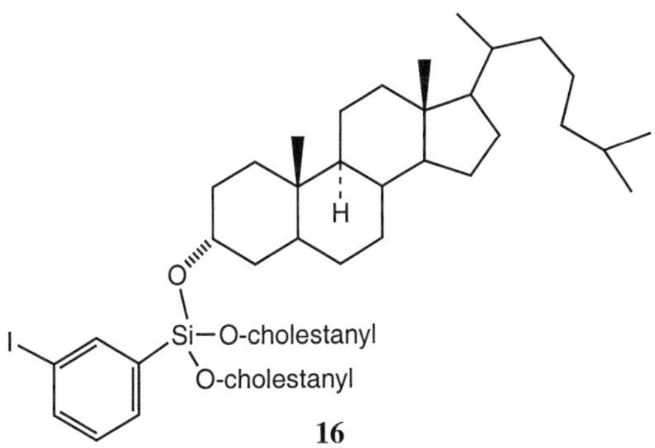

Scheme 6-7 Radical relay functionalization with a triply tethered template and with a tethered pyridine template.

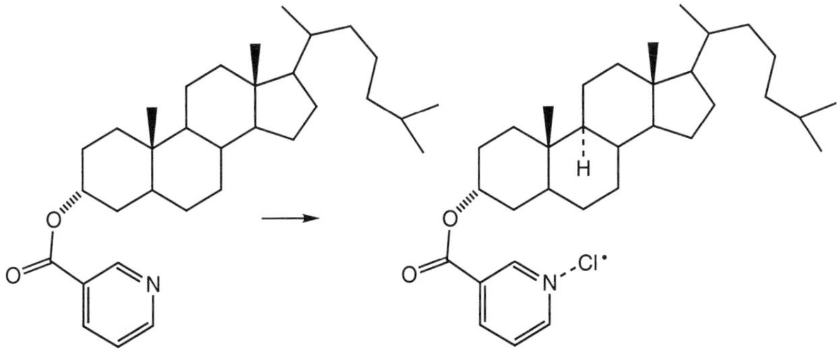

Chlorine atoms would complex with benzene rings, but the geometry and stability were not sufficient to allow a simple phenyl group to be used as the relay template. The same was not true with pyridine rings. Radical relay processes were successful with tethered templates containing pyridine rings, as in **17**. The process involved attachment of the relayed chlorine atom coordinated to the nitrogen atom and in the pyridine plane [50, 51]. The results indicate that the pyridine-based tethered templates were superior to those based on aryl iodides or thioethers. Theoretical calculations and spectroscopic measurements indicate the nature of the complex between a pyridine ring and a chlorine atom [52]. Specifically, the chlorine atom is held to the pyridine nitrogen by a three-electron bond – two pyridine electrons, one chlorine odd electron – which with two bonding and one antibonding electron is weaker than an ordinary covalent bond, and thus able to transfer the chlorine atom to an available hydrogen attached to carbon in the radical relay process.

A further study examined fused pyridine rings, in quinoline and acridine, as tethered templates [53]. Although the chlorine atom complex might have had a different structure from that of the simple pyridine complex, these fused pyridines were also effective templates for the selective chlorination of steroids.

Some variants on the simple template-directed chlorination were also developed. For example, a steroid carrying a tethered iodophenyl group was chlorinated by electrolysis of a solution carrying chloride ion [54]. In this case, the electrolysis furnished Cl_2 in solution to carry a radical relay process and electrolysis also initiated the radical process by one-electron oxidation of the iodophenyl group. As another variant, the radical relay mechanism requires that it be a chlorine atom that attaches to the iodine or pyridine or sulfur to abstract hydrogen, since a complexed bromine atom is not reactive enough, but the new bond to the substrate does not have to be a carbon–chlorine link. That bond is formed by untemplated attack of the substrate carbon radical on a reagent in solution and, with an appropriate sequence of tandem reactions, other atoms can be linked to the substrate.

We showed that radical relay in the presence of phenyliodine dichloride but with an excess of CBr_4 afforded the bromosteroid, with bromine at the point that is normally chlorinated [55]. When $(SCN)_2$ was the additive, rather than CBr_4, an SCN group was formed [55]. The mechanism shown (Scheme 6-8) involves a bromine atom transfer to

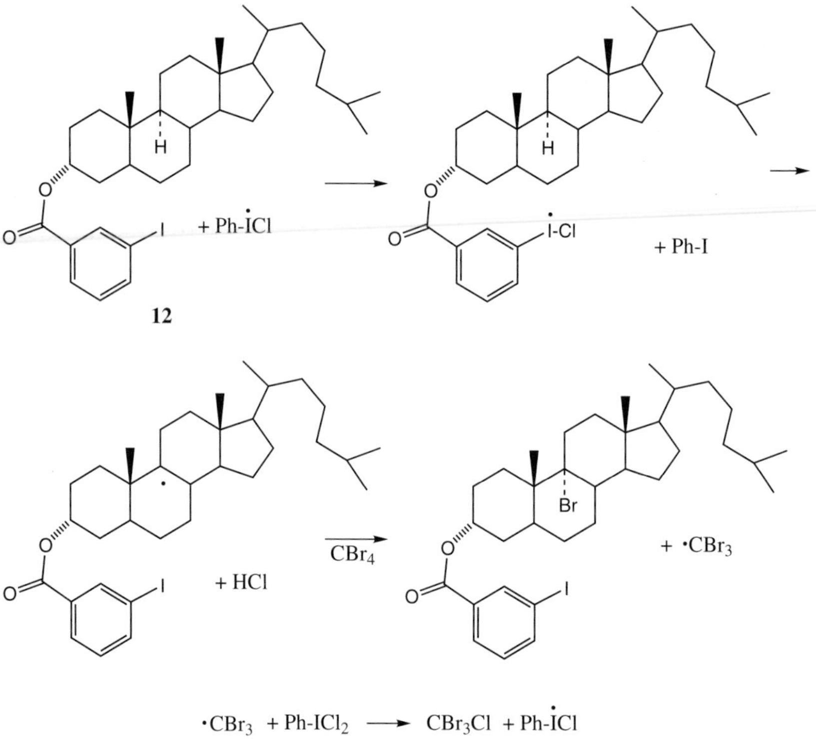
H
O
O
I
+ Ph-İCl
12
H
I-Cl
+ Ph-I
+ HCl
CBr4
Br
+ •CBr3
•CBr3 + Ph-ICl2 → CBr3Cl + Ph-İCl

Scheme 6-8 Mechanism of formation of a bromosteroid by radical relay functionalization.

the steroid radical to form the bromosteroid and $\cdot CBr_3$, but this radical then abstracts Cl from the solution $PhICl_2$ so as to form the phenyl iodochloride radical, which transfers its Cl to the template to permit hydrogen abstraction and complete the cycle. In the thiocyanate case it is the SCN radical that removes Cl from $PhICl_2$ in solution to permit a chlorine radical relay process.

We have largely been describing reactions on steroid substrates, which are conformationally rigid and permit selective functionalizations by appropriate tethered templates. However, when the templates are linked to flexible chains, the results can be used to learn about the conformational preferences of such flexible chains. In one study [56], we examined the positional selectivities of insertion reactions into flexible chains by attached benzophenone units, a process we had also examined earlier [31], and compared the results with those from the intramolecular chlorination of such flexible chains by attached aryliodine dichlorides. The results were complementary. In another study [57] we used long-chain alkyl esters of nicotinic acid in radical relay chlorination, and saw some interesting selectivities reflecting conformational preferences in these nominally flexible cases.

A study [58] on the use of non-tethered templates to direct the radical relay mechanism will be described in a later section of this chapter.

6.2.4   Selective Intramolecular Epoxidations Directed by Removable Tethers

Sharpless [59] had extended earlier observations [60] on the selective epoxidation of the double bonds in allylic alcohols catalyzed by metals that coordinate to the hydroxyl groups. We decided to extend this by using a template to bring a coordinating hydroxyl group near in space to an otherwise remote double bond. Thus we showed that we could promote the epoxidation of a double bond in ring D of a steroid with a template temporarily attached to ring A in **18** (Scheme 6-9) [61]. The reaction was selective for such an accessible double bond even if the substrate had an additional double bond allylic to the point at which the template was attached. In the absence of the template, the allylic dou-

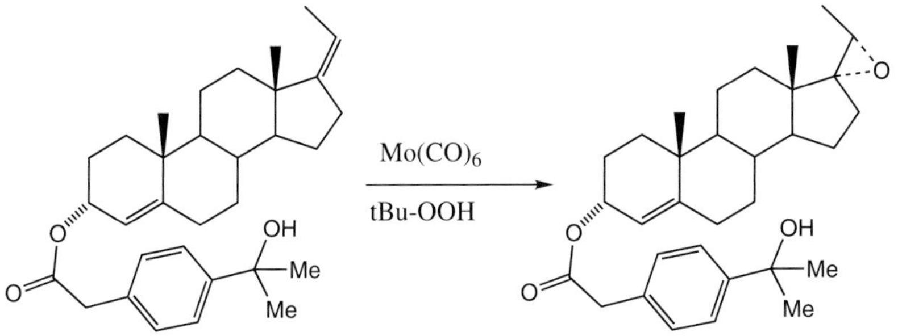

18

Scheme 6-9   Template directed epoxidation.

ble bond was epoxidized, while with the attached template, the remote ring D double bond was the reactive one.

This template directed functionalization of double bonds within reach was also used to determine the conformational preferences of some terpene polyenes [62]. As in the previously described work with attached benzophenone reagents, the selective template-directed epoxidations revealed the details of chain folding in flexible molecules.

6.3   Selective Reactions Directed by Non-covalently Linked Templates

6.3.1   Selective Aromatic Substitution Directed by Cyclodextrin Complexing

The chlorination of anisole by HOCl in aqueous solution produces both *p*- and *o*-chloroanisole, in a ratio of 60:40 [63]. By contrast, in the presence of α-cyclodextrin (cyclohexaamylose) the *para* chlorination dominates, so that with 10 mM cyclodextrin the product is 96% *p*- and 4% *o*-chloroanisole. In water solution anisole binds into the cyclodextrin cavity. The simplest idea would be that this binding blocks chlorination at the *ortho* positions, which are inside the cavity, while the *para* position of the anisole is still exposed to solution. The true situation is more interesting. When the reaction shows 96% selectivity for *para* chlorination, the anisole is only 72% bound, so the bound anisole must be more reactive, not just more selective, than that in solution. The data indicate that *ortho* chlorination is indeed completely blocked for the anisole/cyclodextrin complex, but that the *para* position is 5.6±0.8 times more reactive in the complex than in solution. It was suggested that the chlorine was being delivered to the *para* position of bound anisole by a hydroxyl group of the cyclodextrin, which had been converted to a hypochlorite group (Scheme 6-10). Later work confirmed this idea.

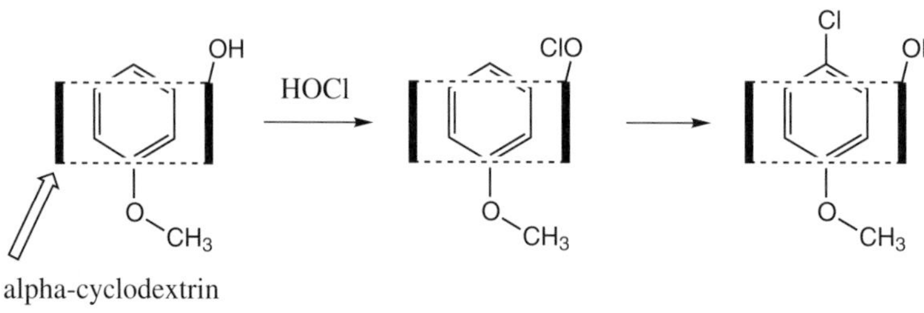

alpha-cyclodextrin

Scheme 6-10   Selective *para*-chlorination of anisole complexed by α-cyclodextrin.

In the full study [64], it was found that a lesser but similar effect was seen with β-cyclodextrin (cycloheptaamylose), which has a larger cavity in which anisole is both more weakly and more flexibly bound. Within this looser complex, *ortho* chlorination is still completely blocked, but *para* chlorination has the same rate as it has in free solution. With 10 mM *β*-cyclodextrin anisole is only 59% bound, not 72%, and the *para/ortho* chlorination ratio becomes only 3.8:1, not 21.6:1 as with *α*-cyclodextrin. Thus, the catalysis of *para* chlorination in the *α*-cyclodextrin complex is promoted by rigid inclusion of the anisole into a tight-fitting cavity.

Kinetic studies revealed something very striking: the rate of chlorination in the complex of anisole with *α*-cyclodextrin had a first-order dependence on HOCl, but in free solution the rate depended on $[HOCl]^2$. Thus, different species perform the chlorination in the two situations. In solution it is known that the $[HOCl]^2$ dependence indicates that the true chlorinating species is Cl_2O, which is formed from two HOCl molecules. HOCl itself is too unreactive for the intermolecular chlorination to proceed rapidly, so the more reactive Cl_2O is used even though there is little of it. However, in the complex with cyclodextrin the HOCl rapidly and reversibly forms a hypochlorite ester with a cyclodextrin hydroxyl group, and this then delivers the chlorine atom to the accessible *para* position of bound anisole. An alkyl hypochlorite is not a strong chlorinating agent, but the intracomplex character of the reaction makes the rate fast enough. As expected, the kinetic advantage of such an intracomplex reaction is greater with the tight anisole complex in *α*-cyclodextrin than with the looser complex in *β*-cyclodextrin. The resemblance of this mechanism to that of the template directed chlorinations described previously is obvious: the OH group of cyclodextrin performs (non-radical) relay chlorine transfer from a solution reagent to the substrate.

When *p*-methylanisole **19** (Scheme-6-11) was used as substrate, only the positions *ortho* to the methoxyl group were chlorinated, whether *α*-cyclodextrin was present or not, but reaction occurred only in free solution. Added *α*-cyclodextrin simply slowed the reaction by complexing the substrate. On the other hand, *p*-cresol **20** was of course also chlorinated *ortho* to the oxygen substituent, but in this case the *α*-cyclodextrin accelerated the reaction. The polar hydroxyl group projects from the cavity so that the *ortho* positions are now within reach of a cyclodextrin hypochlorite group, and intracomplex chlorine delivery occurs.

As a test of this mechanism, the coupling of *p*-benzenediazonium sulfonate **21** with phenol was examined. In solution it couples in the *para* position, but *α*-cyclodextrin

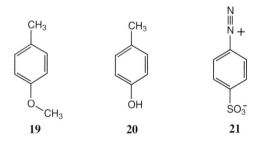

Scheme 6-11   The chlorination of **19** and **20** and the coupling of **21** with phenol was investigated in the presence of *α*-cyclodextrin.

simply inhibits the reaction. The diazonium electrophile cannot be delivered by a cyclo-dextrin hydroxyl group, in contrast with such a delivery of a chlorine atom.

One interesting feature of this work is that the enzyme chlorinase also can chlorinate anisole in water solution, but it gives a random mixture of *ortho* and *para* products [65]. Apparently the enzyme merely makes a chlorinating reagent, probably HOCl, and this then acts on the unbound anisole in free solution. However, the selectivity seen in our anisole complex chlorination is typical of that seen in other enzymes that do indeed direct reactions by geometric control within an enzyme–substrate complex.

The anisole chlorination was examined further with two modified α-cyclodextrin derivatives [66]. In one we prepared the dodecamethyl derivative, in which all the primary hydroxyls and all the C-2 secondary hydroxyls of the α-cyclodextrin were blocked by methyl groups. This was an even more effective catalyst for the direct chlorination of anisole in its *para* position, and less than 1% of *ortho* product could be detected with the catalyst at 10 mM. The extra methyl groups make anisole binding even stronger, accounting for the improved efficiency. Also, this work established that the C-3 hydroxyl group on the secondary face of the cyclodextrin could serve as the catalytic group in chlorine transfer. Of course, if the other hydroxyls were not blocked, as in the unmethylated catalyst, they might also serve to relay chlorine from solution to the bound substrate.

In other work described in the same publication, α-cyclodextrin was converted to a solid polymer by reaction with epichlorohydrin, and this was used in a reactor to perform the flow reaction of anisole with HOCl. The product was completely *para*-chlorinated, and the reactor could be used repeatedly. Thus, the binding of substrate to an immobilized template has potential for practical synthesis with geometric control of selectivity.

Since this early work, other laboratories have reported selective aromatic substitution reactions in cyclodextrin cavities. In most cases the cyclodextrin does not serve as a template, but simply blocks some otherwise reactive spots. These cases have been reviewed [30].

6.3.2 Photochemical Functionalizations by Complexed Benzophenones and Chlorinations Directed by Ion-paired Templates

Benzophenone-4-carboxylic acid **22** complexes with other carboxylic acids in non-polar media by double hydrogen bonding. We examined the functionalization of hexadecanol by converting it to its hemisuccinate **23** and irradiating this in CCl$_4$ solution with benzophenone-4-carboxylic acid (Scheme 6-12) [67]. The products from insertion of the keto carbonyl group into the hexadecane chain were dehydrated and the resulting olefin was ozonized to afford, after saponification to remove the tether, a mixture of ketohexadecanols. The distribution reflected the conformational accessibility of substrate carbons in the tethered complex.

Selectivity was seen when a similar process was performed on the hemisuccinate **24** of 3α,5α-androstanol. The only ketonic product, formed in 21% yield, was 16-keto-3α,5α-androstanol **25**, indicating a rather good selectivity at the most remote carbon from the point of attachment of the removable tether. Because the hydrogen bonding

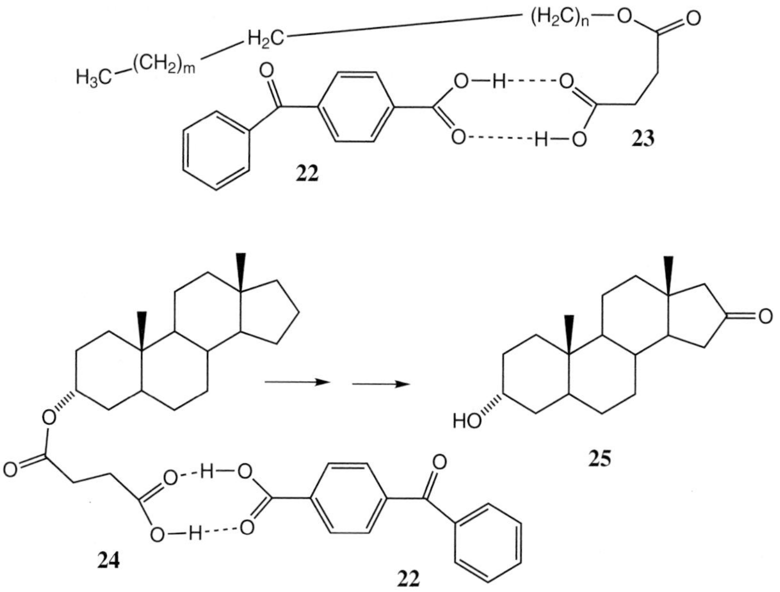

Scheme 6-12 Photochemical functionalizations of alkyl and steroidal esters by complexed benzophenones.

association of reagent with the tether is reversible, the yield was improved by using an excess of reagent to drive the reaction to completion.

Benzophenones carrying carboxylate groups attached *para* either directly or at the end of two-, three-, or four-methylene chains were used to functionalize sodium dodecylsulfate (SDS) and cetyl trimethylammonium bromide (CTAB) chains in micelles, with the hope that the micellar structure might fix chain conformations so that positionally selective functionalizations would occur [68]. It was seen that the chains in the micelles were flexible enough for several positions to be attacked by the triplet benzophenone unit, but very interesting conformational information was obtained. It was seen that the *shorter* benzophenone derivatives attacked *further* into the micellar chains, contrary to simple expectation. This indicates that the chains being attacked are curled back so that their nominally remote ends are actually closer to the micelle surface. This is apparently the way the chains can fill a spherical region: some chains extend all the way to the center of the micelle, while others curl back so as to fill the interchain spaces that would be created if all chains were fully extended. It is the latter curled chains that the benzophenone probes were attacking.

Amphiphiles with two chains per head group can form vesicles, not simple micelles, and we used our probes to determine whether they had greater conformational order and could direct functionalizations more selectively [69]. We found that spherical vesicles showed evidence of conformational disorder comparable to that we had seen in micelles, but that flat vesicles showed considerable order, with attack largely at the expected distance from the surface if the chains were fully extended. Of course such probes may accentuate the amount of disorder in such systems, either by inducing the disorder or by reacting preferentially in disordered regions. However, the overall results of these studies indicated that we had interesting tools for probing structures of micelles and vesicles, but that the flexible chains were not well enough organized to permit synthetically useful selective functionalizations.

To fix the geometries of flexible chains in such reactions, straight-chain α,ω-dicarboxylic acids were used as substrates along with benzophenones that could bind to both carboxyl groups [70]. The benzophenones carried a binding group on the *meta* position of each ring, either trimethylammonium or carboxyl groups. The two cationic trimethylammonium groups in **26** stretched the chain of decanedioic acid dianion **27** across the benzophenone carbonyl, and directed 93% of the functionalization to the two equivalent central methylene groups of the substrate (Scheme 6-13). With the longer dodecanedioic acid there was more randomness, since the chain is not fully extended in the doubly ion-paired complex. The selectivity for the central carbons of the shorter octanedioic acid was slightly less, 81%, when it was complexed by hydrogen bonds to the benzophenonedicarboxylic acid. When there were only single linkage points in related molecules no functionalization occurred under the conditions used, so the double-ended complexing both promoted reaction and directed it selectively in the cases where the substrate and reagent had exactly the correct fit.

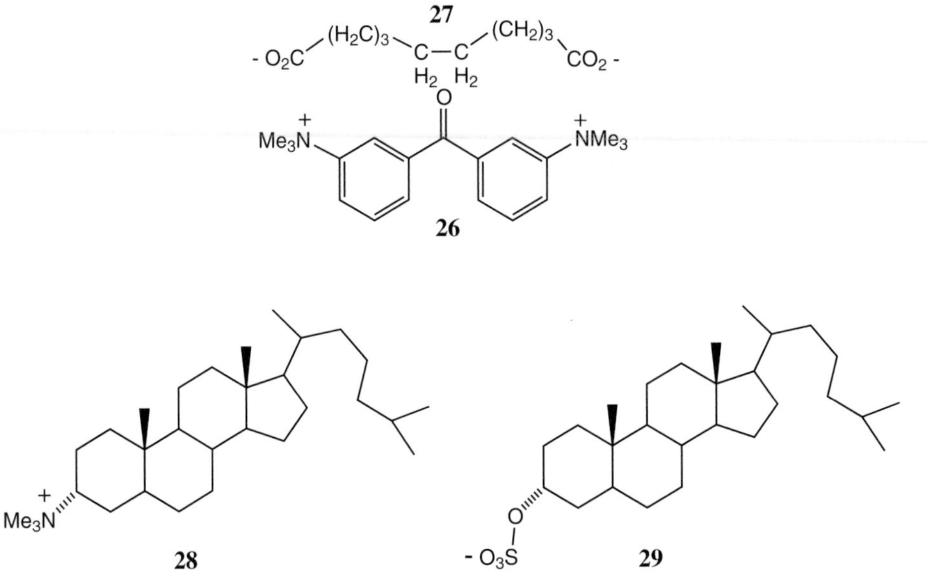

Scheme 6-13   The doubly ion-paired template **26** gives high regioselectivity in the photochemical functionalization of **27**, whereas the single ion-pairing substrates **28** and **29** are not functionalized with high specificity.

Ion pairing was also examined as a way to bind an iodophenyl template to a steroid to direct radical relay chlorination [58]. If successful, such a process could in principle use the template in only a catalytic amount. Cholestane carrying a 3α-trimethylammonium group (compound **28**) was paired with iodophenyl sulfonates or carboxylates, and 3α-cholestanyl sulfate **29** was ion-paired with iodophenyl templates carrying trimethyl-ammonium groups. Indeed, radical relay chlorination did occur with these templates, and not in appropriate control reactions, but the positional selectivities were not as great as with the covalently tethered templates described previously. Apparently ion pairing does not define the relative geometries of template and substrate well enough.

6.3.3 Oxidations Directed by Metalloporphyrin and Metallosalen Templates

Heme (**30**), an iron porphyrin complex (Scheme 6-14), is the catalytic group in many oxidizing enzymes, including the cytochrome P-450s. Thus there has been a serious effort to mimic various aspects of these enzymes using metalloporphyrins. In the gener-

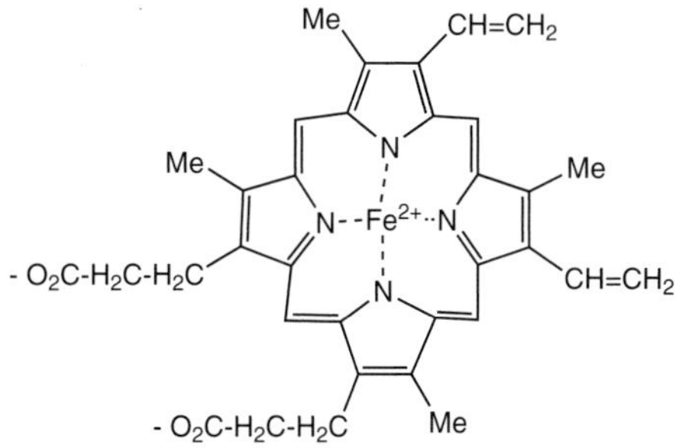

30

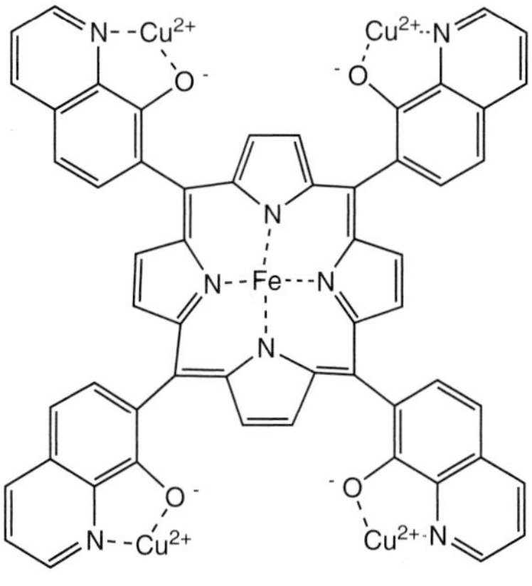

31 with 4 bound Cu^{2+}

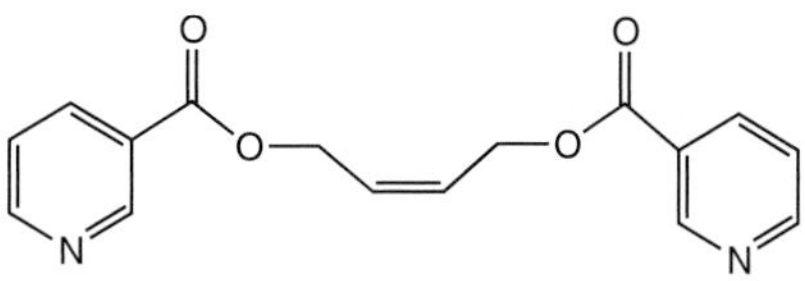

32

Scheme 6-14   Catalytic epoxidation of **32** directed by coordination to the Cu$^{(II)}$ complex **31** in the presence of an oxygen transfer reagent.

al scheme, an iron or manganese derivative of a synthetic porphyrin unit is used to catalyze the transfer of an oxygen atom from an oxidizing reagent such as iodosobenzene, hypochlorite ion, etc., to a substrate. With appropriate substituents to stabilize the porphyrins, some very high catalytic turnovers have been seen. Thus we set out to extend the geometric control ideas described above into this area.

In our first study [71], we synthesized an iron porphyrin **31** carrying 8-hydroxyquinoline units at the four *meso* positions, and used Cu^{2+} to bind substrate **32** carrying two nicotinate ester groups, one at each end (Scheme 6-14). The result was to stretch the substrate across the porphyrin so as to leave a double bond of the substrate directly above the iron in the porphyrin. We saw that the linkage of catalyst and substrate by their mutual binding to the Cu^{2+} ions led to selective epoxidation of the bound substrates relative to analogous cases without such Cu^{2+} linkages. There was also catalytic turnover, as the bound product was replaced by more substrate.

In the same publication, we also reported a similar study using the Mn^{3+} complex **33** (Scheme 6-15) of the tetradentate ligand salen as the catalytic group. There were some differences from the porphyrin results, but auxiliary metal ions were also used to bind the catalyst and substrate in the salen case.

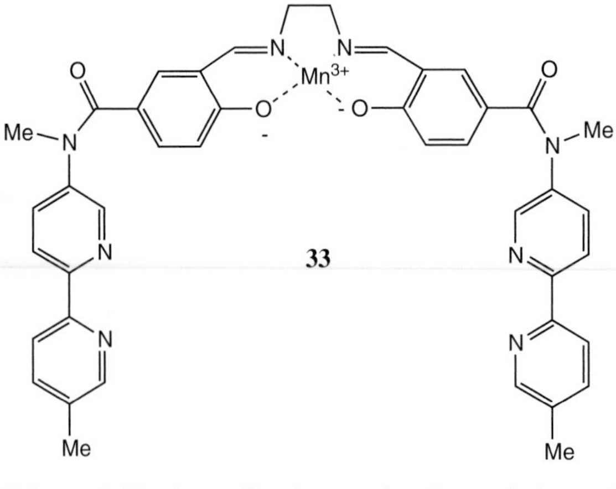

Scheme 6-15 A metallosalen template for catalytic epoxidations.

Not all potential substrates can easily have nicotinate groups attached to both ends. Thus we set out to use hydrophobic binding into cyclodextrins in water as the mechanism for inducing formation of a complex between catalyst and substrate with geometric control and catalytic turnover. In the earliest work we have nonetheless attached groups to the substrates to solubilize them and promote their binding into cyclodextrins.

We made porphyrins carrying four phenyl groups attached to the *meso* positions in which we were able to attach four β-cyclodextrin units to the *para* position of each phenyl (**34**), or two cyclodextrins attached to the para positions of neighboring phenyls (**35**) or of those on opposite sides of the porphyrin unit (**36**) (Scheme-6-16) [72]. The three

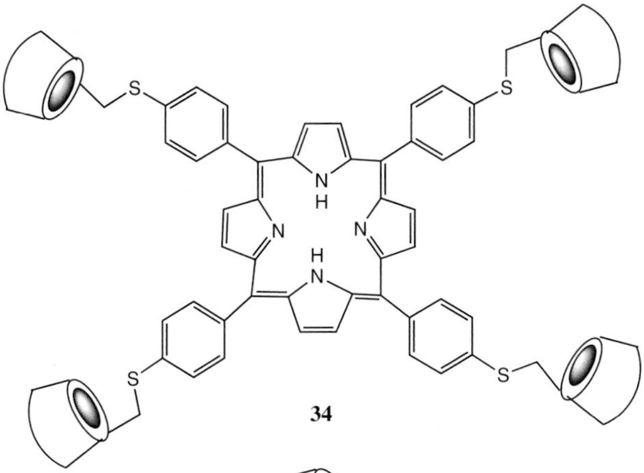

34

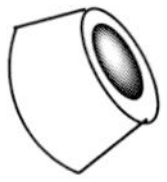

Beta-cyclodextrin

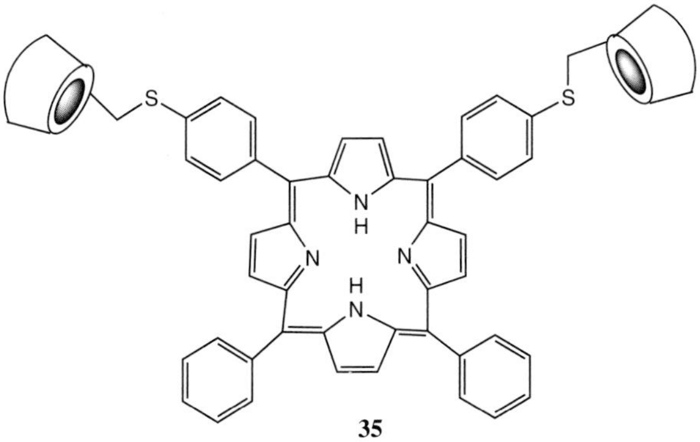

35

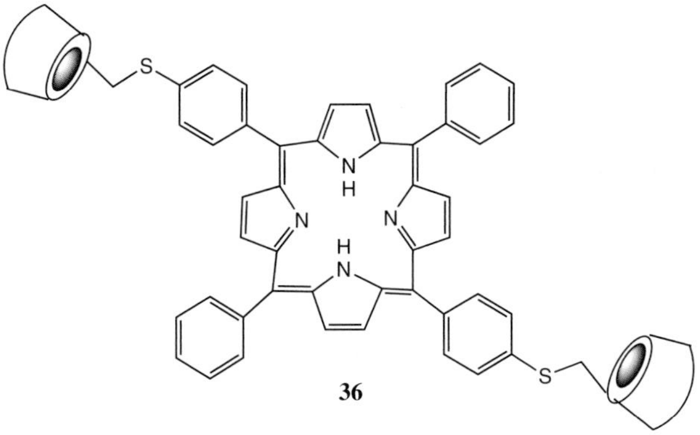

36

Scheme 6-16   Porphyrins
34–36 with appended β-
cyclodextrins catalyze the
epoxidation of double bonds
in substrates with hydro-
phobic termini which are
bound in the cyclodextrin
cavities.

porphyrin derivatives, as their Mn^{3+} complexes, were examined as catalysts for the epoxidation of double bonds in substrates that had hydrophobic ends to bind into cyclodextrin rings in water solution. We found that good catalysis and selectivity were seen when the catalyst carried two cyclodextrins on opposite sides of the ring (Mn^{3+} derivatives of **34** or **36**), but not when they were on neighboring phenyls (Mn^{3+} derivative of **35**). Thus the prediction from molecular models that the substrates would be attacked only if they stretched across the porphyrin ring was borne out. For convenience, the catalyst based on **34** with four cyclodextrins was used for most of the studies, on the assumption that its productive complex was the one in which the substrates would bind to cyclodextrins on opposite sides so as to stretch the substrate across the porphyrin ring. As in the Cu^{2+} linked systems described above, we saw that there was selective turnover epoxidation of those substrates with hydrophobic ends that would bind to the cyclodextrins.

These catalysts were then examined in the selective hydroxylation of saturated unactivated carbons in steroids [73, 74]. Androstane-3,17-diol **37** was converted to a diester **38** carrying *tert*-butylphenyl binding groups and sulfonate solubilizing groups, and oxidized in water solution with iodosobenzene catalyzed by the Mn^{3+}-porphyrin **39** carrying a β-cyclodextrin attached to the *para* position of four phenyls on the porphyrin *meso* positions (Scheme 6-17). The steroid was hydroxylated only on the 6α-position to produce **40**, and with about four turnovers. This was hydrolyzed to the androstanetriol **41**. Control studies in which the *tert*-butylphenyl groups were not present on the substrate showed that there was no oxidation at all under the same conditions, so binding is required. A cholestanol carrying only one binding group was hydroxylated, but to a mixture of products. When the catalyst contained Fe^{2+} rather than Mn^{3+} similar results were obtained, but with much less turnover.

Catalytic turnover is limited by the oxidative destruction of the porphyrin, and prior work showed how to diminish this problem. It had been found [75] that electron-withdrawing groups on the porphyrin carbons stabilized it to oxidation, and in particular porphyrins carrying pentafluorophenyl groups in the *meso* positions were much more stable than those with unsubstituted phenyls. Thus we synthesized a new catalyst **42** with fluorinated phenyls (Scheme 6-18) [76]. The synthesis was particularly convenient, since we found that the readily available porphyrin **43** carrying pentafluorophenyl groups on each *meso* position reacts with β-cyclodextrinthiol to replace a *para* fluorine in each phenyl ring, easily attaching a cyclodextrin in its place and forming **44**. Its Mn^{3+} complex **42** still showed the same selectivity as that of the unfluorinated compound, but now with 187 turnovers before the catalyst was destroyed.

Further work is under way in this area. Substrates are being examined that directly bind with defined geometry into our catalysts, without the need to attach binding groups. Geometries are being varied in order to carry out other selective steroid hydroxylations, including those of direct pharmaceutical interest. Other oxidants, and still more stable catalysts, are being explored. Catalysts with more than two-point binding, and with an ability to hydroxylate unactivated carbons even in the presence of sensitive double bonds, are being examined. The potential of these biomimetic oxidizing catalysts seems quite substantial.

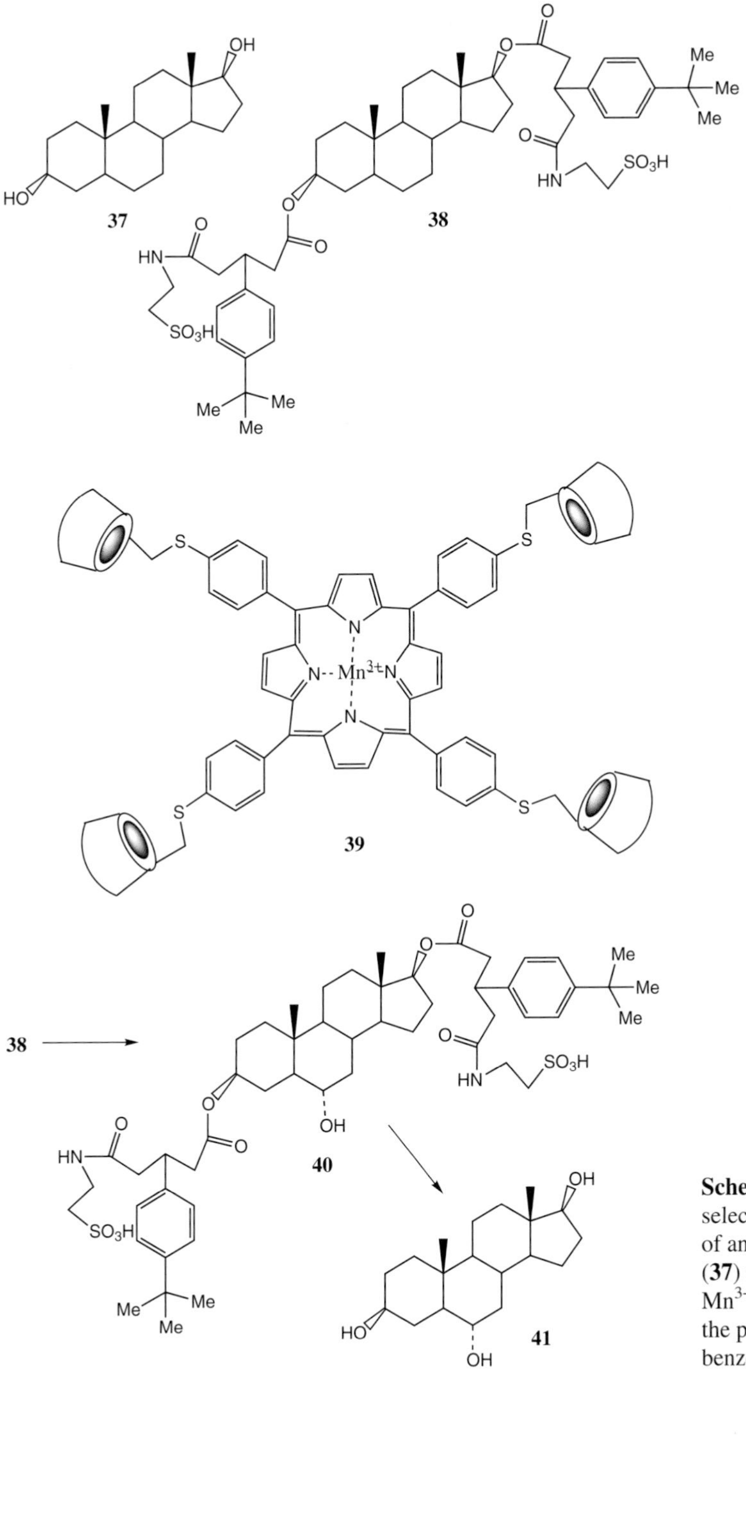

Scheme 6-17   Regio-selective hydroxylation of androstane-3,17-diol (**37**) catalyzed by the Mn^{3+}-porphyrin **39** in the presence of iodoso-benzene.

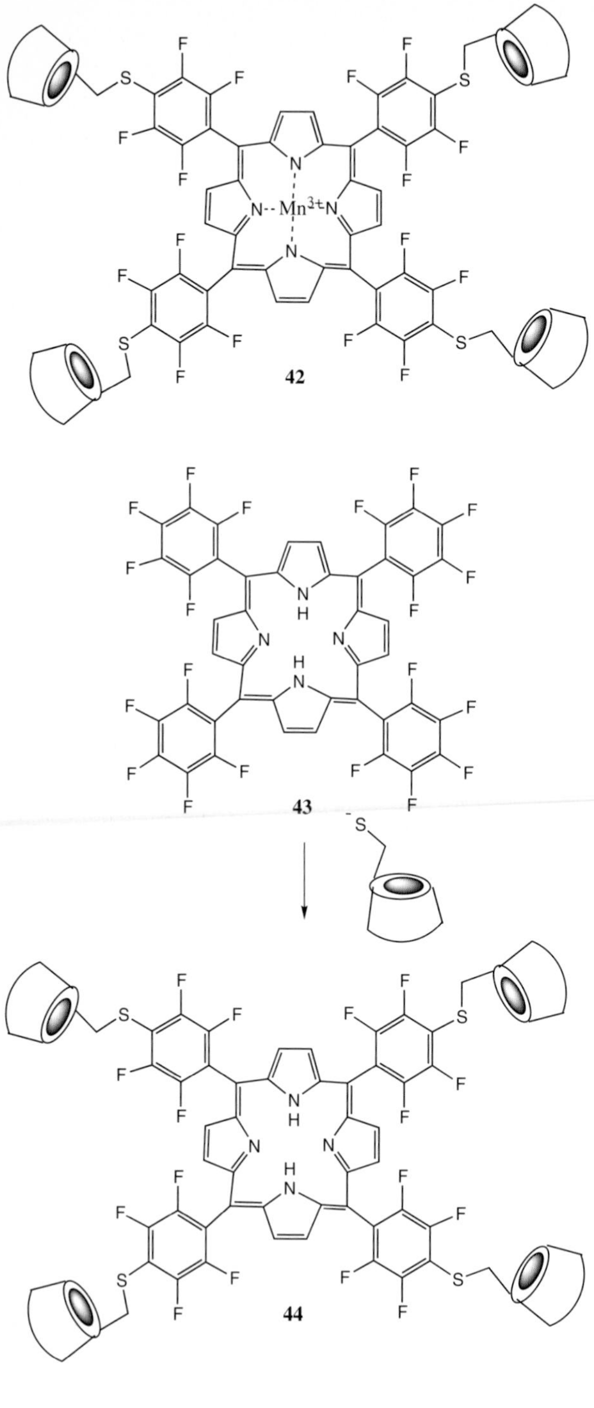

Scheme 6-18 The highly stabilized catalyst **42** is readily prepared from **43** via **44**.

The transfer of oxygen atoms by porphyrin catalysts is of course quite appealing, but it occurred to us some time ago that selective insertion of nitrogen atoms into substrates could also be useful, and accomplished in the same way. Thus we examined the nitrogen analogue of iodosobenzene, tosyliminoiodobenzene **45** [77]. It was able to insert the tosylamine group into cyclohexane solvent to form **46**, catalyzed by manganese or iron tetraphenylporphyrin (Scheme 6-19). In a collaborative study, it was found that even some isozymes of cytochrome P-450 were able to catalyze that reaction [78]. This work has stimulated others to take up the use of tosyliminoiodobenzene and related compounds for nitrogen functionalization reactions.

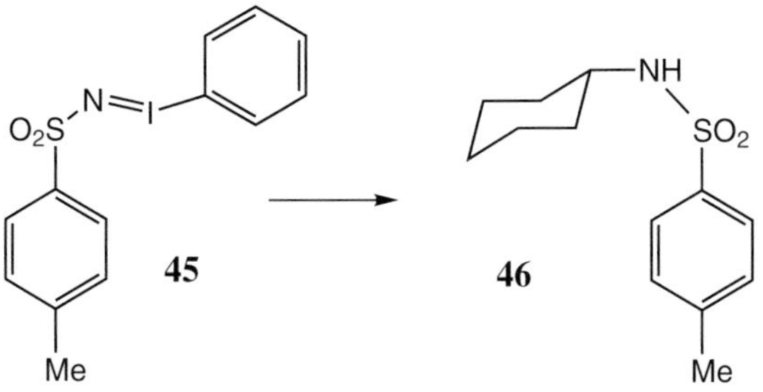

Scheme 6-19   N-atom transfer catalyzed by manganese or iron tetraphenylporphyrin.

6.3.4   Other Reactions Catalyzed by Coordinated Template Catalysts

Intramolecular catalyses of cleavage reactions at carboxylic acid derivatives are well known, and outside the scope of this book. However, cleavage reactions can also be catalyzed by species that bind to the substrate and carry out a selective reaction that depends on the geometry of the mixed complex. Since this is in the spirit of the rest of what has been described, it will be briefly included here. Such reactions imitate enzymatic cleavage processes, so they are "biomimetic" in the same sense that selective functionalizations within mixed complexes are biomimetic [1]. The work using cyclodextrin binding has been extensively reviewed recently [30].

Esters can be cleaved by template catalysts that use a metal ion as both a binding group and part of the catalytic system [79–81]. However, metal ion catalysis has also been extended to cases in which the principal substrate binding involves cyclodextrin inclusion; indeed, the first catalyst described as an artificial enzyme was such an example [82]. A cyclodextrin dimer **47** with a bound metal ion between the two cyclodextrins is a particularly effective hydrolytic catalyst for esters that can bind into both cyclodextrin units (Scheme 6-20) [83, 84].

Esters can also be cleaved by reaction with the hydroxyl group of cyclodextrin within a complex [85–91]. Some very high rates have been observed, and calculations account for the geometric preferences in such reactions.

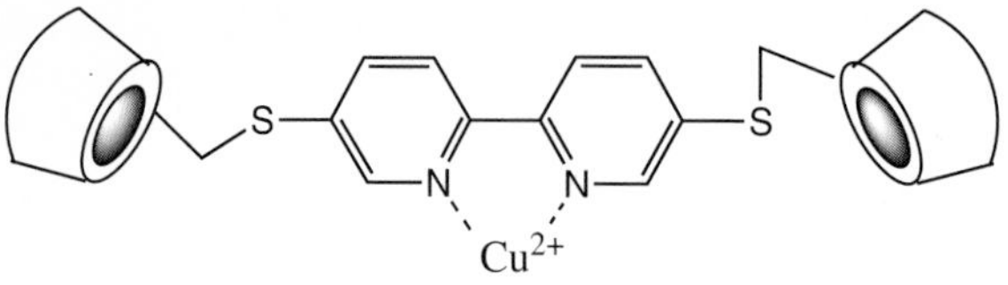

47

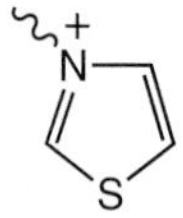

48 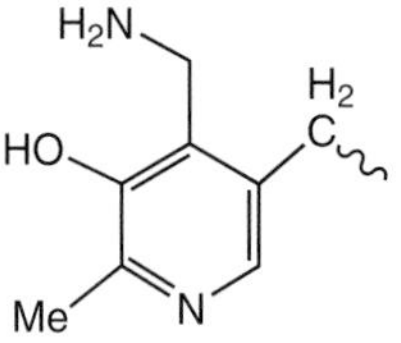   **49** 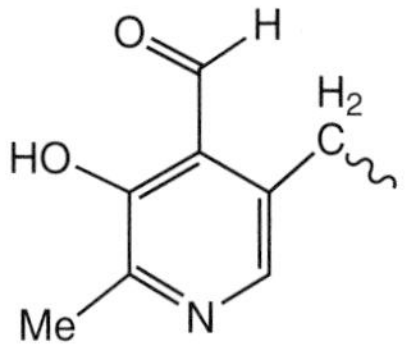   **50**

Scheme 6-20   Coordinated template catalyst **47** and cofactor analogues attached covalently to cyclodextrins in biomimetic catalysis.

Phosphate esters can be cleaved by template catalysts, especially those with cyclodextrin binding groups and linked catalytic groups. Catalysis of the hydrolysis of a bound cyclic phosphate by ribonuclease mimics has been extensively studied [92–98], as has catalysis by enzyme mimics carrying bound metal ions [99–102].

When a coenzyme is linked to a binding group, sometimes with additional catalytic groups, interesting enzyme mimics result. A mimic carrying linked thiazolium ions **48** can perform reactions like those catalyzed in enzymes that use thiamine pyrophosphate [103, 104], while mimics carrying pyridoxamine (**49**) and pyridoxal (**50**) units mimic the amino acid biochemistry that uses pyridoxal phosphate and pyridoxamine phosphate as coenzyme [105–115]. Mimics of enzymes that use coenzyme B-12 have also been made [116, 117], as well as aldolase mimics [118–121]. The double binding of substrates into two cyclodextrin rings that is involved in a number of catalytic systems has also been examined in itself, in studies directed at determining binding constants [122–124]. Remarkably, the chelate effect that leads to very strong binding in these cases was found to be caused by an advantage in *enthalpy*, not entropy, relative to singly binding analogues [123].

The use of tethers and templates in the chemistry described above can furnish valuable information about the conformations of flexible molecules, and in appropriate cases can lead to selective functionalization reactions, including those at unactivated positions. Thus these reactions are "biomimetic," imposing geometric control over the normal substrate reactivity control. It remains for the future to develop such reactions, and catalysts, to the point at which they take their place in the tools of synthesis for research and manufacturing. However, such progress seems likely, judging from the progress that has been made to date.

References

[1] R. Breslow, *Chem. Soc. Rev.* **1972**, *1*, 553–580.

[2] R. Breslow, *J. Steroid Biochem.* Part A, **1979**, *11*, 19–26.

[3] R. Breslow, *Isr. J. Chem.* **1979**, *18*, 187–191.

[4] R. Breslow, *Acc. Chem. Res.* **1980**, *13*, 170–177.

[5] R. Breslow, in D. Dolphin, C. McKenna, Y. Murakami and I. Tabushi (Eds.): *Biomimetic Chemistry*, ACS, Washington, DC **1980**, p. 1–15.

[6] R. Breslow, *Science* **1982**, *218*, 532–537.

[7] R. Breslow, *Chem. Brit.* **1983**, *19*, 126–131.

[8] R. Breslow, in J. L. Atwood, J. E. D. Davies (Eds.): *Inclusion Compounds, Vol. 3*, Academic Press, Orlando, FL, **1984**, pp. 473–508.

[9] R. Breslow, in R. Setton (Ed.): *Chemical Reactions in Organic and Inorganic Constrained Systems*, NATO ASI Ser., Ser. C, D. Reidel, **1986**, pp. 17–28.

[10] R. Breslow, in A. Meister (Ed.): *Adv. Enzymol. Rel. Areas Mol. Biol.,Vol. 58*, John Wiley, **1986**, pp. 1–60.

[11] R. Breslow, *Ann. N. Y. Acad. Sci.* **1986**, *471*, 60–69.

[12] R. Breslow, in Z. Yoshida (Ed.): *New Synthetic Methodology and Functionally Interesting Compounds*, Kodansha, Tokyo, **1986**, pp. 423–439.

[13] R. Breslow, in G. v. Binst (Ed.): *Proceedings of the XVIIIth Solvay Conference on Chemistry*, Springer, Berlin, **1986**, pp. 185–197.

[14] R. Breslow, *Cold Spring Harbor Symposia on Quantitative Biology* **1987**, *52*, 75–81.

[15] R. Breslow, *Proc. Robert A. Welch Found. Conf. Chem. Res.* **1988**, *31*, pp. 73–89.

[16] R. Breslow, *Chemtracts: Org. Chem.* **1988**, *1*, 333–348.

[17] R. Breslow, in C. S. Craik, R. Fletterick, C. R. Matthews, J. Wells (Eds.): *UCLA Symp. Mol. Cell. Biol.* **1990**, *110*, pp. 135–144.

[18] R. Breslow, *Ciba Foundation Symposium* **1991**, *158*, 115–127.

[19] R. Breslow, in B. M. Trost (Ed.): *Comprehensive Organic Synthesis, Vol. 7*, Pergamon, Oxford, **1991**, p. 39–52.

[20] R. Breslow, *Supramol. Chem.* **1992**, 411–428.

[21] R. Breslow, *Isr. J. Chem.* **1992**, *32*, 23–30.

[22] R. Breslow, *Minutes of the Sixth International Symposium on Cyclodextrins*, Editions de Santé, **1992**, pp. 625–630.

[23] R. Breslow, *Supramol. Chem.* **1993**, *1*, 111–118.

[24] R. Breslow, *Pure Appl. Chem.* **1994**, *66*, 1573–1582.

[25] R. Breslow, *Recl. Trav. Chim. Pays-Bas* **1994**, *113*, 493–498.

[26] R. Breslow, *Acc. Chem. Res.* **1995**, *28*, 146–153.

[27] R. Breslow, *Supramol. Chem.* **1995**, *6*, 41–47.

[28] R. Breslow, in C. Chatgilialoglu, V. Snieckus (Eds.): *Chemical Synthesis – Gnosis to Prognosis, NATO ASI Ser., Ser. E*, **1996**, *320*, 113–135.

[29] R. Breslow, *Pure Appl. Chem.* **1998**, *70*, 267–270.

[30] R. Breslow, S. D. Dong, *Chem. Rev.* **1998**, *98*, 1997–2011.

[31] R. Breslow, M. A. Winnik, *J. Am. Chem. Soc.* **1969**, *91*, 3083–3084.

[32] R. Breslow, S. W. Baldwin, *J. Am. Chem. Soc.* **1970**, *92*, 732–734.

[33] R. Breslow, P. Kalicky, *J. Am. Chem. Soc.* **1971**, *93*, 3540–3541.

[34] R. Breslow, S. Baldwin, T. Flechtner, P. Kalicky, S. Liu, W. Washburn, *J. Am. Chem. Soc.* **1973**, *95*, 3251–3262.

[35] R. L. Wife, D. Prezant, R. Breslow, *Tetrahedron Lett.* **1976**, 517–520.

[36] R. Breslow, R. Corcoran, J. A. Dale, S. Liu, P. Kalicky, *J. Am. Chem. Soc.* **1974**, *96*, 1973–1974.

[37] R. Breslow, R. J. Corcoran, B. B. Snider, *J. Am. Chem. Soc.* **1974**, *96*, 6791–6792.

[38] R. Breslow, *Organic Reaction Mechanisms*, W. A. Benjamin, New York, **1969**.

[39] P. White, R. Breslow, *J. Am. Chem. Soc.* **1990**, *112*, 6842–6847.

[40] R. Breslow, B. B. Snider, R. J. Corcoran, *J. Am. Chem. Soc.* **1974**, *96*, 6792–6794.

[41] B. B. Snider, R. J. Corcoran, R. Breslow, *J. Am. Chem. Soc.* **1975**, *97*, 6580–6581.

[42] R. Breslow, R. J. Corcoran, B. B. Snider, R. J. Doll, P. L. Khanna, R. Kaleya, *J. Am. Chem. Soc.* **1977**, *99*, 905–915.

[43] R. Breslow, T. Link, *Tetrahedron Lett.* **1992**, *33*, 4145–4148.

[44] R. Batra, R. Breslow, *Tetrahedron Lett.* **1989**, *30*, 535–538.

[45] U. Maitra, R. Breslow, *Tetrahedron Lett.* **1986**, *27*, 3087–3090.

[46] R. Breslow, U. Maitra, D. Heyer, *Tetrahedron Lett.* **1984**, *25*, 1123–1126.

[47] R. Breslow, R. L. Wife, D. Prezant, *Tetrahedron Lett.* **1976**, 1925–1926.

[48] R. Breslow, T. Guo, *Tetrahedron Lett.* **1987**, *28*, 3187–3188.

[49] R. Breslow, D. Heyer, *J. Am. Chem. Soc.* **1982**, *104*, 2045–2046.

[50] R. Breslow, M. Brandl, J. Hunger, A. D. Adams, *J. Am. Chem. Soc.* **1987**, *109*, 3799–3801.

[51] R. Breslow, A. Adams, M. Brandl, T. Guo, J. Hunger, *Lec. Heterocycl. Chem.* **1987**, *9*, 43–49.

[52] R. Breslow, M. Brandl, J. Hunger, N. J. Turro, K. Cassidy, K. Krogh-Jespersen, J. D. West-brook, *J. Am. Chem. Soc.* **1987**, *109*, 7204–7206.

[53] R. Breslow, D. Wiedenfeld, *Tetrahedron Lett.* **1993**, *34*, 1107–1110.

[54] R. Breslow, R. Goodin, *Tetrahedron Lett.* **1976**, *31*, 2675–2676.

[55] D. Wiedenfeld, R. Breslow, *J. Am. Chem. Soc.* **1991**, *113*, 8977–8978.

[56] R. Breslow, J. Rothbard, F. Herman, M. L. Rodriguez, *J. Am. Chem. Soc.* **1978**, *100*, 1213–1218.

[57] R. Batra, R. Breslow, *Heterocycles* **1989**, *28*, 23–28.

[58] R. Breslow, D. Heyer, *Tetrahedron Lett.* **1983**, *24*, 5039–5042.

[59] K. B. Sharpless, R. C. Michaelson, *J. Am. Chem. Soc.* **1973**, *95*, 6136–6137.

[60] R. Hiatt, in R. C. Augustine, D. L. Trecker (Eds.): *Oxidation, Vol. 2*, Marcel Dekker, New York, **1971**.

[61] R. Breslow, L. M. Maresca, *Tetrahedron Lett.* **1977**, 623–626.

[62] R. Breslow, L. M. Maresca, *Tetrahedron Lett.* **1978**, 887–890.

[63] R. Breslow, P. Campbell, *J. Am. Chem. Soc.* **1969**, *91*, 3085.

[64] R. Breslow, P. Campbell, *Bioorg. Chem.* **1971**, *1*, 140–146.

[65] F. S. Brown, L. P. Hager, *J. Am. Chem. Soc.* **1967**, *89*, 719–720.

[66] R. Breslow, H. Kohn, B. Siegel, *Tetrahedron Lett.* **1976**, 1645–1646.

[67] R. Breslow, P. C. Scholl, *J. Am. Chem. Soc.* **1971**, *93*, 2331–2333.

[68] R. Breslow, S. Kitabatake, J. Rothbard, *J. Am. Chem. Soc.* **1978**, *100*, 8156–8160.

[69] M. F. Czarniecki, R. Breslow, *J. Am. Chem. Soc.* **1979**, *101*, 3675–3676.

[70] R. Breslow, R. Rajagopalan, J. Schwarz, *J. Am. Chem. Soc.* **1981**, *103*, 2905–2907.

[71] R. Breslow, A. B. Brown, R. D. McCullough, P. W. White, *J. Am. Chem. Soc.* **1989**, *111*, 4517–4518.

[72] R. Breslow, X. Zhang, R. Xu, M. Maletic, R. Merger, *J. Am. Chem. Soc.* **1996**, *118*, 11678–11679.

[73] R. Breslow, X. Zhang, Y. Huang, *J. Am. Chem. Soc.* **1997**, *119*, 4535–4536.

[74] R. Breslow, Y. Huang, X. Zhang, J. Yang, *Proc. Natl. Acad. Sci. USA* **1997**, *94*, 11156–11158.

[75]   P. E. Ellis, J. E. Lyons, *Coord. Chem. Rev.* **1990**, *105*, 181–193.
[76]   R. Breslow, B. Gabriele, J. Yang, *Tetrahedron Lett.* **1998**, *39*, 2887–2890.
[77]   R. Breslow, S. H. Gellman, *J. Chem. Soc. Chem. Commun.* **1982**, 1400–1401.
[78]   E. W. Svastits, J. H. Dawson, R. Breslow, S. Gellman, *J. Am. Chem. Soc.* **1985**, *107*, 6427–6428.
[79]   R. Breslow, D. Chipman, *J. Am. Chem. Soc.* **1965**, *87*, 4195–4196.
[80]   R. Breslow, M. Schmir, *J. Am. Chem. Soc.* **1971**, *93*, 4960–4961.
[81]   R. Breslow, D. Berger, D.-L. Huang, *J. Am. Chem. Soc.* **1990**, *112*, 3686–3687.
[82]   R. Breslow, L. E. Overman, *J. Am. Chem. Soc.* **1970**, *92*, 1075–1077.
[83]   R. Breslow, B. Zhang, *J. Am. Chem. Soc.* **1992**, *114*, 5882–5883.
[84]   B. Zhang, R. Breslow, *J. Am. Chem. Soc.* **1997**, *119*, 1676–1681.
[85]   M. F. Czarniecki, R. Breslow, *J. Am. Chem. Soc.* **1978**, *100*, 7771–7773.
[86]   R. Breslow, M. F. Czarniecki, J. Emert, H. Hamaguchi, *J. Am. Chem. Soc.* **1980**, *102*, 762–770.
[87]   G. L. Trainor, R. Breslow, *J. Am. Chem. Soc.* **1981**, *103*, 154–158.
[88]   R. Breslow, G. Trainor, A. Ueno, *J. Am. Chem. Soc.* **1983**, *105*, 2739–2744.
[89]   W. J. le Noble, S. Srivastava, R. Breslow, G. Trainor, *J. Am. Chem. Soc.* **1983**, *105*, 2745–2748.
[90]   H.-J. Thiem, M. Brandl, R. Breslow, *J. Am. Chem. Soc.* **1988**, *110*, 8612–8616.
[91]   R. Breslow, S. Chung, *Tetrahedron Lett.* **1990**, *31*, 631–634.
[92]   R. Breslow, J. B. Doherty, G. Guillot, C. Lipsey, *J. Am. Chem. Soc.* **1978**, *100*, 3227–3229.
[93]   R. Breslow, P. Bovy, C. L. Hersh, *J. Am. Chem. Soc.* **1980**, *102*, 2115–2117.
[94]   E. Anslyn, R. Breslow, *J. Am. Chem. Soc.* **1989**, *111*, 5972–5973.
[95]   E. Anslyn, R. Breslow, *J. Am. Chem. Soc.* **1989**, *111* 8931–8932.
[96]   R. Breslow, E. Anslyn, D.-L. Huang, *Tetrahedron* **1991**, *47*, 2365–2376.
[97]   R. Breslow, *J. Mol. Cat.* **1994**, *91*, 161–174.
[98]   R. Breslow, C. Schmuck, *J. Am. Chem. Soc.* **1996**, *118*, 6601–6605.
[99]   S. Gellman, R. Petter, R. Breslow, *J. Am. Chem. Soc.* **1986**, *108*, 2388–2394.
[100] R. Breslow, S. Singh, *Bioorg. Chem.* **1988**, *16*, 408–417.
[101] R. Breslow, B. Zhang, *J. Am. Chem. Soc.* **1994**, *116*, 7893–7894.
[102] W. H. Chapman, Jr., R. Breslow, *J. Am. Chem. Soc.* **1995**, *117*, 5462–5469.
[103] D. Hilvert, R. Breslow, *Bioorg. Chem.* **1984**, *12*, 206–220.
[104] R. Breslow, E. Kool, *Tetrahedron Lett.* **1988**, *29*, 1635–1638.
[105] R. Breslow, M. Hammond, M. Lauer, *J. Am. Chem. Soc.* **1980**, *102*, 421–422.
[106] R. Breslow, A. W. Czarnik, *J. Am. Chem. Soc.* **1983**, *105*, 1390–1391.
[107] S. C. Zimmerman, A. W. Czarnik, R. Breslow, *J. Am. Chem. Soc.* **1983**, *105*, 1694–1695.
[108] J. Winkler, E. Coutouli-Argyropoulou, R. Leppkes, R. Breslow, *J. Am. Chem. Soc.* **1983**, *105*, 7198–7199.
[109] S. C. Zimmerman, R. Breslow, *J. Am. Chem. Soc.* **1984**, *106*, 1490–1491.
[110] A. W. Czarnik, R. Breslow, *Carbohydrate Res.* **1984**, *128*, 133–139.
[111] W. Weiner, J. Winkler, S. C. Zimmerman, A. W. Czarnik, R. Breslow, *J. Am. Chem. Soc.* **1985**, *107*, 4093–4094.
[112] R. Breslow, A. W. Czarnik, M. Lauer, R. Leppkes, J. Winkler, S. Zimmerman, *J. Am. Chem. Soc.* **1986**, *108*, 1969–1979.
[113] R. Breslow, J. Chmielewski, D. Foley, B. Johnson, N. Kumabe, M. Varney, R. Mehra, *Tetrahedron* **1988**, *44*, 5515–5524.
[114] R. Breslow, J. W. Canary, M. Varney, S. T. Waddell, D. Yang, *J. Am. Chem. Soc.* **1990**, *112*, 5212–5219.

[115] J. T. Koh, L. Delaude, R. Breslow, *J. Am. Chem. Soc.* **1994**, *116*, 11234–11240.
[116] R. Breslow, P. J. Duggan, J. P. Light, *J. Am. Chem. Soc.* **1992**, *114*, 3982–3983.
[117] M. Rezac, R. Breslow, *Tetrahedron Lett.* **1997**, *38*, 5763–5766.
[118] R. Breslow, A. Graff, *J. Am. Chem. Soc.* **1993**, *115*, 10988–10989.
[119] J. M. Desper, R. Breslow, *J. Am. Chem. Soc.* **1994**, *116*, 12081–12082.
[120] R. Breslow, J. Desper, Y. Huang, *Tetrahedron Lett.* **1996**, *37*, 2541–2544.
[121] D.-Q. Yuan, S. D. Dong, R. Breslow, *Tetrahedron Lett.* **1998**, *39*, 7673–7676.
[122] R. Breslow, N. Greenspoon, T. Guo, R. Zarzycki, *J. Am. Chem. Soc.* **1989**, *111*, 8296–8297.
[123] B. Zhang, R. Breslow, *J. Am. Chem. Soc.* **1993**, *115*, 9353–9354.
[124] R. Breslow, S. Halfon, B. Zhang, *Tetrahedron* **1995**, *51*, 377–388.

7 Regio- and Stereoselective Multiple Functionalization of Fullerenes

François Diederich, Roland Kessinger

7.1 Introduction

The synthesis of multiple adducts of C_{60} has attracted much attention in recent years [1]. Investigations of covalent derivatives with varying degrees and patterns of addition make it possible to explore how characteristic fullerene properties are affected by functionalization-induced changes in the conjugated π-chromophore [2]. Furthermore, higher adducts of C_{60} represent an unprecedented family of three-dimensional building blocks for molecular scaffolding targeting advanced fullerene materials and technologies [3]. The direct sequential multiple functionalization of the carbon sphere, however, is problematic in most cases, and isolation of a pure higher adduct usually requires tedious chromatographic isomer separation [4–6]. Moreover, a great diversity of interesting addition patterns is not accessible by sequential derivatization. Therefore, we proposed in 1994 a more general approach to the regioselective formation of multiple adducts of C_{60} by taking advantage of tether-directed remote functionalization techniques [7]. Such strategies had been initially developed by Breslow and co-workers for the selective functionalization of steroids and long-chain alkanes (see Chapter 6) [8]. Here, we illustrate how the application of non-covalent and covalent templates provides access to a great diversity of three-dimensionally functionalized fullerene building blocks. Templated synthesis now serves as a powerful tool for chemists interested in exploiting the outstanding technological potential of fullerenes and their covalent derivatives.

7.2 Anthracenes as Reversible Covalent Templates

Among the most versatile reactions in fullerene chemistry is the Bingel cyclopropanation with 2-bromomalonate in the presence of base [9]. The problematics of regioisomeric product formation in nucleophilic additions to C_{60} is illustrated in the following, using this reaction as an example. By sequential double Bingel addition, Hirsch and co-workers obtained seven (**1**–**7**) out of eight possible bis-adducts named as *cis*-1–3, e_{face} and e_{edge} (these two bis-adducts are identical if the two addends are identical), and *trans*-1–4 (Scheme 7-1) [10, 11], which could be separated by high-performance liquid chromatography (HPLC). Only the *cis*-1 bis-adduct was not formed in detectable quantities due to steric interactions between the addends which are forced into close proximity. The *e* (**3**) and *trans*-3 (**5**) bis-adducts are the major products in the sequential double addition, which is kinetically controlled. The coefficients of the two nearly degenerate lowest unoccupied molecular orbitals (LUMO and LUMO+1) of the monomethanofullerene, which are accepting the electron density from the incoming nucleophile, determine the product ratio of the bis-adducts. These orbitals have their highest coefficients at the *e* and *trans*-3 positions [12, 13].

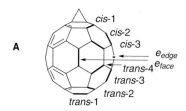

A

cis-1
cis-2
cis-3
e_{edge}
trans-4 e_{face}
trans-3
trans-2
trans-1

B

a)

1-7

1 (*cis*-2);	0.9%
2 (*cis*-3);	2.5%
3 (*e*);	15.5%
4 (*trans*-4);	3.7%
5 (*trans*-3);	12.0%
6 (*trans*-2);	5.3%
7 (*trans*-1);	0.8%

Scheme 7-1   (A) Position notation for bis-adducts of C_{60} according to Hirsch et al. [10]. (B) Products isolated from the second Bingel addition to bis(diethoxycarbonyl)methanofullerene $[C_{61}(COOEt)_2]$ [10]. a) Diethyl 2-bromomalonate, NaH, toluene, 20 °C.

Upon further functionalization, the regioselectivity of the additions becomes remarkably enhanced. A third di(ethoxycarbonyl)methano addend is preferentially introduced (40%) into the equatorial (*e*) position. Starting from *e,e,e*-tris-adduct, a remarkable sequence of stepwise *e* additions, with regard to the previous addends, affords hexakis-adduct **8** [12]. This compound features a pseudo-octahedral addition pattern and can also be directly obtained in 14% yield by reacting C_{60} with 8 equiv of diethyl 2-bromomalonate/1,8-diazabicyclo[5.4.0]undec-7-ene (DBU) [12].

Hirsch and co-workers subsequently discovered that the yield of hexakis-adducts such as **8** or **9** can be greatly increased and their separation from side products substantially facilitated, if 9,10-dimethylanthracene (DMA) is used as a template (Scheme 7-2) [14, 15]. Anthracene derivatives such as DMA are well known to undergo reversible Diels–Alder additions with fullerenes at ambient temperature [16–18]. After addition to the fullerene, the template directs diethyl malonate addends in the Bingel addition regioselectively into *e* positions, ultimately yielding the hexakis-adducts with a pseudo-octahedral, all-*e* addition pattern. The templated activation of *e* 6–6 bonds (bonds between two six-membered rings) is also efficient starting from C_{60} mono-adducts, and several examples are shown in Scheme 7-2. Thus, bis(alkynyl)methanofullerene **10** [19] reacted with diethyl 2-bromomalonate/DBU (8 equiv) and DMA (12 equiv) to provide hexakis-adduct **11** in 28% yield [20]. Without a template, the yield was only 12% and the targeted product was contaminated with an inseparable impurity. Also, mono-adduct **12** was transformed in 35% yield into hexakis-adduct **13** [21], the precursor for the Pt^{II} directed self-assembly of a dinuclear cyclophane of the molecular-square type [22], containing two appended fullerenes. Other examples for the use of this template activation method include the formation of hexakis-adducts **14**–**16** [14, 23, 24], with compound **16** providing an interesting example of a dendrimer with a fully buried fullerene core [25–27].

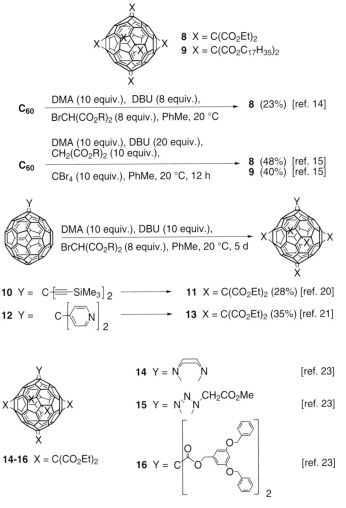

Scheme 7-2 Templated synthesis of hexakis-adducts **8** [14,15], **9** [15], **11** [20], **13** [21], **14** [23], **15** [23] and **16** [24] using 9,10-dimethylanthracene (DMA) as a template. DBU = 1,8-diazabicyclo [5.4.0]undec-7-ene.

A spectacular example for a solid-state template effect was reported by Kräutler et al. [28]. In their elegant preparation of *trans*-1 bis-anthracene adduct **17**, the crystal packing provided the molding effect characteristic of a template. Heating a sample of crystalline mono-adduct **18** at 180 °C for 10 min afforded 48% each of C$_{60}$ and bis-adduct **17**, resulting from a topochemically controlled, regioselective intermolecular anthracene-transfer reaction (Scheme 7-3). The two anthracene addends in **17** were subsequently used as covalent templates in further solution conversions [29]. The two residues regio-selectively directed four new malonate addends regiospecifically into the *e* positions, yielding hexakis-adduct **19**. Thermal removal of the anthracene templates eventually provided tetrakis-adduct **20**, with all four malonate addends aligned along an equatorial belt on the carbon sphere [29, 30].

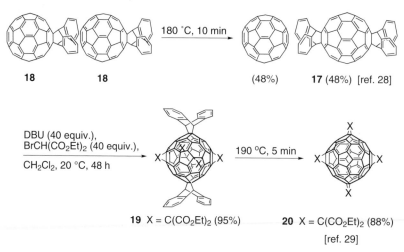

18   **18**   (48%)   **17** (48%) [ref. 28]

DBU (40 equiv.),
BrCH(CO$_2$Et)$_2$ (40 equiv.),
——————————————
CH$_2$Cl$_2$, 20 °C, 48 h

190 °C, 5 min

19 X = C(CO$_2$Et)$_2$ (95%)   **20** X = C(CO$_2$Et)$_2$ (88%)
[ref. 29]

Scheme 7-3   Topochemical solid-state synthesis of bis-adduct **17** [28] and templated formation of tetrakis-adduct **20** [29].

7.3   Tether-directed Remote Functionalizations of C$_{60}$

Prior to the development of tether-directed functionalization methods, regioisomerically pure higher adducts of C$_{60}$ usually were obtained by additions of transition metal complexes [31–33] or radical halogenations [34, 35]. These reactions either occur under thermodynamic control or lead to the precipitation of the least soluble derivative. Iso-merically pure higher adducts of C$_{60}$ sometimes are also readily isolated out of more complex product mixtures [36]. Tether-directed remote functionalization of C$_{60}$ allows the construction of fullerene derivatives with addition patterns that are difficult to obtain by thermodynamically or kinetically controlled reactions with free untethered reagents. Since the description of the first such reaction in 1994 [7], which is the subject of Section 7.3.1, an increasing variety of such regioselective functionalization protocols have been developed, as documented in this overview. Of course, a tether-directed remote functionalization should only be called a template-directed reaction if the tether is removable, and this criterion is met by nearly all the examples described below. The benefits of these synthetic developments for the advancement of fullerene chemistry have been enormous. Thus, the availability of a great diversity of multiple adducts allowed a detailed investigation of the changes in the unique properties of parent C$_{60}$ that occur with increasing degree of covalent functionalization and concomitant reduction of the fullerene chromophore [2, 37–39]. Similarly, effects of the nature of the addition patterns and the nature of the addends on fullerene properties could be probed as well. Examples of properties that change profoundly upon increasing the functionalization of C$_{60}$ are the facile reversible electrochemical reducibility [2, 40, 41], the efficiency in singlet oxygen photosensitization [2, 42, 43], or the reactivity of the carbon sphere against nucleophilic attack [2]. Furthermore, higher adducts of C$_{60}$ are finding increasing application in the construction of supramolecular advanced materials [3]. They also have been of great importance in the exploration of the different origins of fullerene chirality [44–46]. In line with the topic of this monograph, the following chapters will mainly focus on the preparative aspects of tether-directed remote functionalizations and the reader is referred to the cited literature for more details on the physico-chemical investigations made possible by these synthetic methodology developments.

7.3.1 The First Tether-directed Multiple Functionalization of C$_{60}$: Formation of Higher Adducts with Novel Addition Patterns

To achieve the regioselective formation of an *e* bis-adduct of C$_{60}$, methano[60]fullerenecarboxylic acid **21** as an anchor was reacted with the tether-reactive group conjugate **22** (Scheme 7-4) [7, 47]. The particular reactive group in conjugate **22** was chosen since 2-substituted buta-1,3-dienes are known to readily undergo irreversible Diels–Alder additions with C$_{60}$ at 6–6 bonds [48]. The design of the tether was based on molecular model examinations and semi-empirical PM3 calculations (PM3 = parametric method 3) [49, 50] which suggested that mono-adduct **23**, in which conjugate **22** had been attached to anchor **21**, would readily differentiate in a Diels–Alder addition between the desired position e_{face} (Scheme 7-1) and adjacent positions such as the *cis*-3, e_{edge}, and *trans*-4 bonds. In these calculations, the selection criterion was the relative heat of formation of the various possible regioisomeric products; more elaborate computer modeling led to similar predictions [51]. Coupling of **21** to anchor **22** by esterification afforded the wine-red methanofullerene **23** which, upon heating, underwent the predicted Diels–Alder addition at the e_{face} 6–6 bond to give the rather insoluble, orange-brown C_s-symmetrical bis-adduct **24** in 23% yield. Chromatographic analysis revealed that no other regioisomer had formed and that the rather low yield of **24** was due to a rapid $^1O_2$–ene reaction of the fused cyclohexene ring in the presence of light and O$_2$, with the fullerene derivative itself acting as efficient photosensitizer [52, 53]. Fullerene derivatives with fused cyclohexene rings require careful handling under rigorous exclusion of light and air in order to avoid this undesirable side reaction.

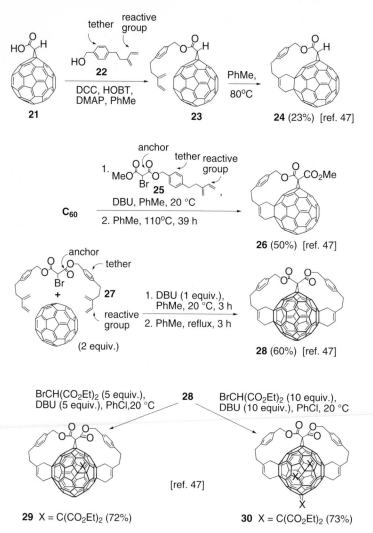

Scheme 7-4 Preparation of bis-adducts **24** [7, 47] and **26** [41, 47] and tris-adduct **28** [7, 47] by tether-directed remote functionalization. A sequence of *e* attacks starting from **28** gives hexakis-adduct **30** with a pseudo-octahedral addition pattern [7, 47]. DCC = *N,N''*-dicyclohexylcarbodi-imide, HOBT = 1-hydroxy-1*H*-benzotriazole, DMAP = 4-(dimethylamino)pyridine.

For the synthesis of a more soluble *e* bis-adduct, the preformed anchor−tether−reactive group conjugate **25** was attached in a Bingel reaction to magenta−purple C_{60} and subsequent heating of the formed wine-red mono-adduct in toluene for 39 h led to the desired brown product **26**, which was isolated in 50% yield as the only regioisomer formed (Scheme 7-4) [41, 47]. Similarly, conjugate **27** underwent the sequence of Bingel attachment followed by double Diels–Alder addition at the two *e* positions on opposite sides of the carbon sphere to yield the orange-brown tris-adduct **28** in 60% yield with complete regioselectivity [47]. The C_{2v}-symmetrical tris-adduct **28** can be readily prepared in multi-gram quantities; its addition pattern cannot be accessed by stepwise, non-tethered reaction sequences. Starting from tris-adduct **28**, sequential malonate additions (excess of 2-bromomalonate, DBU) to *e* 6–6 bonds provided novel tetrakis-(**29**), pentakis-, and, eventually, yellow-colored hexakis-adducts such as **30** (Scheme 7-4) in which the fullerene π-chromophore is reduced to a benzenoid cubic cyclophane-type substructure [54]. The reactivity and physical properties of these higher adducts have been investigated in great detail [47, 55, 56]; also, the corresponding endohedral ^{3}He compounds were prepared and their π-electron ring-current effects studied by ^{3}He-NMR [37].

After advantage had been taken of the *e*-directing cyclohexene rings in the preparation of tetrakis- to hexakis-adducts, it was of interest to develop a method for removing these rings together with the *p*-xylylene tethers, thereby transforming the tether-directed remote functionalization into a true template-directed reaction. However, a direct removal of the cyclohexene rings in tetrakis-adduct **29** or hexakis-adduct **30** by retro-Diels–Alder reaction was not successful [57]. Therefore, an elegant alternative procedure, introduced by Rubin and co-workers [52, 53], was applied. When a solution of **30** containing C_{60} as $^1O_2$ sensitizer was irradiated while a stream of O_2 was bubbled through, the $^1O_2$–ene reaction (the Schenk reaction [58]) at the two cyclohexene rings yielded a mixture of isomeric allylic hydroperoxides **31** with endocyclic double bonds (Scheme 7-5) [2, 30]. In-situ reduction of **31** with PPh$_3$ gave a mixture of isomeric allylic alcohols **32** which was subsequently heated in toluene together with toluene-4-sulfonic acid (TsOH) and dimethyl acetylenedicarboxylate (DMAD). Under these conditions, **32** was dehydrated to the corresponding bis(cyclohexa-1,3-diene) derivative which, via a Diels–Alder retro-Diels–Alder sequence, afforded tetrakis-adduct **33** in 42% overall yield starting from **30**. Transesterification with K_2CO_3 in EtOH/THF yielded the octakis(ethyl ester) **20**, which was characterized by X-ray crystallography. By a similar approach, the novel C_{2v}-symmetrical tris-adduct **35** was prepared starting from pentakis-adduct **34** [2]. Compounds such as **20** are of interest since physical studies showed that the location of multiple all-methano addends along an equatorial belt, as in **20**, leads to the smallest possible perturbation of the fullerene π-chromophore. Thus, the electronic properties of **20** do not differ much from those in C_{60} whereas those of tetrakis-adducts with the functional groups placed all around the carbon sphere differ substantially from those of the parent fullerene [2, 30]. Furthermore, tetrakis-adduct **20** provides a versatile intermediate for further molecular nanoscaffolding. It contains two reactive 6–6 bonds at the two poles activated by the four malonate addends in *e* positions around the equator. Thus, addition of 20 equiv of dialkynyl bromide **36** afforded hexakis-adduct **37** and, after deprotection, tetrakis-ethynylated **38** as an interesting novel building block for molecular construction [57].

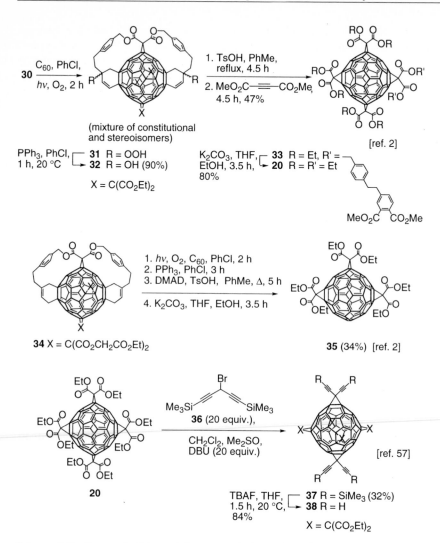

Scheme 7-5 Removal of the cyclohexene moieties and *p*-xylylene tethers in **30** and **34** provides tetrakis-adduct **20** [2, 30] as precursor to the tetraethynylated hexakis-adduct **38** [57] and tris-adduct **35** [2], respectively.

7.3.2 The Bingel Macrocyclization

The simplest, most versatile method for the preparation of covalent bis-adducts of C_{60} with high regio- and diastereoselectivity is the macrocyclization between C_{60} and bismalonate derivatives in a double Bingel reaction (Scheme 7-6) [25, 59, 60]. As a general protocol, diols are first transformed into bis(ethyl malonyl) derivatives. Subsequently, the corresponding bis(bromomalonate)s are prepared and subjected to the Bingel reaction with C_{60} in toluene in the presence of DBU (variant A). By the even shorter variant B, bis(iodomalonate)s are prepared in situ [59, 62], and the one-pot reaction of C_{60}, bis(ethyl malonate), and DBU in toluene at 20 °C generates the macrocyclic bis-adducts in yields usually between 20 and 40% and high regio- and diastereoselectivity.

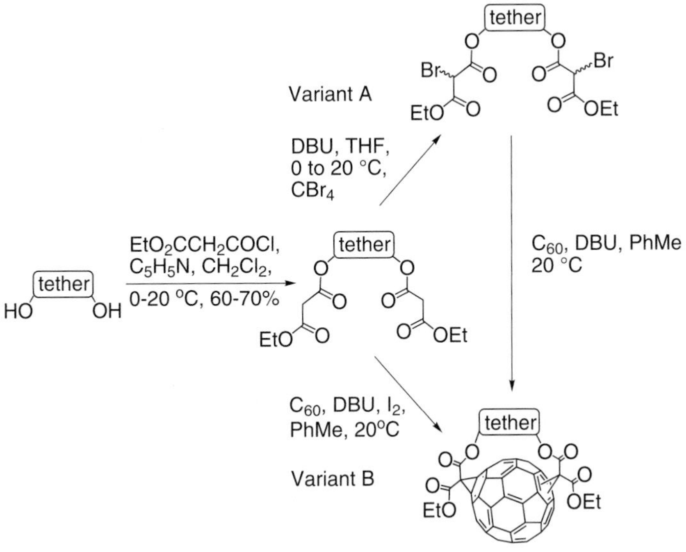

Scheme 7-6   General protocols for the Bingel macrocyclization reaction [25, 59, 60].

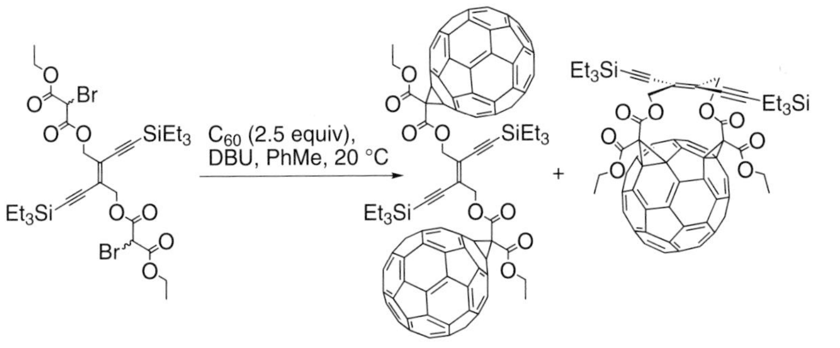

40   **39** (C_2, 33%)   (±)-**41** (C_1,16%) [ref. 60]

Scheme 7-7   First observation of the Bingel macrocyclization reaction [60].

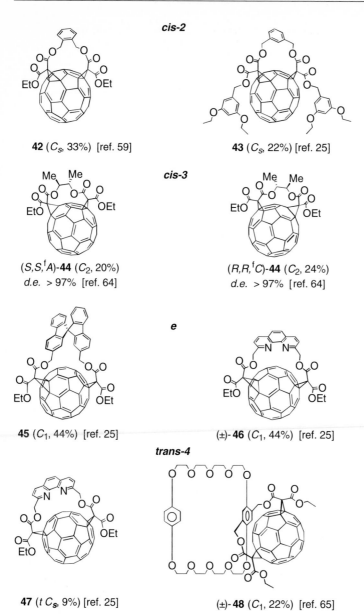

cis-2

42 (C_S, 33%) [ref. 59]

43 (C_S, 22%) [ref. 25]

cis-3

$(S,S,{}^fA)$-**44** (C_2, 20%)
d.e. > 97% [ref. 64]

$(R,R,{}^fC)$-**44** (C_2, 24%)
d.e. > 97% [ref. 64]

e

45 (C_1, 44%) [ref. 25]

(±)-**46** (C_1, 44%) [ref. 25]

trans-4

47 ($t\,C_S$, 9%) [ref. 25]

(±)-**48** (C_1, 22%) [ref. 65]

Scheme 7-8   Examples of C_{60} bis-adducts directly produced by the Bingel macrocyclization reaction (Scheme 7-6). For each compound, the molecular symmetry and the preparation yield are given.

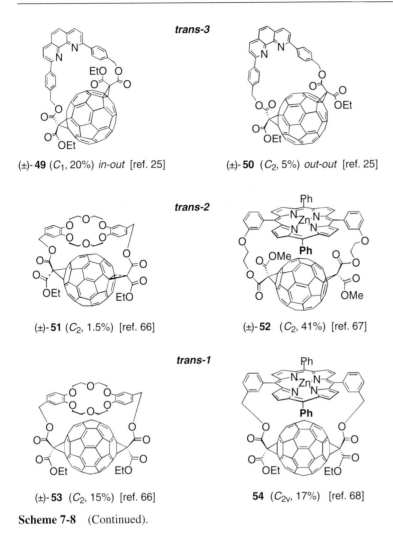

trans-3

(±)-**49** (C_1, 20%) *in-out* [ref. 25]

(±)-**50** (C_2, 5%) *out-out* [ref. 25]

trans-2

(±)-**51** (C_2, 1.5%) [ref. 66]

(±)-**52** (C_2, 41%) [ref. 67]

trans-1

(±)-**53** (C_2, 15%) [ref. 66]

54 (C_{2v}, 17%) [ref. 68]

Scheme 7-8   (Continued).

The Bingel macrocyclization was first observed during attempts to prepare poly(tri-acetylene) oligomers [63] with laterally pendant C_{60} spheres as advanced materials with promising electronic and optical properties. When (E)-hex-3-ene-1,5-diyne **39**, a direct precursor, after silyl group deprotection, to the desired PTA oligomers, was prepared by double Bingel reaction between two C_{60} molecules and bis(bromomalonate) **40**, the cyclophane-type C_1-symmetrical *cis*-2 bis-adduct (±)-**41**, resulting from Bingel macro-cyclization, was obtained with complete diastereoselectivity as a significant side product (Scheme 7-7) [59, 60]. This observation led to a first comprehensive study [60] which demonstrated the enormous versatility of the new tether-directed bis-functionalization. With the exception of *cis*-1, all possible bis-addition patterns (Scheme 7-1) have been obtained today and some interesting examples (**42–54**) [25, 59, 60, 64–68] for the various functionalization patterns are presented in Scheme 7-8.

7.3.2.1   Formation of *cis*-2 bis-adducts

In theory, each macrocyclic isomer could be formed as a mixture of diastereoisomers, depending on how the EtOCO residues at the two methano bridge C-atoms are oriented with respect to each other (*in–in, in–out,* and *out–out* stereoisomerism) [69]. However, with the exception of the *in–out* isomer (±)-**49** (Scheme 7-8), *out–out* stereoisomers have been obtained exclusively until now. The *out–out* geometry of *cis*-2 bis-adduct **42** with an *o*-xylylene tether was proven by X-ray crystallography [59].

The Bingel macrocyclization has been increasingly exploited in the construction of molecular building blocks for supramolecular systems. For this purpose, *cis*-2 bis-adducts, such as **43** with a *m*-xylylene tether (Scheme 7-8), have been particularly useful. Thus, bis-adducts **55** [70], a fullerene–glycodendron conjugate, and **56** [71] were shown to form stable Langmuir monolayers at the air–water interface (Scheme 7-9).

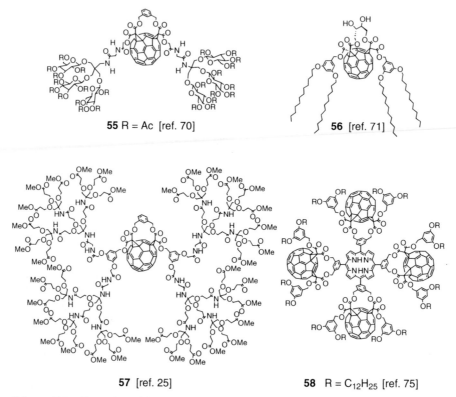

55 R = Ac [ref. 70]   **56** [ref. 71]

57 [ref. 25]   **58**   R = C$_{12}$H$_{25}$ [ref. 75]

Scheme 7-9   Examples of functional supramolecular systems containing C$_{60}$ *cis*-2 bis-adducts prepared by the Bingel macrocyclization.

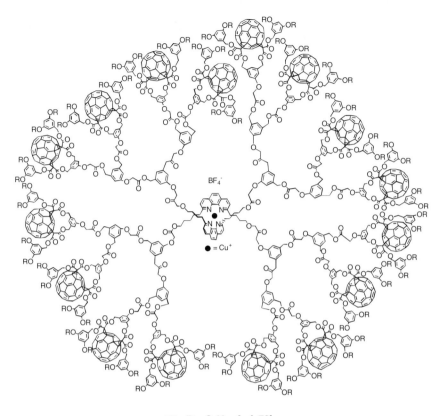

59 R = C$_8$H$_{17}$ [ref. 73]

Scheme 7-9 (Continued).

Other examples include the fullerene dendrimer **57** [61] with a molecular mass of 7344 Da, dendrimer **59** featuring a bis(1,10-phenanthroline)copper(II) core and 16 peripheral fullerene groups (molecular mass 31601 Da) [72, 73], and porphyrin derivatives with two [74] and four (**58**) [75] appended fullerenes.

Higher adducts are also accessible in a regiospecific way. Thus, starting from *cis*-2 bis-adduct **61**, which was prepared by Bingel macrocyclization of **60**, a clipping reaction yielded tetrakis-adduct **62** with an all-*cis*-2 addition pattern along an equatorial belt (Scheme 7-10) [25]. It is evident that by a judicious combination of tethers, a rich variety of higher adducts of C$_{60}$ with original addition patterns should become accessible. A first example suggests that the malonates can also be replaced by other groups. Thus, the tethered bis(β-keto ester) **63** was used to prepare the *cis*-2 bis-adduct **64** [76a] and the addition of tethered bismalonamides found application in the formation of a fullerene–calixarene conjugate [76b]. It is clear that the full scope of the Bingel macrocyclization remains yet to be explored.

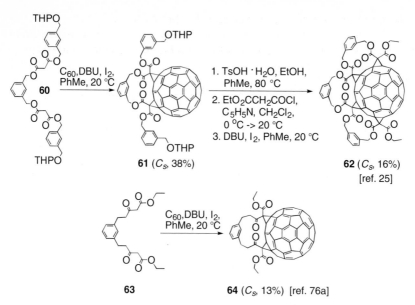

Scheme 7-10   Extensions of the Bingel macrocyclization: preparation of a higher adduct of C₆₀ [25] and use of tethered bis(β-keto esters) instead of tethered bismalonates [74]. THP = 3,4,5,6-tetrahydro-2*H*-pyran-2-yl.

7.3.2.2   Formation of *cis*-3 bis-adducts

Since *cis*-3, *trans*-3, and *trans*-2 bis-adducts of C₆₀ with identical addends are chiral as a result of inherently chiral functionalization patterns [10, 44–46], it was of interest to explore whether Bingel macrocyclizations with bismalonates bridged by non-racemic tethers would provide an enantioselective synthesis of these compounds. An overall enantioselective synthesis of optically active C₆₀ bis-adducts had been achieved previously by asymmetric Sharpless bis-osmylation [77]; however, this sequential bis-functionalization lacks the regioselectivity of the Bingel macrocyclization, and therefore requires tedious regioisomer separations.

A sequence of a highly diastereoselective Bingel macrocyclization using a non-racemic tether, followed by removal of the tether via transesterification, indeed provided an enantioselective synthesis of optically active *cis*-3 bis-adducts in which the chirality results exclusively from the addition pattern [25, 46, 59]. Starting from (*R,R*)-**65** and (*S,S*)-**65**, which were prepared from the corresponding optically pure diols, the two enantiomeric *cis*-3 bis-adducts (*R,R*,ᶠ*A*)-**66** and (*S,S*,ᶠ*C*)-**66** were obtained with high diastereoselectivity (diastereoisomeric excess d.e. > 97%) (Scheme 7-11). In each macrocyclization, two diastereoisomeric *out–out cis*-3 bis-adducts are possible due to the chiral addition pattern; however, the high asymmetric induction by the optically active tether in the second intramolecular Bingel addition led to the formation of (*R,R*,ᶠ*A*)-**66** and (S,S,^{f}C)-**66** only. The simple stereochemical descriptors ^{f}C and ^{f}A (f = fullerene, C = clockwise, A = anticlockwise) have been introduced to specify the configurations of chiral fullerenes and fullerene derivatives with a chiral functionalization pattern [44]. Similarly, the *cis*-3 adducts (R,R,^{f}C)-**44** and (S,S,^{f}A)-**44** (Scheme 7-8) were formed with a d.e. exceeding 97% (HPLC) starting from the corresponding optically active tethered bismalonates [64]. Transesterification of (R,R,^{f}A)-**66** and (S,S,^{f}C)-**66** yielded the *cis*-3 tetraethyl esters (^{f}A)-**2** and (^{f}C)-**2** with an enantiomeric excess (e.e.) higher than 99% [(^{f}A)-**2**] and 97% [(^{f}C)-**2**] (HPLC), reflecting the e.e.s of the corresponding commercial starting diols. The absolute configurations of these optically active fullerene derivatives could be assigned from their calculated circular dichroism (CD) spectra [46]. The fact that the tethers in (R,R,^{f}A)-**66** and (S,S,^{f}C)-**66** could be readily removed by transesterification makes them true covalent templates.

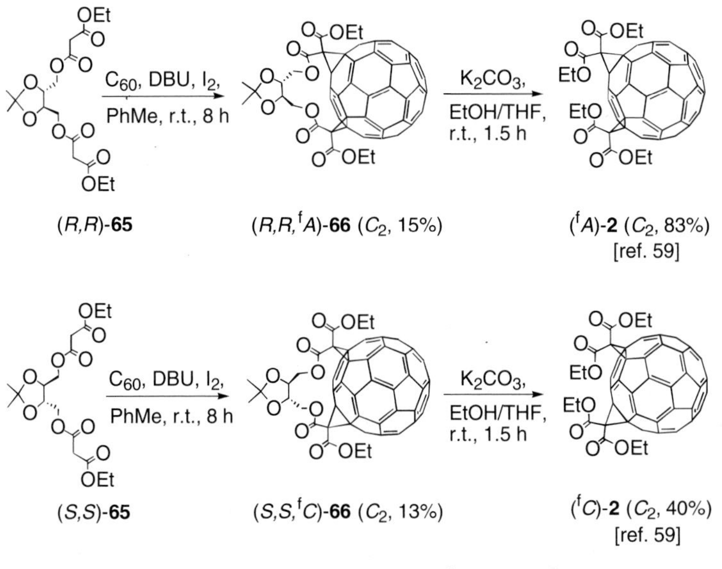

Scheme 7-11   Enantioselective synthesis of (^{f}A)-**2** and (^{f}C)-**2** by diastereoselective tether-directed bis-cyclopropanation of C$_{60}$, followed by transesterification [25, 46, 59].

Of substantial interest were the chiroptical properties of these enantiomeric C$_{60}$ derivatives with chiral functionalization patterns: the mirror-image CD spectra of (^{f}C)-**2** and (^{f}A)-**2** displayed large Cotton effects between 250 and 750 nm with $\Delta\varepsilon$ values approaching 150 cm^2 mmol^{-1} [25, 59]. These intensities of the CD bands are of similar magnitude to those measured for inherently chiral fullerenes (such as D$_2$-C$_{76}$ [78] and their derivatives [79]) but are much larger than those measured for non-racemic fullerene derivatives in which the chirality only results from the addends [25].

7.3.2.3   Other bis-functionalization patterns

The *e* bis-adducts **45** (Scheme 7-8) and *ent*-**45** (not shown) provide examples for optical-ly active fullerene derivatives in which the chirality originates only from the optically active tether, and the intensities of their CD bands ($\Delta\varepsilon$ values < 30 cm^2 mmol^{-1}) are smaller than those of ($^f C$)-**2** and ($^f A$)-**2** (Scheme 7-11) with inherently chiral functional-ization patterns [25].

The Bingel macrocyclization is not completely regioselective in each case [25, 59]. As an example, the use of a tether derived from 1,10-phenanthroline-2,9-dimethanol yield-ed both *e* bis-adduct (±)-**46** and *trans*-4 bis-adduct **47**.

By the Bingel macrocyclization, organic chromophores or receptor sites can be pre-cisely positioned in close proximity to the fullerene surface, thus offering the potential for inducing changes in the physical properties of the carbon allotrope. This was exploit-ed in the construction of the C$_{60}$-containing catenane (±)-**67**·4PF$_6$ starting from the *trans*-4 crown ether (±)-**48** [65]. The formation of (±)-**67**·4PF$_6$ by self-assembly, follow-ing the elegant strategy developed by Stoddart and co-workers [80], is depicted in Scheme 7-12. With the fullerene moiety acting as an electron acceptor, (±)-**67**·4PF$_6$ fea-tures an unprecedented intramolecular A−D−A−D−A stack (A = acceptor, D = donor).

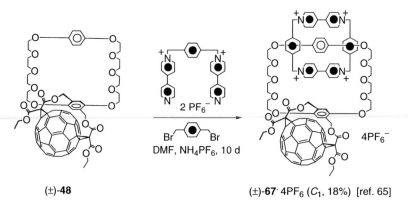

(±)-**48**   (±)-**67**·4PF$_6$ (C_1, 18%) [ref. 65]

Scheme 7-12   Preparation of (±)-**67**·4PF$_6$ by self-assembly [65].

The Bingel macrocyclization with a bismalonate bridged by a large tether derived from 2,9-bis[4-(hydroxymethyl)phenyl]-1,10-phenanthroline provided, as already men-tioned in Section 7.3.2.1, the only example so far of the formation of both stereoisomer-ic *in−out trans*-3 ((±)-**49**) and *out−out trans*-3 ((±)-**50**) bis-adducts [25]. Steric consid-erations based on molecular model examinations and computer modeling indeed had indicated that the large tether would lead to the formation of the two diastereoisomers with similar calculated steric strain energy. The formation of *in−out* (±)-**49** was fully confirmed by transesterification under cleavage of the tether leading to the known tetra-ethyl ester (±)-**5** [10].

Substantial perturbations of the electronic structure of the fullerene were observed when an alkali metal cation was bound to the crown ether moieties in *trans*-2 bis-adduct (±)-**51** or *trans*-1 bis-adducts (±)-**53** [66]. In these compounds, the crown ether receptor site is positioned closely and tightly on the fullerene surface. Upon complexation of alkali metal ions to this site, the first fullerene-based electrochemical reduction process becomes significantly facilitated due to the stabilization of the formed fullerene radical anion by the cationic center bound in close proximity. This was the first ever observed effect of cation complexation on the redox properties of the carbon sphere in fullerene–crown ether conjugates [81–86]. By transesterification of (±)-**53** with Cs_2CO_3 in anhydrous 1-hexanol/THF, the tether was removed under formation of the corresponding D_{2h}-symmetrical tetrakis(hexyl ester) (34%) [66], thereby providing a versatile entry into diverse molecular scaffolding using *trans*-1 bis-adducts from C_{60}.

Porphyrin-fullerene conjugates attract wide attention for their intramolecular energy and electron transfer properties [87, 88]. By attachment of the porphyrin to two points on the C_{60} surface, the interchromophoric spatial relationship in the cyclophane-type molecular dyads *trans*-2 (±)-**52** [67] and *trans*-1 **54** [68] (Scheme 7-8), which controls both energy and electron transfer, is rigorously defined. In the two systems, as well as in the fullerene–porphyrin conjugate **58** [75] (Scheme 7-9), the close proximity between fullerene and porphyrin chromophore leads to a nearly complete quenching of the porphyrin luminescence, presumably as a result of efficient energy transfer between the porphyrin donor and the fullerene acceptor.

7.3.3   Formation of Bis-adducts by Double Diels–Alder Addition of Tethered Bis(buta-1,3-diene)s to C_{60}

The double Diels–Alder addition [89, 90] of a tethered bis(buta-1,3-diene) to C_{60} has been used for the regioselective formation of a *trans*-1 bis-adduct by Rubin and his group in their highly original program targeting the creation of large holes in the shell of C_{60} and the formation of endohedral metallofullerenes [91]. These researchers prepared the tethered bis(cyclobutene) **68** as a mixture of stereoisomers (Scheme 7-13). Upon refluxing **68** with C_{60} in toluene, ring-opening of the cyclobutenes occurred and the resulting 1,3-diene underwent cycloaddition regioselectively to afford the *trans*-1 derivative **69** in 30% yield. Upon treatment with TsOH, the tether is eliminated under formation of a fullerene containing two fused cyclohexa-1,3-dienes, which each undergo a rearrangement sequence [92] consisting of a photochemical [4 + 4] cycloaddition followed by a thermal [2 + 2 + 2] retro-cycloaddition to yield the mixture of stereoisomers **70** and **71** which have not yet been separated.

Nishimura and co-workers used bis(*o*-quinodimethanes) connected by α,ω-dioxamethylene tethers for the regioselective bis-functionalization by double Diels–Alder addition (Scheme 7-14) [93]. The tethered bis(*o*-quinodimethanes) were intermediately formed by heating the bis(bromomethyl) derivatives **72a–d** together with C_{60} in toluene in the presence of KI and [18]crown-6 [90]. With an O–$(CH_2)_2$–O tether (**72a**), the *cis*-2 and *cis*-3 bis-adducts **73a** and (±)-**74a** were isolated in 10% and 8% yield, respec-

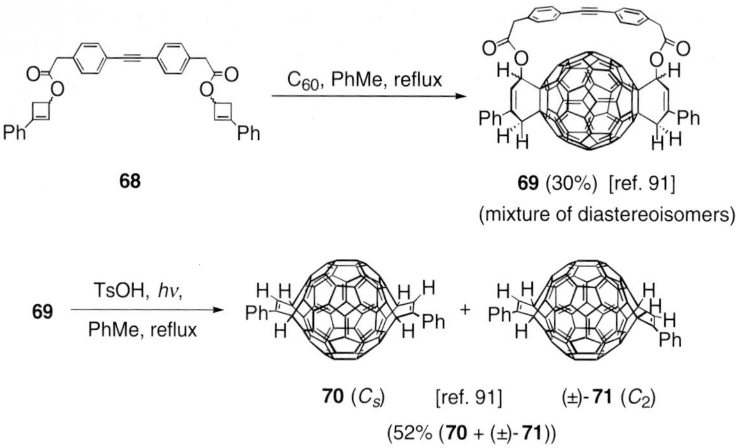

68

69 (30%) [ref. 91]
(mixture of diastereoisomers)

C_{60}, PhMe, reflux

69 $\xrightarrow{\text{TsOH, }hv,}$ PhMe, reflux

70 (C_S)   [ref. 91]   (±)-**71** (C_2)

(52% (**70** + (±)-**71**))

Scheme 7-13 Preparation of *trans*-1 bis-adduct **69** and elimination of the tether followed by rearrangement to the mixture of stereoisomers **70** and (±)-**71** [91].

tively. Compound **72b**, with an $O-(CH_2)_3-O$ tether, showed a higher regioselectivity, and the *cis*-2 bis-adduct **73b** was isolated in 20% yield in addition to *cis*-3 bis-adduct (±)-**74b** as a minor product (9% yield). Interestingly, the reaction of **72c**, with an $O-(CH_2)_4-O$ tether, displayed hardly any regioselectivity whereas the conversion of **72d**, with the longest $O-(CH_2)_5-O$ tether, proceeded with high regioselectivity and yielded the *e* bis-adduct (±)-**75** in 30% yield with the yield of other regioisomers, which were not isolated, not exceeding a total of 2–3%.

The tethers were subsequently readily removed by ether cleavage with BBr_3 in benzene, providing the corresponding bis(phenols) **76**, (±)-**77**, and (±)-**78** in nearly quantitative yield. The racemic mixtures (±)-**74a**, (±)-**74b**, (±)-**77**, and (±)-**78** were successfully resolved by HPLC on a chiral stationary phase (CHIRALCEL OD®) [94]. In agreement with other findings [25, 45, 59], the CD spectra of non-racemic **74a** and **77** featured very intense Cotton effects due to the strong chiroptical contributions of their chiral functionalization patterns. In contrast, the functionalization patterns in **74b** and **78** are achiral and their enantiomers displayed much weaker CD bands. With tetrabromide **79**, featuring an optically active tether, the double Diels–Alder addition yielded one single diastereoisomeric *cis*-3 bis-adduct, presumably **80**, with complete diastereoselectivity besides *cis*-2 bis-adduct **81** as the major product (Scheme 7-14).

An interesting templated bis-functionalization was reported by Shinkai and co-workers, who introduced two boronic acid groups regioselectively via Diels–Alder addition onto the fullerene using saccharides as "imprinting" templates (Scheme 7-15) (for the concept of molecular imprinting, see Chapter 2) [95,96]. The formation of cyclic boronate esters between boronic acids and carbohydrates has in recent years been elegantly applied by Shinkai and his group to carbohydrate recognition, transport, and sensing [97]. They now use saccharides as templates to "imprint" into the surface of fullerenes two boronic acid groups, with their relative position on the surface of the carbon sphere being directly related to the structure of the templating saccharide. After cleavage of the boronate esters, the resulting bis(boronic acid)-substituted fullerene should provide a specific recognition site for the sugar that served as the initial sugar template.

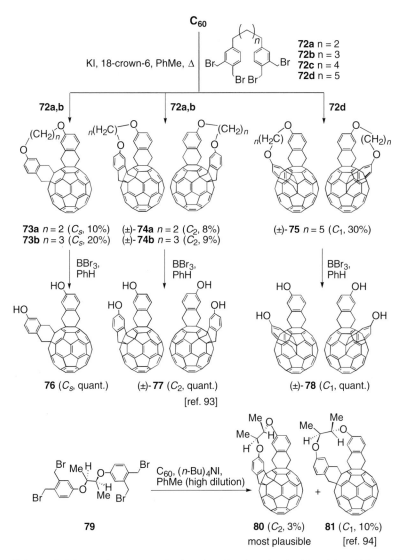

Scheme 7-14   Regioselective bis-functionalizations by double Diels–Alder addition of tethered bis(*o*-quinodimethanes) to C₆₀ [93, 94].

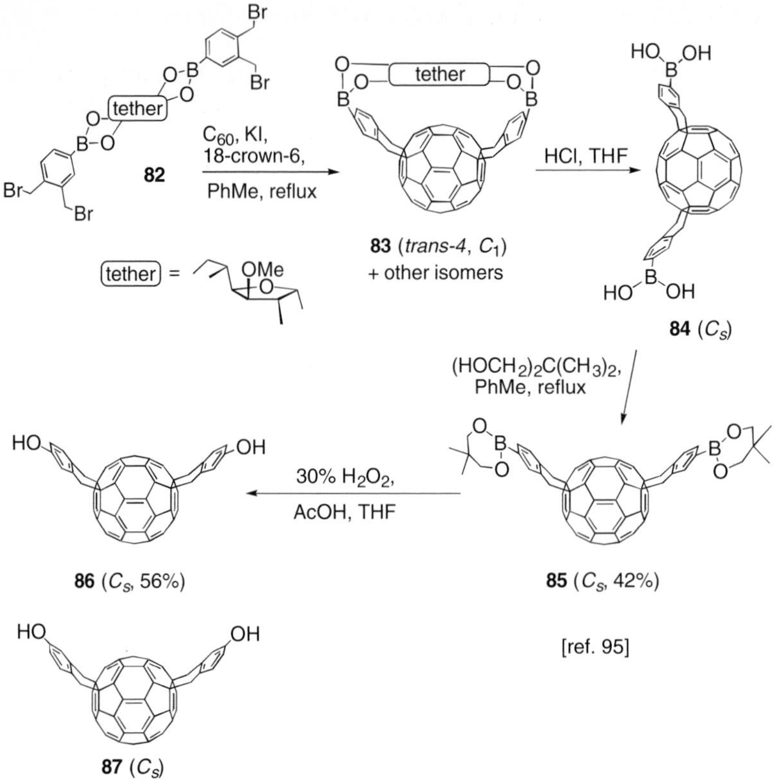

Scheme 7-15   Regioselective bis-functionalization of C_{60} using a saccharide as imprinting template [95, 96].

Thus, compound **82** containing 3-*O*-methyl-D-glucofuranose as a template underwent a double Diels–Alder addition with C_{60} providing as the major product the *trans*-4 bis-adduct **83** [95]. Removal of the saccharide template yielded boronic acid **84**, which is a potentially selective receptor for the initial template saccharide molecule. Preliminary recognition studies indeed showed that **84** retains the memory for the initial template. Protection of the boronic acid residues afforded **85** (42% yield, HPLC) which was transformed into diol **86** by treatment with H_2O_2. Scheme 7-15 shows the structure of an alternative *trans*-4 diol **87** with C_s symmetry that could potentially also be formed in the reaction sequence starting from **82**; an unambiguous assignment of the exact structure of **83** and the following bis-adducts **84**–**86** remains yet to be reported. There is little doubt that the concept of molecular imprinting in homogeneous solution on surfaces such as those of a fullerene bears exceptional fundamental and technological potential.

7.3.4   Double [3+2] Cycloaddition of Tethered Vinylcarbenes

Nakamura and co-workers synthesized the oligomethylene-tethered bis(cyclopropenone acetals) **88a–c** to generate by thermolysis minute amounts of nucleophilic vinylcarbenes which subsequently underwent two-fold [3+2] cycloadditions to 6–6 bonds of C$_{60}$ (Scheme 7-16) [98, 99]. Thus, reagent **88a**, with a –(CH$_2$)$_3$– tether, afforded the C_s-sym-

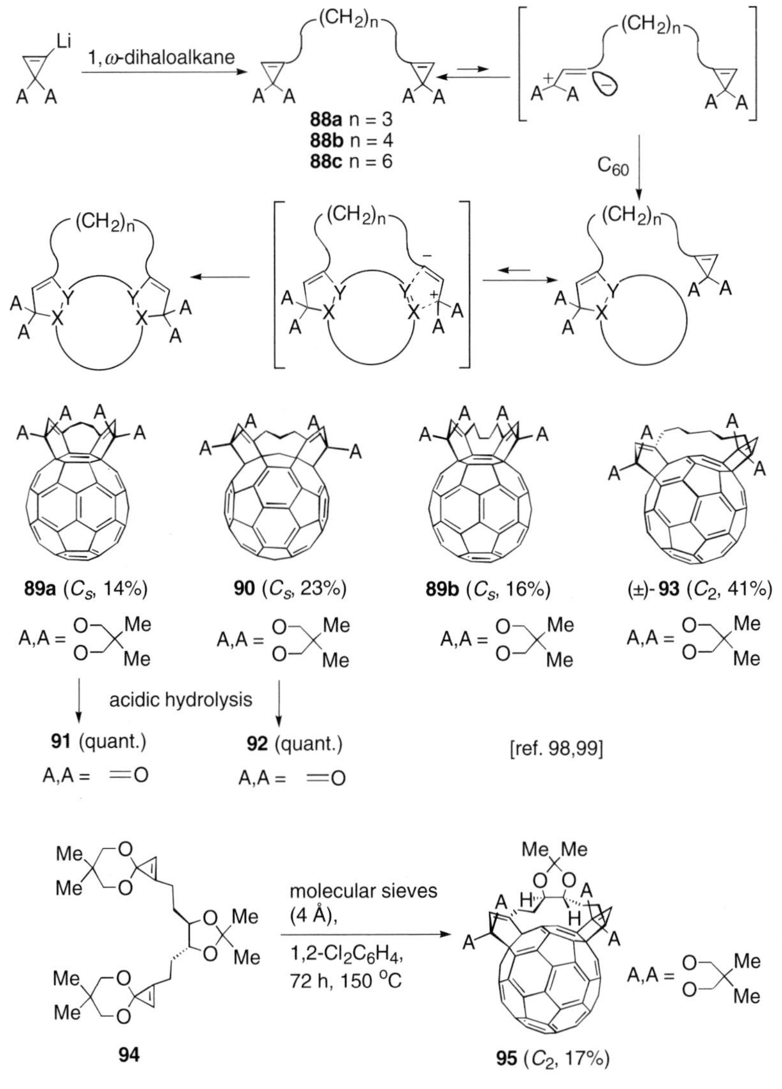

88a n = 3
88b n = 4
88c n = 6

89a (C_s, 14%)   **90** (C_s, 23%)   **89b** (C_s, 16%)   (±)-**93** (C_2, 41%)

A,A = O⟩Me / O⟩Me   A,A = O⟩Me / O⟩Me   A,A = O⟩Me / O⟩Me   A,A = O⟩Me / O⟩Me

acidic hydrolysis

91 (quant.)   **92** (quant.)   [ref. 98,99]
A,A = =O   A,A = =O

94   molecular sieves (4 Å), 1,2-Cl$_2$C$_6$H$_4$, 72 h, 150 °C   **95** (C_2, 17%)   A,A = O⟩Me / O⟩Me

Scheme 7-16   Regioselective bis-functionalization by double [3+2] cycloaddition of tethered vinylcarbenes to C$_{60}$ [98, 99].

metrical *cis*-1 bis-adduct **89a** in 14% yield together with the C_s-symmetrical *cis*-2 bis-adduct **90** which was formed as the major regioisomer in 23% yield. Upon acidic hydrolysis of the acetal groups, the corresponding diketones **91** and **92** were obtained in quantitative yield. Interestingly, **88b**, with the longer $-(CH_2)_4-$ tether, provided the C_s-symmetrical *cis*-1 bis-adduct **89b** as the only formed regioisomer in 16% yield. Apparently, the conformational strain of the tether moiety in the transition state of the second addition, rather than product stability, controls the regioselectivity of the attack, and computational studies were undertaken to elucidate the origin of this selectivity [99]. The reagent with the longest $-(CH_2)_6-$ tether, **88c**, afforded in high yield (41%) the C_2-symmetrical *cis*-3 bis-adduct (±)-**93**, which could also be quantitatively transformed by acid hydrolysis into the corresponding diketone. Preliminary results obtained with **94**, with an optically active tether, showed that a single non-racemic *cis*-3 stereoisomer (**95**) was obtained in a highly diastereoselective way.

7.3.5   Double [3+2] Cycloadditions of Tethered Bis-azides

The addition of azides, followed by extrusion of dinitrogen, leading to aza-bridged fullerenes, provides a versatile method for the functionalization of C_{60}. Numerous reports have appeared since this conversion was first described in the literature [100], and the reader is referred to a recent review by Hirsch [101]. In order to control the regioselectivity of bis-adduct formation, tethered bis-azides have been used by Luh and co-workers (Scheme 7-17) [102–104]. When bis-azides **96a/b**, tethered by a $-(CH_2)_2-$ or $-(CH_2)_3-$ bridge, were reacted with C_{60}, bis-aza-fullerenes **97a/b** were obtained in which two aza bridges are positioned at two open 6–5 junctions of a fluorene moiety of C_{60} (Scheme 7-17) [102]. When C_{60} was treated with 2,2-dibenzyl-1,3-diazidopropane (**98**) in refluxing chlorobenzene for 8 h, the three products **99**, **100**, and **101** were isolated in 18, 24, and 11% yields, respectively [104]. Whereas compound **101**, similarly to **97**, possesses two aza bridges at the open 6–5 junctions of a fluorene moiety in C_{60}, the structures of the two other compounds are remarkably different. The bis-azide had undergone [3+2] cycloaddition to two neighboring 6–6 bonds in a six-membered ring (*cis*-1 addition pattern) to yield **100**. Compound **99** also features a *cis*-1 addition pattern with two fused aziridine rings at neighboring 6–6 bonds in a six-membered ring. Thermolysis of **100** in refluxing benzene gave a 40:54:6 mixture of **99**, **101**, and C_{60}, respectively, in quantitative yield, clearly showing that compound **100** is an intermediate in the formation of **99** and **101**. Mechanisms for these interesting conversions were proposed by Luh and co-workers [105]. Finally, the reaction of C_{60} with the optically active bis-azides **102a/b** yielded the enantiomerically pure bis-aza-fullerenes **103a/b** with an addition pattern as in **97a/b** [103].

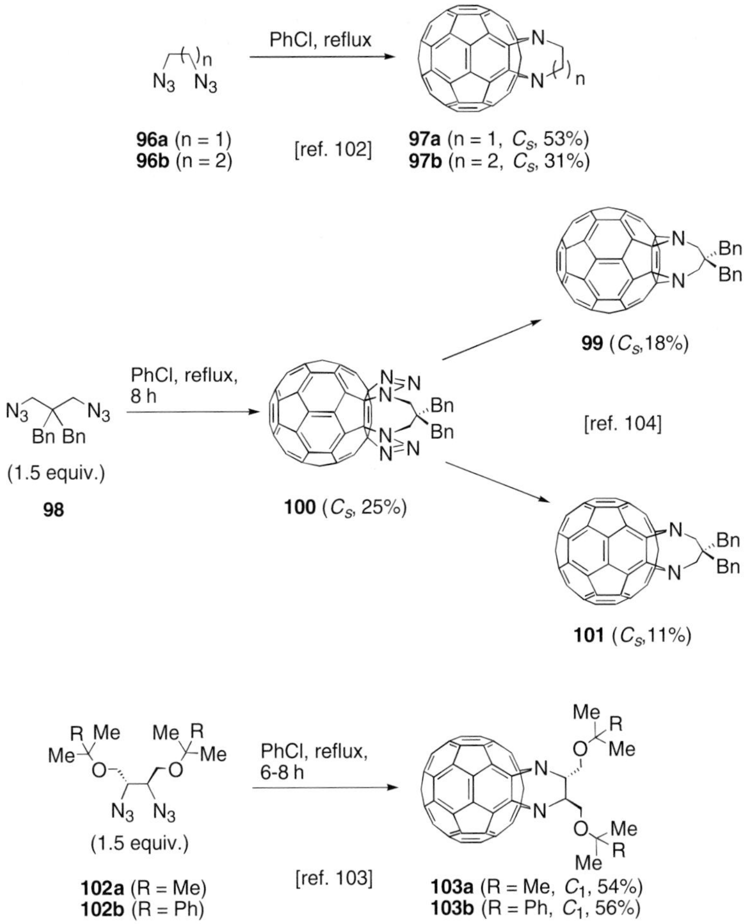

Scheme 7-17   Regioselective bis-addition of tethered bis-azides to C_{60} [102–104].

7.4 Conclusions

Templated synthesis approaches are currently revolutionizing the covalent chemistry of fullerenes. They provide elegant, imaginative protocols to highly functionalized fullerene building blocks for three-dimensional molecular scaffolding. There is much room for innovation in this emerging area. The variety of reactions that can be used to attack the fullerene is large and the number of structural and functional templates seemingly unlimited. A particularly exciting aspect of this research is related to the high stereoselectivity observed in the preparations of chiral *cis*-3 bis-adducts described above. With non-racemic tethers acting as chiral auxiliaries, highly diastereoselective bis-additions occur and, after tether removal, overall enantioselective syntheses of doubly functional-ized fullerenes are accomplished. It is becoming clear that the curved surface of the car-bon sphere imposes substantial stereoelectronic restraints which restrict the number of energetically favorable reaction pathways and transition states in tethered bis-function-alizations. By imprinting the precise alignment of functional groups and ligands for met-al ions onto fullerene surfaces, new generations of reagents and catalysts for asymmetric reactions in homogenous solution may emerge. Covalent derivatives of fullerenes prom-ise fascinating applications as advanced materials in diverse advanced technologies such as energy and information storage and transport, sensorics, or microfabrication. Only the continued invention of functionalization methodology such as that described in this chapter will ensure that these desirable objectives are reached in the future.

7.5   Experimental Procedures

7.5.1   Preparation of Hexakis-adduct 11 using DMA as a Template (Scheme 7-2) [20]

A solution of **10** [19] (145 mg, 0.156 mmol) and DMA (323 mg, 1.56 mmol) in dry toluene (14.5 mL) was degassed and subsequently stirred at 20 °C for 2 h. Diethyl 2-bromomalonate (210 µL, 1.24 mmol) and DBU (280 µL, 1.87 mmol) were successively added, and the dark red solution was stirred for 5 d at 20 °C, then diluted with CH_2Cl_2 (100 mL), washed with H_2O (3 × 10 mL), and dried ($MgSO_4$). Evaporation of the sol-vent, flash chromatography (2×; SiO_2H, cyclohexane/AcOEt 7 : 3, then SiO_2, CH_2Cl_2/cyclohexane 4 : 1) gave **11** (75 mg, 28%) as a yellow solid. M.p. > 240 °C.

7.5.2   Preparation of Bis-adduct 24 by Reaction of Methano[60]fullerenecarboxylic Acid 21 with the Tether–Reactive-group Conjugate 22 (Scheme 7-4) [47]

To **21** (100 mg, 0.128 mmol) in PhBr (20 mL) was added sequentially DCC (26.5 mg, 0.128 mmol), HOBT (17.4 mg, 0.128 mmol), DMAP (15.7 mg, 0.128 mmol), and **22** (48.4 mg, 0.256 mmol). After stirring at 20 °C for 19 h, the mixture was loaded directly onto column and chromatographed (SiO_2, PhMe/hexane 2 : 1), yielding a wine-red solu-tion of mono-adduct **23** which was evaporated to remove the hexane and then diluted with toluene to a total volume of 500 mL. The resulting solution was deoxygenated by purging with Ar for 10 min, further degassed by two freeze–pump–thaw cycles and then heated under Ar to 80 °C for 44 h. After concentration to 100 mL, the orange-red solu-tion was chromatographed under exclusion of air and light (SiO_2, PhMe/hexane 2 : 1) to afford one main product which was recrystallized from CS_2/MeOH then from CS_2/hex-ane and dried to give **24** (27.7 mg, 23%) as a black solid. M.p. > 270 °C.

7.5.3   Preparation of Tris-adduct 28 by Addition of the Anchor–Tether–Reactive-group Conjugate 27 to C_{60} (Scheme 7-4) [47]

To a solution of C_{60} (2.753 g, 3.82 mmol) and **27** (1.0 g, 1.91 mmol) in toluene (700 mL), DBU (320 mg, 2.10 mmol) in toluene (50 mL) was added over 5 min. After 3 h, the mixture was concentrated to 100 mL, diluted with hexane (100 mL), and chromatographed (SiO_2, toluene/hexane 1:1 then 2:1) to give the wine-red monomethanofullerene. After evaporation of hexane, toluene was added to give a total volume of 2.5 L, then the solution was deoxygenated by purging with Ar for 1.5 h and heated to reflux for 36 h. Concentration and column chromatography (SiO_2, toluene), followed by recrystallization from $CHCl_3$/MeOH and drying gave **28** (1.335 g, 60%) as a brown solid. M.p. > 270 °C.

7.5.4   Formation of Tetrakis-adduct 20 by Removal of the Tether in Hexakis-adduct 30 and Transesterification (Scheme 7-5) [2]

A solution of **30** (200 mg, 0.122 mmol) and C_{60} (102 mg, 0.142 mmoL) in chlorobenzene (160 mL) was irradiated in a Pyrex photochemical reactor at 20 °C for 2 h with a medium-pressure Hg lamp, while a stream of O_2 was bubbled through. The mixture containing **31** ($R_f = 0.43$, TLC, SiO_2, CH_2Cl_2/AcOEt 9:1) was then transferred into a flask and deoxygenated by inserting a stream of Ar, and PPh_3 (320 mg, 1.221 mmol) in chlorobenzene (5 mL) was added. After stirring at 20 °C for 1 h under Ar, plug filtration (SiO_2) with toluene to remove chlorobenzene and C_{60}, then with CH_2Cl_2/AcOEt 19:1 and then 9:1, yielded **32** which was precipitated out of CH_2Cl_2 by addition of hexane: 183 mg (90%). TsOH (54 mg, 0.1285 mmol) and DMAD (135 mg, 0.950 mmol) were added to a solution of the isomeric mixture **32** (159 mg, 0.095 mmol) in deoxygenated toluene (200 mL), and the mixture was refluxed for 4.5 h in the dark. Column chromatography (SiO_2, CH_2Cl_2 followed by CH_2Cl_2/AcOEt 49:1 then 24:1) yielded **33** as a yellow-green product, that was precipitated from CH_2Cl_2 by addition of hexane: 86 mg (47%) of a bronze solid. To a solution of **33** (30 mg, 0.0156 mmol) in anhydrous EtOH/THF 1:1 (30 mL) was added K_2CO_3 (155 mg, 1.123 mmol), and the mixture was stirred at 20 °C under Ar for 3.5 h. After filtration and evaporation, column chromatography (SiO_2, CH_2Cl_2/AcOEt) afforded **20** which was precipitated from CH_2Cl_2 by addition of hexane: 17 mg (80%) of a bronze solid. M.p. > 270 °C.

7.5.5   Bingel Macrocyclization: Synthesis of *cis*-2 Bis-adduct 42 Starting from Benzene-1,2-dimethanol (Scheme 7-8) [25]

Ethyl 3-chloro-3-oxopropanoate (ethyl malonyl chloride, 2.3 mL, 18.09 mmol) was added to a stirred solution of benzene–1,2-dimethanol (1.00 g, 7.24 mmol) and pyridine (1.5 mL, 18.09 mmol) in CH_2Cl_2 (100 mL) at 0 °C. The solution was allowed to slowly warm to 20 °C (over 1 h) and then stirred for 3 h. The mixture was washed with saturated aqueous NH_4Cl solution (2×), dried ($MgSO_4$), and evaporated to dryness. Column chromatography (SiO_2, CH_2Cl_2/MeOH 97:3) yielded 1,2-bis{[(ethoxycarbonyl)acetoxy]methyl}benzene (1.81 g, 68%) as a colorless oil. DBU (0.4 mL, 2.496 mmol) was added at 20 °C to a solution of C_{60} (300 mg, 0.416 mmol), I_2 (211 mg, 0.832 mmol) and 1,2-bis{[(ethoxycarbonyl)acetoxy]methyl}benzene (152 mg, 0.416 mmol) in toluene (600 mL). The solution was stirred for 5 h. The crude material was filtered through a short plug (SiO_2) eluting first with toluene (to remove unreacted C_{60}) and then with CH_2Cl_2. Column chromatography (SiO_2, CH_2Cl_2/hexane 2:1) and recrystallization (hexane/$CHCl_3$) provided **42** (151 mg, 33%) as a dark red solid.

7.5.6 Enantioselective Synthesis of *cis*-3 Bis-adduct (^fC)-2 by Bingel Macrocylization of (*S,S*)-66 with C_{60} Followed by Transesterification (Scheme 7-11) [25]

DBU (0.25 mL, 1.662 mmol) was added at 20 °C to C_{60} (200 mg, 0.277 mmol), I_2 (155 mg, 0.609 mmol), and (–)-(4*S*,5*S*)-bis{[(ethoxycarbonyl)acetoxy]methyl}-2,2-dimethyl-1,3-dioxolane (*S,S*)-**65** (119 mg, 1.662 mmol) in toluene (400 mL), and the mixture was stirred for 8 h. The crude material was filtered through a short plug (SiO_2), eluting first with toluene to remove unreacted C_{60} and then with CH_2Cl_2/MeOH 97:3. Column chromatography (SiO_2) eluting with CH_2Cl_2/hexane 9:1 yielded the corresponding *cis*-2 bis-adduct which was precipitated from $CHCl_3$/hexane (61 mg, 20%), and eluting with CH_2Cl_2/MeOH gave *cis*-3 bis-adduct (*S,S,^fC*)-**66** which was precipitated from CH_2Cl_2/hexane (41 mg, 13%) to give a brown solid. K_2CO_3 (260 mg, 1.886 mmol) was added to (*S,S,^fC*)-**66** (21 mg, 0.019 mmol) in THF/EtOH 1:1 (28 mL), and the mixture was stirred at 20 °C for 1.5 h, then filtered and evaporated to dryness. Column chromatography (SiO_2, toluene) followed by precipitation from CH_2Cl_2/hexane yielded (^fC)-**2** (8 mg, 40%) as a brown solid. M.p. > 250 °C.

Acknowledgment

We thank the Swiss National Science foundation for continuing generous support of fullerene research and Prof. Jules Rebek, Jr., at the Skaggs Institute for Chemical Biology, The Scripps Research Institute, La Jolla, for hosting F.D. for his sabbatical, during which this article was written.

References

[1] F. Diederich, R. Kessinger, *Acc. Chem. Res.* **1999**, *32*, 537–545.
[2] F. Cardullo, P. Seiler, L. Isaacs, J.-F. Nierengarten, R. F. Haldimann, F. Diederich, T. Mordasini-Denti, W. Thiel, C. Boudon, J.-P. Gisselbrecht, M. Gross, *Helv. Chim. Acta* **1997**, *80*, 343–371.

[3] F. Diederich, M. Gómez-López, *Chimia* **1998**, *52*, 551–556.

[4] F. Diederich, C. Thilgen, *Science (Washington, D.C.)* **1996**, *271*, 317–323.

[5] S. Samal, S. K. Sahoo, *Bull. Mater. Sci.* **1997**, *20*, 141–230.

[6] A. Hirsch, *Top. Curr. Chem.* **1999**, *199*, 1–65.

[7] L. Isaacs, R. F. Haldimann, F. Diederich, *Angew. Chem.* **1994**, *106*, 2434–2437; *Angew. Chem., Int. Ed. Engl.* **1994**, *33*, 2339–2342.

[8] R. Breslow, *Acc. Chem. Res.* **1995**, *28*, 146–153.

[9] C. Bingel, *Chem. Ber.* **1993**, *126*, 1957–1959.

[10] A. Hirsch, I. Lamparth, H. R. Karfunkel, *Angew. Chem.* **1994**, *106*, 453–455; *Angew. Chem., Int. Ed. Engl.* **1994**, *33*, 437–438.

[11] F. Djojo, A. Herzog, I. Lamparth, F. Hampel, A. Hirsch, *Chem. Eur. J.* **1996**, *2*, 1537–1547.

[12] A. Hirsch, I. Lamparth, T. Grösser, H. R. Karfunkel, *J. Am. Chem. Soc.* **1994**, *116*, 9385–9386.

[13] A. Hirsch, I. Lamparth, G. Schick, *Liebigs Ann.* **1996**, 1725–1734.

[14] I. Lamparth, C. Maichle-Mössmer, A. Hirsch, *Angew. Chem.* **1995**, *107*, 1755–1757; *Angew. Chem., Int. Ed. Engl.* **1995**, *34*, 1607–1609.

[15] X. Camps, A. Hirsch, *J. Chem. Soc., Perkin Trans. 1* **1997**, 1595–1596.

[16] J. A. Schlueter, J. M. Seaman, S. Taha, H. Cohen, K. R. Lykke, H. H. Wang, J. M. Williams, *J. Chem. Soc., Chem. Commun.* **1993**, 972–974.

[17] M. Tsuda, T. Ishida, T. Nogami, S. Kurono, M. Ohashi, *J. Chem. Soc., Chem. Commun.* **1993**, 1296–1298.

[18] K. Komatsu, Y. Murata, N. Sugita, K. Takeuchi, T. S. M. Wan, *Tetrahedron Lett.* **1993**, *34*, 8473–8476.

[19] P. Timmerman, H. L. Anderson, R. Faust, J.-F. Nierengarten, T. Habicher, P. Seiler, F. Diederich, *Tetrahedron* **1996**, *52*, 4925–4947.

[20] P. Timmerman, L. E. Witschel, F. Diederich, C. Boudon, J. P. Gisselbrecht, M. Gross, *Helv. Chim. Acta* **1996**, *79*, 6–20.

[21] T. Habicher, J.-F. Nierengarten, V. Gramlich, F. Diederich, *Angew. Chem.* **1998**, *110*, 2019–2022; *Angew. Chem., Int. Ed. Engl.* **1998**, *37*, 1916–1919.

[22] P. J. Stang, *Chem. Eur. J.* **1998**, *4*, 19–27.

[23] I. Lamparth, A. Herzog, A. Hirsch, *Tetrahedron* **1996**, *52*, 5065–5075.

[24] X. Camps, H. Schönberger, A. Hirsch, *Chem. Eur. J.* **1997**, *3*, 561–567.

[25] J.-F. Nierengarten, T. Habicher, R. Kessinger, F. Cardullo, V. Gramlich, J.-P. Gisselbrecht, C. Boudon, M. Gross, *Helv. Chim. Acta* **1997**, *80*, 2238–2276.

[26] C. J. Hawker, K. L. Wooley, J. M. J. Fréchet, *J. Chem. Soc., Chem. Commun.* **1994**, 925–926.

[27] M. Brettreich, A. Hirsch, *Tetrahedron Lett.* **1998**, *39*, 2731–2734.

[28] B. Kräutler, T. Müller, J. Maynollo, K. Gruber, C. Kratky, P. Ochsenbein, D. Schwarzenbach, H.-B. Bürgi, *Angew. Chem.* **1996**, *108*, 1294–1296; *Angew. Chem., Int. Ed. Engl.* **1996**, *35*, 1204–1206.

[29] R. Schwenninger, T. Müller, B. Kräutler, *J. Am. Chem. Soc.* **1997**, *119*, 9317–9318.

[30] F. Cardullo, L. Isaacs, F. Diederich, J.-P. Gisselbrecht, C. Boudon, M. Gross, *Chem. Commun.* **1996**, 797–799.

[31] P. J. Fagan, J. C. Calabrese, B. Malone, *Acc. Chem. Res.* **1992**, *25*, 134–142.

[32] A. L. Balch, J. W. Lee, B. C. Noll, M. M. Olmstead, *Inorg. Chem.* **1994**, *33*, 5238–5343.

[33] H.-F. Hsu, J. R. Shapley, *J. Am. Chem. Soc.* **1996**, *118*, 9192–9193.

[34] F. N. Tebbe, R. L. Harlow, D. B. Chase, D. L. Thorn, G. C. Campbell, Jr., J. C. Calabrese, N. Herron, R. J. Young, Jr., E. Wasserman, *Science (Washington D. C.)* **1992**, *256*, 822–825.

[35] A. G. Avent, P. R. Birkett, C. Christides, J. D. Crane, A. D. Darwish, P. B. Hitchcock, H. W. Kroto, M. F. Meidine, K. Prassides, R. Taylor, D. R. M. Walton, *Pure Appl. Chem.* **1994**, *66*, 1389–1396.

[36] Y. Murata, M. Shiro, K. Komatsu, *J. Am. Chem. Soc.* **1997**, *119*, 8117–8118.

[37] M. Rüttimann, R. F. Haldimann, L. Isaacs, F. Diederich, A. Khong, H. Jiménez-Vásquez, R. J. Cross, M. Saunders, *Chem. Eur. J.* **1997**, *3*, 1071–1076.

[38] D. M. Guldi, H. Hungerbühler, K.-D. Asmus, *J. Phys. Chem.* **1995**, *99*, 9380–9385.

[39] D. M. Guldi, K.-D. Asmus, *J. Phys. Chem. A* **1997**, *101*, 1472–1481.

[40] L. Echegoyen, L. E. Echegoyen, *Acc. Chem. Res.* **1998**, *31* 593–601.

[41] C. Boudon, J.-P. Gisselbrecht, M. Gross, L. Isaacs, H. L. Anderson, R. Faust, F. Diederich, *Helv. Chim. Acta* **1995**, *78*, 1334–1344.

[42] C. S. Foote, *Top. Curr. Chem.* **1994**, *169*, 347–363.

[43] T. Hamano, K. Okuda, T. Mashino, M. Hirobe, K. Arakane, A. Ryu, S. Mashiko, T. Nagano, *Chem. Commun.* **1997**, 21–22.

[44] C. Thilgen, A. Herrmann, F. Diederich, *Helv. Chim. Acta* **1997**, *80*, 183–199.

[45] C. Thilgen, A. Herrmann, F. Diederich, *Angew. Chem.* **1997**, *109*, 2362–2374; *Angew. Chem., Int. Ed. Engl.* **1997**, *36*, 2268–2280.

[46] H. Goto, N. Harada, J. Crassous, F. Diederich, *J. Chem. Soc., Perkin Trans. 2* **1998**, 1719–1723.

[47] L. Isaacs, F. Diederich, R. F. Haldimann, *Helv. Chim. Acta* **1997**, *80*, 317–342.

[48] B. Kräutler, M. Puchberger, *Helv. Chim. Acta* **1993**, *76*, 1626–1631.

[49] J. J. P. Stewart, *J. Comput. Chem.* **1989**, *10*, 209–220.

[50] F. Diederich, L. Isaacs, D. Philp, *J. Chem. Soc., Perkin Trans. 2* **1994**, 391–394.

[51] S. H. Friedman, G. L. Kenyon, *J. Am. Chem. Soc.* **1997**, *119*, 447–448.

[52] Y.-Z. An, G. A. Ellis, A. L. Viado, Y. Rubin, *J. Org. Chem.* **1995**, *60*, 6353–6361.

[53] Y.-Z. An, A. L. Viado, M.-J. Arce, Y. Rubin, *J. Org. Chem.* **1995**, *60*, 8330–8331.

[54] P. Seiler, L. Isaacs, F. Diederich, *Helv. Chim. Acta* **1996**, *79*, 1047–1058.

[55] L. Isaacs, P. Seiler, F. Diederich, *Angew. Chem.* **1995**, *107*, 1636–1639; *Angew. Chem., Int. Ed. Engl.* **1995**, *34*, 1466–1469.

[56] R. F. Haldimann, F.-G. Klärner, F. Diederich, *Chem. Commun.* **1997**, 237–238.

[57] F. Cardullo, *ETH Dissertation No.* 12409, Zürich, **1997**.

[58] G. O. Schenk, H. Eggert, W. Denk, *Liebigs Ann. Chem.* **1953**, *584*, 177–198.

[59] J.-F. Nierengarten, V. Gramlich, F. Cardullo, F. Diederich, *Angew. Chem.* **1996**, *108*, 2242–2244; *Angew. Chem., Int. Ed. Engl.* **1996**, *35*, 2101–2193.

[60] J.-F. Nierengarten, A. Herrmann, R. R. Tykwinski, M. Rüttimann, F. Diederich, C. Boudon, J.-P. Gisselbrecht, M. Gross, *Helv. Chim. Acta* **1997**, *80*, 293–316.

[61] T. Habicher, *ETH Dissertation No.* 12965, Zürich, **1998**.

[62] C. Bingel, presentation at the meeting *New Perspectives in Fullerene Chemistry and Physics*, Rome (Italy), **1994**.

[63] R. E. Martin, U. Gubler, C. Boudon, V. Gramlich, C. Bosshard, J.-P. Gisselbrecht, P. Günter, M. Gross, F. Diederich, *Chem. Eur. J.* **1997**, *3*, 1505–1512.

[64] R. Kessinger, F. Diederich, unpublished results.

[65] P. R. Ashton, F. Diederich, M. Gómez-López, J.-F. Nierengarten, J. A. Preece, F. M. Raymo, J. F. Stoddart, *Angew. Chem.* **1997**, *109*, 1611–1614; *Angew. Chem., Int. Ed. Engl.* **1997**, *36*, 1448–1451.

[66] J.-P. Bourgeois, L. Echegoyen, M. Fibbioli, E. Pretsch, F. Diederich, *Angew. Chem.* **1998**, *110*, 2203–2207; *Angew. Chem., Int. Ed. Engl.* **1998**, *37*, 2118–2121.

[67] E. Dietel, A. Hirsch, E. Eichhorn, A. Rieker, S. Hackbarth, B. Röder, *Chem. Commun.* **1998**, 1981–1982.

[68] J.-P. Bourgeois, F. Diederich, L. Echegoyen, J.-F. Nierengarten, *Helv. Chim. Acta* **1998**, *81*, 1835–1844.

[69] R. W. Alder, S. P. East, *Chem. Rev.* **1996**, *96*, 2097–2111.

[70] F. Cardullo, F. Diederich, L. Echegoyen, T. Habicher, N. Jayaraman, R. M. Leblanc, J. F. Stoddart, S. Wang, *Langmuir* **1998**, *14*, 1955–1959.

[71] J.-F. Nierengarten, C. Schall, J.-F. Nicoud, B. Heinrich, D. Guillon, *Tetrahedron Lett.* **1998**, *39*, 5747–5750.

[72] J.-F. Nierengarten, D. Felder, J.-F. Nicoud, *Tetrahedron Lett.* **1999**, *40*, 269–272.

[73] J.-F. Nierengarten, D. Felder, J.-F. Nicoud, *Tetrahedron Lett.* **1999**, *40*, 273–276.

[74] J.-F. Nierengarten, L. Oswald, J.-F. Nicoud, *Chem. Commun.* **1998**, 1545–1546.

[75] J.-F. Nierengarten, C. Schall. J.-F. Nicoud, *Angew. Chem.* **1998**, *110*, 2037–2040; *Angew. Chem., Int. Ed. Engl.* **1998**, *37*, 1934–1936.

[76] (a) J.-F. Nierengarten, D. Felder, J.-F. Nicoud, *Tetrahedron Lett.* **1998**, *39*, 2747–2750; (b) A. Soi, A. Hirsch, *New. J. Chem.* **1998**, 1337–1339.

[77] J. M. Hawkins, A. Meyer, M. Nambu, *J. Am. Chem. Soc.* **1993**, *115*, 9844–9845.

[78] R. Kessinger, J. Crassous, A. Herrmann, M. Rüttimann, L. Echegoyen, F. Diederich, *Angew. Chem.* **1998**, *110*, 2022–2025; *Angew. Chem., Int. Ed. Engl.* **1998**, *37*, 1919–1922.

[79] A. Herrmann, F. Diederich, *Helv. Chim. Acta* **1996**, *79*, 1741–1756.

[80] D. Philp, J. F. Stoddart, *Angew. Chem.* **1996**, *108*, 1242–1286; *Angew. Chem., Int. Ed. Engl.* **1996**, *35*, 1154–1196.

[81] M. Kawaguchi, A. Ikeda, S. Shinkai, *J. Chem. Soc., Perkin Trans. 1* **1998**, 179–184.

[82] P. S. Baran, R. R. Monaco, A. U. Kahn, D. I. Schuster, S. R. Wilson, *J. Am. Chem. Soc.* **1997**, *119*, 8363–8364.

[83] F. Arias, L. A. Godínez, S. R. Wilson, A. E. Kaifer, L. Echegoyen, *J. Am. Chem. Soc.* **1996**, *118*, 6086–6087.

[84] D. A. Leigh, A. E. Moody, F. A. Wade, T. A. King, D. West, G. S. Bahra, *Langmuir* **1995**, *11*, 2334–2336.

[85] U. Jonas, F. Cardullo, P. Belik, F. Diederich, A. Gügel, E. Harth, A. Herrmann, L. Isaacs, K. Müllen, H. Ringsdorf, C. Thilgen, P. Uhlmann, A. Vasella, C. A. A. Waldraff, M. Walter, *Chem. Eur. J.* **1995**, *1*, 243–251.

[86] J. Osterodt, M. Nieger, P.-M. Windscheif, F. Vögtle, *Chem. Ber.* **1993**, *126*, 2331–2336.

[87] H. Imahori, K. Yamada, M. Hasegawa, S. Taniguchi, T. Okada, Y. Sakata, *Angew. Chem.* **1997**, *109*, 2740–2742; *Angew. Chem., Int. Ed. Engl.* **1997**, *36*, 2626–2629.

[88] D. Carbonera, M. Di Valentin, C. Corvaja, G. Agostini, C. Giacometti, P. A. Liddell, D. Kuciauskas, A. L. Moore, T. A. Moore, D. Gust, *J. Am. Chem. Soc.* **1998**, *120*, 4398–4405.

[89] Y. Rubin, S. Khan, D. I. Freedberg, C. Yeretzian, *J. Am. Chem. Soc.* **1993**, *115*, 344–345.

[90] P. Belik, A. Gügel, J. Spickermann, K. Müllen, *Angew. Chem.* **1993**, *105*, 95–97; *Angew. Chem., Int. Ed. Engl.* **1993**, *32*, 78–80.

[91] (a) Y. Rubin, *Chem. Eur. J.* **1997**, *3*, 1009–1016. (b) W. Qian, Y. Rubin, *Angew. Chem.* **1999**, *111*, 2504–2508; *Angew. Chem. Int. Ed.* **1999**, *38*, 2356–2360.

[92] M.-J. Arce, A. L. Viado, Y.-Z. An, S. I. Khan, Y. Rubin, *J. Am. Chem. Soc.* **1996**, *118*, 3775–3776.

[93] M. Taki, S. Sugita, Y. Nakamura, E. Kasashima, E. Yashima, Y. Okamoto, J. Nishimura, *J. Am. Chem. Soc.* **1997**, *119*, 926–932.

[94] M. Taki, Y. Nakamura, H. Uehara, M. Sato, J. Nishimura, *Enantiomer* **1998**, *3*, 231–239.

[95] T. Ishi-i, K. Nakashima, S. Shinkai, *Chem. Commun.* **1998**, 1047–1048.

[96] T. Ishi-i, K. Nakashima, S. Shinkai, K. Araki, *Tetrahedron* **1998**, *54*, 8679–8686.

[97] T. D. James, K. R. A. S. Sandanayake, S. Shinkai, *Angew. Chem.* **1996**, *108*, 2038–2050; *Angew. Chem., Int. Ed. Engl.* **1996**, *35*, 1910–1922.

[98] E. Nakamura, H. Isobe, H. Tokuyama, M. Sawamura, *Chem. Commun.* **1996**, 1747–1748.

[99] H. Isobe, H. Tokuyama, M. Sawamura, E. Nakamura, *J. Org. Chem.* **1997**, *62*, 5034–5041.

[100] M. Prato, Q. C. Li, F. Wudl, V. Lucchini, *J. Am. Chem. Soc.* **1993**, *115*, 1148–1150.

[101] A. Hirsch, *Principles of Fullerene Reactivity*, in [6], pp. 1–65.

[102] L.-L. Shiu, K.-M. Chien, T.-Y. Liu, T.-I. Lin, G.-R. Her, T.-Y. Luh, *J. Chem. Soc., Chem. Commun.* **1995**, 1159–1160.

[103] C. K.-F. Shen, K.-M. Chien, C.-G. Juo, G.-R. Her, T.-Y. Luh, *J. Org. Chem.* **1996**, *61*, 9242–9244.

[104] C. K.-F. Shen, H.-h. Yu, C.-G. Juo, K.-M. Chien, C.-R. Her, T.-Y. Luh, *Chem. Eur. J.* **1997**, *3*, 744–748.

[105] For a possible mechanism, see: E. U. Wallenborn, R. F. Haldimann, F.-G. Klärner, F. Diederich, *Chem. Eur. J.* **1998**, *4*, 2258–2265.

8 Template Controlled Oligomerizations

Ken S. Feldman, Ned A. Porter, Jennifer R. Allen

8.1 Introduction

The synthesis of repeating molecular units with control of stereochemistry has been the subject of much research in the natural products arena, and significant advances have emerged from both traditional ionic chemistry and free-radical chemistry [1]. For example, (–)-lardolure (**1**), the aggression pheromone of *Lardoglyphus konoi*, contains four identical units, as emphasized by the abbreviated and equivalent representation **1'**, Scheme 8-1. Both enantiomers of lardolure have been synthesized using a linear approach by the groups of Mori and Yamamoto [2, 3]. In addition, controlled-length oligomers can be found within the backbone of the polyene macrolide antibiotics, naturally occurring 1,3-polyols that constitute an important class of clinically valuable antifungal agents. A recent synthesis of filipin III (**2**) exemplifies the stepwise, iterative approach that defines current art in the construction of targets containing these repeating molecular segments [4].

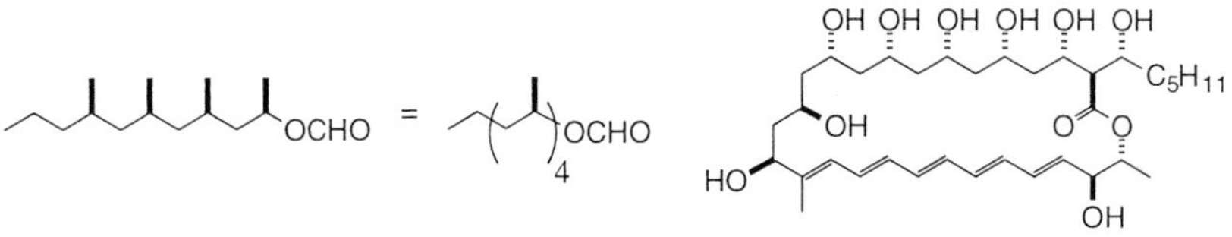

Scheme 8-1   Representative oligomeric natural products.

An attractive and powerful feature of free-radical addition sequences is their ability to create multiple carbon–carbon bonds in a single reaction [5]. In order for this methodology to be successful in the synthesis of repetitive molecular fragments such as those noted above, two general considerations must be addressed: (1) control of stereochemistry, and (2) control of the number of iterative units which become incorporated.

There has been intense interest in the control of stereochemistry of free-radical addition reactions by use of chiral auxiliaries [6] and, even more recently, using enantioselective catalysis [7–12]. As a result, the current understanding of stereochemical control has clearly progressed to the point where this obstacle can be overcome. There have also been reports of strategies that simplify telomer distribution in free-radical oligomerizations [13]. While simple adjustment of the monomer/chain transfer reagent ratio can provide a crude measure of dispersity control, this approach is only satisfactory for the lowest telomers.

Two distinct but related strategies that rely on templates to control the number of monomers incorporated into an oligomeric product can be envisioned. One of these approaches, shown in Scheme 8-2, relies on templated radical macrocyclization reactions to control telomer size [14, 15]. This strategy requires attachment of all of the monomer units to the template backbone and uses macrocyclization, which faces competition from intermolecular chain transfer, to control the telomer length. The chain transfer agent $T-I'$ (i.e., telomerization terminator) is not attached to the template.

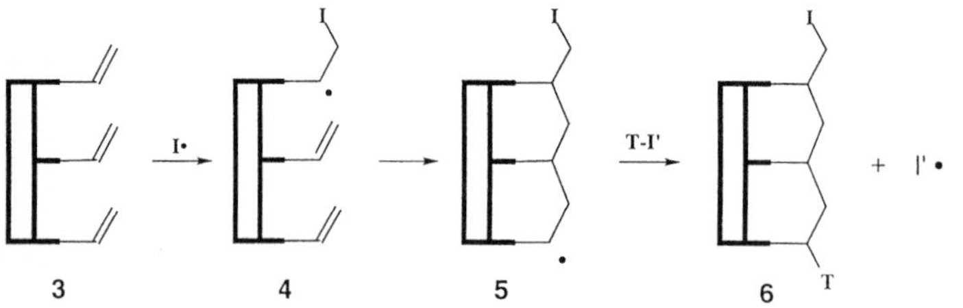

Scheme 8-2   Oligomerization control via template-bound monomers. I· = radical initiator, $T-I'$ = chain transfer agent.

The second approach, shown in Scheme 8-3, relies on attachment of both the radical initiating and radical terminating functionalities, but none of the monomer units, onto a rigid backbone [16–18]. Upon initiation, intermolecular chain growth proceeds until the growing telomer spans the length of the template backbone and intramolecular termination (macrocyclization) becomes feasible. Thus, the distance between the initiating end and the terminating end defines the number of monomer addition steps that can occur before termination.

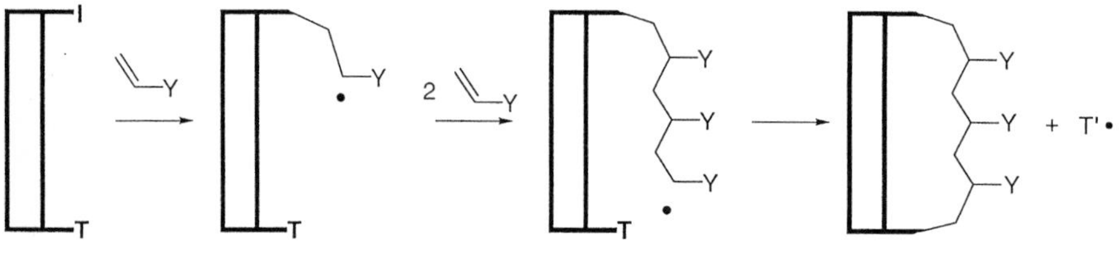

Scheme 8-3   Oligomerization control via template-bound initiator and terminator. I = radical initiator, T = radical terminator.

8.2 Controlled Oligomerizations with Tethered Monomers

One approach to oligomer control in a free-radical polymerization utilizes bound monomers and relies on templated radical macrocyclization reactions. Successful execution of this strategy requires that *cyclotelomerization* effectively compete with intermolecular chain transfer. Scheme 8-2 in Section 8.1 depicts this chemistry schematically wherein radical addition (A), cyclization (C), and chain transfer (T) provide an $n = 3$ telomer. The key macrocyclizations (cyclotelomerizations) must precede chain transfer. These transformations are well precedented by systematic investigations of free-radical macrocyclizations that appeared in the 1980s [19–23] and by the seminal contributions of Kämmerer, Scheme 8-4 [24–34].

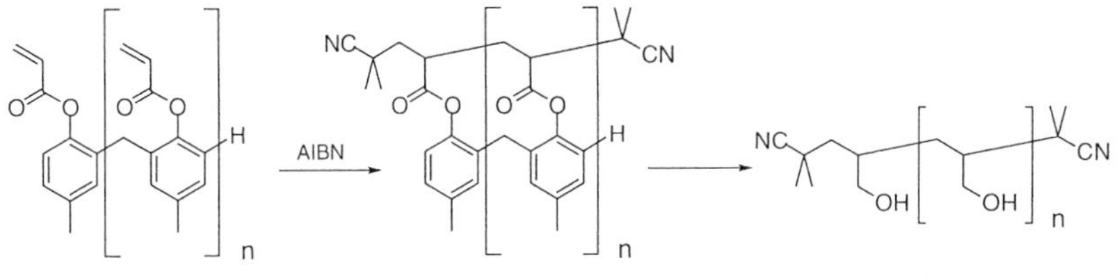

Scheme 8-4   Kämmerer's strategy for template control of acrylate oligomerization.

8.2.1 Templates for $n = 2$ Telomers

An investigation into the preparation of $n = 2$ telomers was successful in showing that the ACT strategy with templates of type **14** is a viable means for producing isotactic 1,3 stereocenters as exemplified in the production of **16** ($n = 2$) (Scheme 8-5). The oxazolidine unit has documented success as a stereocontrol element in acyclic radical reactions [35–37], and thus its incorporation into this template provides, in effect, a chiral auxiliary to control the configuration of new stereogenic centers formed in the sequence.

Exposure of **14** ($m = 3$, R = *tert*-Bu) to cyclohexyl iodide, allyltributylstannane, and AIBN leads to a macrocycle **15** with two new stereogenic centers. The allyl group provides additional functionality for further transformations and also creates a new stereocenter in the process. In order to effect the desired macrocyclization, addition of the first-formed radical to the proximal acrylamide moiety must be faster than addition to the chain transfer agent allyltributylstannane, a requirement that can be fulfilled under appropriate reaction conditions. Premature chain transfer in this particular system, under conditions that discourage bimolecular reaction between two templates, leads to two simple $n = 1$ products (vide infra).

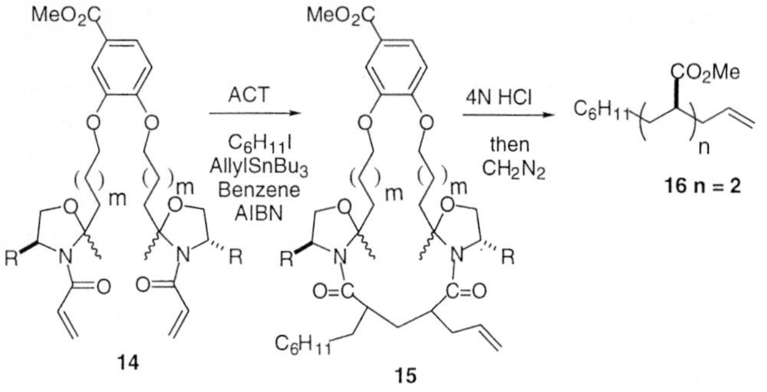

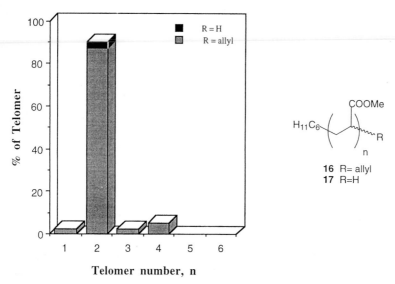

Scheme 8-5 The ACT reaction with templates of type **14** leading to *n* = 2 telomers.

The combination of components within the covalent template assembly **14** that gave the best overall performance was identified through systematic variation in the length of the tether separating the template from the monomer, the configuration at the site of the auxiliary attachment, the auxiliary blocking group R, and the rigid base compound. It was found that each of these variables had a profound influence on the performance of the template assembly in obtaining *n* = 2 telomers **16**. Impressively, in these studies **16** (*n* = 2) could be obtained as 87% of all of the telomers formed with a diastereoselectivity as high as 20 : 1. The telomer distribution histogram for reaction of **14** (*m* = 3, R = *tert*-Bu) is shown in Scheme 8-6.

Scheme 8-6 Telomer distribution for reaction of template **14** (*m* = 3, R = *tert*-Bu) with allyltributylstannane and cyclohexyl iodide.

8.2.2 Templates for *n* = 4 Telomers

It is of some interest to inquire if this template strategy can be applied to the construction of telomers with *n* > 2. This point is of some practical importance since the smaller oligomer units can be approached by efficient synthetic sequences that do not rely on templated chemistry. As the telomer length increases, non-template strategies become tedious and time-consuming, while the template approaches may provide a more direct route to oligomers of defined length. A strategy for achieving *n* = 4 telomers is shown in Scheme 8-7. A template that has four attached monomers can serve, via intramolecular macrocyclizations, to control the telomer length. After addition of an initiating radical to one of the monomers, three consecutive macrocyclizations followed by chain transfer would yield the desired *n* = 4 polycycle **21**. Any deviation from the sequence (e.g., three consecutive cyclizations followed by chain transfer) would not lead to the desired *n* = 4 telomers. Competing processes that are available include failure to cyclize prior to chain transfer, intramolecular chain transfer by a hydrogen atom abstraction process, or intermolecular addition reactions of templates. It is recognized that the last deviation gives rise to the possibility of obtaining telomers higher than *n* = 4, while premature chain transfer gives telomers smaller than *n* = 4.

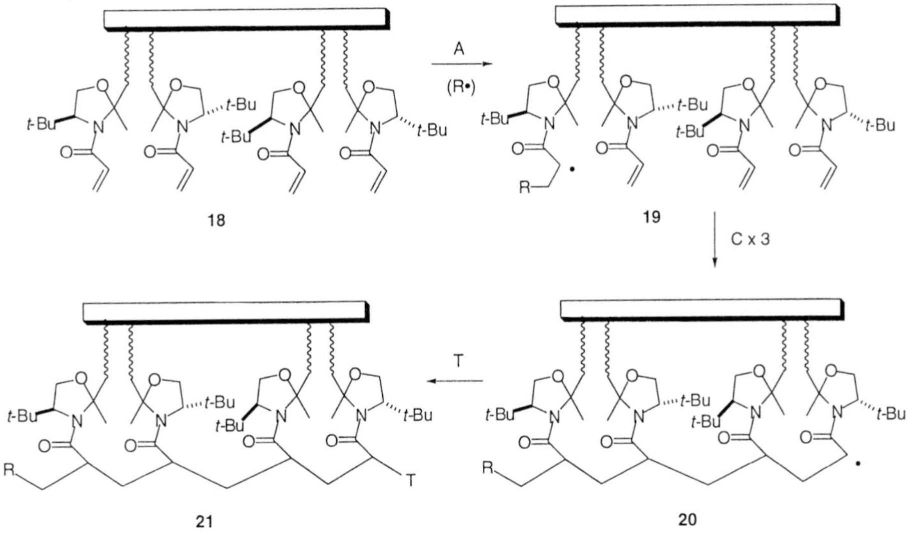

Scheme 8-7   An approach to *n* = 4 telomers with control of stereochemistry.

In analogy to the *n* = 2 template **14**, template **24** was prepared according to Scheme 8-8. This compound was prone to light initiated polymerization when left concentrated and therefore was stored as a stock solution typically on the order of 1.0 M in benzene. The *n* = 4 base was synthesized starting from protocatechuic acid. Bis-ester formation was achieved by protection of the phenolic functionalities (TBDMS) followed by conversion of the free acid to the corresponding acid chloride [38] and coupling with ethylene glycol. Deprotection of all four silyl ethers with HF gave the tetraphenol 22. Tetraetherification of this tetraphenol with an excess of the known oxazolidine acrylamide 23 gave 24 in 45% yield.

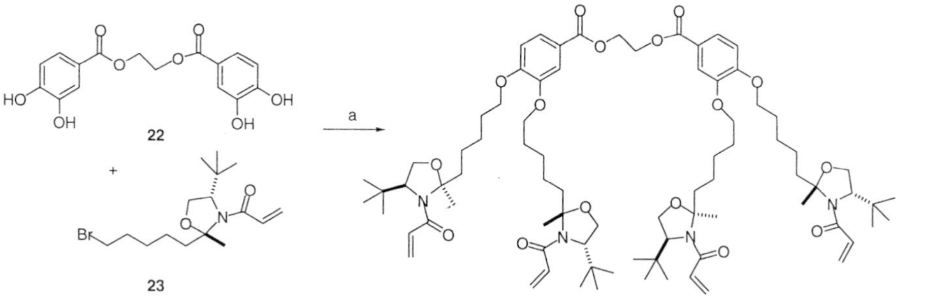

Scheme 8-8 Synthesis of the $n=4$ template **24**. Key: (a) K_2CO_3, catalytic [18]crown-6, 30% DMSO, acetone, reflux, 18 h.

The ACT sequence with template **24** was expected to give a complex mixture of products. Success in obtaining $n=4$ telomers by this sequence is obviously dependent on the balance between the rate of chain transfer, the rate of cyclization, and the rate of intermolecular addition. Formation of the $n=4$ product requires three consecutive macrocyclizations prior to chain transfer. If the rate of chain transfer is too high, premature quenching of intermediate radicals will lead ultimately to smaller telomers. It is clear that cyclization can be promoted, in part, by lowering the concentrations of allyltin (Sn) and template (T), provided that intermolecular reactions remained suppressed.

For analysis, the crude ACT reaction mixtures were subjected to a standard telomer assay as follows [14, 15]. The tin products were removed by chromatography and treatment with KF. The resulting macrocyclic mixtures were then hydrolyzed under acidic conditions to yield the corresponding acrylic acid telomers. Hydrolysis of the desired $n=4$ macrocycle is shown in Scheme 8-9. Esterification of the reaction mixture with diazomethane gave the methyl acrylate telomers **26** that were then readily identified by gas chromatography/mass spectrometry.

A simple template-free methyl acrylate telomerization was conducted using the cyclohexyl iodide/allyltributyltin system in order to identify retention times for the telomers **26** from the ACT reactions. Telomerization of methyl acrylate (900 mM) with 120 mM cyclohexyl iodide and 300 mM allyltributyltin gave 27% of the $n=4$ telomers, Scheme 8-10. Under the best GC conditions found, only four different $n=4$ telomers, out of a possible 16, were observed in this product mixture. This experiment gives a baseline value for telomer distribution obtained without the use of a template. It is clear that simple template-free telomerization is not useful for the preparation of an oligomer of a specific length.

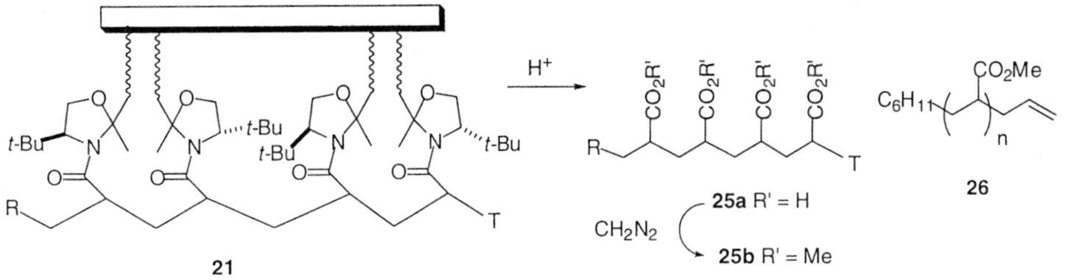

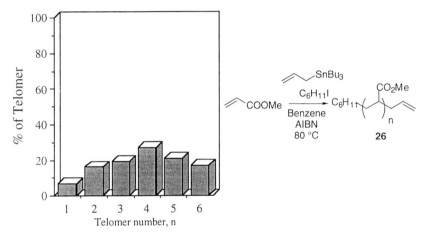

Scheme 8-9   Assay procedures for analysis of ACT reactions giving $n=4$ products.

Scheme 8-10   Histogram for template-free methyl acrylate telomerization with $900^{MA}/120^I/300^{Sn}$.

The reagent ratios chosen for initial studies with template **24** were based on previous work on the templated ACT reaction: 2.5 mM in **24**, 80.0 mM in cyclohexyl iodide (I), and 200.0 mM in allyltin (abbreviated from this point forward as $2.5^T/80^I/200^{Sn}$). The product histogram acquired for this reaction, following the standard telomer assay, is shown in Scheme 8-11 a. The results obtained were promising in some aspects. First, the desired $n=4$ product was the major telomer formed, at 35%. Secondly, only two $n=4$ isomers in a ratio of 8.5:1 were detected by GC analysis. For comparison, the $n=2$ isomers that are easily separated were found in a ratio of 17:1. Lastly, no telomers **26** higher than $n=4$ were detected, an observation consistent with suppression of bimolecular reactions. Based on these results, ACT with **24** was investigated under conditions with lower ratios of both template/initiator and chain transfer agent/initiator. The results obtained from ACT under $1.0^T/80^I/100^{Sn}$ conditions are shown in Scheme 8-11 b. In contrast to the product distribution obtained under the conditions described in Scheme 8-11 a, more of the longer telomers is formed at the expense of the shorter ones. In addition, 50% of all of the telomers formed had $n=4$. Under these conditions, four $n=4$ stereoisomers were detected by GC analysis in a ratio of $1.0:2.1:21.9:1.0$. One complication in this reaction, however, was the production of telomers that were terminated by hydrogen atom abstraction instead of by addition to the chain transfer agent allyltributylstannane. These products are shown in the product histograms as the darker portions in the stacked columns and, in this reaction, make up 10% of the total oligomers formed. The formation of these H-terminated species was also seen in the previous studies with template **14** and presumably results from an intramolecular process [39].

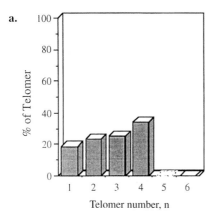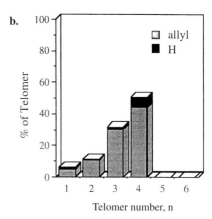

Scheme 8-11   Product histogram for ACT of **24** under (a) $2.5^T/80^I/200^{Sn}$; (b) $1.0^T/80^I/100^{Sn}$ conditions.

A further attempt was made to optimize tetramerization with this template system. However, the results for the ACT of **24** with a $2.5^T/80^I/100^{Sn}$ ratio of components were rather disappointing. The major telomer formed under these conditions was the $n=3$ species and not the desired $n=4$ oligomer. In addition, a small amount of the higher $n=5$ and 6 telomers were seen in these experiments, but no products resulting from hydrogen atom abstraction were detected. Clearly, adjusting the chain transfer agent's concentration to maximize cyclotelomerization and minimize side reactions is difficult with this system.

The feasibility and practicality of the ACT sequence in obtaining diastereomerically pure $n=4$ telomers were investigated using the optimum conditions determined for template **24**. The possibility of isolation of the desired $n=4$ template from all of the other reaction products by the use of HPLC was explored. In a standard template-free stereorandom methyl acrylate telomerization, three peaks are observed for the $n=3$ telomers by normal-phase HPLC and three more broad peaks are separated for the $n=4$ telomers. Each of these latter peaks represents a mixture of isomers as determined by GC analysis.

An ACT reaction with 225 mg of **24** under $1.0^T/80^I/100^{Sn}$ conditions was performed, the resulting mixture was subjected to the telomer assay procedure, and the oligomeric products were then purified by HPLC (Scheme 8-12). Only one peak was observed for the $n=3$ and $n=4$ telomers, although the $n=4$ peak exhibited a slight shoulder. Gas chromatography, high-resolution mass spectrometry and two-dimensional ^{1}H-NMR spectroscopy all revealed that the compound isolated, **27**, was a single $n = 4$ diastereomer. HETCOR analysis showed this compound to be all isotactic, as described below and shown in **27**. The isolated 10 mg of **27** represents a 20% overall yield starting from **24**. In addition, 6 mg of an all-isotactic $n = 3$ telomer was isolated.

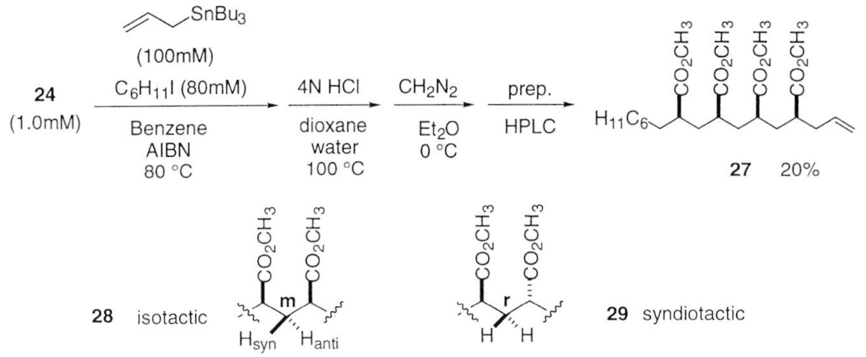

Scheme 8-12 ACT reaction maximizing formation of the $n = 4$ isotactic telomer.

Much literature precedent supports the assignment of tacticity in methyl acrylate polymers using NMR techniques [40, 41]. In the ^{1}H-NMR spectrum, the shift of the methylene protons is sensitive to dyad stereochemistry. For example, in an isotactic (*meso*) dyad **28**, the methylene protons are chemically non-equivalent and appear as two separate sets of signals, whereas in a syndiotactic (racemic) dyad **29**, the methylene protons are equivalent. The ^{1}H-NMR spectrum of **27** showed multiplets at 1.89 and 1.5 ppm due to the two diastereotopic methylene protons of the isotactic dyad. The rest of the spectrum is consistent with the structure of the $n = 4$ tetrad **27**. A racemic dyad structure would have been expected to give resonances of intermediate shift to that of the two resonances observed for the telomer **27**. This evidence strongly implies that **27** has the all-isotactic configuration shown in Scheme 8-12.

Two-dimensional techniques (HETCOR) also give information about triad and tetrad configuration [42]. The ^{1}H multiplets at 1.89 and 1.5 ppm couple to three secondary carbon atoms at 34.94, 34.23, and 33.42 ppm in the ^{13}C spectrum. These methylene carbons do not show crosspeaks with any other region in the proton spectra. Thus, these results show that the major isomer isolated from the ACT reaction of **24** is all isotactic. Although this characterization does not allow determination of the absolute stereochemistry of **27**, the tentative assignment shown is consistent with expectations based upon previous work done with chiral oxazolidine auxiliaries.

The results presented herein show that the ACT strategy to obtain higher telomers has some promise. ACT sequences with templates designed to obtain $n = 4$ telomers have been successful, as the desired product is isolated from these reactions in 20% overall yield. It is clear, however, from the concentration studies that optimization of this system is complex, most expectedly because of the requirement for three consecutive macro-cyclizations. Making these macrocyclizations more efficient would offer a potential solution and could make the ACT sequence more useful.

8.2.3   Ether-based Templates

Templates based on ether tethers also have been examined. Improvements in product yield may be anticipated from the reduced steric congestion in the cyclization step due to the substitution of an –O linkage for a –CH$_2$ linkage [43]. In addition, such oxygenated tethers might serve as crown-ether type substrates, sequestering ions that in turn serve to promote macrocyclization. The beneficial effect on macrocyclization rates by complex-ation of this type has precedent in the work of Mandolini and Illuminati [44].

The synthesis of the requisite glycol tethers is shown in Scheme 8-13 and starts from diethylene glycol (**30**). Hydroxyl monoprotection within **30** as a *tert*-butyldimethylsilyl ether was followed by oxidation of the remaining OH unit to give the corresponding aldehyde **31**. This oxidation reaction was very problematic in that all other oxidants examined gave the ester corresponding to oxidation of the hemiacetal product resulting from starting alcohol attack on product aldehyde. The Dess–Martin reagent [45–47], however, was outstanding in obtaining **31** without the formation of side products. Reac-tion of **31** with methylmagnesium bromide gave the carbinol, and subsequent oxidation with PCC gave the ketone **32**. This ketone was converted to the requisite oxazolidine acrylamides **33** and **34** by the same procedure as was utilized in the synthesis of **14**. These diastereomeric species were easily separable using flash column chromatography. The first-eluting tethers typically have the methyl and auxiliary *anti* to one another whereas the second eluting isomer has the alternative *syn* arrangement.

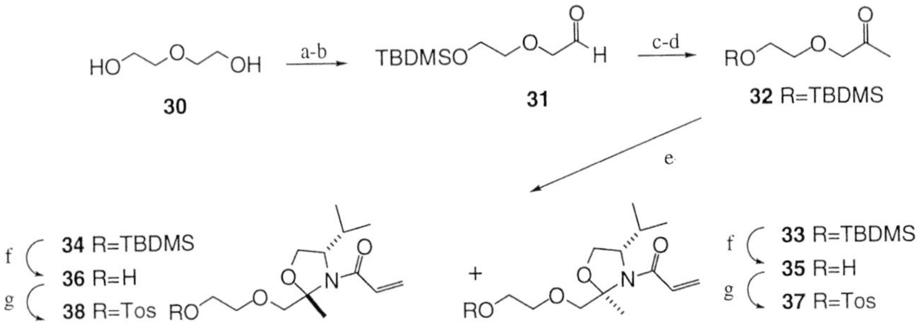

Scheme 8-13   Synthesis of ether-based tethers. Key: (a) TBDMS-Cl, imidazole, DMF, 93%; (b) Dess-Martin reagent, CH$_2$Cl$_2$, 100%; (c) MeMgBr, Et$_2$O, 0 °C, 81%; (d) PCC, CH$_2$Cl$_2$, mol sieves, 90%; (e) *S*-valinol, benzene, cat. TsOH, 80 °C then acryloyl chloride, Et$_3$N, cat. DMAP, 69% com-bined; (f) Bu$_4$NF, THF, 0 °C, 100%; (g) TsCl, pyridine, CH$_2$Cl$_2$, 85%.

The corresponding $n = 4$ polyether template **40** was synthesized according to Scheme 8-14. Coupling of the methyl ester of protocatechuic acid (**39**) with two equivalents of the diastereomerically pure tosylate **38** gave **40** in 83% yield.

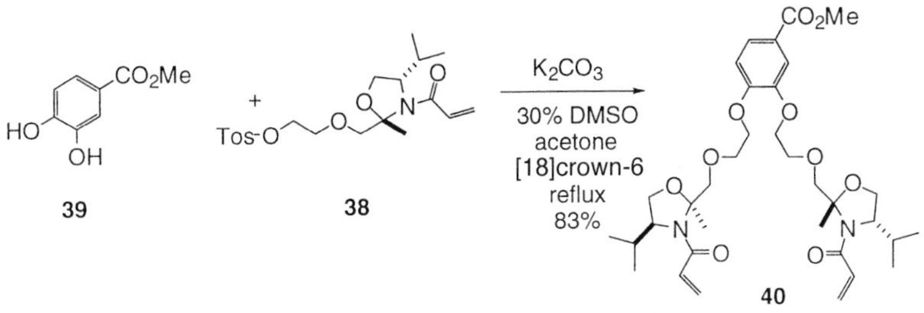

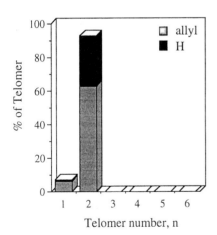

Scheme 8-14   Preparation of template **40** with ether-based tethers.

It should be noted that the ACT sequence and the standard telomer assay were employed to study oligoselectivity with this template and thus the products analyzed are the same methyl acrylate telomers identified in the previous studies. The ACT reaction of **40** under standard conditions with cyclohexyl iodide and allyltributyltin ($2.5^{T}/80^{I}/200^{Sn}$), without the aid of a counter-ion, showed interesting results. The product histogram obtained with **40** after telomer assay is presented in Scheme 8-15.

Scheme 8-15   Product histogram for ACT of **40** under $2.5^{T}/80^{I}/200^{Sn}$ conditions.

Although 96% of the telomers formed were the desired $n = 2$ species, a significant amount (30%) were terminated by hydrogen atom abstraction to give **42** rather than by reaction with allyltributyltin to furnish **16** ($n = 2$) (Scheme 8-16). While intervention of this competing chain transfer process has been seen in previous work (vide supra), in this sequence it appears to be extreme. Modifying the tethers to include an oxygen atom apparently promotes a more facile intramolecular hydrogen atom abstraction.

Scheme 8-16   Proposed H-atom abstraction in the chain transfer sequence.

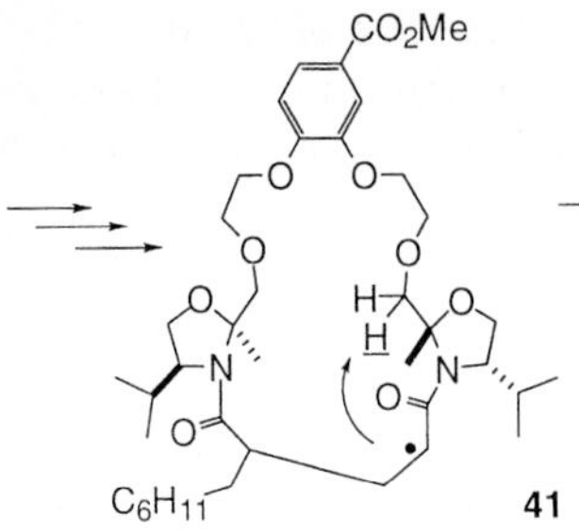

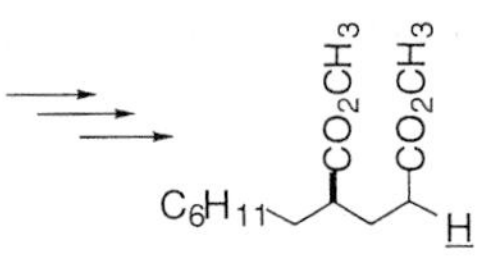

A possible source of the abstracted hydrogen atom is indicated in Scheme 8-16 [48]. A 1,5-abstraction as shown is analogous to "back-biting" reactions found in polymer chemistry and therefore it would not be surprising if these abstractions were operating in the ACT reaction. Chain transfer of the intermediate carbon radical in **41** to the methylene adjacent to the tether oxygen would continue the chain, and work-up would lead to product **42**. The hydrogen atom source has not, however, been unambiguously identified.

As a first impression, these results were somewhat promising because they showed that replacing a carbon atom with an oxygen atom in the tether did enhance the macro-cyclization. For example, in the analogous five-atom system with carbon (e.g., **14**, $m = 3$, R = *tert*-Bu), the same conditions resulted in 80.5% of the $n = 2$ telomer compared to 95% in the system with the glycol tethers. This enhancement is not surprising because of the reduced transannular interactions in the glycol system. The drawback in this system remains, however, the significant amount of hydrogen atom abstraction rather than chain transfer (30%). Complexing an appropriate ion within the macrocycle might alter these product ratios. Telomerizations carried out in the presence of potassium tetraphenylborate and other tetraarylborate salts did not result in any change in the product histograms, nor were any changes observed in the ^{1}H-NMR spectrum of the starting template **40**, even at high temperatures. A systematic investigation of these templates with a variety of Lewis acids may lead to an appropriate system for this free-radical transformation. This work remains for the future.

8.3   Controlled Oligomerizations with a Polynorbornane-based Template

The template approach to controlling oligomerization described in the preceding section requires prior attachment of the monomer units to a (semi-)rigid backbone. An alternative strategy relies on template-bound initiating and terminating functionalities that spatially define the reactivity loci for simple unattached monomers (Scheme 8-3). Upon initiation, intermolecular chain-growth polymerization proceeds until the growing telomer bridges the template backbone and intramolecular termination via macrocyclization becomes feasible. Thus, the size of the template-bound oligomer produced by this sequence should, in principle, correlate with the size of the initiator–terminator gap of the template. This approach to oligoselective polymerization is described in this section.

8.3.1 Template Design and Synthesis

The synthesis of rigid rod templates with the general structure **7** is predicated upon two key observations: (1) Norbornadiene (NBD, **43**) can be polymerized under nickel(0) catalysis to give predominantly *trans-anti-trans* disposed oligomers [49], and (2) norbornene is an effective dienophile in anthracene-based Diels–Alder reactions [50]. From this foundation, templates with appropriately positioned functionality featuring either 13, 17, or 24 Å gaps spanning putative initiator and terminator units can be assembled. In each case, issues of stereoisomer separation and stereochemical assignment must be addressed before oligomerization studies start. In addition, reagent compatibility while selectively manipulating chemically disparate initiator and terminator moieties, and an overarching concern about solubility of intermediates and the final templates, all demand attention. The resolution of these issues en route to the preparation of a family of (NBD)$_n$-based templates of potential use for controlling oligomer length in polymerization reactions is described below.

The nickel(0)-mediated polymerization of norbornadiene developed by Pruett and Rick at Union Carbide represents one of the simplest and most efficient protocols for accessing functionalized, rigid rod molecules of approximately nanometer length. This process could, in principle, provide a broad mixture of oligomeric polynorbornadiene products. However, concentration/solvent regimes can be identified under which a physical property, solubility, imposes a stop-message on the polymerization. Thus, reaction of a concentrated solution of norbornadiene in 1,4-dioxane with 0.2 mol% of NiCOD$_2$ precatalyst leads to a solid residue from which the *trans* dimer **44**, the *trans-anti-trans* trimer **45** and the *trans-anti-trans-anti* tetramer **46** can be isolated in pure form by sequential fractional sublimation and then recrystallization from isopropanol (*i*-prOH) (Scheme 8-17). The final recrystallization removes small amounts of diastereomeric oligomers. It is noteworthy that the dimer/trimer/tetramer suite of products accounts for 94% of the starting norbornadiene. NBD tetramer insolubility severely limits further reaction. Structurally similar rigid rod spacers have been prepared through an iterative linear synthesis by Warrener and Butler [51, 52]. These species are formed as single isomers by this stepwise approach without the need for a tedious separation, but at the cost of overall efficiency. For the planned oligomerization scouting studies, rapid access to

Scheme 8-17   NiCOD$_2$-catalyzed norbornadiene oligomerization.

large amounts (>1 g) of template was deemed essential, and so the Union Carbide group's "shotgun" oligomerization approach was preferred.

A high-lying HOMO for the strained alkene in norbornene and, presumably, the higher oligomers **44–46**, makes it an attractive partner for an electron-deficient diene such as methyl 9-anthracenecarboxylate (**47**) (Scheme 8-18). High yields of a 1:1 stereoisomeric mixture of the double Diels–Alder cycloaddition products **48/49** and **50/51** emerged from combination of **45** and **46** with **47**, respectively [17]. In both cases, the diastereomeric products were separated cleanly by chromatography. The C_s **48** and C_2 **49** (NBD)$_3$ isomers could be distinguished by their characteristic ^{1}H-NMR signals (**48**: $H_a \neq H_b$, δ 1.73, 1.63; **49**: $H_a = H_b$, δ 1.67). The (NBD)$_4$ templates **50** and **51** were more difficult to differentiate, but eventual recourse to the derived Mosher esters **52** and **53**, respectively, permitted unambiguous assignment. Thus, the diester **52** prepared from the C_2 symmetric diester **50** exists as two diastereomers that could be separated by HPLC, whereas its counterpart **53**, derived from the C_i-symmetry diester **51**, is a single compound that displays two distinct OCH$_3$ signals.

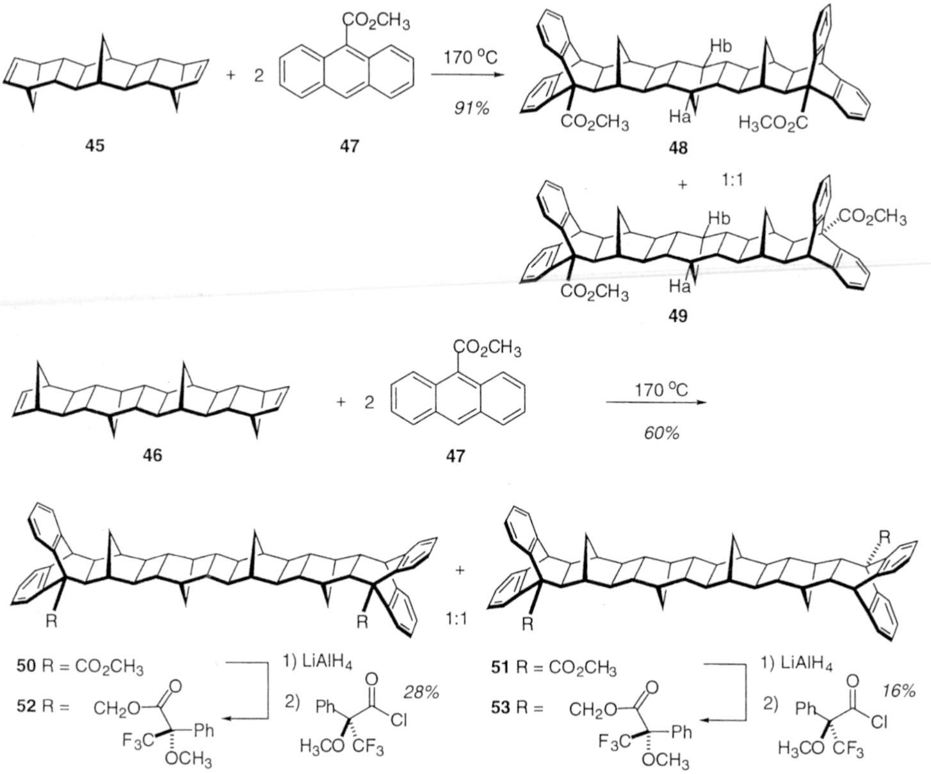

Scheme 8-18   Template synthesis via double Diels–Alder cycloaddition of anthracene endcaps to norbornadiene oligomers.

Access to an even longer template formally derived from an NBD hexameric spacer can be achieved via the mono Diels–Alder adduct **54** (Scheme 8-19). This species, available through incomplete reaction of **45** with **47**, can be photodimerized under the influence of $(CuOTf)_2$ to furnish the $(NBD)_6$-containing templates **55/56** in modest yield as an inseparable mixture of diastereomers. No further studies were performed with this material, as a pure sample of the required C_2 isomer was not forthcoming.

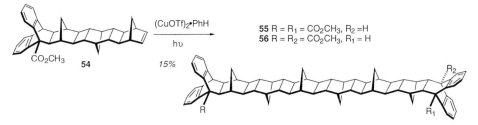

$(CuOTf)_2 \cdot PhH$

$h\upsilon$

15%

55 R = R$_1$ = CO$_2$CH$_3$, R$_2$ =H
56 R = R$_2$ = CO$_2$CH$_3$, R$_1$ = H

CO$_2$CH$_3$ **54**

Scheme 8-19   Synthesis of an $(NBD)_6$-based template by photodimerization of a norbornene precursor.

The majority of the exploratory template functionalization reactions were pursued with the C_s-symmetric $(NBD)_3$-based template **48**. A series of related templates designed to arrest free-radical alkene polymerization, **63–66**, were prepared from diester **48** through straightforward chemistry (Scheme 8-20) [17]. A trichloroacetyl initiator was chosen based on the documented utility of this species in methacrylate polymerizations, especially graft polymerizations [53]. The choice of terminator moiety required some deliberation, given the absolutely critical role that this unit will play in stopping the polymerization. The methacrylate derivatives **63–66** were selected for two attributes that might predispose this template construct towards oligoselectivity: (1) They should be intrinsically more reactive to a methacryloyl radical than the monomer methacrylate, and (2) they feature an effective radical leaving group appropriately positioned to serve as a "stop-message" upon radical addition to the alkene terminus.

Similar functionalization chemistry with the $(NBD)_4$-derived diester **50** afforded the trichloroacetyl, phenylthiomethacrylate-bearing template **67** whose initiator–terminator pair spans an approximately 17 Å gap (Scheme 8-20). Single-crystal X-ray analysis of the phenylthio-containing species **63** provided a detailed glimpse of the fully functionalized template (**63′**, Scheme 8-20). The anthracene-derived endcaps appear bowed slightly away from the apical *exo* hydrogens of the proximal norbornane units, as predicted by the molecular mechanics-based structural model **63″**. These distortions are unlikely to have any consequence for the planned chemistry.

Template-imposed oligoselectivity in (anionic) group transfer polymerization was also of interest with this template system. A suitable initiator–terminator pair for effecting this chemistry was proposed to be the trimethylsilylacetate/fluoromethacrylate species shown in the $(NBD)_3$-derived template **69** (Scheme 8-21). Under fluoride catalysis,

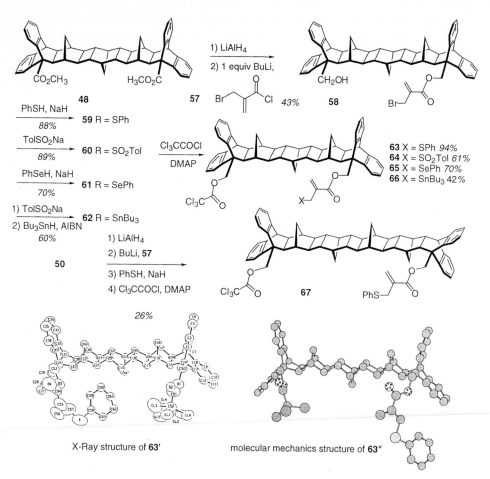

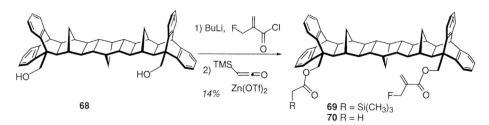

Scheme 8-20 Synthesis of initiator–terminator functionalized templates for controlling the free radical polymerization of activated alkenes.

Scheme 8-21 Synthesis of an initiator–terminator functionalized template for controlling the group transfer polymerization of methacrylate.

silylacetates have been reported to initiate the group transfer polymerization of methacrylates [54]. Termination of the silicon transfer oligomerization was anticipated to result from living-end silyl ketene acetal addition to the template-bound fluoromethacrylate unit, leading to fluoride ejection. The construction of these initiator and terminator units followed the strategy described earlier and is detailed in Scheme 8-21.

8.3.2   Measuring the Initiator–Terminator Gap in Molecular Terms

The measured span between initiator and terminator attachment points (e.g., 13 Å with **48**) does not accurately represent the distance that must be traversed by a growing polymer chain as it bridges the initiator–terminator gap. Both the conformational preferences and the steric demands of the growing substituted hydrocarbon chain will surely conspire to trace a non-least-motion path of substantially greater length. Thus, prediction of oligomer size for any given template will not be straightforward. One approach to empirically measuring the "molecular size" of the initiator–terminator span utilizes bis-lactonization reactions with the $(NBD)_3$ template diol **68**. In this experiment, the diol **68** is condensed with either the bis-acid chlorides **71–76** (DMAP catalysis) or with the corresponding bis-acids **77–82** (Steglich/Keck conditions [55]) to furnish bis-lactones **84** (Scheme 8-22). Assuming that the rate of initial ester formation (e.g., **83**) is independent of the chain length n, the yields of bis-lactone **84** will scale with the rate of macrolacton-

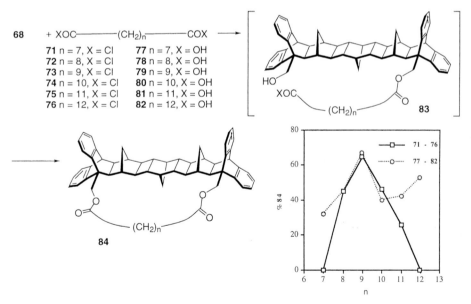

Scheme 8-22   Bis-lactonization experiments to "measure" the effective initiator–terminator gap in the $(NBD)_3$-derived template.

ization. In turn, this macrolactonization rate will be sensitive to the chain dynamics and chain length of each activated diacid. Thus, to a first approximation, the highest-yielding macrolactonization should reveal the optimal chain length for the corresponding alkene oligomerization process. The data are presented in Scheme 8-22. Under both sets of reaction conditions, a clear preference for macrocyclization when $n=9$ is observed. This chain length translates into the expectation that three olefinic monomer units should optimally fit between the initiator and terminator units of templates **63–66**.

8.3.3   Model Studies for Optimizing Terminator Performance

Preliminary scouting experiments were conducted prior to actual oligomerization studies in order to explore the effect of the radical leaving group X (cf. **86–91**) in the key coupling step (Scheme 8-23). The halogen-containing model terminators were not effective partners for a chloroacetate-derived radical. In contrast, all of the chalcogen-based terminators performed with acceptable efficiency. Examination of relative radical leaving group abilities [56] (Scheme 8-23) does not lend any support to hypotheses that attempt to correlate reaction efficiency with X˙ departure rates. While these experiments permitted exclusion of halogen-based terminators from further consideration, at this juncture the basis for choosing between the chalcogen variants was not evident. A final model experiment (Scheme 8-23) revealed that the trichloroacetate template moiety **85** and the chalcogen-based terminator template **90** indeed serve as an effective initiator–terminator pair for the uncontrolled oligomerization of methyl methacrylate.

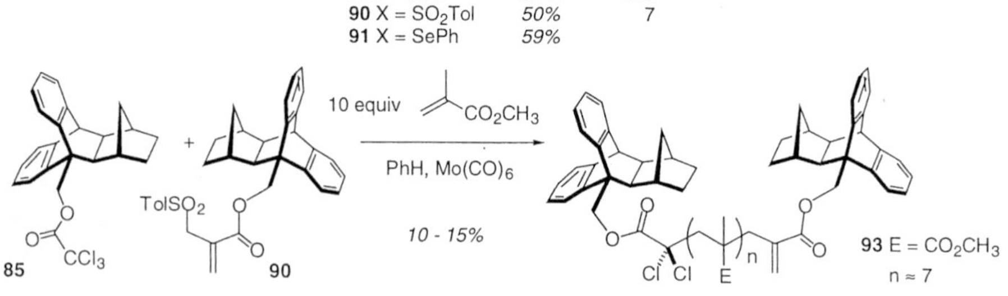

	rel. rate [56]
86 X = Cl 0%	1
87 X = Br 0%	65
88 X = SnBu3 51%	
89 X = SPh 69%	110
90 X = SO2Tol 50%	7
91 X = SePh 59%	

Scheme 8-23   Model experiments designed to test the effectiveness of various terminating functionalities in free-radical polymerizations.

8.3.4 Template Controlled Oligomerization Studies

These encouraging preliminary results prompted extensive optimization studies of the template controlled oligoselective free-radical polymerization of methyl methacrylate (MMA) using the family of templates **63–65** (Scheme 8-24) [16]. Early experiments with the sulfonyl terminator system **64** led, for the first time, to isolation of small quantities of macrocyclized material **94** – template controlled oligomerization had been achieved! Initial explorations of the effect of concentration, reagent ratio, and temperature variations on product **94** yield for the sulfonyl-based system led to the following optimized parameters: a 10-fold excess of monomer MMA and a 1.2-fold excess of $Mo(CO)_6$ over template in benzene (10 mM in template) at 150 °C. These optimized conditions resulted in isolation of ca. 16% of the desired MMA-trimer-bound template **94** along with 25% of recovered template **64** and an equal amount of uncontrolled oligomer. Solvent characteristics also play an important, if ill-defined, role in maximizing the yield of macrocyclized material. Limited template solubility necessarily constrained the solvent choices, but successful template controlled oligomerization of MMA with template **64** proceeded in benzene, ethyl acetate, and isopropanol. Benzene appeared to be the solvent of choice among these options. Further examination of the influence of leaving group X on controlled oligomer formation revealed a substantial improvement with the thiophenyl-based species **63**, whereas the PhSe analogue **65** offered little advantage over the sulfonyl case. Lower temperatures and longer reaction times also contributed to improved yields of macrocyclized product, and eventually as much as 47% of the MMA trimer species **94** could be isolated when the reaction was run under optimized conditions. A brief look at the effect of solvent variation on macrocycle **94** yield with the

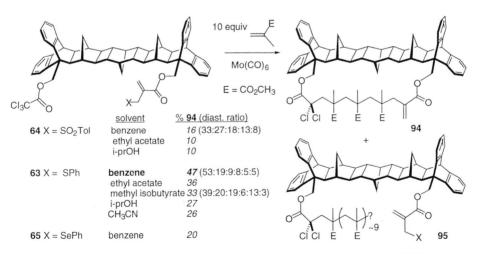

Scheme 8-24 Oligoselective free-radical polymerization of MMA with the $(NBD)_3$-derived template: effects of variation in solvent and radical leaving group.

superior thiophenyl-bearing template **63** provided results consistent with the earlier studies using **64**. Careful examination of the cyclization mixtures did not afford any evidence for formation of macrocyclized material with other than three MMA monomers bridging the initiator–terminator gap. Inspection of this $(MMA)_3$ linker segment reveals that the $n = 9$ prediction from the macrolactonization experiments was precisely borne out.

The stereochemical outcome of this oligoselective polymerization is of some interest. In principle, eight stereoisomeric macrocycles **94** can be formed. However, HPLC analysis of the cyclized material revealed that only six of these possibilities are represented in the product mixture. In benzene as solvent, over half of the product mixture is a single stereoisomer, whereas in methyl isobutyrate as solvent the diastereomers are more evenly distributed. Preliminary attempts to ascertain the relative stereochemistry of the major isomer within **94** via DNOE NMR measurements did not allow unambiguous assignment. Without this structural information in hand, further speculation on the relationship between chain stereochemistry and cyclization efficiency within **99** (see Scheme 8-27) is not warranted. Nevertheless, there must be some influence, given the non-statistical distribution of isomers. In comparison, the ^{1}H-NMR spectrum of the pMMA portion of uncontrolled oligomer **95** is superimposable with that of atactic (i.e., random stereochemistry) pMMA.

Exploration of the template controlled free-radical oligomerization of other activated olefins began with standard monomers utilized in bulk polymer synthesis and the template **63**. Vinyl acetate and acrylonitrile led only to uncontrolled polymerization, while vinylene carbonate did not react under the standard experimental conditions. More exotic monomers, such as vinyl trifluoroacetate and *tert*-butyl acrylate, were also unsuccessful. Only methyl acrylate polymerization was arrested by template **64** to provide the macrocyclized product **96** in modest yield as a mixture of five diastereomers (Scheme 8-25). Subsequent studies with the more effective thiophenyl-bearing template **63** at lower temperatures improved this yield to 35%. The diastereomer distribution was reminiscent of the methyl methacrylate-derived product, although no stereochemical assignments were made in this case either.

Attempts to extend the template mediated free-radical oligomerization of MMA to the longer $(NBD)_4$-derived template **67** did not afford unequivocal results (Scheme 8-26).

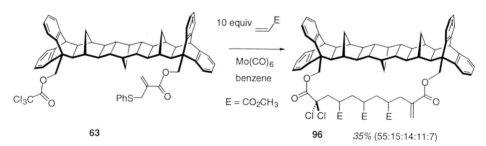

Scheme 8-25   Oligoselective free-radical polymerization of methyl acrylate with the $(NBD)_3$-derived template **63**.

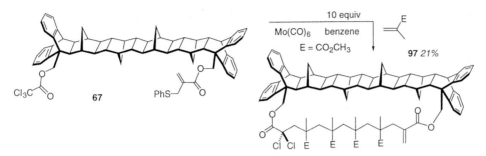

Scheme 8-26 Oligoselective free-radical polymerization of MMA with the (NBD)$_4$-derived template **67**.

Treatment of template **67** with 10 equivalents of MMA and 1.2 equivalents of Mo(CO)$_6$ in benzene at 70 °C furnished a 21% yield of material which appeared by ^{1}H-NMR spectroscopy to consist of a mixture of diastereomeric template-bound MMA tetramers. Unfortunately, all attempts to confirm this molecular composition by mass spectral analysis were not fruitful.

Finally, the ability of the silylacetate, fluoromethacrylate-bearing template **69** to enforce oligoselectivity on the group transfer polymerization of MMA was examined. However, no fluoride-initiated polymerization conditions could be identified which led to template-bound oligomer. Typically, off-template initiation (F$^-$ + MMA?) furnished uncontrolled polymer. In these cases, the template **69** suffered simple desilylation to deliver the acetate **70**. Perhaps a template-bound silyl ketene acetal initiator would have performed in a more desirable manner, but that species remained elusive.

8.3.5 Mechanistic Studies: Probing Bimolecular Termination and Solvent Effects

The maximum yield of the MMA trimerization product **94** from template **63** was 47%. In addition to this macrocyclic product, ca. 10% of unreacted template was recovered. The fate of the remaining ca. 40% of the original template was of some concern. An additional 20–25% of the template appeared to have a pMMA chain extending by about nine monomer units, on average (^{1}H-NMR integration), from the initiator, and an intact terminator (cf. **100**). The terminal functionality on the pMMA chain could not be identified, but no evidence for termination by the phenylthiomethacrylate moiety of a second template (i.e., bimolecular termination) could be found. At present, candidates for this radical quenching species include Cl and PhS (from Ph$_2$S$_2$ formed as the reaction progresses). Thus, the intervention of bimolecular termination as a significant yield-limiting process is called into question.

This curious observation refocuses concerns about yield loss to a critical choice that a template-bound pMMA radical **99** faces: macrocyclization to afford the controlled-length oligomer **94** vs. addition of a fourth monomer to lead ultimately to **100** (Scheme 8-27). Furthermore, the energetics of these diverse reaction channels seem to be influenced by some characteristic of the solvent, as the aromatic solvent benzene apparently differentially favors the former path over the latter one compared to other non-aromatic solvents. Additional insight into the role (or lack thereof) of bimolecular termination in limiting macrocycle yield, and also into the effect of solvent on the options available to radical **99**, was sought through the series of terminator challenge experiments discussed below.

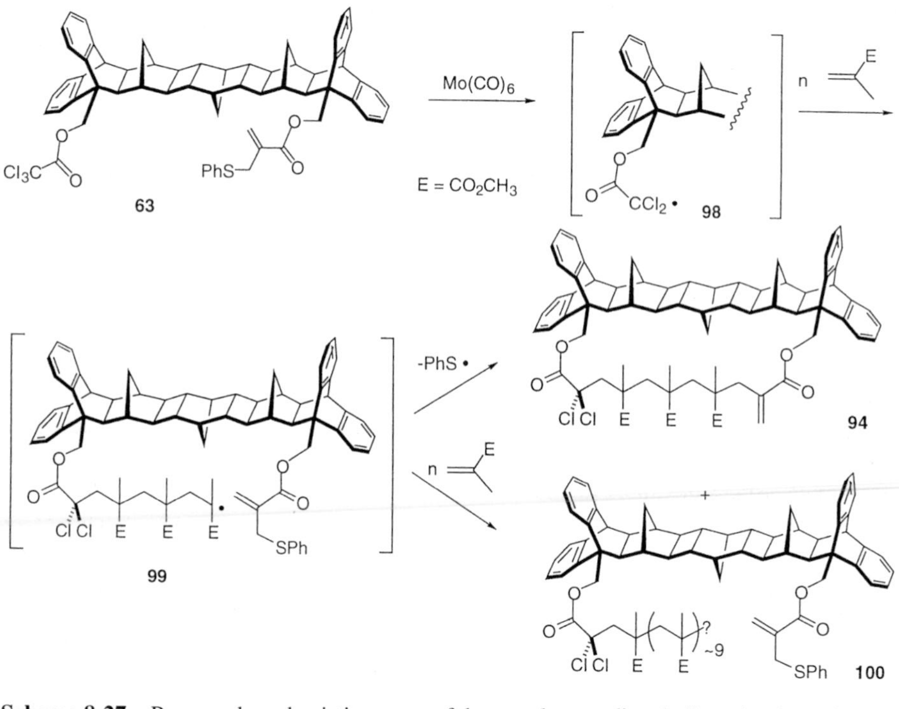

Scheme 8-27 Presumed mechanistic course of the template mediated oligoselective polymerization of MMA.

Varying amounts of two competing terminators, the simple methyl ester **106** and the template model ester **107**, were introduced into the standard MMA oligomerization experiment with template **63** (Scheme 8-28) [18]. Both of these exogenous terminators were of comparable intrinsic reactivity in combination with a simple methacryloyl radical, as seen by the model reaction **101** + (**102** + **103**) → **104** + **105** shown in Scheme 8-28. The yield of macrocycle **94** was monitored as a function of the amount of **106** or **107** added in both the optimum solvent benzene and, independently, in a solvent chosen to resemble the growing pMMA chain, methyl isobutyrate.

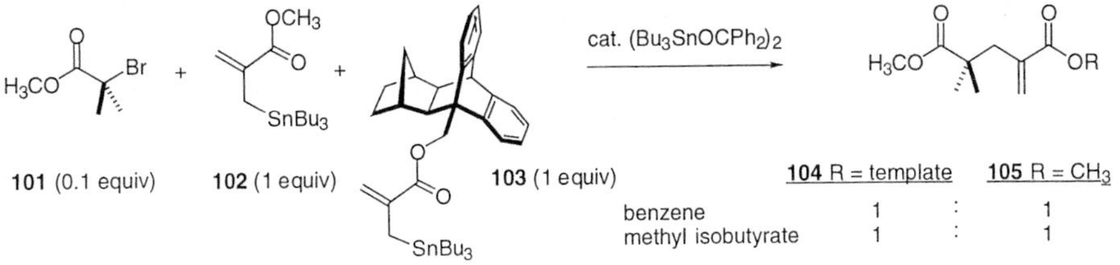

OCH₃ → OCH_3

H₃CO → H_3CO

Br

101 (0.1 equiv)   102 (1 equiv)   SnBu₃

103 (1 equiv)

cat. (Bu₃SnOCPh₂)₂ → cat. $(Bu_3SnOCPh_2)_2$

H_3CO OR

SnBu₃

	104 R = template	105 R = CH₃
benzene	1 :	1
methyl isobutyrate	1 :	1

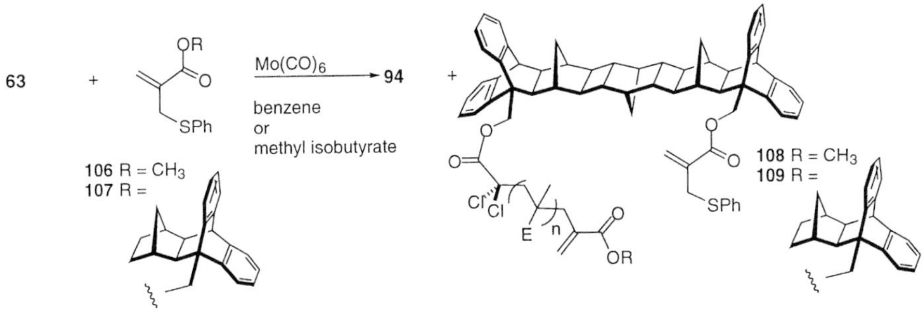

63   +

OR

SPh

106 R = CH₃
107 R =

Mo(CO)₆ → $Mo(CO)_6$ 94   +

benzene
or
methyl isobutyrate

Cl   Cl

E   n

OR

SPh

108 R = CH₃
109 R =

Scheme 8-28   Terminator challenge experiments: relative reactivity of the two probes **106** and **107** (as modeled by **102** and **103**), and presumed role as bimolecular terminators in a controlled oligomerization experiment.

These four experiments were designed to test two related hypotheses: (1) in an actual template system, bimolecular termination is suppressed as a consequence of the demanding steric environment about the terminator moiety. Observation of greater macrocycle yield with the bulkier terminator **107** present compared with the smaller analogue **106** in either solvent, given that these species have equivalent intrinsic reactivity with a small radical, would support this premise. (2) More pronounced solvent–solute interactions in benzene as compared to non-aromatic solvents contribute to favoring intramolecular termination (→ **94**) over bimolecular monomer addition (→ **100**) for radical **99**. Observation of relatively greater macrocycle yield as either terminator concentration is increased in benzene compared to the same reaction in methyl isobutyrate is consistent with this hypothesis.

The results of these studies are presented in complementary graphical form in Schemes 8-29 and 8-30. Scheme 8-29 displays the consequences of increasing terminator concentration on macrocycle yield independently for each solvent. In both benzene and methyl isobutyrate, the yields of macrocycle **94** were greater when **107** was present compared with **106**. Presumably, the bulkier model template terminator **107** is less effective than the smaller methyl ester analogue **106** at intercepting radical intermediates en route to **94**. Given the intrinsic reactivity equivalency between these two species with an unencumbered radical (Scheme 8-28), it is reasonable to conclude that the bulk of the template–oligomer construct **99** and the bulk of the terminator moiety in **107** are responsible for the differential reactivity observed. The differences are not large in either solvent (~ 10%), but they can contribute to the surprising lack of substantial bimolecular termination in the polymerizing system.

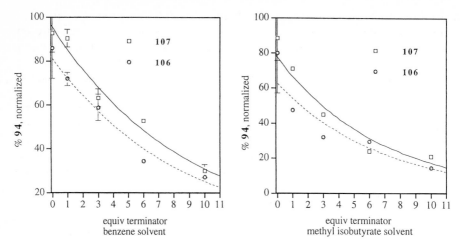

Scheme 8-29   Template macrocycle **94** yield as a function of added terminator **106** or **107** in benzene and, independently, in methyl isobutyrate solvent.

These data can be recast to expose the differing effects of solvent on yield suppression with each terminator (Scheme 8-30). With either **106** or **107**, the yield of macrocycle **94** was greater in benzene than in methyl isobutyrate. Thus, intermolecular trapping of an intermediate template-bound radical is less significant in benzene than in methyl isobutyrate. In these experiments, the differences are large (~20%) and suggest that a template-bound radical (e.g., **99**) is not as accessible to either added terminator, **106** or **107**, in benzene as it is in methyl isobutyrate.

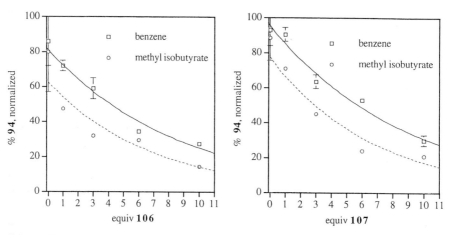

Scheme 8-30   Template macrocycle **94** yield as a function of solvent (benzene or methyl isobutyrate) in the presence of increasing amounts of the added terminators **106** and **107**.

Taken together, these results support an interpretation of the template controlled free-radical oligomerization of MMA in benzene that cites a solvent-induced steric effect in favoring intramolecular termination over bimolecular reaction. Specifically, a solvophobic effect [57–60] may force the growing hydrocarbon-like pMMA chain against the hydrocarbon-like template in benzene. This solvent enforced proximity of reactive entities in **99** may then contribute to an unexpected acceleration of macrocyclization compared with the alternative of further monomer addition. Consistent with this hypothesis, Endo et al. have shown that the relative energy of interaction between two phenyl rings is substantially larger than the energy of interaction of a phenyl ring with a hydrocarbon chain (Table 8-1) [60]. Thus, benzene's affinity for itself may lead to "sequestering" of the hydrocarbon portions of **99**, and in so doing enhance intramolecular reaction compared with a non-aromatic solvent.

Table 8-1   Relative interaction energies between a phenyl ring and various hydrocarbon species (the more negative the value, the stronger the interaction) [60]

Solvent	Relative energy of interaction with a phenyl ring (kcal/mol)
Anisole	−1.51
Benzene	−1.32
Cyclohexane	−0.44
n-Hexane	−0.04
n-Octane	0.00

Finally, the diastereomer distribution for macrocycle **94** was examined in each of the experiments reported in Schemes 8-29 and 8-30. While some variation was observed from run to run, no meaningful trends could be discerned. There is no evidence that certain stereoisomers of radical **99** are intrinsically slower to cyclize and hence would be more susceptible to trapping by **106** or **107**.

8.4   Concluding Remarks

The successful demonstration of oligoselective free-radical polymerizations of acrylate and methacrylate monomers by the template constructs **14, 24, 40, 63**, and **64** establishes the feasibility of this venture. In each case, optimization of both reaction parameters and template attributes led to very low-dispersity oligomers incorporating by design either $n=2$, $n=3$, or $n=4$ monomer units. Yields were tolerable, given the number of new C–C bonds formed and the complexity of the transformations. Furthermore, differing strategies for stereocontrol revealed that stereoselective *and* oligoselective polymerizations are within reach. These preliminary experiments hint at the possibility that future studies may advance these techniques to the point where they will be valuable assets in the synthesis of stereodefined repeating molecular segments of target molecules.

8.5   Experimental Procedures

8.5.1   (2R,4S)- and (2S,4S)-2-(5-Bromopentyl)-2-methyl-3-acryloyl-4-*tert*-butyl oxazolidine (23)

7-Bromo-2-heptanone (208 mg, 1.08 mmol) and *tert*-leucinol (189 mg, 1.61 mmol) were dissolved in benzene (10 mL). Catalytic TsOH was added, and the mixture was refluxed, with a Dean–Stark trap containing molecular sieves, for 18 h. The reaction solution was cooled to rt, solid NaHCO$_3$ was added, and the mixture was poured into saturated NaHCO$_3$ solution (200 mL) and extracted with ether (2×200 mL). The organic extracts were dried over Na$_2$SO$_4$, filtered, and condensed to a pale yellow oil which was dissolved in CH$_2$Cl$_2$ (5.0 mL) and used in the next step. The solution was cooled to 0 °C, triethylamine (0.24 mL, 1.7 mmol) and DMAP (3 mg) were added, followed by freshly distilled acryloyl chloride (0.123 mL, 1.51 mmol). The mixture was allowed to warm to rt over 12 h, diluted with CH$_2$Cl$_2$ (50 mL) and washed with saturated NaHCO$_3$ solution (3×10 mL). The organic layer was dried over Na$_2$SO$_4$ and condensed to a yellow oil which was purified by column chromatography (10% EtOAc/hexane) to give a combined yield of 195 mg (52%) of two diastereomers. First eluting diastereomer: R_f 0.20 (20% EtOAc/hexanes); ^{1}H-NMR δ 6.51 (dd, J = 16.6, 10.0 Hz, 1H), 6.33 (dd, J = 16.6, 1.9 Hz, 1H), 5.60 (dd, J = 10.0, 1.9 Hz, 1H), 4.40 (d, minor rotamer), 3.99 (d, 1H), 3.89 (m, 1H), 3.75 (d, 1H), 3.45 (t, 2H), 2.42 (m, 1H diastereotopic), 1.85 (m, 2H), 1.68 (s, 3H), 1.6–1.2 (b, 5H), 0.95 (s, 9H), 0.92 (s, minor rotamer); ^{13}C-NMR δ 164.90, 130.69, 127.36, 98.24, 65.38, 64.80, 35.89, 34.65, 33.82, 32.68, 28.29, 27.67, 27.32, 23.98; GC/CIMS (CH$_4$/NH$_3$) *m/z* 347 (MH$^+$). Second-eluting diastereomer: R_f 0.10 (20% EtOAc/hexanes); ^{1}H-NMR δ 6.51 (dd, J = 16.6, 10.0 Hz, 1H), 6.34 (dd, J = 16.6, 1.9 Hz, 1H), 5.63 (m, 1H), 4.1–3.7 (set of 3 multiplets, 3H), 3.38 (t, 2H), 2.58 (m, 1H diastereotopic), 1.85 (m, 2H), 1.51 (s, 3H), 1.6–1.2 (b, 5H), 0.95 (s, 9H); ^{13}C-NMR δ 165.05, 130.76, 127.30, 98.04, 65.12, 64.64, 39.05, 35.70, 33.79, 32.61, 28.42, 27.68, 23.58, 19.69; GC/CIMS (CH$_4$/NH$_3$) *m/z* 347 (MH$^+$).

8.5.2   1,2-Bis-[3,4-bis-(5-(2R-methyl-3-acryloyl-4S-*tert*-butyl-2-oxazolidinyl)-pentyloxy)]-ethane (24)

The following procedure is representative for the synthesis of the n = 4 templates. Tetrol 22 (110 mg, 0.33 mmol) and 23 were dissolved in acetone (3.5 mL) and distilled DMSO (1.5 mL). The solution was charged with K$_2$CO$_3$ (273 mg, 1.98 mmol) and catalytic [18]crown-6 and brought to reflux for 18 h. The reaction mixture was cooled, extracted with ether (2×50 mL) and washed with saturated NaCl solution (2×100 mL). The organic extracts were pooled, dried over Na$_2$SO$_4$, filtered, and condensed to a viscous oil which was purified by column chromatography (50% EtOAc/hexane) to give 24 as 204 mg (44%) of a colorless oil: R_f 0.32 (50% EtOAc/hexanes); ^{1}H-NMR δ 7.61 (dd, J = 8.1, 1.5 Hz, 2H), 7.49 (d, J = 1.5 Hz, 2H), 6.81 (d, J = 8.4 Hz, 2H), 6.52 (dd, J = 16.6, 9.9 Hz, 4H), 6.33 (dd, J = 16.6, 1.8 Hz, 4H), 5.64 (dd, J = 9.9, 1.8 Hz, 4H), 4.59 (s, 4H),

4.00–3.87 (m, 16H), 3.79–3.69 (m, 4H), 1.80–1.53 (b, 16H), 1.71 (s, 12H), 1.45–1.30 (b, 16H), 0.93 (s, 36H, major), 0.90 (s, minor); ^{13}C-NMR δ 166.25, 164.91, 153.26, 148.42, 130.77, 129.72, 127.30, 123.81, 114.24, 111.90, 98.31, 69.07, 68.70, 64.79, 62.41, 39.63, 35.91, 34.78, 29.15, 27.69, 27.35, 24.05. Analysis: Calcd. for $C_{80}H_{122}O_{16}N_4$: C, 68.83; H, 8.81; N, 4.01. Found: C, 69.08; H, 8.71; N, 3.74%.

8.5.3 Typical conditions for ACT reaction ($2.5^A/80^I/200^{Sn}$)

From a benzene stock solution of template **14** ($m = 3$, R = *tert*-Bu), 0.0625 mmol (87 mg) of **14** was delivered to a reaction flask charged with cyclohexyl iodide (0.258 mL, 2.0 mmol, 32 equiv) and allyltributylstannane (1.55 mL, 5.0 mmol, 80 equiv) and the reaction diluted to a total volume of 25 mL in benzene. The reaction flask was purged with argon for 30 min, and then the mixture was brought to reflux. An initial portion of AIBN (2 mg, 0.0125 mmol, 0.2 equiv) was added upon reflux, a second portion was added after 2 h, and reflux was maintained for another 10 h. The mixture was cooled to rt and the solvent carefully removed under reduced pressure. The residual oil was subjected to column chromatography using gradient elution (100% hexanes → 5%, 10%, 15%, 25%, 50%, 75% → 100% EtOAc), and the isolated products were dissolved in ether (50 mL) and stirred overnight over an aqueous 10% KF solution (50 mL). The layers were separated, and the organic layer was dried over MgSO$_4$, condensed to a mixture of oils and solids, redissolved in EtOAc (5.0 mL), refiltered, and condensed to an oil.

8.5.4 Telomer assay

The isolated oil was mixed with 1 M HCl (3.0 mL) and *p*-dioxane (3.0 mL) and refluxed at 100 °C for 12 h. The reaction mixture was poured into 1 M KOH (100 mL) and washed with ether (50 mL). The aqueous layer was cooled in an ice bath and acidified with conc. HCl (approximately 10 mL), followed by stirring for 2 h with ether (50 mL) while saturating the aqueous layer with NaCl. The layers were separated, and the aqueous layer was further extracted with ether (2×25 mL). The pooled ethereal layers were dried over MgSO$_4$, filtered, condensed to a total volume of approximately 5.0 mL, and treated with an excess of diazomethane (generated from 0.5 g Diazald). Excess diazomethane was quenched with a minimum amount of glacial acetic acid and the solution was washed with saturated NaHCO$_3$, dried over MgSO$_4$, and filtered. The filtrate was analyzed by GC (1.0 min at 80 °C, ramp at 15 °C/min to 300 °C, and hold for 5.0 min; calibrated vs. an authentic mixture of methyl acrylate telomers) and assignments were confirmed by GC/MS.

8.5.5 Synthesis of Template Diester 48

A solution of norbornadiene (**43**) (85.4 g, 927 mmol) and NiCOD$_2$ (1.3 g, 4.73 mmol) in 100 mL of 1,4-dioxane was stirred at rt for 22 h, leading to deposition of a white precipitate. A second charge of norbornadiene (85.4 g, 927 mmol) was added, and the heterogeneous reaction mixture was warmed to 50–55 °C for 48 h and then at 40 °C for 16 h. Concentration of the reaction solution and fractional sublimation of the residue at 0.05 Torr afforded 96.6 g of dimer **44** at 60–70 °C (57%), 59.2 g of trimer **45** at 155–165 °C (34%), and 5.8 g of tetramer **44** at 240 °C (3%). The trimer was recrystallized from hot isopropanol to furnish 47.8 g of **45** as fluffy white crystals; m.p. 198–201 °C (lit. [49] m.p. 201 °C); IR (CDCl$_3$) 3035, 1610 cm^{-1}; ^{1}H-NMR (200 MHz, CDCl$_3$) δ 5.93 (s, 4H), 2.59 (s, 4H), 1.89–1.12 (m, 16H); ^{13}C-NMR (90 MHz, CDCl$_3$) δ 135.3, 44.2, 41.8, 41.6, 40.6, 28.9; MS m/z (relative intensity) 276 (100, M$^+$).

A solution of NBD trimer **45** (11 g, 40 mmol), methyl 9-anthracenecarboxylate (**47**) (20 g, 85 mmol), and 3-*tert*-butyl-4-hydroxy-5-methylphenyl sulfide (20 mg, 0.06 mmol) in 50 mL of benzene was heated to 170 °C in a steel-walled medium-pressure reactor. After 200 h, the solution was cooled, concentrated in vacuo, and the residue was purified by flash chromatography on SiO$_2$ with 3 : 1 benzene/hexane as eluent to afford 12.5 g (43%) of the C_s template diester **48** and 13.5 g (47%) of the template C_2 diester **49**. Both solids decomposed upon heating (**48**: 256 °C, **49**: 338 °C). **48**: IR (CDCl$_3$) 1730 cm^{-1}; ^{1}H-NMR (200 MHz, CDCl$_3$) δ 7.51–6.96 (m, 16H), 4.17 (d, $J=2.2$ Hz, 2H), 4.05 (s, 6H), 2.04–1.05 (m, 22H), –0.61 (d, $J=10.2$ Hz, 2H); ^{13}C-NMR (90 MHz, CDCl$_3$) δ 172.7, 143.7, 142.7, 141.6, 139.1, 126.5, 126.2, 126.0, 125.5, 124.3, 124.0, 123.0, 121.8, 58.8, 51.9, 49.4, 48.6, 48.4, 47.1, 47.0, 42.6, 42.30, 42.28, 41.5, 41.3, 41.2, 29.0, 27.6; FABMS m/z (relative intensity) 747 (22, M–H$^+$).

8.5.6   Controlled Oligomerization of Methyl Methacrylate with Template 63

A solution of **63** (80 mg, 0.078 mmol), Mo(CO)$_6$ (25 mg, 0.10 mmol), and MMA (84 µL, 0.78 mmol) in 8 mL of benzene in a heavy-walled vessel equipped with a Teflon stopcock was deoxygenated by three freeze–pump–thaw cycles and placed in a 70 °C oil bath for nine days. At that time, the reaction was poured into 5 mL of 1 M HCl, and the aqueous layer was extracted with 4 × 10 mL of CH$_2$Cl$_2$. The combined organic layers were washed with brine and dried over Na$_2$SO$_4$. Filtration, concentration, and purification of the residue by flash chromatography on SiO$_2$ with 2% Et$_2$O/CH$_2$Cl$_2$, 4% Et$_2$O/CH$_2$Cl$_2$, 6% Et$_2$O/CH$_2$Cl$_2$ as eluent gave 43 mg (47%) of the template-bound MMA trimer **94** as a mixture of six isomers. This material could be further purified by HPLC (SiO$_2$, 2% Et$_2$O/CH$_2$Cl$_2$ eluent) to furnish six separate diastereomers of **94**. **94** (major diastereomer): IR (CCl$_4$) 1725 cm^{-1}; ^{1}H-NMR (400 MHz, CDCl$_3$) δ 7.2–6.9 (m, 16H), 6.13 (d, $J=1.0$ Hz, 1H), 5.80 (d, $J=11.4$ Hz, 1H), 5.74 (d, $J=11.3$ Hz, 1H), 5.34 (s, 1H), 4.91 (d, $J=11.7$ Hz, 1H), 4.88 (d, $J=12.0$ Hz, 1H), 4.17 (m, 2H), 3.65 (s, 3H), 3.60 (s, 3H), 3.53 (s, 3H), 3.06 (d, $J=14.4$ Hz, 1H), 2.48 (d, $J=14.9$ Hz, 1H), 2.41 (d, $J=13.4$ Hz, 1H), 2.34 (d, $J=13.2$ Hz, 1H), 2.07 (d, $J=14.2$ Hz, 1H), 2.00 (d, $J=14.2$ Hz, 1H), 1.9–1.2 (m, 24H), 1.00 (s, 3H), 0.93 (s, 3H), 0.92 (s, 3H), –0.32 (m, 2H); ^{13}C-NMR (100 MHz, CDCl$_3$) δ 177.1, 175.9, 175.6, 167.9, 165.9, 145.1, 144.9, 143.6, 142.8, 142.7, 142.6, 141.3, 140.8, 137.3, 128.3, 126.3, 126.2, 126.0, 125.7, 125.6, 125.5, 125.4, 124.6, 124.5, 123.3, 121.0, 120.8, 120.5, 83.0, 66.4, 62.5, 56.6, 53.2, 52.7, 52.2, 51.9, 51.6, 48.4, 48.2, 48.1, 47.9, 47.8, 47.7, 47.6, 47.5, 47.4, 47.2, 46.9, 45.6, 44.5,

44.3, 42.6, 42.3, 42.1, 42.0, 41.9, 41.4, 39.8, 39.6, 29.0, 28.5, 28.3, 17.7, 16.7, 16.2; FABMS *m/z* (relative intensity) 1172, 1171, 1170, 1169, 1168 (6, 9, 8, 9, 2 M^+, M^++1).

Acknowledgments

The dedicated efforts of talented co-workers at Penn State (Dr. Yoon Lee, John Bobo, Dr. Gregory Tewalt, Dr. William Shiang, Paul Weinreb and Susan Ensel) and at Duke (Greg Miracle and Scott Cannizzaro) are gratefully acknowledged, as is financial support from the National Science Foundation (Penn State and Duke) and the National Institutes of Health (Duke).

References

[1] D. P. Curran, N. A. Porter, B. Giese, *Control of Stereochemistry in Free Radical Reactions; Concepts, Guidelines, and Synthetic Applications*, VCH: Weinheim, **1995**.

[2] K. Mori, S. Kuwahara, *Tetrahedron* **1986**, *42*, 5539–5544.

[3] M. Kaino, Y. Naruse, K. Ishihara, H. Yamamoto, *J. Org. Chem.* **1990**, *55*, 5814–5815.

[4] T. I. Richardson, S. D. Rychnovsky, *J. Am. Chem. Soc.* **1997**, *119*, 12360–12361.

[5] D. P. Curran, "Radical Addition Reactions," and "Radical Cyclization Reactions and Sequential Radical Reactions." In *Comprehensive Organic Synthesis*; B. M. Trost, I. Fleming, Eds.; Pergamon: Elmsford, NY, **1991**; *Vol. 4*, Chapters 4.1 and 4.2.

[6] N. A. Porter, D. P. Curran, B. Giese, *Acc. Chem. Res.* **1991**, *24*, 296–304, and references therein.

[7] J. H. Wu, R. Radinov, N. A. Porter, *J. Am. Chem. Soc.* **1995**, *117*, 11029–11030.

[8] M. P. Sibi, J. Ji, J. H. Wu, S. Gürtler, N. A. Porter, *J. Am. Chem. Soc.* **1996**, *118*, 9200–9201.

[9] D. Nanni, D. P. Curran, *Tetrahedron–Asymmetry* **1996**, *8*, 2417–2422.

[10] M. Blumenstein, K. Schwarzkopf, J. O. Metzger, *Angew.Chem.* **1997**, *36*, 235–236.

[11] M. P. Sibi, J. Ji, *J. Org. Chem.* **1997**, *62*, 3800–3801.

[12] M. Murakata, T. Jono, Y. Mizuno, O. Hoshino, *J. Am. Chem. Soc.* **1997**, *119*, 11713–11714.

[13] For a historical perspective, see: G. S. Miracle, S. C. Cannizzaro, N. A. Porter, *ChemTracts* **1993**, 147–171.

[14] N. A. Porter, G. S. Miracle, S. C. Cannizzaro, R. L. Carter, A. T. McPhail, L. Liu, *J. Am. Chem. Soc.* **1994**, *116*, 10255–10266.

[15] G. S. Miracle, S. C. Cannizzaro, N. A. Porter, *J. Am. Chem. Soc.* **1992**, *114*, 9683–9684.

[16] K. S. Feldman, Y. B. Lee, *J. Am. Chem. Soc.* **1987**, *109*, 5850–5851.

[17] K. S. Feldman, J. S. Bobo, S. M. Ensel, Y. B. Lee, P. H. Weinreb, *J. Org. Chem.* **1990**, *55*, 474–481.

[18] K. S. Feldman, J. S. Bobo, G. L. Tewalt, *J. Org. Chem.* **1992**, *57*, 4573–4574.

[19] N. A. Porter, D. R. Magnin, B. T. Wright, *J. Am. Chem. Soc.* **1986**, *108*, 2787–2788.

[20] N. A. Porter, V. H.-T. Chang, *J. Am. Chem. Soc.* **1987**, *109*, 4976–4981.

[21] N. A. Porter, V. H.-T. Chang, D. R. Magnin, B. T. Wright, *J. Am. Chem. Soc.* **1988**, *110*, 3554–3560.

[22] N. A. Porter, "Stereochemical and Regiochemical Aspects of Free Radical Macrocyclizations." In *Organic Free Radicals*; Proceedings of the Fifth International Symposium; H. Fischer, H. Heimgartner, Eds.; Springer: Berlin, **1988**; p. 187.

[23] N. A. Porter, B. Lacher, V. H.-T. Chang, D. R. Magnin, *J. Am. Chem. Soc.* **1989**, *111*, 8309–8310.

[24] H. Kämmerer, S. Ozaki, *Makromol. Chem.* **1966**, *91*, 1–9.

[25] H. Kämmerer, A. Jung, *Makromol. Chem.* **1967**, *101*, 284–295.

[26] H. Kämmerer, A. Jung, J. S. Shukla, *Makromol. Chem.* **1967**, *107*, 259–261.

[27] W. Kern, H. Kämmerer, *Pure Appl. Chem.* **1967**, *15*, 421–433.

[28] H. Kämmerer, N. Önder, *Makromol. Chem.* **1968**, *111*, 67–73.

[29] H. Kämmerer, J. S. Shukla, *Makromol. Chem.* **1968**, *116*, 62–71.

[30] H. Kämmerer, J. S. Shukla, G. Scheuermann, *Makromol. Chem.* **1968**, *116*, 72–77.

[31] H. Kämmerer, G. Hegemann, *Makromol. Chem.* **1970**, *139*, 17–21.

[32] H. Guéniffey, H. Kämmerer, C. Pinazzi, *Makromol. Chem.* **1973**, *165*, 73–81.

[33] H. Kämmerer, G. Hegemann, *Makromol. Chem.* **1982**, *183*, 1435–1444.

[34] H. Kämmerer, G. Hegemann, *Makromol. Chem.* **1984**, *185*, 635–646.

[35] N. A. Porter, J. D. Bruhnke, W.-X. Wu, I. J. Rosenstein, R. A. Breyer, *J. Am. Chem. Soc.* **1991**, *113*, 7788–7790.

[36] N. A. Porter, W.-X. Wu, A. T. McPhail, *Tetrahedron Lett.* **1991**, *32*, 707–710.

[37] N. A. Porter, I. J. Rosenstein, R. B. Breyer, J. D. Bruhnke, W.-X. Wu, A. T. McPhail, *J. Am. Chem. Soc.* **1992**, *114*, 7664–7676.

[38] K. E. Harding, C. Tseng, *J. Org. Chem.* **1978**, *43*, 3972–3977.

[39] The source of H-atoms has not been identified, but the position α to the phenolic ether would be reactive toward electrophilic or ambiphilic radicals.

[40] H. Girard, P. Monjol, *C. R. Hebd. Seances. Acad. Sci., Ser. C (Paris)* **1974**, 553–555.

[41] K. Matsuzaki, T. Uryu, A. Ishida, M. Takeuchi, T. Ohki, *Polym. Sci. Part A-1* **1967**, *5*, 2167–2177.

[42] N. A. Porter, T. Allen, R. B. Breyer, *J. Am. Chem. Soc.* **1992**, *114*, 7676–7683, and references cited therein.

[43] A. Philippon, M. Degueil-Castaing, A. L. J. Beckwith, B. Maillard, *J. Org. Chem.* **1998**, *63*, 6814–6819.

[44] G. Illuminati, L. Mandolini, *Acc. Chem. Res.* **1981**, *14*, 95–102.

[45] D. B. Dess, J. C. Martin, *J. Am. Chem. Soc.* **1991**, *113*, 7277–7287.

[46] R. E. Ireland, L. Lui, *J. Org. Chem.* **1993**, *58*, 2899–2903.

[47] S. D. Meyer, S. L. Schreiber *J. Org. Chem.* **1994**, *59*, 7549–7552.

[48] Radical stabilization energies for radicals α to ethers have been estimated at 12 kcal/mol. F. G. Bordwell, X.-M. Zhang, M. S. Alnajjar, J. A. Franz, *J. Org. Chem.* **1995**, *60*, 4976–4979.

[49] R. L. Pruett, E. A. Rick, US Patent 3 440 294, April 22, 1969.

[50] U. Azzena, S. Cossu, O. De Lucchi, G. Melloni, *Gazz. Chim. Ital.* **1989**, *119*, 357–358.

[51] R. N. Warrener, I. G. Pitt, D. N. Butler, *J. Chem. Soc., Chem. Commun.* **1983**, 1340–1341.

[52] R. N. Warrener, D. N. Butler, R. A. Russell, *SynLett* **1998**, 566–573.

[53] C. H. Bamford, R. W. Dyson, G. C. Eastmond, D. Whittle, *Polymer* **1969**, *10*, 759–770.

[54] D. Y. Sogah, W. R. Hertler, O. W. Webster, G. M. Cohen, *Macromolecules* **1987**, *20*, 1473–1488.

[55] E. P. Boden, G. E. Keck, *J. Org. Chem.* **1985**, *50*, 2394–2395.

[56] P. J. Wagner, J. H. Sedon, M. J. Lindstrom, *J. Am. Chem. Soc.* **1978**, *100*, 2579–2580.

[57] M. H. Abraham, *J. Am. Chem. Soc.* **1982**, *104*, 2085–2094.

[58] T. Dunams, W. Hoekstra, M. Pentaleri, D. Liotta, *Tetrahedron Lett.* **1988**, *29*, 3745–3748.

[59] R. Breslow, T. Guo, *J. Am. Chem. Soc.* **1988**, *110*, 5613–5617.

[60] M. M. Itoh, J-i. Kato, S. Takagi, E. Nakashiro, T. Sato, Y. Yamada, H. Saito, T. Namiki, I. Takamura, K. Wakatsuki, T. Suzuki, T. Endo, *J. Am. Chem. Soc.* **1988**, *110*, 5147–5152.

9 Templated or Not Templated, That is the Question: Analysis of Some Ring Closure Reactions

Alois Fürstner

9.1 Introduction

Searching for the entries "template" or "templated" in the common chemical databases makes clear that these expressions are widely used but very ill defined. If we understand a template as a molecular device which shapes a given substrate and thereby provides bias for a certain pathway over a competing one, and which is removed once it has exerted its function, we will have no problem in calling many ring closure reactions "templated" ones. A prototype example of a cyclization using a covalently linked template is the Corey–Nicolaou method for macrolide synthesis (Scheme 9-1) [1]. This transformation takes advantage from the polyfunctional character of thiopyridyl esters which (i) activate the acid part, (ii) help to bring the reacting sites in proximity by an acid–base interaction, and (iii) activate the –OH group *if* a reactive conformation is adopted. Similarly classical is crown ether formation by intramolecular substitution reactions, where a cation of appropriate size coordinates to the heteroatoms of the incipient ring and thereby allows cyclizations to occur at concentrations that are much higher than usual for systems of this size (Scheme 9-2) [2]. Finally, the cyclotrimerization of butadiene pioneered by Wilke and co-workers should be mentioned as an illustrative example of a catalytic, templated ring closure which crucially depends upon the assembly of the monomer units within the coordination sphere of a nickel template that changes its oxidation state and coordination number while running through the catalytic cycle (Scheme 9-3) [3].

In generalizing the latter examples one may be tempted to assume that all metal-assisted cyclization and cross coupling reactions are templated processes. This viewpoint, however, not only sets an impractically wide scope for a review, it may even be incorrect. Therefore, the present article does not try to cover all potentially "templated" ring closure reactions [4], but presents just a very limited and personal selection of metal-assisted or metal-catalyzed cyclizations based on C–C coupling reactions. They are analyzed with the intention to deduce more accurately what the phenomenological effect of templating refers to.

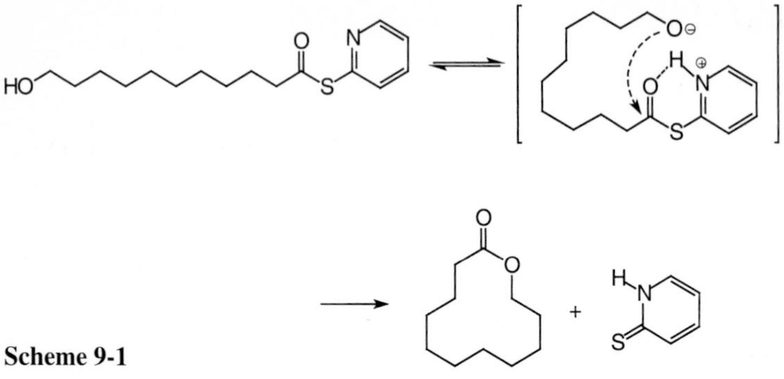

Scheme 9-1

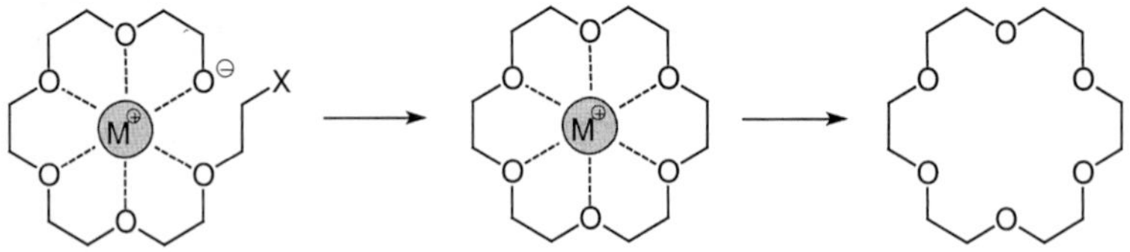

Scheme 9-2

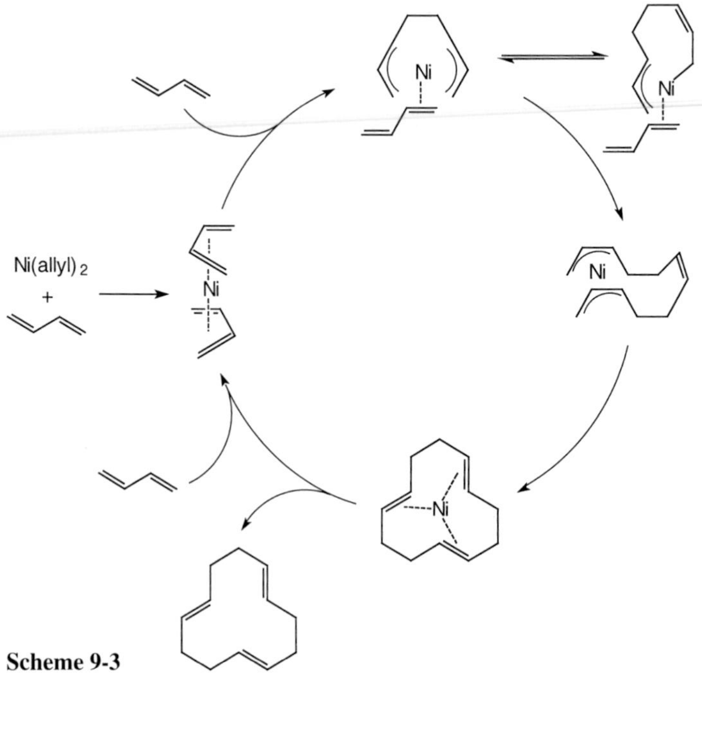

Scheme 9-3

9.2 Possible Pitfalls

It is common knowledge that the preparation of medium-sized rings is a formidable challenge because unfavorable entropic and enthalpic factors impede their formation [5]. Therefore, many successful entries into this series rely on templates which lock the flexible substrate in a suitable conformation for ring closure and thereby increase the probability of encounters of the reacting sites. With this notion in mind, the surprising efficiency of the catalytic ring expansion depicted in Scheme 9-4, a key step in a recent total synthesis of the immunosuppressive alkaloid streptorubin B [6], might be ascribed to the well-known affinity of Pt^{II} toward alkenes and alkynes which fosters the desired formation of the 10-membered ring in a cooperative manner. This analysis, though seemingly reasonable, is probably incorrect. During an in-depth study of the reaction mechanism it was found that the same transformation can also be effected in respectable yield with protons (!) instead of PtCl$_2$. Therefore one must conclude that the efficiency of this particular ring expansion mainly results from the geometrical constraints of the substrate rather than from a preorganization mediated by the catalyst. If any template effect is operative, it accounts only for the slightly higher efficiency of the PtCl$_2$-catalyzed over the proton-induced pathway.

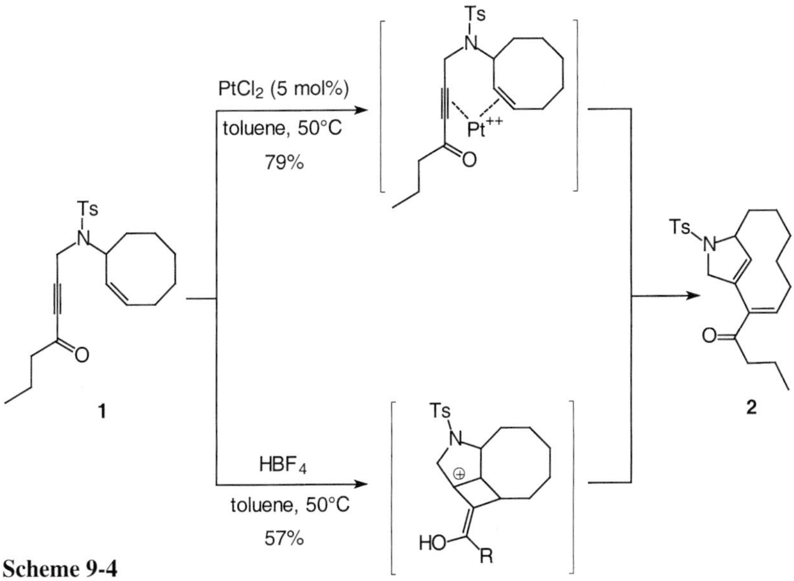

Scheme 9-4

Even in cases where the topology of a given substrate renders cyclization a priori rather unlikely, it is potentially misleading to invoke a template effect without appropriate control experiments. An illustrative example was reported by Negishi and co-workers (Scheme 9-5) [7]. Intramolecular carbopalladation reactions of ω-haloallenes provide a potentially general method for the synthesis of common, medium, and large rings. There-fore the efficiency of the conversion of substrate **3** into the eight-membered ring com-pound **6** could be attributed to coordination-induced proximity: oxidative insertion of Pd^0 into the aryl halide provides an arylpalladium(II) species which coordinates the allene entity and thereby facilitates the C−C-bond formation despite of the unfavorable size of the incipient ring.

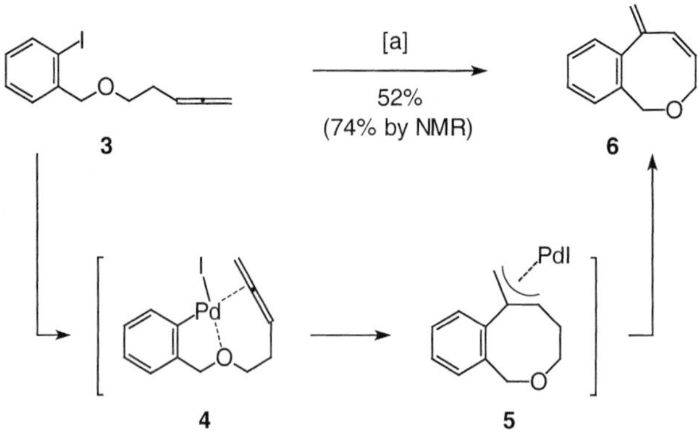

[a] PdCl$_2$(PPh$_3$)$_2$ (5 mol%), K$_2$CO$_3$, EtOH/DMF, 100°C

Scheme 9-5

This scenario, however, does not explain why the cyclization of the corresponding *ω*-haloalkene **7** is terribly inefficient, although it might follow a similar course (Scheme 9-6); simple dehalogenation of the substrate rather than ring closure becomes the major pathway. This outcome indicates that factors other than the template effect of Pd^{II} foster the allene case, which are not operative with the alkene. One may speculate whether kinetic parameters such as the generally higher reactivity of allenes toward organopalla-dium reagents caused by their higher strain energies, or if thermodynamic features such as the stability of the resulting *π*-allylpalladium species **5** formed by an irreversible C−C-coupling process drive the conversion of **3 → 6**.

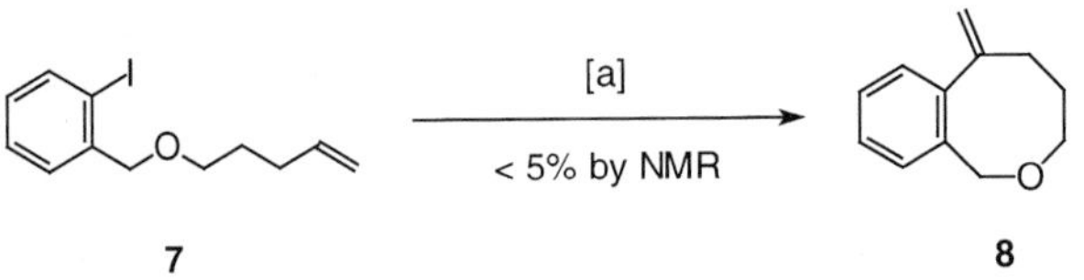

[a] PdCl$_2$(PPh$_3$)$_2$ (5 mol%), K$_2$CO$_3$, EtOH/DMF, 100°C

Scheme 9-6

These examples pinpoint the notions that

(i) one must be very prudent in ascribing the performance of ring closure reactions to a template effect if the substrate contains any element of structural preorganization. The same caution applies – maybe even to a larger extent – to all cyclizations involving kinetically favorable ring sizes [8];
(ii) control experiments are mandatory in order to distinguish between a possible template effect and other, more conventional incentives for ring closure.

In searching for templated cyclization reactions it is therefore best to look at substrates which are as flexible as possible in order to rule out any potentially misleading influence of simple geometrical constraints. For the same reason, macrocycle syntheses rather than the formation of five- to seven-membered rings are generally more suitable for studying a possible template effect, since the proximity of the reacting sites in the required precursors is a priori less likely [9]. Therefore the following discussion is restricted to some examples of ring closure reactions by metal-induced C–C-bond formation which meet these criteria and for which direct rather than presumptive evidence is available that a template effect is in fact operative.

9.3   Titanium-induced Intramolecular Carbonyl Coupling Reactions

The reductive coupling of carbonyl compounds by means of low-valent titanium [Ti], a process which is generally referred to as the McMurry reaction in honor of one of its pioneers, has attracted considerable attention over recent decades [10]. One of its major advantages stems from the fact that this transformation makes it possible to form carbo- and heterocycles of any ring size, including medium and macrocyclic ones. This feature is witnessed by a considerable number of elegant syntheses of natural and non-natural products and has often been ascribed to the template effect exerted by the polar surface of the low-valent titanium reagents used. Several pieces of evidence – in fact – support this argument.

Specifically, McMurry et al. [11] and Baumstark et al. [12] have systematically investigated the intramolecular coupling of a homologeous series of conformationally flexible dicarbonyl compounds by means of [Ti] derived from either $TiCl_3$/Li or $TiCl_3$/LiAlH$_4$. These early studies have shown that the yield of the cycloalkenes formed is essentially independent of the ring size (Table 9-1), an outcome that is in clear contrast to the profile of most other classical cyclization reactions. For comparison, the Dieckmann condensation of diesters or the Thorpe–Ziegler dinitrile method fail to afford medium-sized rings in reasonable yields and perform only poorly in the macrocyclic series [13, 15]. Even the acyloin condensation [14], which was one of the first reasonable entries into products with 8–11 ring atoms, does not show the same striking uniformity as the titanium mediated process.

Table 9-1 Effect of the ring size on intramolecular, low-valent titanium-induced carbonyl coupling reactions.

$$R^1\text{CO-}(CH_2)_n\text{-CO-}R^2 \xrightarrow{[Ti]} (CH_2)_n\text{C}=\text{C}\begin{smallmatrix}R^1\\R^2\end{smallmatrix}$$

R^1	R^2	n	Yield (%)	Ref.
Ph	Ph	2	61	[12]
		3	62	[12]
		4	60	[12]
		6	61	[12]
		7	53	[12]
		8	49	[12]
		10	61	[12]
H	H	10	76	[11]
		12	71	[11]
		14	85	[11]
Ph	Me	4	79	[11]
C_5H_{11}	C_5H_{11}	6	67	[11]
		7	68	[11]
C_4H_9	C_4H_9	8	75	[11]
		9	76	[11]
C_3H_7	C_3H_7	11	65	[11]
Et	Et	12	75	[11]
Ph	H	13	80	[11]
Me	Me	14	90	[11]
		20	83	[11]

The experiments mentioned above have been carried out under high dilution, i.e., conditions which are known to favor cyclization over competing intermolecular reactions [15, 16]. Even more surprising are reports on McMurry-type macrocyclizations which deliberately disregard the "Ziegler–Ruggli dilution principle" [15, 16] but still provide good to excellent results. One of the most striking examples is the formation of a 36-membered ring depicted in Scheme 9-7, where the conformationally flexible substrate **9** is converted into product **10** in 90% yield when exposed to a refluxing slurry of commercial titanium powder activated by TMSCl even if its concentration is as high as 0.02 M [17]. This outcome indicates an exceptional preference for ring closure which must be attributed to a preorganization of the diketone by the chemically activated titanium surface [18].

Similar observations have been made during a study on the formation of crownophanes such as **14** (Scheme 9-8) which proceed smoothly even in rather concentrated solutions [19]. In this particular case, photochemical control experiments provide strong evidence that the titanium mediated process benefits from a significant template effect. Whereas the photoinduced intramolecular pinacolization of diketone **11** to diol **12** in 0.002 M solution is rather inefficient ($\approx 29\%$) [20], treatment of the same substrate with

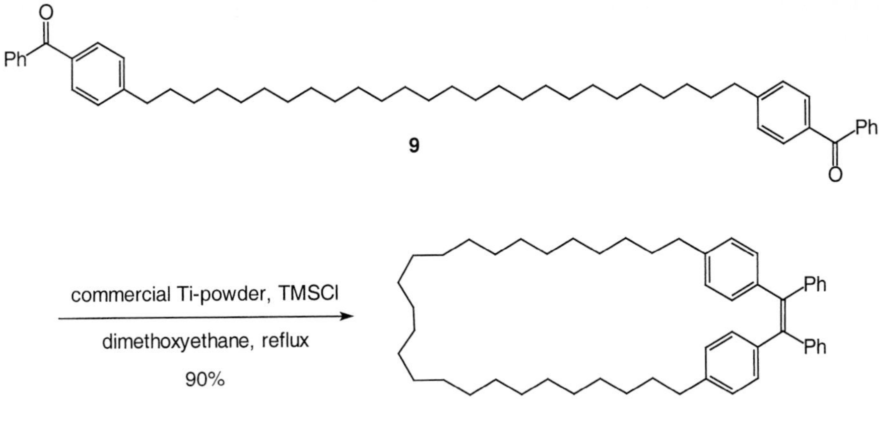

Scheme 9-7

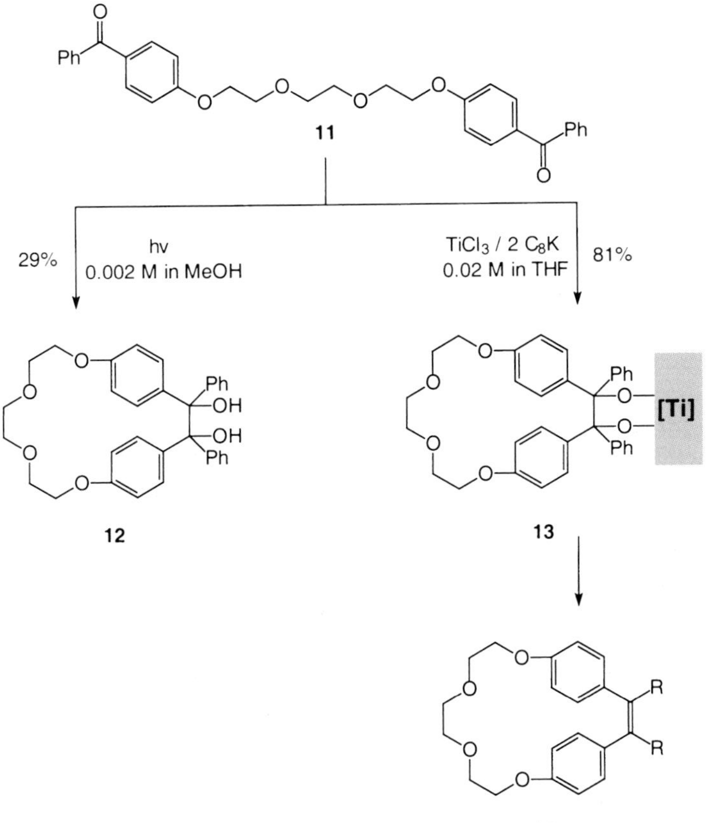

Scheme 9-8

[Ti] leads to the desired 20-membered ring **14** in 57–83% yield by deoxygenation of an intermediate titanium pinacolate **13** at concentrations ≥ 0.02 M! One may, however, object that this outcome is not due to a template effect of [Ti] but rather to the considerable amounts of (alkali) metal salts such as KCl, LiCl, $MgCl_2$, $ZnCl_2$, or $AlCl_3$ accumulating in the reaction mixture during the preparation of the [Ti] reagent from $TiCl_3$ or $TiCl_4$ and Li, K, Mg, Zn, $LiAlH_4$, etc.; these salts may preorganize the polyether substrate as shown in Scheme 9-9 and thereby facilitate the cyclization process.

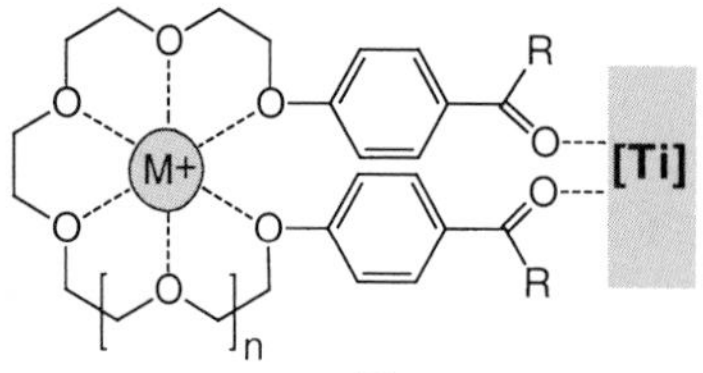

15

Scheme 9-9

A series of experiments, however, using (i) different [Ti] species, including titanium activated under "salt-free" conditions (Ti/TMSCl) [17, 18], and (ii) different substrates with and without polyether chains, made it possible to distinguish between the effect of [Ti] and that of the admixed salts. From the results summarized in Table 9-2 it must be deduced that [Ti] is largely responsible for the smooth cyclizations, whereas the admixed salts play only a minor role. The much more intricate question, however, of how this striking property of [Ti] can be explained on the molecular level, is largely unsolved. The heterogeneous character of McMurry reactions makes spectroscopic investigations rather difficult, and all mechanistic studies are further hampered by the fact that [Ti] exhibits a much more complicated inorganic chemistry than was previously anticipated [21].

One can, however, safely pretend that the pronounced oxophilicity of titanium is the chemical basis for the observed template effect. As mentioned above, the coupling of carbonyl compounds, which proceeds via ketyl radical anions as the actual intermediates, occurs smoothly in an inter- as well as an intramolecular manner and is largely unaffected by the ring size formed. In stark contrast, the titanium mediated coupling of allylic alcohols is only productive as an intermolecular process, whereas the intramolecular variant fails to afford reasonable amounts of macrocyclic products (Scheme 9-10) [22]. This difference can probably be explained by assuming that free allyl radicals are formed in the latter case which cannot stick to [Ti] as they lack the oxygen atom of their ketyl analogues. Rather than being preorganized in a favorable conformation for ring closure on the oxophilic surface of the reagent, they diffuse freely in solution and get reduced by hydrogen abstraction from the ethereal medium.

Table 9-2 Probing the template effect of low-valent-titanium [Ti] and admixed alkali halides [19].

Product	[Ti]	Admixed salt	Yield (%)
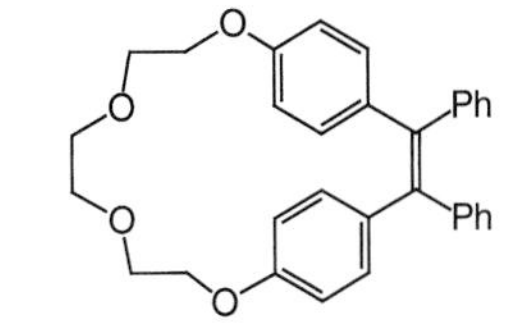	TiCl$_3$/2 C$_8$K	KCl	81
	TiCl$_3$/2 Na	NaCl	83
	Ti/TMSCl	–	57
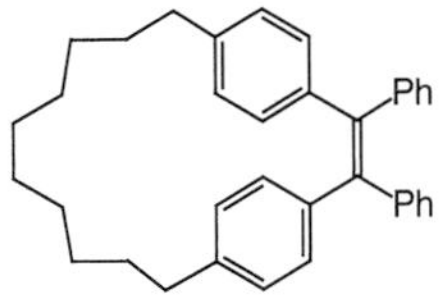	TiCl$_3$/2 C$_8$K	KCl	86
	TiCl$_3$/2 Na	NaCl	86
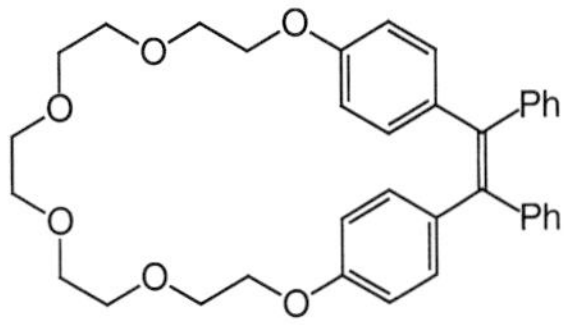	TiCl$_3$/2 C$_8$K	KCl	63
	Ti/TMSCl	–	60

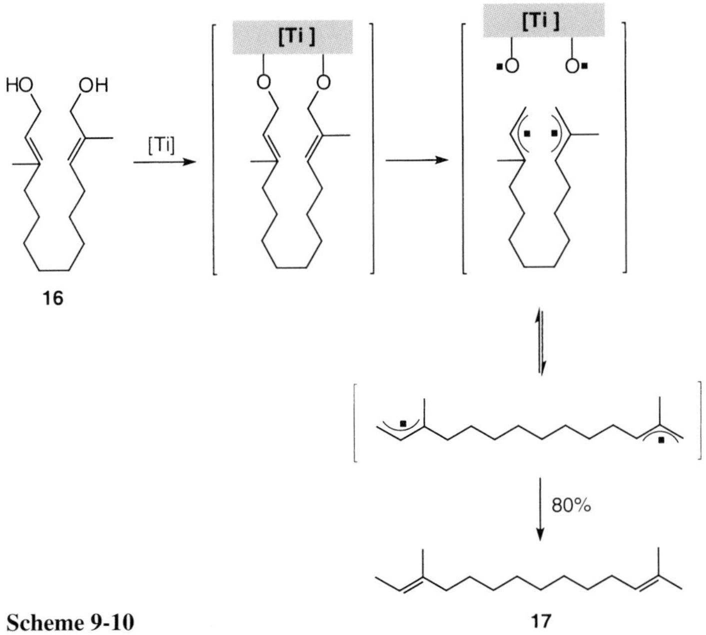

Scheme 9-10

9.4   Zinc versus Samarium Mediated Reformatsky Reactions

According to a general definition, the Reformatsky reaction can be taken as subsuming all enolate formations by oxidative addition of a metal or a low-valent metal salt into a carbon–halogen bond activated by a vicinal carbonyl group, followed by reaction of the enolates thus formed with an appropriate electrophile [23]. The insertion of metallic zinc into α-halo esters is historically the first and still the most widely used form of this process although Zn may be beneficially replaced but other metals and reducing metal salts in certain cases.

A particular advantage of the Reformatsky reaction stems from the fact that the site of reaction is strictly determined by the halogen moiety. This feature can be advantageously used for regioselective enolate formations in polycarbonyl compounds which are difficult to achieve by proton abstraction methods. As a consequence, this transformation has seen a renaissance over recent decades and has found many elegant uses, particularly in intramolecular variants [23].

The traditional zinc mediated process, though well suited for the formation of five- and six-membered rings, performs rather poorly in the more challenging macrocyclic series [24]. This limitation, can be overcome by replacing Zn by SmI_2 as the reagent [25]. Comparative data using the conformationally flexible substrate **18** show that the latter provides excellent yields whereas Zn does not afford preparatively useful results (Scheme 9-11). This outcome probably stems from the high oxophilicity and the flexible coordination geometry of the Sm^{3+} counterion to the enolate, which acts as a template and coordinatively binds the aldehyde group in a much tighter transition state than Zn^{2+} does. Applications to the synthesis of complex natural products such as ferrulactone [26], octalactin A21 [27], and Taxol® [28] (Scheme 9-12) witness the truly remarkable efficiency of such intramolecular SmI_2 mediated Reformatsky variants. From the examples compiled in this and the previous sections one must conclude that the *affinity* of certain transition metals towards oxygen-containing functional groups can be macroscopically expressed as a template effect in ring closure reactions.

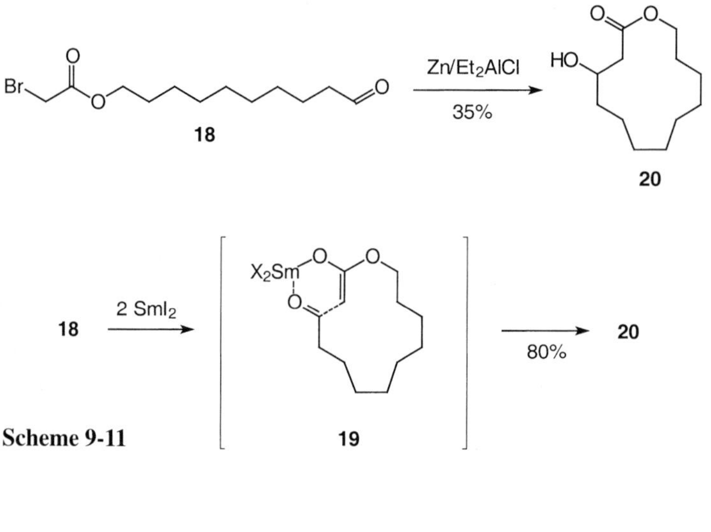

Scheme 9-11

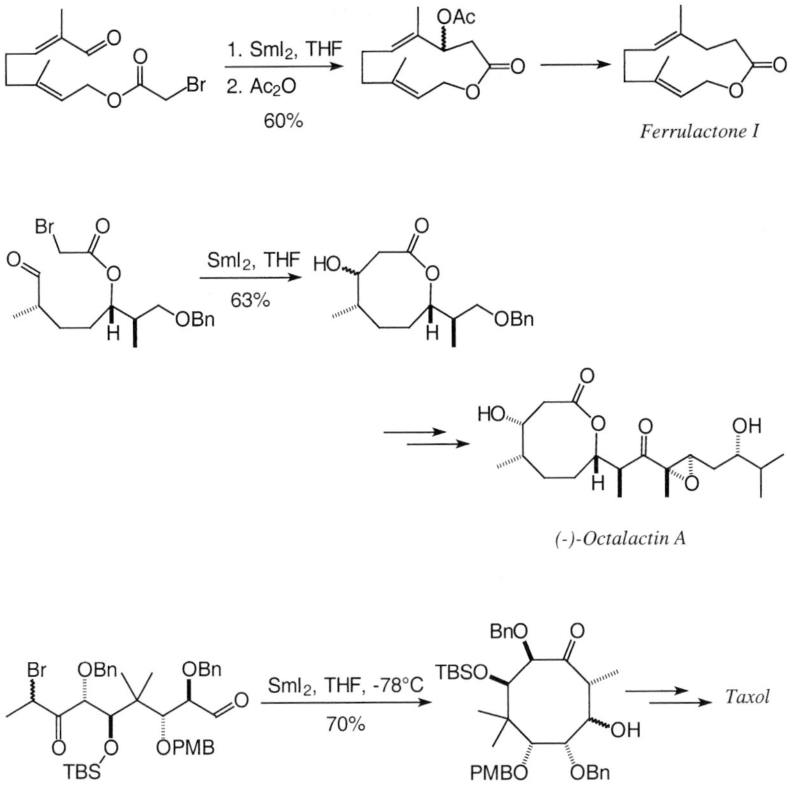

Ferrulactone I

(-)-Octalactin A

Scheme 9-12

9.5 Macrocycle Syntheses by Ring Closing Metathesis (RCM)

Formally speaking, the cyclization of a dicarbonyl compound to a cycloalkene via a McMurry coupling closely resembles the ring closing metathesis (RCM) of a 1,ω-diene, although these reactions have nothing in common as to their mechanisms (Scheme 9-13) [29, 30]. Whereas the former has recently been recognized as a very efficient entry into macrocycles due to the inherent template effect of [Ti] discussed in Section 9-3, the latter transformation has long been thought to be inappropriate for the formation of large rings, except for systems which are a priori strongly biased for ring closure by a rigid conformation of the diene substrate [31].

This assumption, however, has been totally revised in recent years: surprisingly enough, RCM turned out to be among the best entries into macrocycles even with conformationally flexible diene substrates, provided that a few basic parameters are properly assessed [32]. These key features indicate that a subtle template effect is responsible for the effectiveness and scope of RCM-based ring closure reactions.

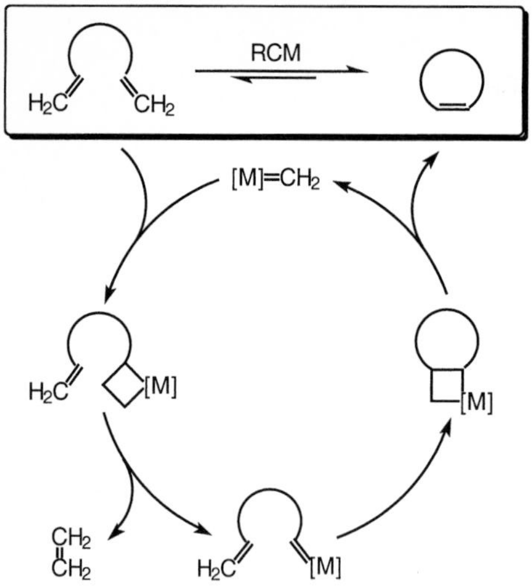

Scheme 9-13

Specifically, the entirely different behavior of dienes **21** and **23** (Scheme 9-14) shows that the presence of a polar functional group in the substrate is a fundamental requirement for productive macrocyclization [32]. In addition, the distance of the polar substituents from the dienes to be metathesized turned out to be another decisive factor, as can be gleaned from Scheme 9-15 [32].

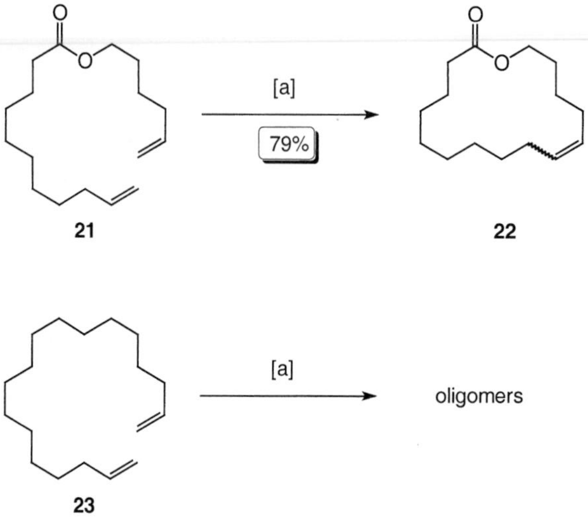

[a] (Cy$_3$P)$_2$Cl$_2$Ru=CHCH=CPh $_2$, 4 mol%

Scheme 9-14

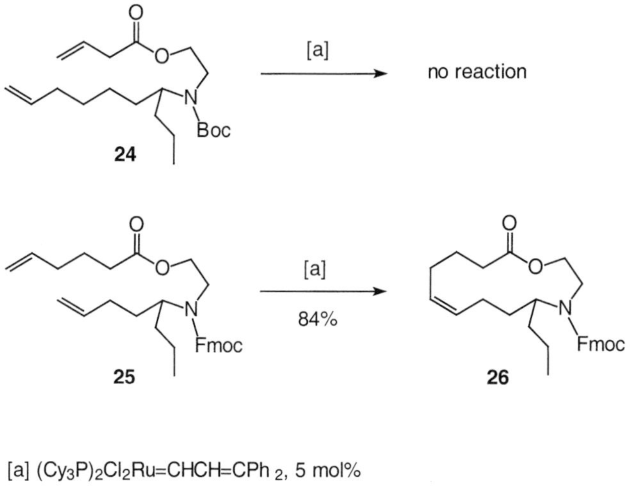

[a] (Cy$_3$P)$_2$Cl$_2$Ru=CHCH=CPh$_2$, 5 mol%

Scheme 9-15

These observations are best rationalized in terms of coordination of the emerging carbene and/or metallacyclic intermediates of the metathesis reaction with the polar "relay"
substituents of the substrate [32]. Such a Lewis-acid/Lewis-base interaction assists in
assembling the reacting sites within the coordination sphere of, e.g., the ruthenium catalyst and hence provides the necessary bias for cyclization as *formally* depicted in Scheme
9-17. If, however, such an intermediate becomes too stable, as might be the case for some
five- or six-membered chelate structures (e.g., structures **B** and **C** or similar), the catalyst
is sequestered in an unproductive form and the conversion is likely to cease. Yet, the
stability of the assumed intermediates is not only defined by geometric parameters such
as distance and relative orientation, but will also be very strongly influenced by the affinity of the "soft" Ru center to the heteroatom. This explains why the presence of sul-
fur(II), amino and cyano groups usually stops RCM events catalyzed by standard Grubbs-
type complexes such as (PCy$_3$)$_2$Cl$_2$Ru=CHR (R=Ph, CH=CPh$_2$) (Scheme 9-16) [33,
34d].

Although the experimental data summarized above provide convincing evidence that
a subtle and *transient* template effect resulting from a finely tuned Lewis-acid/Lewis-
base interaction is operative in productive RCM-based macrocyclizations, it should be
emphasized that a detailed spectroscopic characterization of the key intermediates
involved has not yet been achieved. Therefore structure **A** represents just a mnemotic
device indicating that *polarity, distance,* and *affinity* are key issues in such transformations but does not imply any structural information [32].

Although this "relay model" is rather simple, it provides a safe guidance for retrosynthetic planning and its predictive capacity has been successfully tested in numerous
applications. Scheme 9-18 compiles some of the targets which have been obtained in our
laboratory so far [34], which are supplemented by many additional examples from other

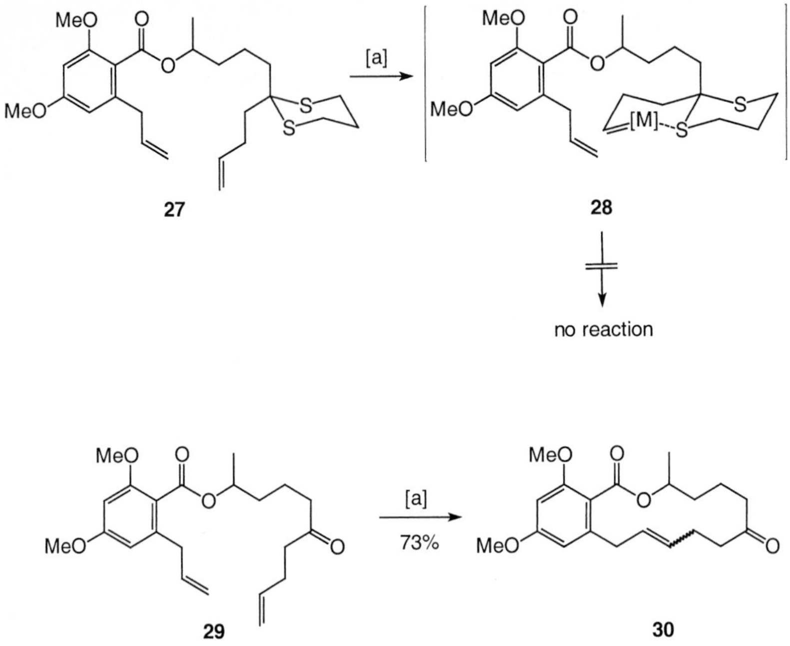

[a] (Cy₃P)₂Cl₂Ru=CHCH=CPh ₂ (cat.)

Scheme 9-16

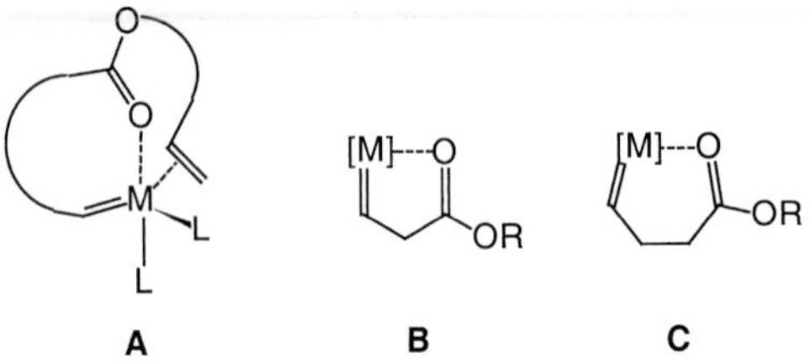

Scheme 9-17

groups [35]. The synthesis of (−)-gloeosporone is particularly informative, as it forced us to address and overcome a potential limitation of RCM-based macrocyclization processes [36].

(−)-Gloeosporone **31** is a germination self-inhibitor produced by the fungus *Colletotrichum gloeosporoides,* which has attracted attention because of its interesting structural and biological properties [37]. While all previous syntheses of **31** are based on macrolactonization strategies, RCM might enable a conceptually new and unprecedentedly short approach to this macrolide as depicted in Scheme 9-19.

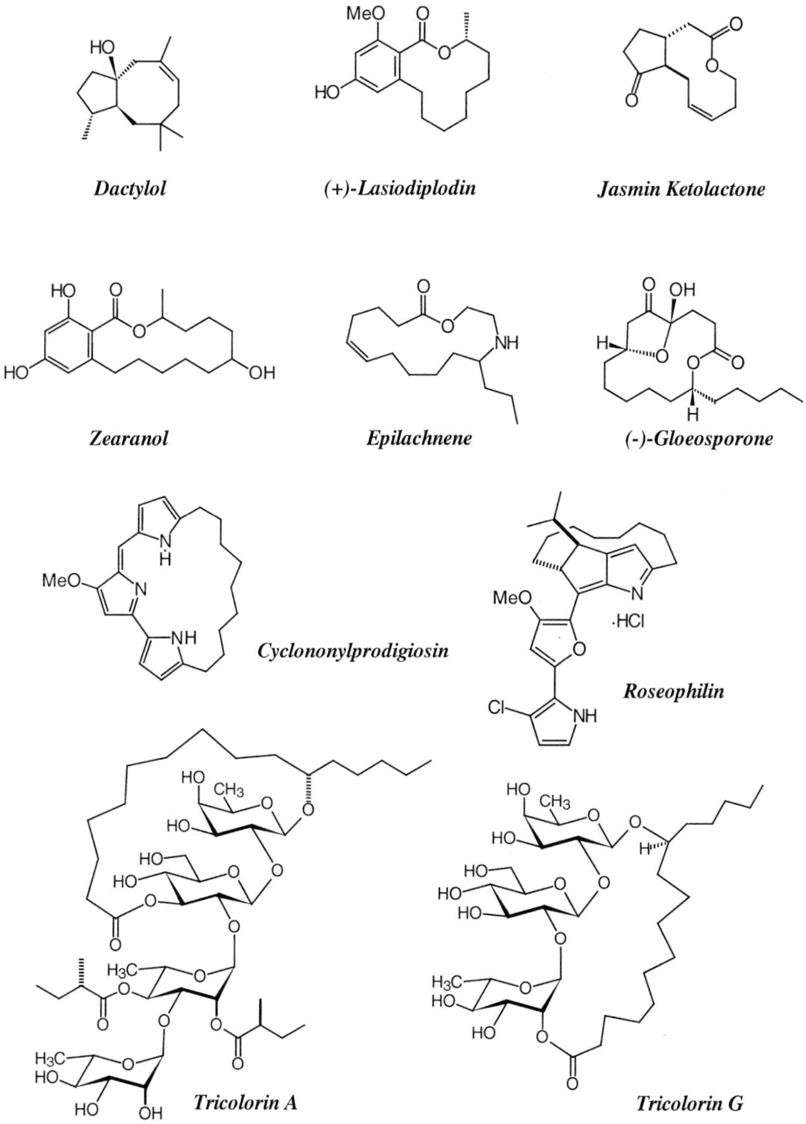

Dactylol   **(+)-Lasiodiplodin**   **Jasmin Ketolactone**

Zearanol   **Epilachnene**   **(-)-Gloeosporone**

Cyclononylprodigiosin   **Roseophilin**

Tricolorin A   **Tricolorin G**

Scheme 9-18

A more detailed assessment of this retrosynthetic scheme, however, reveals a potential difficulty, if the "relay model" outlined above is valid. Assuming that the catalyst attacks the 4-pentenoate entity of the envisaged cyclization precursor **34**, a fairly stable six-membered chelate structure may be obtained which can intercept the catalyst in an unproductive form.

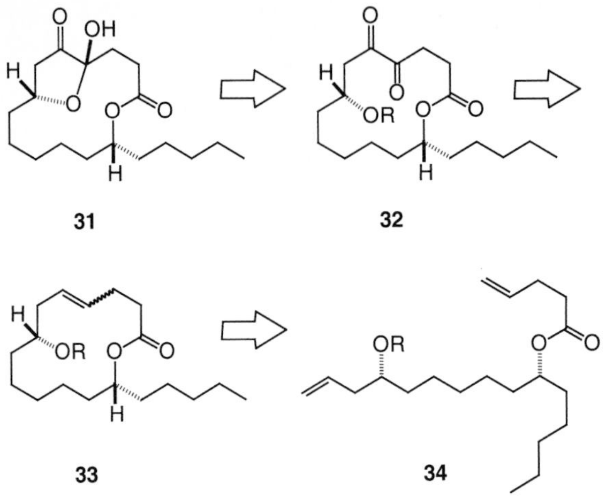

Scheme 9-19

In fact, diene **34**, which is easily obtained in enantiomerically pure form from cycloheptene, fails to afford any cyclized product on exposure to catalytic amounts of the ruthenium carbene $(PCy_3)_2Cl_2Ru=CHPh$. We reasoned that it is necessary to destabilize the likely unproductive chelate responsible for this outcome and/or to activate the catalyst. Either goal might be achieved by running the reaction in the presence of a Lewis acid which competes with the evolving carbene for the Lewis-basic ester group and/or will interact with $(PCy_3)_2Cl_2Ru=CHPh$, e.g., by abstraction of a basic phosphine ligand. The additive, however, must not destroy the catalyst, should cause a minimum of acid catalyzed side reactions, and – most importantly – should undergo a *kinetically labile* coordination to the ester.

We identified $Ti(OiPr)_4$ as a candidate which meets these stringent criteria. Gratifyingly, exposure of diene **34** to a mixture of catalytic amounts of $(PCy_3)_2Cl_2Ru=CHPh$ and substoichiometric amounts of $Ti(OiPr)_4$ in refluxing CH_2Cl_2 effects a smooth cyclization reaction (Scheme 9-20)! The resulting product **35** is then oxidized to *vic*-diketone **36** with $KMnO_4$ in Ac_2O, which on cleavage of the silyl protecting group during the aqueous workup cyclizes to (–)-gloeosporone **31**. This approach delivers enantiomerically pure **31** in only eight synthetic operations with excellent overall yield starting from cycloheptene; moreover, the chosen route is highly flexible and can be easily adapted to the synthesis of analogues of this bioactive target.

Although the precise mode of action of the binary RCM catalyst (i.e., $(PCy_3)_2Cl_2Ru=CHPh+Ti(OiPr)_4$) remains to be elucidated, its key role within the gloeosporone synthesis gives further credence to the notion that a transient and kinetically labile chelate complex acts as a crucial template which shapes the diene precursor and thereby provides internal bias for ring closure.

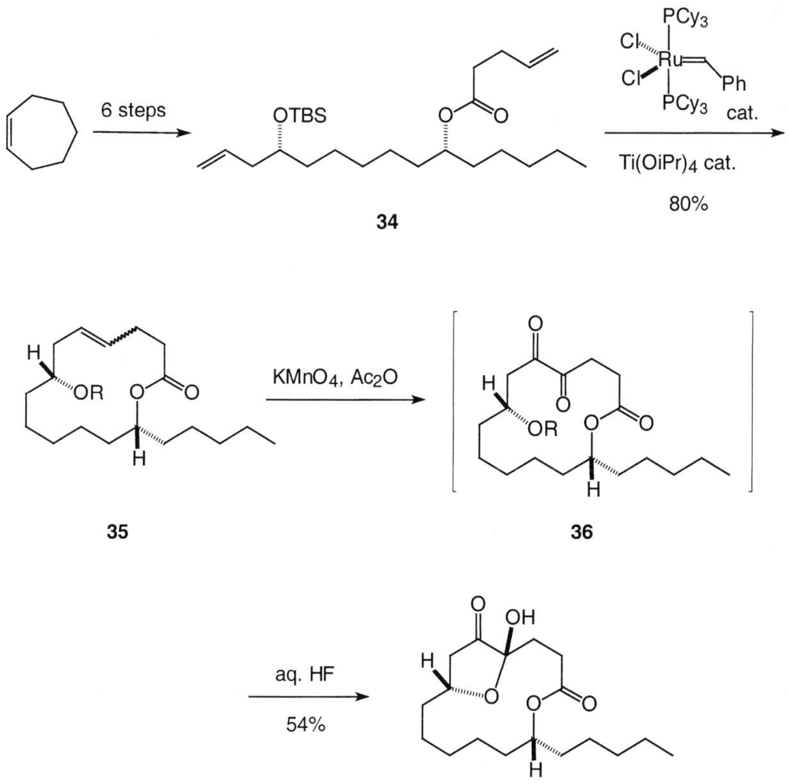

Scheme 9-20

9.6 Macrocycles via Tsuji–Trost Allylation

The comparison of intramolecular carbopalladation reactions of allenes and alkenes outlined in Schemes 9-5 and 9-6 illustrates that not every transition metal catalyzed ring closure necessarily involves a template effect. Others, however, clearly benefit from it. A prototype example is the palladium catalyzed cycloisomerization of alkenyl epoxides carrying distal pre-nucleophiles [38, 39], representing one variant of the famous Tsuji–Trost allylation [40].

This particular transformation cleverly exploits the polyfunctional character of the reacting units. It starts with an oxidative insertion of Pd^0 into the allylic moiety leading to the formation of a dipolar π-allylpalladium species. This step usually occurs with particular ease since the strain of the epoxide ring is released. The resulting alkoxide adjacent to the π-allyl unit acts as a base and deprotonates the pre-nucleophile *if* the substrate adopts a suitable conformation. Charge attraction then keeps the emerging nucleophile in the proximity of the electrophilic π-allylpalladium entity, thereby greatly enhances the chance of reactive encounters, and strongly favors the formation of the desired cyclic product. This intrinsic stabilization of the properly shaped conformer is macroscopically expressed in the ease of cyclization ascribed to a "template effect". An additional advantage of this methodology stems from the fact that the hydroxyl group liberated during the oxidative insertion step directs the entering nucleophile to the distal position of the π-allylpalladium complex and therefore efficiently controls the regioselectivity of the reaction [38–40].

It should be mentioned, however, that high dilution is still mandatory in order to obtain high yields [15, 16], which can be achieved either by low feeding rates of the substrate into the solution containing the catalyst or by exploiting the "pseudo-dilution effect" of polymer bound Pd^0 [41]. This experimental requirement shows that the palladium center itself does not shape the substrate but just stabilizes the reactive conformation once it is adopted (Scheme 9-21). Therefore the observed "template effect" must be largely attributed to electrostatic attraction and/or favorable ion pairing properties of the intermediates.

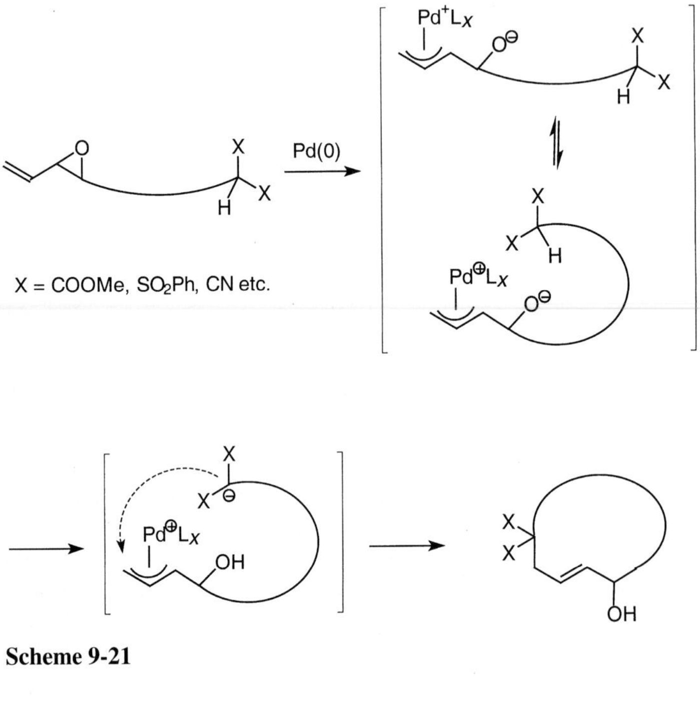

Scheme 9-21

The key step of a recent total synthesis of roseophilin **41**, a structurally unique alkaloid with promising cytotoxic activities, constitutes a characteristic example illustrating the performance of this process (Scheme 9-22) [42]. Thus, treatment of compound **38** with catalytic amounts of Pd(PPh$_3$)$_4$ and dppe in refluxing THF leads to the formation of the highly functionalized 13-membered ring **39**, which can be converted into the macro-tricyclic ketopyrrole core **40** of the target in a few operations. It is interesting to note that the same compound has also been made by a ring closing metathesis route [43], with the yields of the palladium- and the RCM-based approaches being almost identical.

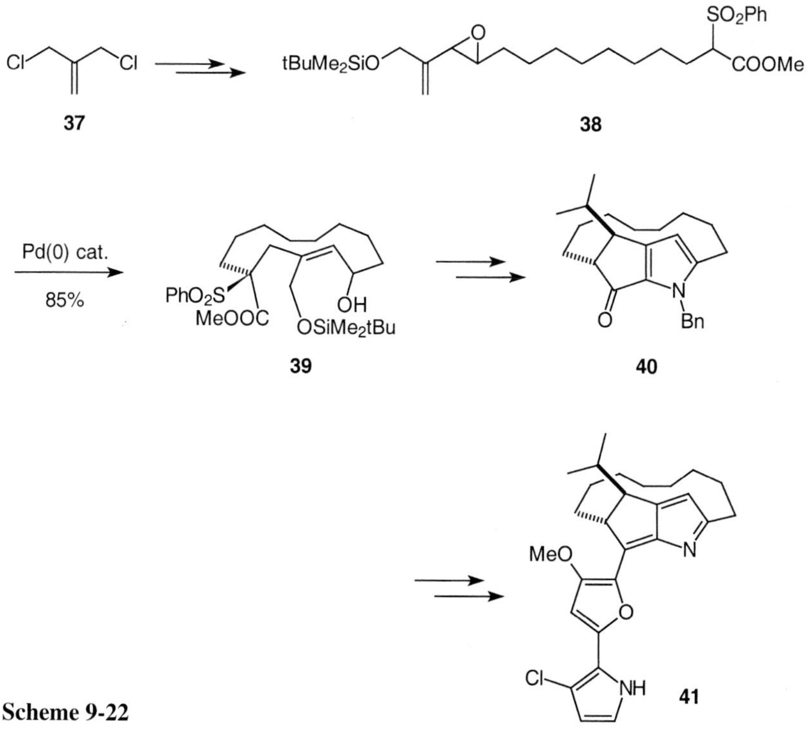

Scheme 9-22

A conceptually closely related approach to medium-sized and large rings has been recently reported, in which the cyclization is triggered by a hydropalladation of an allene rather than by oxidative addition of Pd⁰ into a vinyl epoxide (Scheme 9-23) [44]. In this case, however, the addition of an external base as well as of a proton source are necessary in order to generate the nucleophilic entity and the required palladium hydride catalyst, respectively. 4-Dimethylaminopyridine (4-DMAP) and HOAc turned out to be the best combination for this purpose. One may assume that the hydroacetate formed exerts a bifunctional catalysis and simultaneously acts as an effective template that shapes the conformation of the cyclization precursor. This interpretation is particularly tempting as such reactions proceed smoothly even at rather high concentrations. A straightforward synthesis of the 10-membered product **44**, a promising endopeptidase inhibitor, illustrates the convenience of this elegant new methodology (Scheme 9-24) [44, 45].

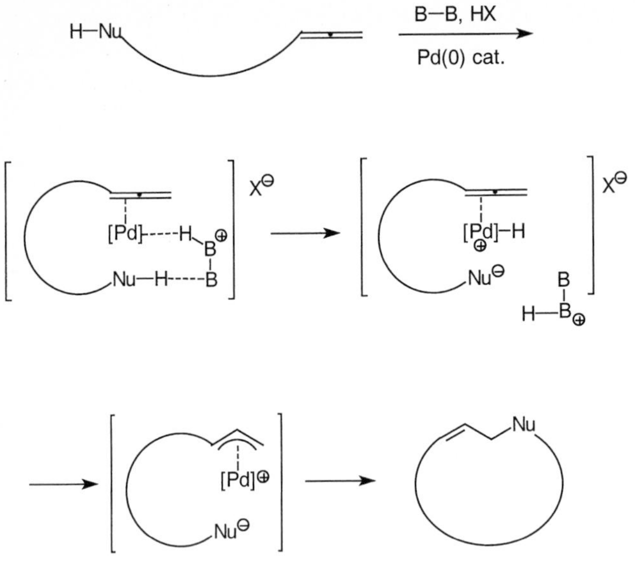

B—B: denotes a compound with 2 basic sites

Scheme 9-23

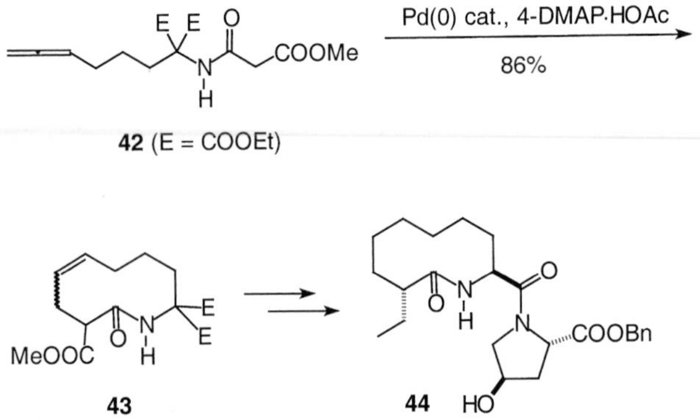

42 (E = COOEt)

43   **44**   HO

Scheme 9-24

9.7 Summary

Although the few examples compiled in this review provide by no means a comprehensive picture, their analysis highlights a few general notions concerning templated ring closure reactions. They show beyond doubt that a "template effect" is not a uniform phenomenon and no simplistic global picture can be drawn. The effect may be thermodynamic or kinetic in nature and may derive from pronounced chemical affinities, from ion pairing and/or electrostatic attraction of the reacting sites, from coordination induced proximity, from the polarizability of a metal cation or from a specific coordination geometry, etc. Moreover, it can be rather difficult to distinguish between a "template effect" and other yet more conventional incentives for ring closure. This is particularly true if the substrate contains structural elements that favor cyclization over competing intermolecular reactions (e.g., rigid backbones, multiple chiral centers affecting its conformation, strong internal hydrogen bonds, *gem*-dimethyl groups) [46] or if ring closure is a priori kinetically highly favorable. Therefore a less prolific use of the terminology "template" seems appropriate which should always be based on a set of appropriate control experiments ruling out more trivial arguments.

9.8   Experimental Procedures

9.8.1   Synthesis of 15,16-Diphenyl-1,4,7,10-tetraoxa(10.2)[20]-paracyclophan-15-ene (14, Scheme 9-8) by an Intramolecular McMurry Coupling under "Salt Free" Conditions [19]

A suspension of commercial titanium powder (325-mesh) (2.06 g, 43 mmol) in DME and chlorotrimethylsilane (TMSCl) (6 mL) was refluxed for 48 h under Ar. A solution of diketone 11 (298 mg, 0.584 mmol) in DME (8 mL) was added at once and reflux was continued until TLC showed complete conversion of the substrate (ca. 8 h). The slurry was cooled to ambient temperature, diluted with THF (20 mL), and filtered through a pad of silica, the insoluble residues were washed with THF (20 mL in several portions), the combined filtrates were evaporated, and the crude product was purified by flash chromatography (silica, hexanes/ethyl acetate = 2 : 1) to afford product 15 as colorless, hygroscopic crystals (158 mg, 57%), m.p. (DSC) = 184.6 °C.

9.8.2   Synthesis of Oxacyclohexadec-11-en-2-one (22, Scheme 9-14) via RCM [32]

A solution of diene 21 (298 mg, 1.12 mmol) in CH_2Cl_2 (100 mL) was slowly added via a dropping funnel to a solution of $(PCy_3)_2Cl_2Ru=CHCH=CPh_2$ (40 mg, 4 mol%) in CH_2Cl_2 (100 mL) under Ar. After being stirred for 30 h at ambient temperature, the solvent was evaporated and the crude product purified by flash chromatography (silica, hexanes/ethyl acetate = 100 : 1) affording lactone 22 (E/Z = 46 : 54) as a colorless syrup (219 mg, 79%).

9.8.3   Palladium-catalyzed Alkenyl Epoxide Cyclization: Synthesis of Macrocycle 39 (Scheme 9-22) [42]

To a refluxing solution of Pd(PPh$_3$)$_4$ (23 mg, 0.02 mmol) and dppe (16 mg, 0.04 mmol) in THF (100 mL) was added a solution of substrate 38 (108 mg, 0.20 mmol) in THF

(40 mL) dropwise over a period of 6 h. The reaction mixture was refluxed for an additional 10 h, cooled to ambient temperature, and extracted with H_2O/ethyl acetate, and the organic layer was dried over Na_2SO_4. After evaporation of the solvent, the crude product was purified by flash chromatography (SiO_2, hexanes/ethyl acetate = 3 : 1) affording product **39** (92 mg, 85%, mixture of diastereoisomers) as a pale yellow foam.

References

[1] (a) E. J. Corey, K. C. Nicolaou, L. S. Melvin, *J. Am. Chem. Soc.* **1975**, *97*, 653–654 and 654–655; (b) K. C. Nicolaou, *Tetrahedron* **1977**, *33*, 683–710.

[2] The first proposal that a template effect is operative in such cases has been made by Greene, although a perusal of Pederson's original paper suggests that its seminal crown ether synthesis also relies on this effect. Cf.: (a) C. J. Pedersen, *J. Am. Chem. Soc.* **1967**, *89*, 7017–7036; (b) R. N. Greene, *Tetrahedron Lett.* **1972**, 1793–1796. For a comprehensive treatise of the template effect in crown ether syntheses see [4] and literature cited therein.

[3] (a) P. W. Jolly, G. Wilke, *The Organic Chemistry of Nickel*, Vol. 2, Academic Press, New York, **1975**; (b) G. Wilke, *Angew. Chem.* **1988**, *100*, 189–211; *Angew. Chem., Int. Ed Engl.* **1988**, *27*, 185–206.

[4] (a) For a recent comprehensive treatise of templated syntheses of macrocyclic polyethers and related host compounds see: *Template Synthesis* (N. V. Gerbeleu, V. B. Arion, J. Burgess, Eds.), Wiley–VCH, Weinheim, **1999** and literature cited therein. (b) See also: R. Hoss, F. Vögtle, *Angew. Chem.* **1994**, *106*, 389–398; *Angew. Chem. Int. Ed. Engl.* **1994**, *33*, 375–384.

[5] Reviews: (a) G. Rousseau, *Tetrahedron* **1995**, *51*, 2777–2849; (b) N. A. Petasis, M. A. Patane, *Tetrahedron* **1992**, *48*, 5757–5821; (c) G. A. Molander, *Acc. Chem. Res.* **1998**, *31*, 603–609.

[6] A. Fürstner, H. Szillat, B. Gabor, R. Mynott, *J. Am. Chem. Soc.* **1998**, *120*, 8305–8314.

[7] S. Ma, E. Negishi, *J. Am. Chem. Soc.* **1995**, *117*, 6345–6359.

[8] For general discussions of the effect of ring size on cyclization see: (a) E. L. Eliel, S. H. Wilen, L. N. Mander, *Stereochemistry of Organic Compounds*, Wiley, New York, **1994**; (b) J. E. Baldwin, *J. Chem. Soc., Chem. Commun.* **1976**, 734–736 and 738–741.

[9] For general reviews on macrocycle syntheses see: (a) P. Knops, N. Sendhoff, H. B. Mekelburger, F. Vögtle, *Top. Curr. Chem.* **1992**, *161*, 1–36; (b) Q. Meng, M. Hesse, *Top. Curr. Chem.* **1992**, *161*, 107–176; (c) M. Bartra, F. Urpi, J. Vilarrasa in *Recent Progress in the Chemical Synthesis of Antibiotics and Related Microbial Products*, Vol. 2 (Lukacs, G., Ed.), Springer, Berlin, **1993**, pp. 1–55; (d) C. J. Roxburgh, *Tetrahedron* **1995**, *51*, 9767–9822; (e) H. Stach, M. Hesse, *Tetrahedron* **1988**, *44*, 1573–1590; (f) M. A. Tius, *Chem. Rev.* **1988**, *88*, 719–732.

[10] For recent reviews see: (a) A. Fürstner, B. Bogdanovic, *Angew. Chem.* **1996**, *108*, 2583–2609; *Angew. Chem., Int. Ed. Engl.* **1996**, *35*, 2442–2469; (b) A. Fürstner in *Transition Metals for Organic Synthesis* (M. Beller, C. Bolm, Eds.), Wiley–VCH, Weinheim, **1998**, pp. 381–401; (c) T. Lectka in *Active Metals. Preparation, Characterization, Applications* (A. Fürstner, Ed.), VCH, Weinheim, **1996**, pp. 85–131; (d) J. E. McMurry, *Chem. Rev.* **1989**, *89*, 1513–1524; (e) M. Ephritikhine, *Chem. Commun.* **1998**, 2549–2554.

[11] J. E. McMurry, M. P. Fleming, K. L. Kees, L. R. Krepski, *J. Org. Chem.* **1978**, *43*, 3255–3266.

[12] A. L. Baumstark, C. J. McCloskey, K. E. Witt, *J. Org. Chem.* **1978**, *43*, 3609–3611.

[13] J. P. Schaefer, J. J. Bloomfield, *Org. React.* **1967**, *15*, 1–203.

[14] Reviews: (a) J. J. Bloomfield, D. C. Owsley, J. M. Nelke, *Org. React.* **1976**, *23*, 259–403; (b) K. Rühlmann, *Synthesis* **1971**, 236–253; (c) K. T. Finley, *Chem. Rev.* **1964**, *64*, 573–589 and literature cited.

[15] (a) K. Ziegler in *Houben-Weyl, Methoden der Organischen Chemie, Vol. 4(2)*, Thieme, Stuttgart, **1955**, pp. 729–822; (b) P. Ruggli, *Liebigs Ann. Chem.* **1912**, *392*, 92–100; (c) Review: L. Rossa, F. Vögtle, *Top. Curr. Chem.* **1983**, *113*, 1–86.

[16] It should be emphasized that the expression "high dilution" is ambiguous terminology, although quite frequently used by preparative chemists to denote a slow feeding rate of the substrate into the reaction mixture (e.g., via syringe pumps). A more accurate description of cyclization reactions refers to the "effective molarity parameter (EM)", which is a quantitative measure of the propensity of a given substrate to cyclization. For an in-depth treatise see the following reference and literature cited therein: L. Mandolini, *Adv. Phys. Org. Chem.* **1986**, *22*, 1–111; see also Chapter 1.

[17] A. Fürstner, A. Hupperts, *J. Am. Chem. Soc.* **1995**, *117*, 4468–4475.

[18] For a study on the morphological basis for the activation of commercial titanium by chlorosilanes see: A. Fürstner, B. Tesche, *Chem. Mater.* **1998**, *10*, 1968–1973.

[19] (a) A. Fürstner, G. Seidel, C. Kopiske, C. Krüger, R. Mynott, *Liebigs Ann.* **1996**, 655–662; (b) for an investigation on the structure and ion-pairing behavior of these crownophanes see: F. Barbosa, V. Péron, G. Gescheidt, A. Fürstner, *J. Org. Chem.* **1998**, *63*, 8806–8814.

[20] M. D. Ferguson, J. A. Butcher, *Tetrahedron Lett.* **1993**, 3719–3722.

[21] It has long been assumed that metallic Ti is formed on treatment of $TiCl_3$ or $TiCl_4$ with reducing agents such as Li, K, Mg, Zn, $LiAlH_4$, etc. and that the different reactivity of individual samples stems from differences in the texture and size of the particles. Recent investigations, however, prove beyond doubt that this simplified view is incorrect and that low-valent rather than metallic titanium samples are formed which differ in their composition and reactivity depending on their mode of preparation. For leading references see: (a) L. E. Aleandri, S. Becke, B. Bogdanovic, D. J. Jones, J. Rozière, *J. Organomet. Chem.* **1994**, *472*, 97–112; (b) L. E. Aleandri, B. Bogdanovic, A. Gaidies, D. J. Jones, S. Liao, A. Michalowicz, J. Rozière, A. Schott, *J. Organomet. Chem.* **1993**, *459*, 87–93; (c) B. Bogdanovic, A. Bolte, *J. Organomet. Chem.* **1995**, *502*, 109–121.

[22] J. E. McMurry, M. G. Silvestri, M. P. Fleming, T. Hoz, M. W. Grayston, *J. Org. Chem.* **1978**, *43*, 3249–3254.

[23] (a) A. Fürstner, *Synthesis* **1989**, 571–590; (b) A. Fürstner, in *Organozinc Reagents. A Practical Approach* (P. Knochel, P. Jones, Eds.), Oxford University Press, Oxford, **1999**, pp. 287–305; (c) M. W. Rathke, in *Comprehensive Organic Synthesis, Vol. 2* (B. M. Trost, I. Fleming, Eds.), Pergamon, Oxford, **1991**, pp. 277–299; (d) M. W. Rathke, *Org. React.* **1975**, *22*, 423–460.

[24] K. Maruoka, S. Hashimoto, Y. Kitagawa, H. Yamamoto, H. Nozaki, *J. Am. Chem. Soc.* **1977**, *99*, 7705–7707.

[25] (a) T. Tabuchi, K. Kawamura, J. Inanaga, M. Yamaguchi, *Tetrahedron Lett.* **1986**, *27*, 3889–3890; (b) J. Inanaga, Y. Yokoyama, Y. Handa, M. Yamaguchi, *Tetrahedron Lett.* **1991**, *32*, 6371–6374.

[26] T. Moriya, Y. Handa, J. Inanaga, M. Yamaguchi *Tetrahedron Lett.* **1988**, *29*, 6947–6948.

[27] S. Inoue, Y. Iwabuchi, H. Irie, S. Hatakeyama, *Synlett* **1998**, 735–736.

[28] (a) T. Mukaiyama, I. Shiina, H. Iwadare, M. Saitoh, T. Nishimura, N. Ohkawa, H. Sakoh, K. Nishimura, Y. Tani, M. Hasegawa, K. Yamada, K. Saitoh, *Chem. Eur. J.* **1999**, *5*, 121–161; (b) See also: I. Shiina, K. Uoto, N. Mori, T. Kosugi, T. Mukaiyama, *Chem. Lett.* **1995**, 181–182.

[29] For general reviews on RCM see: (a) R. H. Grubbs, S. Chang, *Tetrahedron* **1998**, *54*, 4413–4450; (b) A. Fürstner, *Top. Organomet. Chem.* **1998**, *1*, 37–72; (c) M. Schuster, S. Blechert, *Angew. Chem.* **1997**, *109*, 2125–2144; *Angew. Chem., Int. Ed. Engl.* **1997**, *36*, 2037–2056; (d) A. Fürstner, *Top. Catal.* **1997**, *4*, 285–299; (e) R. H. Grubbs, S. J. Miller, G. C. Fu, *Acc. Chem. Res.* **1995**, *28*, 446–452.

[30] (a) Mechanistic discussion of metathesis in general: J. L. Hérisson, Y. Chauvin, *Makromol. Chem.* **1970**, *141*, 161–176; (b) K. J. Ivin, J. C. Mol, *Olefin Metathesis and Metathesis Polymerization*, 2nd edn., Academic Press, New York, **1997**.

[31] This view has been launched, e.g., in [29e]; see also: (a) H.-G. Schmalz, *Angew. Chem.* **1995**, *107*, 1981–1984; *Angew. Chem., Int. Ed. Engl.* **1995**, *34*, 1833–1836; (b) U. Koert, *Nachr. Chem. Techn. Lab.* **1995**, *43*, 809–814.

[32] (a) A. Fürstner, K. Langemann, *J. Org. Chem.* **1996**, *61*, 3942–3943; (b) A. Fürstner, K. Langemann, *Synthesis* **1997**, 792–803.

[33] (a) Y.-S. Shon, T. R. Lee, *Tetrahedron Lett.* **1997**, *38*, 1283–1286; (b) S. K. Armstrong, B. A. Christie, *Tetrahedron Lett.* **1996**, *37*, 9373–9376; (c) J. L. Mascarenas, A. Rumbo, L. Castedo, *J. Org. Chem.* **1997**, *62*, 8620–8621.

[34] (a) Dactylol: A. Fürstner, K. Langemann, *J. Org. Chem.* **1996**, *61*, 8746–8749; (b) Lasiodiplodin: A. Fürstner, N. Kindler, *Tetrahedron Lett.* **1996**, *37*, 7005–7008; (c) Jasmin Ketolactone: A. Fürstner, T. Müller, *Synlett* **1997**, 1010–1012; (d) Epilachnene: see [32b]; (d) Zeranol: A. Fürstner, G. Seidel, N. Kindler, *Tetrahedron*, **1999**, *55*, 8215–8230; (e) Tricolorin A: A. Fürstner, T. Müller, *J. Org. Chem.* **1998**, *63*, 424–425; (f) Roseophilin: A. Fürstner, T. Gastner, H. Weintritt, *J. Org. Chem.*, **1999**, *64*, 2361–2366; (g) Tricolorin G: A. Fürstner, T. Müller, *J. Am. Chem. Soc*, in press; (h) Cyclononylprodigiosin: A. Fürstner, J. Grabowski, in preparation.

[35] For leading references on other RCM-based macrocyclization reactions see: (a) S. J. Miller, H. E. Blackwell, R. H. Grubbs, *J. Am. Chem. Soc.* **1996**, *118*, 9606–9614; (b) Z. Yang, Y. He, D. Vourloumis, H. Vallberg, K. C. Nicolaou, *Angew. Chem.* **1997**, *109*, 170–173; *Angew. Chem., Int. Ed. Engl.* **1997**, *36*, 166–168; (c) K. C. Nicolaou, Y. He, D. Vourloumis, H. Vallberg, Z. Yang, *Angew. Chem.* **1996**, *108*, 2554–2556; *Angew. Chem., Int. Ed. Engl.* **1996**, *35*, 2399–2401; (d) D. Schinzer, A. Limberg, A. Bauer, O. M. Böhm, M. Cordes, *Angew. Chem.* **1997**, *109*, 543–544; *Angew. Chem., Int. Ed. Engl.* **1997**, *36*, 523–524; (e) P. Bertinato, E. J. Sorensen, D. Meng, S. J. Danishefsky, *J. Org. Chem.* **1996**, *61*, 8000–8001; (f) Z. Xu, C. W. Johannes, A. F. Houri, D. S. La, D. A. Cogan, G. E. Hofilena, A. H. Hoveyda, *J. Am. Chem. Soc.* **1997**, *119*, 10302–10316; (g) S. F. Martin, Y. Liao, H.-J. Chen, M. Pätzel, M. N. Ramser, *Tetrahedron Lett.* **1994**, 6005–6008; (h) A. K. Ghosh, K. A. Hussain, *Tetrahedron Lett.* **1998**, 1881–1884; (i) E. Magnier, Y. Langlois, *Tetrahedron Lett.* **1998**, *39*, 837–840; (j) H. E. Blackwell, R. H. Grubbs, *Angew. Chem.* **1998**, *110*, 3469–3472; *Angew. Chem., Int. Ed. Engl.* **1998**, *37*, 3281–3284; (k) K. Arakawa, T. Eguchi, K. Kakinuma, *J. Org. Chem.* **1998**, *63*, 4741–4745; (l) T. Nishioka, Y. Iwabuchi, H. Irie, S. Hatakeyama, *Tetrahedron Lett.* **1998**, *39*, 5597–5600; (m) W. P. D. Goldring, A. S. Hodder, L. Weiler, *Tetrahedron Lett.* **1998**, *39*, 4955–4958; (n) H. El Sukkari, J. P. Gesson, B. Renoux, *Tetrahedron Lett.* **1998**, *39*, 4043–4046; (o) R. M. Williams, J. Liu, *J. Org. Chem.* **1998**, *63*, 2130–2132; (p) Y. Gao, P. Lane-Bell, J. C. Vederas, *J. Org. Chem.* **1998**, *63*, 2133–2143.

[36] A. Fürstner, K. Langemann, *J. Am. Chem. Soc.* **1997**, *119*, 9130–9136.

[37] Elucidation of the correct structure: (a) W. L. Meyer, W. B. Schweizer, A. K. Beck, W. Scheifele, D. Seebach, S. L. Schreiber, S. E. Kelly, *Helv. Chim. Acta* **1987**, *70*, 281–291. For previous syntheses see: (b) G. Adam, R. Zibuck, D. Seebach, *J. Am. Chem. Soc.* **1987**, *109*, 6176–6177; (c) S. L. Schreiber, S. E. Kelly, J. A. Porco, T. Sammakia, E. M. Suh, *J. Am.*

Chem. Soc. **1988**, *110*, 6210–6218; (d) D. Seebach, G. Adam, R. Zibuck, W. Simon, M. Rouilly, W. L. Meyer, J. F. Hinton, T. A. Privett, G. E. Templeton, D. K. Heiny, U. Gisi, H. Binder, *Liebigs Ann.* **1989**, 1233–1240; (e) N. R. Curtis, A. B. Holmes, M. G. Looney, N. D. Pearson, G. C. Slim, *Tetrahedron Lett.* **1991**, *32*, 537–540; (f) S. Takano, Y. Shimazaki, M. Takahashi, K. Ogasawara, *J. Chem. Soc. Chem. Commun.* **1988**, 1004–1005; (g) M. Matsu-shita, M. Yoshida, Y. Zhang, M. Miyashita, H. Irie, T. Ueno, T. Tsurushima, *Chem. Pharm. Bull.* **1992**, *40*, 524–527.

[38] For the development of the intermolecular version of this reaction and applications thereof see: (a) B. M. Trost, G. A. Molander, *J. Am. Chem. Soc.* **1981**, *103*, 5969–5972; (b) J. Tsuji, H. Kataoka, Y. Kobayashi, *Tetrahedron Lett.* **1981**, *22*, 2575–2578; (c) B. M. Trost, M. A. Ceschi, B. König, *Angew. Chem.* **1997**, *109*, 1562–1564; *Angew. Chem., Int. Ed. Engl.* **1997**, *36*, 1486–1489.

[39] For applications to cyclization reactions see: (a) B. M. Trost, R. W. Warner, *J. Am. Chem. Soc.* **1982**, *104*, 6112–6114; (b) B. M. Trost, R. W. Warner, *J. Am. Chem. Soc.* **1983**, *105*, 5940–5942; (c) A. S. Kende, I. Kaldor, R. Aslanian, *J. Am. Chem. Soc.* **1988**, *110*, 6265–6266; (d) B. M. Trost, J. T. Hane, P. Metz, *Tetrahedron Lett.* **1986**, *27*, 5695–5698; (e) B. M. Trost, T. S. Scanlan, *J. Am. Chem. Soc.* **1989**, *111*, 4988–4990; (f) B. M. Trost, B. A. Vos, C. M. Brzezowski, D. P. Martina, *Tetrahedron Lett.* **1992**, *33*, 717–720.

[40] Reviews: (a) B. M. Trost, *Angew. Chem.* **1989**, *101*, 1199–1219; *Angew. Chem., Int. Ed. Engl.* **1989**, *28*, 1173–1192; (b) J. Tsuji, *Palladium Reagents and Catalysts*, Wiley, New York, **1995**.

[41] For a discussion of the "pseudodilution effect" reached with polymer-bound Pd⁰ see [39 a, f] and references cited therein.

[42] (a) A. Fürstner, H. Weintritt, *J. Am. Chem. Soc.* **1998**, *120*, 2817–2825; (b) A. Fürstner, H. Weintritt, *J. Am. Chem. Soc.* **1997**, *119*, 2944–2945.

[43] For the metathesis route to roseophilin see [34 f]; for yet another RCM approach to this tar-get see: S. H. Kim, I. Figueroa, P. L. Fuchs, *Tetrahedron Lett.* **1997**, *38*, 2601–2604.

[44] B. M. Trost, P.-Y. Michellys, V. J. Gerusz, *Angew. Chem.* **1997**, *109*, 1837–1839; *Angew. Chem., Int. Ed. Engl.* **1997**, *36*, 1750–1753.

[45] For related methodology see inter alia. Y. Yamamoto, M. Al-Masum, N. Asao, *J. Am. Chem. Soc.* **1994**, *116*, 6019–6020.

[46] For comprehensive treatises on conformational effects see: (a) R. Göttlich, B. C. Kahrs, J. Krüger, R. W. Hoffmann *Chem. Commun.* **1997**, 247–257 and literature cited therein; (b) see also [8 a].

10 Use of the Temporary Connection in Organic Synthesis

Liam R. Cox, Steven V. Ley

10.1 Introduction

Classical methods of bond formation invariably involve intermolecular processes. A number of problems, however, are associated with this type of bond-forming event. These often include unacceptably low levels of regio- and stereoselectivity. One approach which helps to circumvent some of these problems is to temporarily intramolecularize [1] a reaction. This exploits the advantages associated with intramolecular transformations, while still affording an end-product which is the net result of the desired intermolecular process.

The advantages of intra- over intermolecular transformations are manifold. By increasing the proximity between reacting centers, activation entropy is reduced. This facilitates bond formation, often allowing the intramolecular reaction to proceed under milder conditions than its intermolecular counterpart. Intramolecularization also affects the kinetics of the process, which often equates to an increase in the rate of the desired reaction relative to reagent-consuming side reactions: in kinetically controlled reactions, this improves the yield of the required product at the expense of troublesome side products. By bringing together the reacting centers about a uniting template, additional conformational constraints are imposed on the reacting system resulting in more defined transition states (T. S.) and concomitant increased levels of regio- and stereoselectivity in the end-products.

One of the more successful methods of achieving intramolecularization is to use the so-called temporary connection [2]. Using this concept, the reactants are covalently attached to one another through an atom or series of atoms, which function as a disposable tether. For this approach to be successful, a number of design criteria need to be satisfied. Introduction of the tether must be facile. It should be of sufficient length and flexibility so as not to hinder reaction – by increasing strain in the T. S. – but it should not be so long that unfavorable entropic factors become dominant once more. Similarly it is also undesirable for the tether to be so short and rigid that approach of the reacting centers is hindered, or causes excessive buildup of strain in the T. S. The tether should also be appropriately designed to act as a template about which the reacting groups are pre-organized to react through a defined T. S. This can lead to high levels of stereo- and regiocontrol which may enhance, or in some cases reverse, the inherent controlling bias in the corresponding intermolecular reaction. Upon completion of the reaction, the tether should be readily removed or allow further manipulation in subsequent transformations (Figure 10-1).

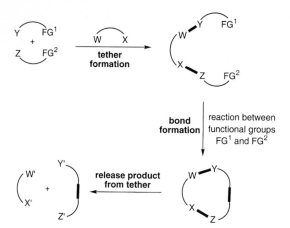

bond formation | reaction between functional groups FG^1 and FG^2

release product from tether

Figure 10-1   Use of the temporary connection in bond formation.

Incorporating asymmetry into the tether also creates the potential for effecting stereochemical transcription where stereochemical information can be relayed in a defined manner from existing sites to the newly generated stereogenic center(s).

This chapter will concentrate on examples of templated synthesis where the template takes the form of a temporary, covalent tether. The intermediate molecule containing both reacting species is in all cases *isolable* (or potentially so). This distinguishes this intra molecularization approach from substrate-directed stereoselective synthesis. In this case, the template, most frequently a metal, is used in a much more transient sense allowing the potential for developing catalytic systems, which is obviously not possible using a covalently bound template.

The guiding concept of pre-organizing reagents by covalent attachment to a template is quite general. It has been successfully applied to a wide range of important reaction types using an array of tethers. This alone bears testimony to the utility of this approach in facilitating reaction and providing regio- and stereocontrol. This chapter will not provide an exhaustive list of tethered reaction systems [3]. Instead, we have chosen to concentrate on highlighting the variety of tethers and transformations which have benefited from the concept of using a temporary connection. Emphasis will be laid on areas where this tethering approach has provided unique advantages over the analogous intermolecular reaction.

The examples of temporary tethering strategies discussed in Section 10.2 all involve cycloaddition reactions, which is where this approach to controlling both regio- and stereoselectivity has found most success. The rest of this chapter will endeavor to illustrate the diversity of reaction types which have also benefited from the use of a temporary connection to bind together reacting fragments. In most cases, the reactions will involve tethers which have been already described. Consequently their preparation and removal will not be discussed in detail; rather we will concentrate on highlighting the advantages of using the tethering approach for achieving stereo- and regiocontrol over the analogous intermolecular systems.

10.2 Cycloaddition Reactions

10.2.1 [4 + 2] The Diels–Alder Reaction

The reaction between a 4π-(diene) and a 2π-system (dienophile), otherwise known as the Diels–Alder reaction, is undoubtedly one of the most important reactions for the construction of six-membered rings. Since up to four adjacent stereogenic centers can be generated in a single step, this reaction has become particularly valuable in synthesis. A wide variety of systems have been investigated and in the case of polarized diene/dienophiles, excellent levels of stereo- and regiocontrol can be observed. However levels of regiocontrol, and sometimes stereocontrol, drop significantly in less polarized systems, rendering the reaction less useful.

By joining the diene and dienophile together through a tether, the intramolecular variant provides an entry into polycyclic systems [4]. Significantly, the intramolecular Diels–Alder (IMDA) reaction differs from its intermolecular counterpart in that levels of stereocontrol, and especially regiocontrol, are usually appreciably higher, even for non-polarized systems. The tether linking the reacting π-systems increases steric and geometrical constraints by minimizing, for example, transannular and $A^{1,3}$-interactions. This serves to reduce the number of possible low-energy T. S.s, accounting for the improvements in both regio- and stereocontrol.

Two variants of the IMDA reaction may be recognized. In the first, the so-called Type I IMDA, the dienophile is joined to a terminal position of the diene, producing relatively strain-free bicyclic products. In the Type II IMDA [5], the dienophile is linked to an internal position of the diene. Cyclization in this case results in the formation of a more strained system containing a bridgehead double bond (Figure 10-2).

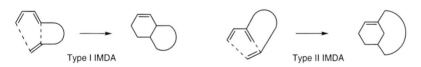

Figure 10-2   The two variants of the IMDA.

If the all-carbon tether is replaced by a connection which may be removed after reaction, then the IMDA can be used to construct six-membered rings with high levels of stereo- and regiocontrol, which would otherwise be impossible or very difficult to prepare using the analogous intermolecular reaction. A wide variety of tethering systems have been examined and proved to be successful in the formation of highly functionalized cyclohexanes.

10.2.1.1 Type I IMDA reactions

Silicon tethers

Silyl acetals

Silyl acetals are attractive tethering agents owing to the ease with which they may be removed after reaction. Craig and co-workers were the first to report the use of such a connecting group in an IMDA reaction [6]. Triene **1** was prepared by the addition of equimolar amounts of dienol **2** and hydroxy enoate **3** to a cooled solution of Ph_2SiCl_2 and Et_3N in CH_2Cl_2 (Scheme 10-1). The desired mixed silyl acetal **1** was isolated in moderate yield after purification by column chromatography on Florisil®, together with small amounts of the symmetrical acetals resulting from double addition of each starting alcohol. Attempted formation of the dimethylsilyl analog proved problematic owing to its greatly increased sensitivity towards hydrolysis.

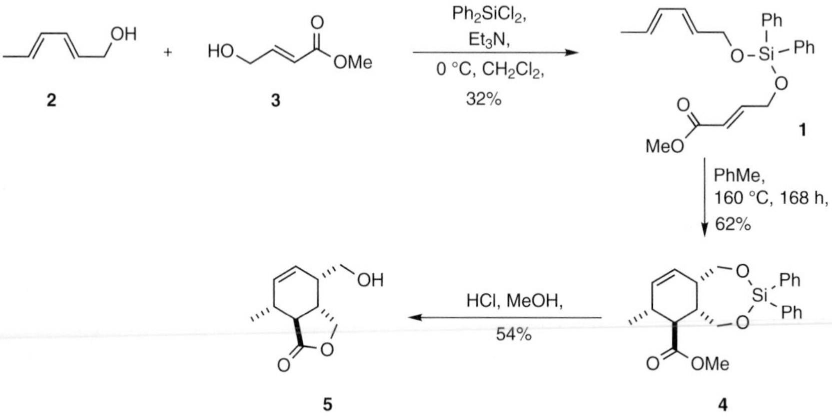

Scheme 10-1   Thermolysis of silyl acetal **1** provides a single IMDA product.

Despite the moderate yields in preparing the starting triene, the subsequent IMDA reaction proved very successful. Thermolysis of **1** in toluene at 160 °C for 168 h afforded a 13:87 mixture of the starting material and a *single*, bicyclic product **4** (Scheme 10-1). Cyclization was completely regio- and stereoselective and may be accounted for by invoking an *exo* T. S. (Figure 10-3). Cleavage of the silyl tether was effected by acidic methanolysis and provided hydroxylactone **5** as a single stereoisomer. The structure of **5** was confirmed by X-ray crystallographic analysis.

The importance of the tether in enabling such high stereo- and regiocontrol was further exemplified upon investigation of the analogous intermolecular reaction [6a, d]. Exposing silyl ethers **6** (diene) and **7** (dienophile) to similar reaction conditions provided a mixture of all four possible regio- and stereoisomers in the ratio 3:2:2:1 (Scheme 10-2).

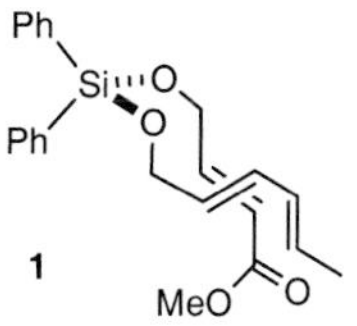

Figure 10-3 Triene **1** reacts through an *exo* T. S.

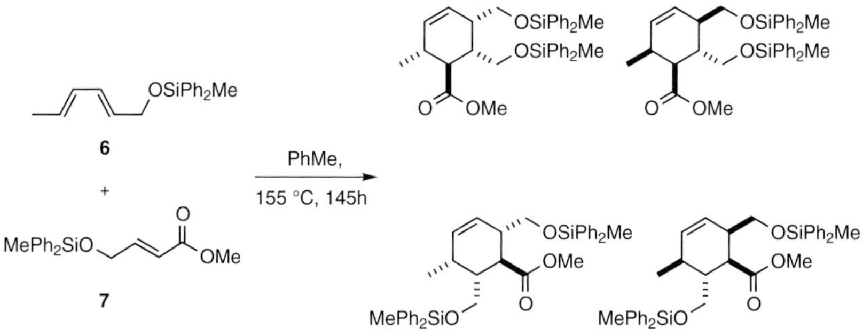

Scheme 10-2 Intermolecular reaction of **6** and **7** is neither stereo- nor regioselective.

Perhaps the greatest drawback with this tethering approach is the relatively poor yield in the formation of the triene IMDA precursor **1**, although the reaction was not optimized. Fortin and co-workers have reported an improved procedure for the efficient formation of unsymmetrical di-*tert*-butylsilyl acetals using $^t\text{Bu}_2\text{Si(Cl)OTf}$, which is readily prepared from the corresponding chlorosilane [7]. Successive incorporation of the dienophile and diene is possible by exploiting the different reactivity of silyl triflates and chlorides towards nucleophilic substitution (Scheme 10-3).

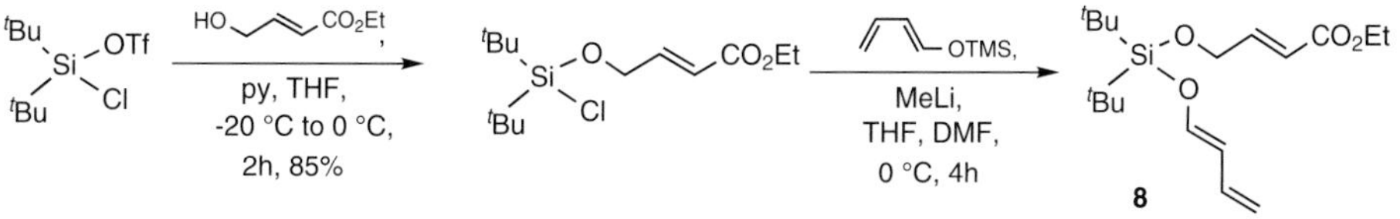

Scheme 10-3 Successive displacement of triflate and chloride allows preparation of mixed di-*tert*-butylsilyl acetals.

Thermolysis of triene **8** in xylene at 160–180 °C afforded a 9:1 mixture of cyclization products in excellent yield, with the major product **9** resulting from an *endo* T. S. Regiocontrol, however, was complete with the formation of exclusively the "head to tail" isomer (Figure 10-4).

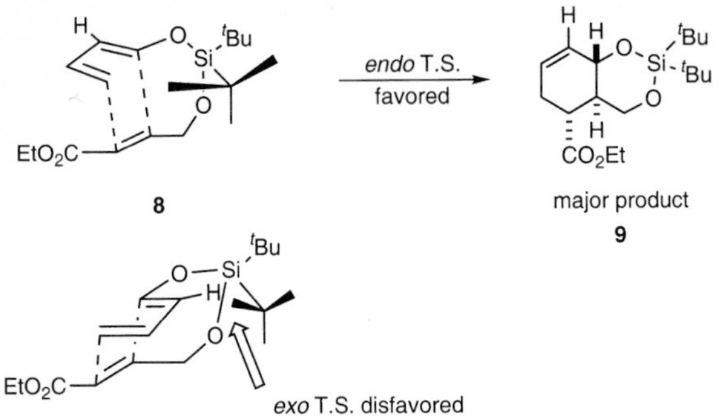

Figure 10-4   Steric interactions between the tether and diene are responsible for the observed stereoselectivity.

This result provides a good example in which the presence of the tether reverses the inherent regiochemical bias of the diene/dienophile system. Both the diene and dienophile are strongly polarized systems. As a result, the corresponding intermolecular reaction proceeds with excellent regiocontrol to afford exclusively the "head to head" isomer as a roughly 1:1 mixture of *endo* and *exo* isomers [7]. The tether not only exerts a complete regiochemical reversal, but also improves the level of stereoselectivity. Steric interactions between one of the bulky ^{t}Bu groups in the tether and the diene account for the adoption of an *endo* T. S. (Figure 10-4).

One of the best tests for any synthetic methodology is in its application to the preparation of biologically important compounds. Posner et al. recently reported the use of a silyl acetal tether in an IMDA reaction for the preparation of the 2-fluoroalkyl analog **10** of 1,25-dihydroxyvitamin D$_3$ **11** (Figure 10-5) [8].

A Lewis acid-promoted, IMDA cycloaddition between an electron-poor pyrone and an electron-rich silyl enol ether was used to prepare the A ring. The requisite triene **12** was synthesized from alcohol **13** and chlorodiisopropylsilyl enol ether **14**. Subsequent ZnBr$_2$-promoted cyclization of **12** allowed slow conversion to the Diels–Alder adducts **15** and **16** in 46% and 12% yields, respectively (Scheme 10-4). The major, *endo* adduct **15** was subsequently elaborated into fluoroalkyl-vitamin D$_3$ analog **10**.

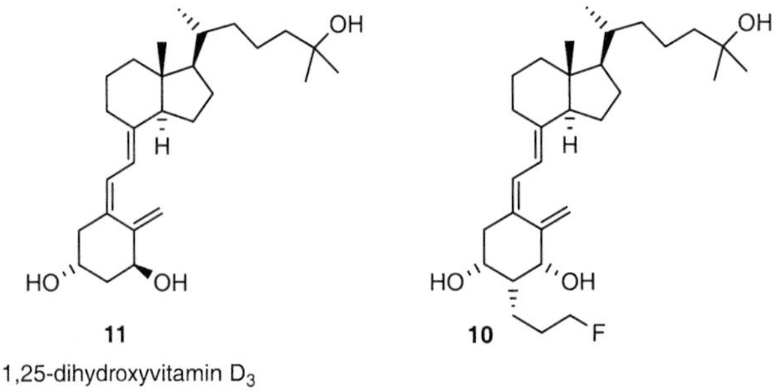

11

1,25-dihydroxyvitamin D$_3$

10

Figure 10-5   Analogs of 1,25-dihydroxyvitamin D$_3$ are important biological targets.

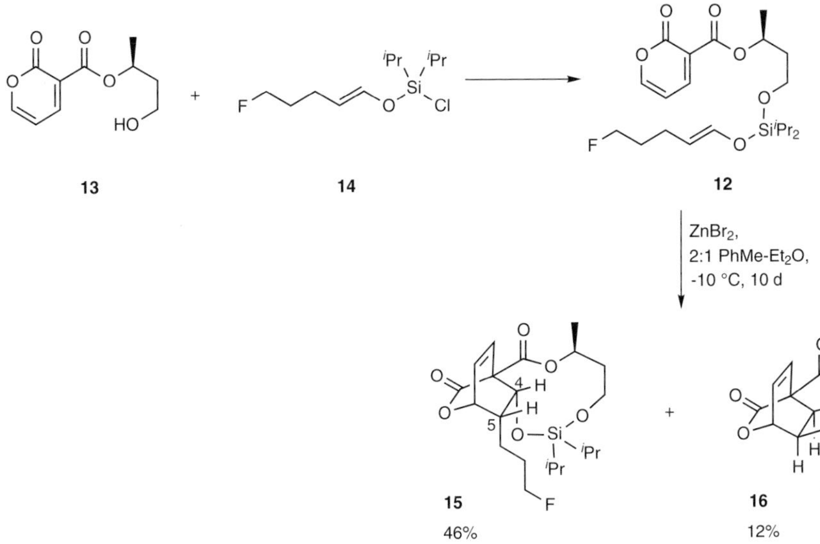

Scheme 10-4 Cyclization proceeds stereoselectively in a stepwise fashion.

The key feature in this IMDA approach was the incorporation of a stereogenic center into the tether, allowing the possibility to control the *absolute* stereochemical outcome of the cyclization. The use of (S)-1,3-butanediol in the tether gave the best results – cyclization proceeded with excellent asymmetric induction affording only two (*exo* and *endo* products) out of the possible four 4,5-*cis* diastereoisomers. It is also noteworthy that this cyclization must proceed in a stepwise fashion since the starting dienophile has a *trans* geometry while the products possess a *cis* substitution pattern at C(4) and C(5).

Vinyl and dienyl silyl ethers

The use of silyl acetals as tethering groups in the IMDA reaction is often hampered by the relatively poor yields observed in the formation of unsymmetrical systems. However, the preparation of the triene precursors can be facilitated if only one of the π-systems is linked through a silyl ether connection, while its reacting partner is attached *directly* to the silicon tether [9].

Dialkylvinylsilyl halides are readily prepared, for example, by hydrosilylation of acetylenes; indeed, the most simple dimethyl- and diphenylvinylsilyl chlorides are now commercially available. Substitution of the halide with the appropriate hydroxydiene then generates the desired triene precursor. Vinyl silyl ethers are also much more stable than silyl acetals, especially towards hydrolysis; nevertheless, a number of mild methods are still available for subsequent removal of the tether.

Stork et al. prepared the two isomeric vinyl silanes **17** and **18,** differing only in the double bond geometry of the dienophile [10]. Cyclization was readily effected in excellent yield on heating a benzene solution for 4–6 h at 80 °C. In both cases, the short length of the connecting tether enforced absolute control of regiochemistry. Cyclization also proceeded with complete stereocontrol through an *endo* addition mode, as evidenced by the formation of the *cis*- and *trans*-fused bicycles **19** and **20** from **18** and **17**, respectively (Scheme 10-5).

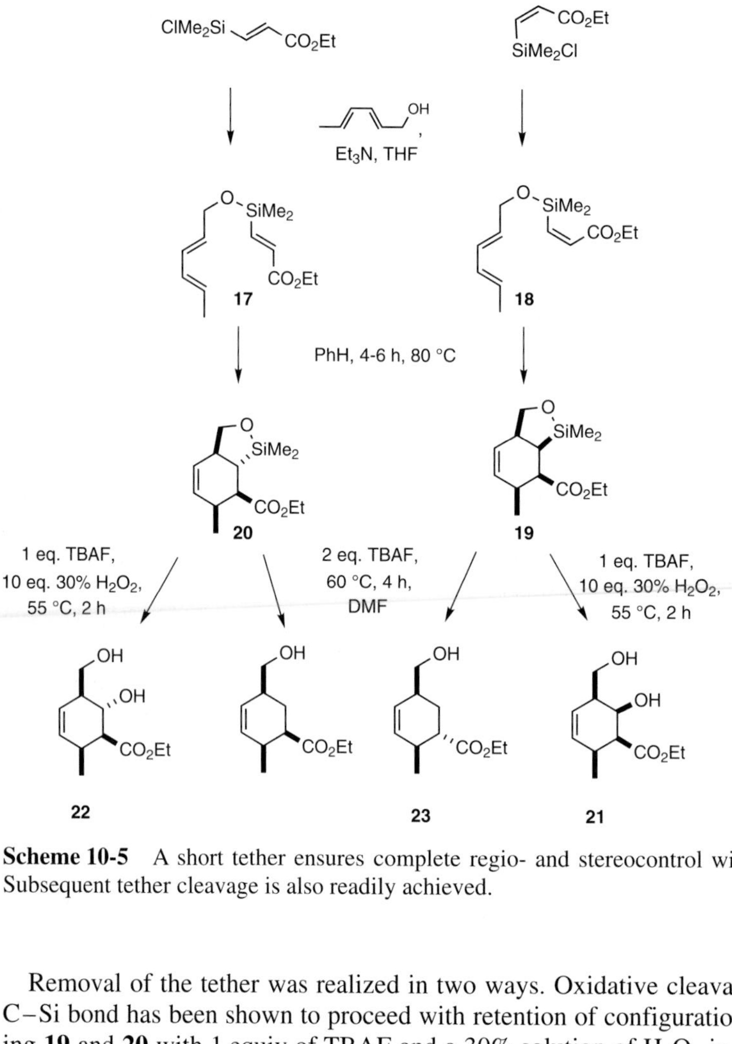

Scheme 10-5   A short tether ensures complete regio- and stereocontrol with vinyl silyl ethers. Subsequent tether cleavage is also readily achieved.

Removal of the tether was realized in two ways. Oxidative cleavage of an activated C–Si bond has been shown to proceed with retention of configuration [11]. Thus treating **19** and **20** with 1 equiv of TBAF and a 30% solution of H_2O_2 in DMF at 55 °C for 2 h afforded the diastereoisomeric diols **21** and **22**, respectively, both in 80% yield. Alternatively hydrodesilylation could be achieved using 2 equiv of TBAF in DMF at 60 °C for 4 h. In the case of the *cis*-fused, bicyclic Diels–Alder adduct **19**, the basic conditions of the reaction led to complete epimerization of the ester substituent onto the less hindered convex face, and the isolation of exclusively alcohol **23** (Scheme 10-5).

Sieburth and Fensterbank have studied the effect of tether length on the efficiency of vinyl silyl ether IMDA reactions [12a]. Trienes with tethers of three or four atoms reacted with equally high efficiency providing good yields of the TMS-substituted alcohols upon workup with MeLi (Scheme 10-6). Extending the tether length by one additional atom, however, resulted in a much slower reaction, presumably owing to entropic factors, and after 20 h at 190 °C appreciable quantities of the triene starting material remained (Scheme 10-6).

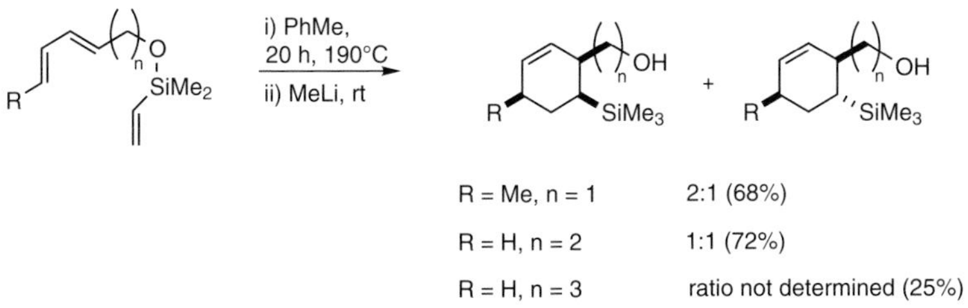

R = Me, n = 1	2:1 (68%)
R = H, n = 2	1:1 (72%)
R = H, n = 3	ratio not determined (25%)

Scheme 10-6   Effect of tether length on the efficiency of IMDA reactions.

In the cases of dimethylsilyl IMDA precursors, the *exo/endo* selectivity was poor. However, this ratio could be readily and quite dramatically influenced by varying the alkyl substituents on the silicon template [12]. Thus with dienol **24**, tether formation with dimethylvinylsilyl chloride and subsequent IMDA reaction afforded a 4:1 mixture of *exo/endo* products (Scheme 10-7). The ratio could be further improved to 10:1 by using a diphenylsilyl tether, and when bulky 'Bu groups were used, a single stereoisomer, resulting from *exo* addition, was observed. This example once more illustrates the potential for tuning the stereoselectivity of the reaction by varying the steric interactions with the tether.

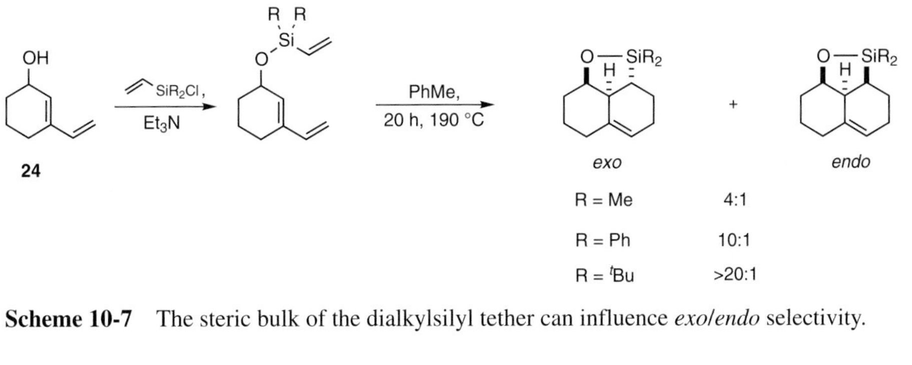

R = Me	4:1
R = Ph	10:1
R = 'Bu	>20:1

Scheme 10-7   The steric bulk of the dialkylsilyl tether can influence *exo/endo* selectivity.

Dienylsilyl ethers have been less widely investigated in spite of the fact that this tethering pattern provides one of the first examples of a silyl-tethered IMDA reaction [13, 14]. Tamao, Ito, and Kobayashi demonstrated that 1,7-diynes undergo stereospecific Ni⁰-catalyzed hydrosilylation generating the corresponding 1,2-dialkylidenecyclohexanes possessing a (Z)-vinylsilane moiety [13]. A variety of silanes could be used to effect this transformation, among them (diethylamino)dimethylsilane. The resulting dienylamino silane **25** allowed further functionalization by nucleophilic displacement of the labile amino functionality with cinnamyl alcohol (Scheme 10-8). Triene **26** then underwent a smooth IMDA reaction on heating a xylene solution at reflux for 40 h. Oxidative cleavage of the C–Si bond afforded diol **27** as a single stereo- and regioisomer.

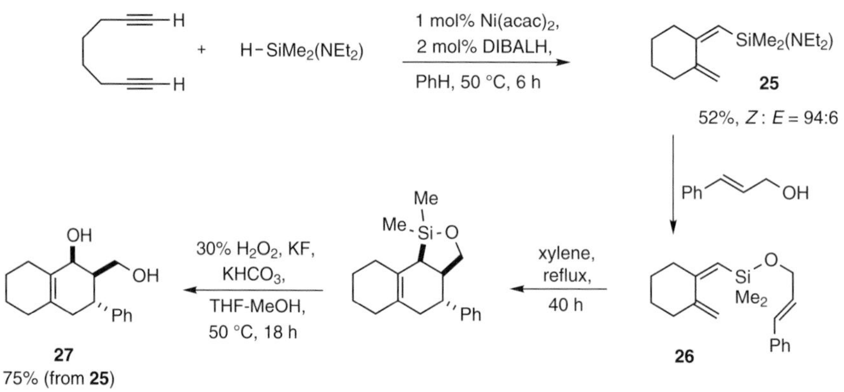

Scheme 10-8   Dienylsilyl ethers can also be used in IMDA reactions.

Acetal tethers

The use of an acetal to tether the diene and dienophile has been investigated by Craig and co-workers [15]. Preparation of the unsymmetrical acetal precursor was best achieved via the corresponding enol ether. For example, the isopropylidene-linked triene **28** was synthesized in two steps from acetate **29**. Tebbe methylenation of **29** afforded enol ether **30**, which was used directly in a Pd^{II}-catalyzed addition of the dienophile **3** to form **28** (acid-catalyzed addition with *p*TSA proved inferior). Thermolysis of **28** in toluene at 165 °C allowed a remarkably facile and completely regiospecific cyclization to take place. Acidic methanolysis of the Diels–Alder adducts resulted in cleavage of the tether and isolation of a 2.7:1 mixture of lactones **31** and **32** in 72% yield (Scheme 10-9).

The reactivity of triene **28** was greatly increased compared to the analog bearing a single methyl group in the acetal tether, which required 27 h for complete reaction [14]. This may be attributed to the greater degree of alkyl substitution at the acetal center reducing the population of unreactive, distal conformers which experience unfavorable non-bonding interactions between the diene and/or dienophile and the additional methyl group.

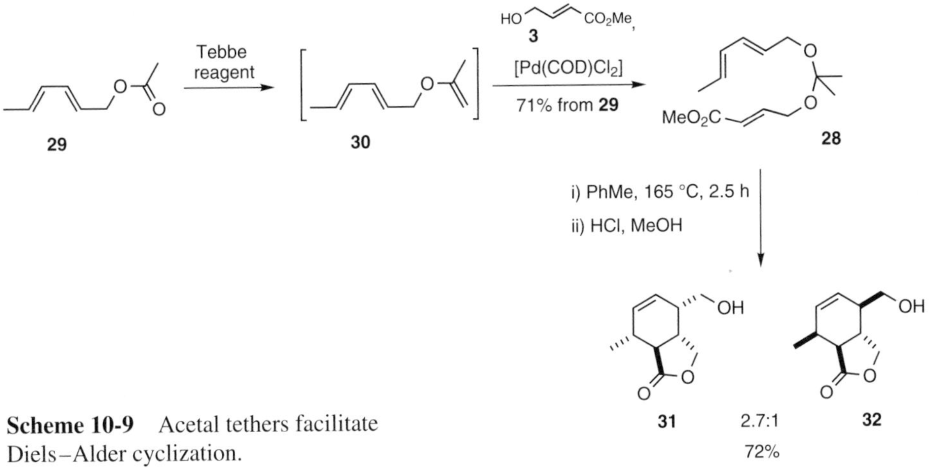

Scheme 10-9 Acetal tethers facilitate Diels–Alder cyclization.

Comparison with the analogous silyl acetal-tethered systems also reveals a number of interesting features. The Diels–Alder precursors (e.g., **28**) linked through a carbon-acetal tether are appreciably more reactive than the analogous silyl acetals (vide supra). This is most probably a result of the shorter C–O bond length (1.43 Å compared with 1.64 Å for Si–O) increasing the local concentration of the dienophile at the reacting diene. On the downside, the use of an acetal tether results in much poorer stereocontrol when compared with the almost complete *exo* selectivity observed with the corresponding silyl acetals: the reduced length of the carbon acetal tether provides a more compact T. S., in which steric interactions between the *exo* diene and dienophile methylene group are increased, disfavoring this addition mode [15].

Boeckman et al. used an acetal tether in a racemic synthesis of the cyclohexene subunit of tetronolide **33** [16]. Although it had previously been recognized that a Diels–Alder strategy could provide rapid entry into this highly functionalized six-membered ring, some of the earlier reported intermolecular reactions suffered from relatively poor reactivity and regio- and stereocontrol [17]. It was anticipated that an intramolecular variant would overcome some of these problems (Scheme 10-10).

Formation of the acetal-tethered triene **34** was achieved by selective nucleophilic displacement of the secondary alkyl bromide in **35** with hydroxydiene **36**. Subsequent thermolysis at 145 °C for 70 h resulted in completely regiospecific cycloaddition with the formation of two out of the possible four diastereoisomers **37** and **38** in a 2.5–3:1 ratio (Scheme 10-11). The presence of a stereogenic center in the tether provides the source of stereocontrol – the products derive from *exo* and *endo* T. S.s in which the bromomethyl group in the tether occupies a pseudo-equatorial position (Scheme 10-11). Small differences in non-bonding interactions between the tether and pyruvate ester most probably account for the slight *exo* selectivity. The separated major product was further elaborated, and finally the tether was cleaved under reductive conditions to provide the advanced cyclohexene intermediate **39**.

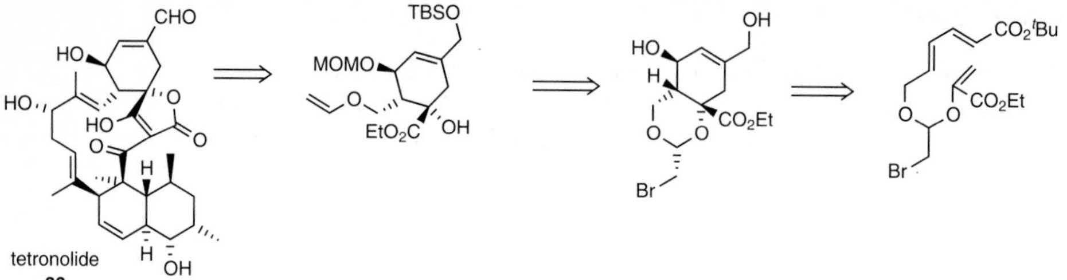

tetronolide
33

Scheme 10-10 An IMDA strategy can be used to prepare the cyclohexene subunit of tetronolide **33**.

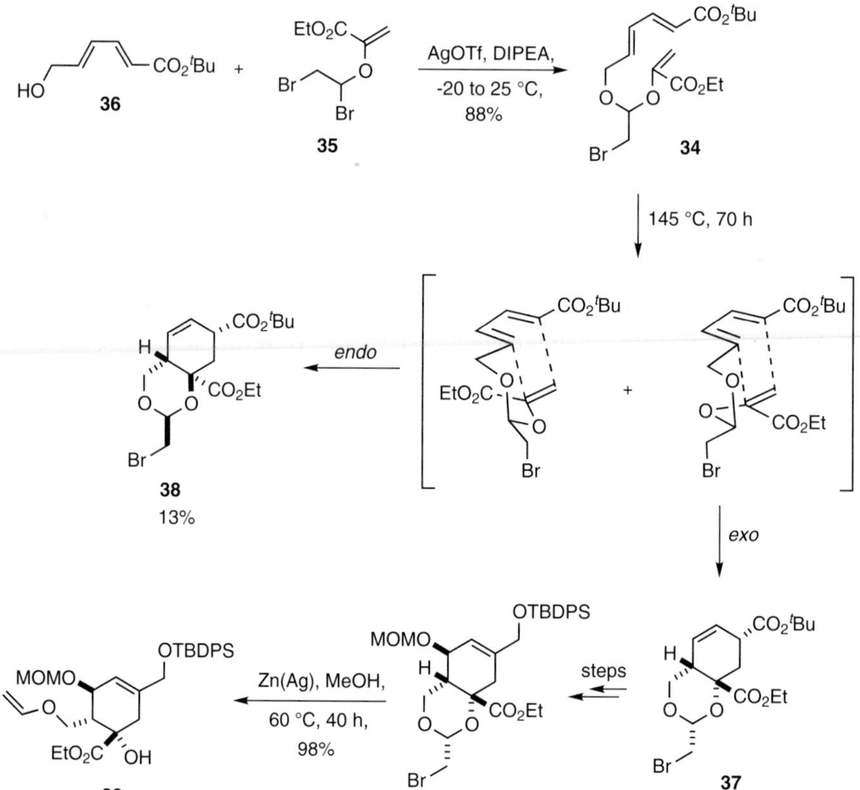

36 **35** **34** **38** 13% **39** **37** 38%

Scheme 10-11 Cyclization of **34** proceeds with complete regiocontrol and moderate stereoselectivity.

Ether tethers

An early example of an IMDA reaction using an ether linkage to connect the diene and dienophile was provided by Funk et al. in their preparation of the hexahydronaphthalene fragment of the important hypocholesterolemic agent, compactin **40** (Scheme 10-12) [18].

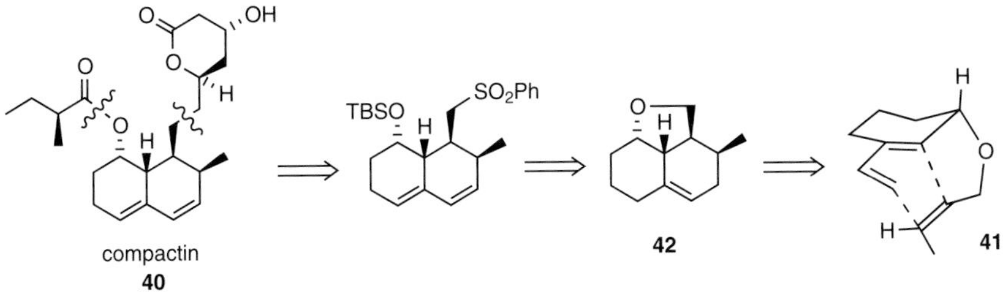

Scheme 10-12   An IMDA reaction can be used to prepare the lower half of compactin.

Formation of the requisite triene **41** was readily achieved by alkylation of hydroxy-diene **24** with 1-bromobut-2-yne (Scheme 10-13). The stereospecificity of the concerted Diels–Alder reaction required a *cis*-configured dienophile to react through the sterically less demanding *exo* T. S. to afford the desired product **42**. A chemoselective hydroboration/protonolysis sequence on propargylic ether **43** allowed access to the desired *cis* olefin **41** in good yield. Subsequent thermolysis at 165 °C in toluene afforded a 4 : 1 mixture of stereoisomers with the major product being the desired *exo* isomer **42**. Unactivated primary alkyl ethers are rarely used as tethers, owing to the relatively harsh reaction con-

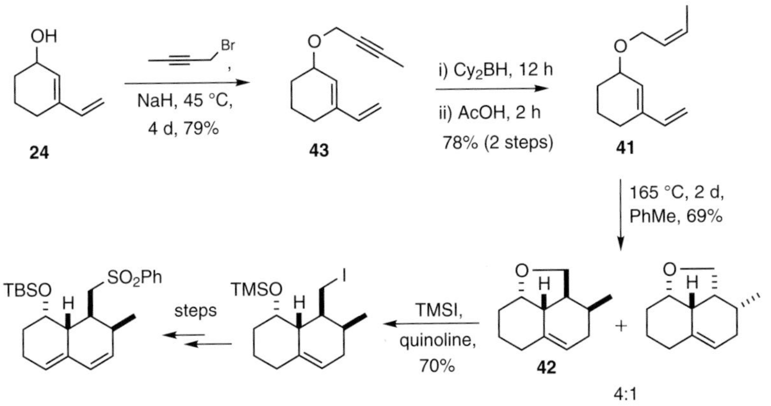

Scheme 10-13   Ether-linked triene **41** has been elaborated to the hexahydronaphthalene portion of compactin.

ditions required for their subsequent cleavage. In this case, however, the use of a more labile ester tether proved problematic in the Diels–Alder step owing to a facile elimination side reaction [18]. Fortunately the ether tether, which did not suffer this side reaction in the cycloaddition reaction, was readily cleaved regioselectively using TMSI in the presence of quinoline (Scheme 10-13).

Craig and co-workers have investigated the use of benzylic and tertiary alkyl ethers as temporary tethers in IMDA reactions [19]. Both these functionalities are more readily cleaved under a variety of conditions. A convergent and highly flexible approach was used to prepare the triene precursors. Thus treating acetone or benzaldehyde with TMS ethers under TMSOTf activation afforded the corresponding acetals, which were then used in a Sakurai coupling reaction with 5-trimethylsilylpenta-1,3-diene. The two steps were best accomplished in a one-pot process providing excellent yields of the ether-linked trienes (Scheme 10-14).

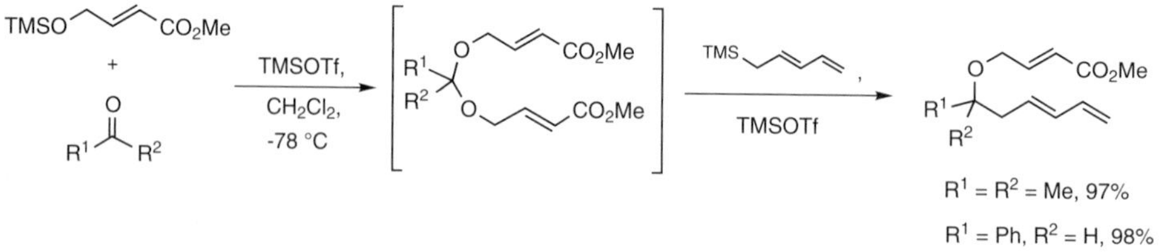

Scheme 10-14   Ether-linked trienes have been prepared in a convergent and flexible manner.

Cyclization studies were carried out in sealed tubes in toluene solutions at 165 °C and provided excellent yields of the corresponding bicyclic cycloadducts in 5–9 h. Reactions of tertiary alkyl ethers showed moderate to complete *trans* selectivity. This selectivity was observed even for inherently *cis*-directing systems and may be attributed to the *gem*-dimethyl group in the ether tether: comparison of the reacting conformations leading to *cis*- and *trans*-fused decalins reveals, in the *cis* case, an unfavorable non-bonding interaction between the axial methyl group and H–C(3) of the diene (Scheme 10-15).

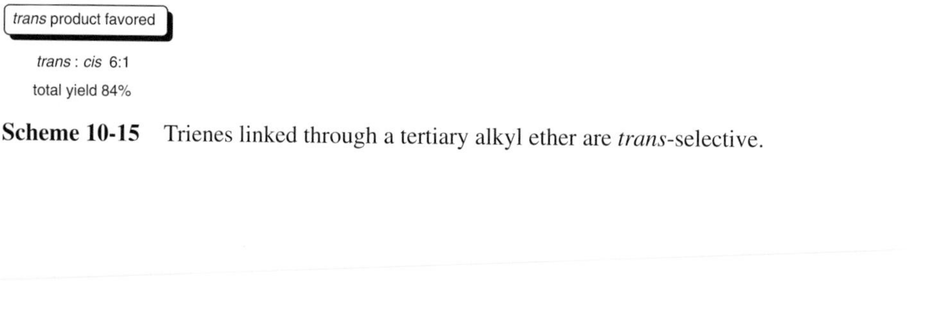

Scheme 10-15   Trienes linked through a tertiary alkyl ether are *trans*-selective.

The presence of an additional stereocenter in the ether linkage of PhCHO-derived trienes provides the possibility for additional stereocontrol. Thus thermolysis of triene **44** afforded a 2 : 1 mixture of decalin products **45** and **46** in excellent yield. These arise respectively from *endo* and *exo* addition of the diene to the *same* diastereoface of the dienophile via a T. S. in which the phenyl substituent in the tether is pseudo-equatorially disposed (Scheme 10-16).

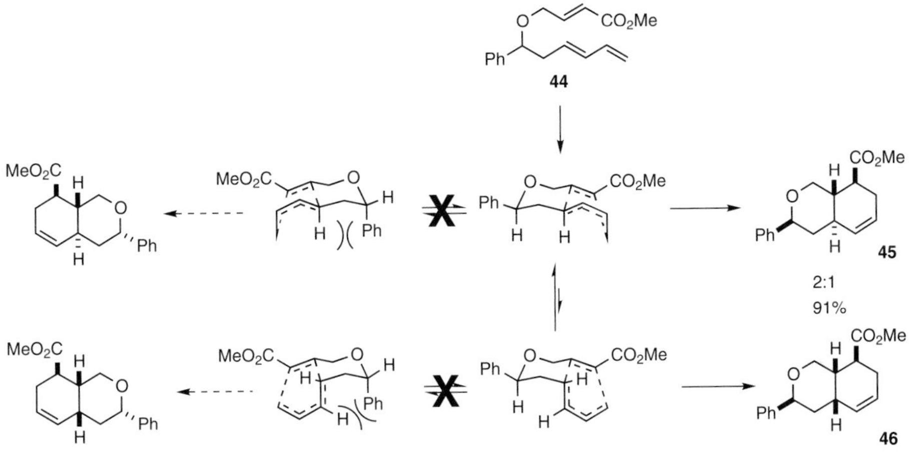

Scheme 10-16   Triene **44** linked through a benzylic ether tether also undergoes a stereoselective IMDA reaction.

As a result of the excellent diastereocontrol observed in these reactions, it was envisaged that the use of chiral ether linkages might impart a degree of absolute stereocontrol on the IMDA reaction [19b]. Preliminary results were encouraging. For example, the menthone-derived triene **47** provided a 10 : 1 mixture of two (out of a possible four) diastereoisomers with proposed structures **48** and **49** in 94% yield (Scheme 10-17).

Cleavage of the tethers was achieved in a number of ways affording a variety of products, ripe for further elaboration (Scheme 10-18).

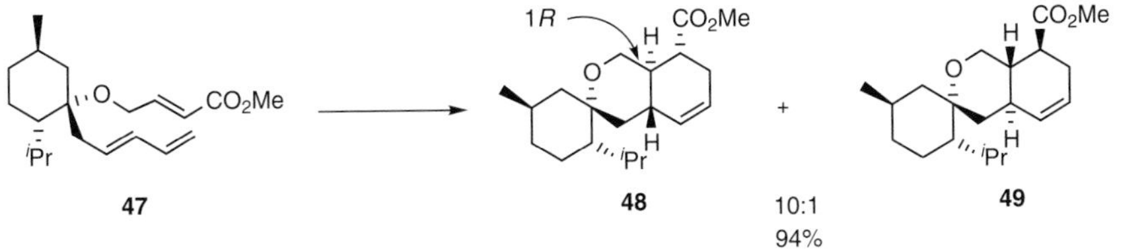

Scheme 10-17   Asymmetry in the ether linkage can impart absolute stereocontrol on the cycloaddition.

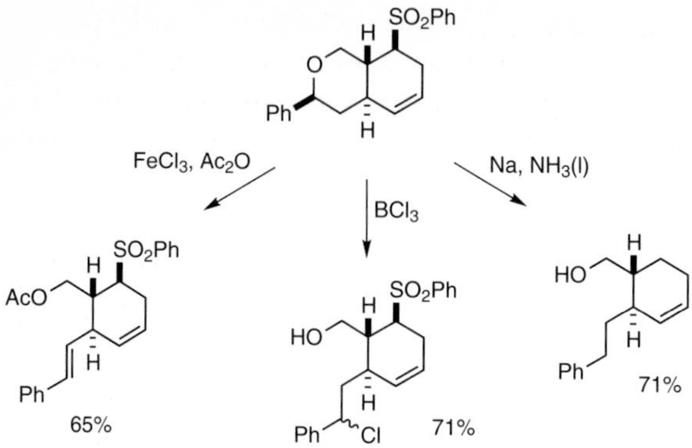

Scheme 10-18   Cleavage of the ether tethers after cycloaddition.

Boron tethers

The use of a boron atom as a covalent template for intramolecularizing the Diels–Alder reaction has been relatively little investigated [20–22]. However Narasaka et al. have reported the use of a boronate template to control regio- and stereoselectivity in an IMDA reaction between anthrone **50** and methyl 4-hydroxybut-2-enoate **3** [20a]. Formation of the boronate tether was achieved by simply heating an equimolar mixture of phenylboronic acid with the diene and dienophile at reflux in pyridine with azeotropic removal of water. After 5 h, a single cycloadduct **51** was obtained in 81% yield. The tether was readily removed oxidatively affording the corresponding diol **52** in excellent yield (Scheme 10-19).

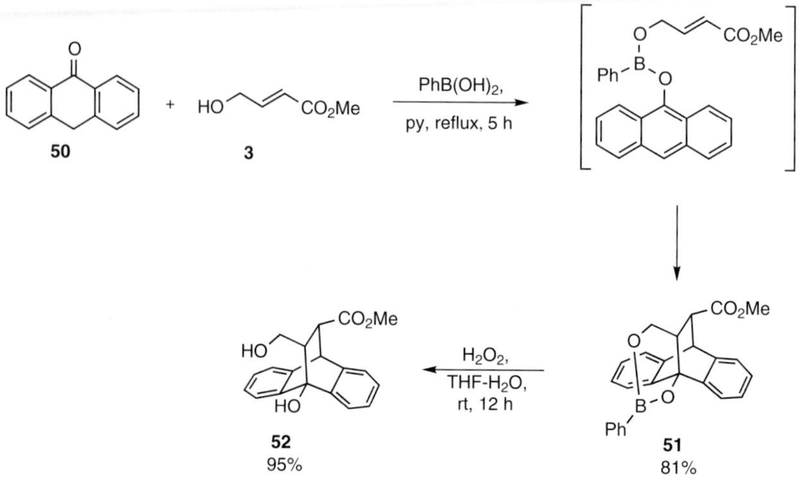

Scheme 10-19   A boronate tether has been used to control the regio- and stereoselectivity of a Diels–Alder reaction.

Although the boronate-tethered triene intermediate was not isolated, its formation can be implied by the lack of any reaction under the same conditions in the absence of PhB(OH)$_2$. The tether also enforces a complete reversal in the regioselectivity of the reaction compared with the analogous intermolecular reactions of acrylates with anthrone [23].

The same group also applied boron reagents to trap reactive dienes and used them in a subsequent IMDA reaction [20b]. *o*-Quinodimethanes are useful dienes since they allow rapid access to the tetralin skeleton. However, they are fairly unstable and only undergo cycloaddition with reactive dienophiles; in the presence of less reactive dienophiles, the diene isomerizes or dimerizes preferentially.

It was envisaged that a suitable boron reagent could be used to intercept an α-hydroxy-*o*-quinodimethane and form a mixed boronate with a hydroxyl-containing dienophile. Tether formation would not only ensure a completely regioselective cycloaddition, but would also prevent isomerization of the reactive diene moiety. Thexylborane was found to be a useful source of the boron template. A solution of the borane, cyclobutanol **53** and dienophile **3** in THF was stirred for 1 h to allow tether formation. The solvent was then exchanged for *m*-xylene and the reaction mixture heated at reflux for 3 h. Cycloadduct **54** was isolated in 59% yield after hydrolytic workup as an unreported mixture of stereoisomers (Scheme 10-20). The opposite regioisomeric product was not observed, indicating the participation of a boron-tethered intermediate. The intermolecular reaction would, once again, be expected to provide the opposite regioisomer to that formed.

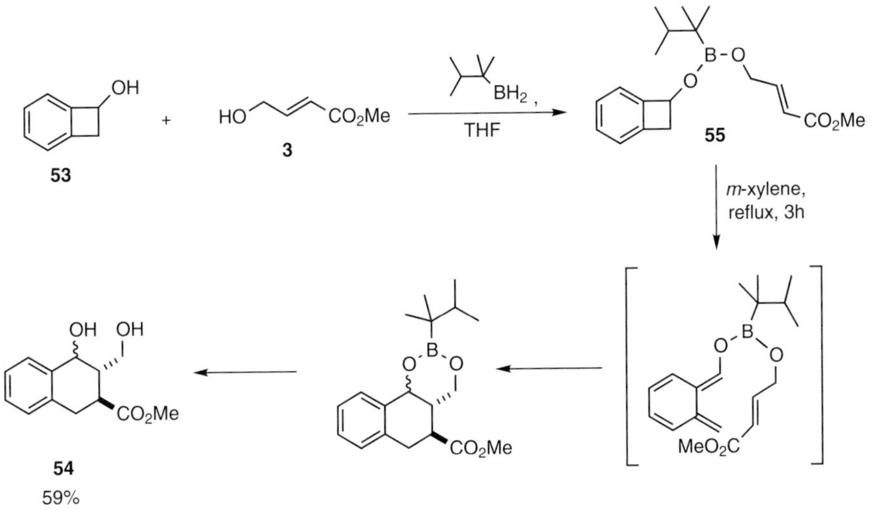

Scheme 10-20   Linking the benzocyclobutane and dienophile through a boron tether to form **55** allows interception of the reactive diene formed on ring opening for use in a regioselective IMDA reaction.

Inspired by these results, Nicolaou et al. used a temporary boronate tether to overcome the undesired regiochemical bias in the intermolecular Diels–Alder reaction of diene **56** and dienophile **57**, required for their approach to the C-ring of taxol **58** [21a, b, 24]. Thermolysis of a benzene solution of **56** and **57** in the presence of PhB(OH)$_2$ under dehydrating conditions provided the cycloadduct **59** in 79% yield (77% conversion) after transacetalization of the boron tether with 2,2-dimethylpropane-1,3-diol and intramolecular acyl transfer to the less strained [4.3.0]bicyclic system. This was further elaborated to an advanced intermediate in the ultimately successful synthesis of taxol (Scheme 10-21) [21c].

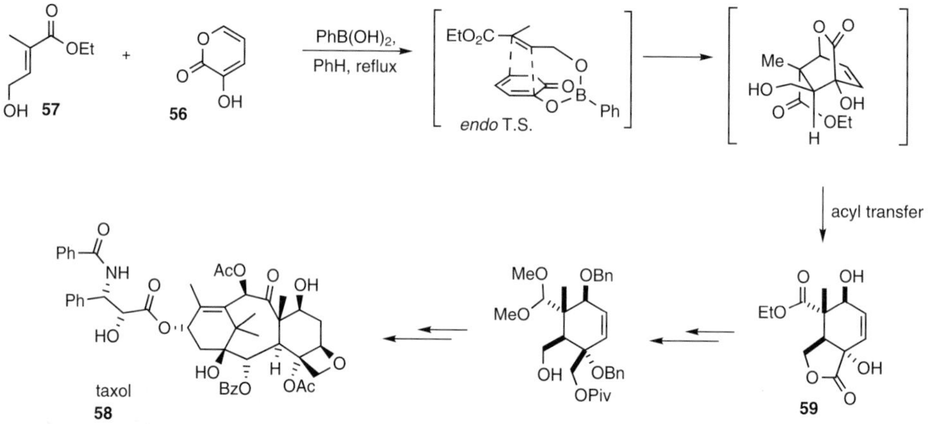

Scheme 10-21   Nicolaou et al. used a boron-tethered cycloaddition to construct the C-ring of taxol.

Sulfonate tethers

Metz et al. have investigated the use of sulfonates as covalent tethers in IMDA reactions [25]. The sulfonate group not only acts as a covalent tether, but if directly attached, can also be used to activate the dienophile. Vinyl sulfonyl chloride is readily prepared [26], and nucleophilic displacement of the chloride functionality by a hydroxy-containing diene effects tether formation. Although the triene precursor can be isolated, the sulfonate functionality is hydrolytically sensitive and generally it is more convenient to carry out the IMDA reaction directly. Vinyl sulfonates exhibit high reactivity in Diels–Alder reactions and cyclization sometimes proceeds at or below room temperature. In the case of triene **60**, a highly diastereo- and regioselective IMDA reaction occurred within 2–3 h at 0 °C [25b] with the observed *exo* cycloadduct **61** being formed as the thermodynamic product. A two-step oxidative cleavage protocol was employed to effect tether cleavage: α-lithiation of the sultone with ^{n}BuLi followed by trapping with 2-methoxy-4,4,5,5-tetramethyl-1,3,2-dioxaborolane generated the corresponding boronate, which was oxidatively decomposed with *m*CPBA affording γ-hydroxyketone **62** in excellent yield (Scheme 10-22). The net result of this reaction is a completely regio- and stereoselective, formal [4 + 2] cycloaddition of ketene to a cyclopentadiene.

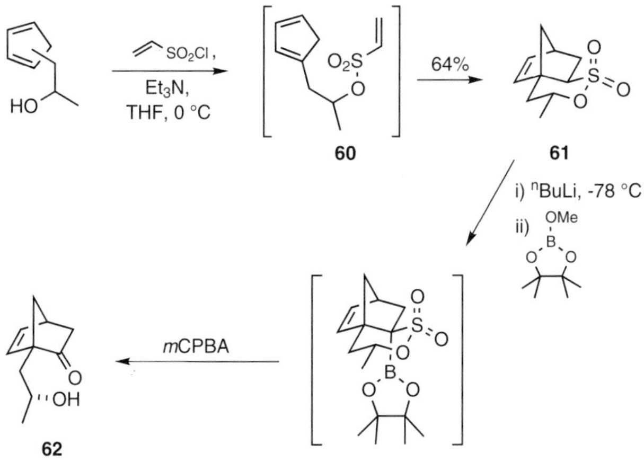

Scheme 10-22   A sulfonate tether affords cycloadducts stereo- and regioselectively.

Recently, Winterfeldt and co-workers used a similar strategy as a key step in the preparation of myltaylenol **63** [27]. It was envisaged that diene **64**, readily prepared in enantiopure form, could be used in an IMDA reaction with a dienophile, temporarily attached via the hydroxy functionality, to incorporate the remaining two stereogenic centers. Tethering the dienophile would control the facial selectivity of the cycloaddition with attack proceeding onto the sterically more hindered, but desired, α-face of the diene. Interestingly, although a number of tethering strategies were investigated, only the sulfonate produced satisfactory results. Treatment of **64** with vinyl sulfonyl chloride proceeded uneventfully affording sulfonate **65** in excellent yield. Subsequent thermolysis in toluene at 110 °C for 20 h ensured complete conversion to cycloadduct **66** in which the predicted α-attack of the diene was realized (Scheme 10-23). An oxidative cleavage strategy was required for further manipulation to the target. However, the sterically very hindered center at the sultone carbon necessitated a modification of the conditions reported by Metz [25b]. After appreciable experimentation it was found that molecular oxygen could be used to trap the lithiated sultone generated using ^{s}BuLi at −78 °C [28]. A large excess of the base was crucial for efficient conversion to the desired hydroxyketone **67**. This was further elaborated into the first enantioselective synthesis of myltaylenol.

Metz et al. have also demonstrated the use of sulfonate tethers in the preparation of the bicyclic sesquiterpene lactone, ivangulin **68** [25f]. The sulfonate derived from vinyl sulfonyl chloride and diene **69** underwent smooth IMDA reaction affording a mixture of *exo* cycloadducts **70** and **71** in a 1:1.4 ratio. Heating this mixture in toluene at reflux allowed equilibration, via a retro Diels–Alder-Diels–Alder process, to an improved ratio of 9.5:1 in favor of **71**, which is thermodynamically favored owing to the equatorially disposed methyl substituent (Scheme 10-24). This major cycloadduct was separated and further elaborated to advanced intermediate **72** in which all three stereogenic centers had been successfully incorporated. This time, cleavage of the sultone was achieved reductively using dissolving-metal conditions, affording tetrasubstituted olefin **73** as the major product, which was readily manipulated into ivangulin **68**.

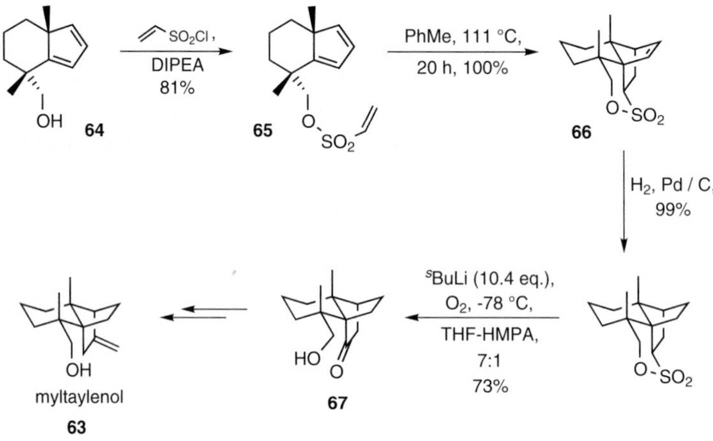

Scheme 10-23   A sulfonate tether strategy was used in Winterfeldt's synthesis of myltaylenol.

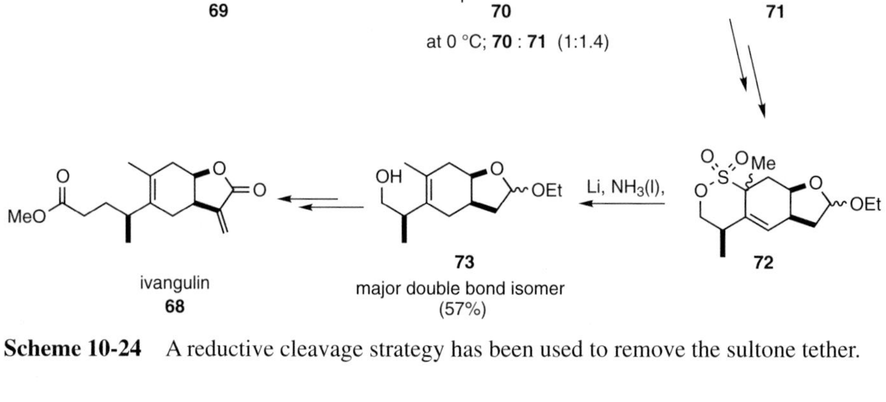

Scheme 10-24   A reductive cleavage strategy has been used to remove the sultone tether.

10.2.1.2 Type II IMDA reactions

The Type II variant of the IMDA reaction usually requires more forcing conditions to effect cyclization than its more frequently encountered Type I analog. This is a direct consequence of the higher-energy T. S. associated with the formation of products containing a strained, bridgehead double bond. This intramolecular variant of the Diels–Alder reaction has been most intensively studied by Shea and co-workers who have shown that it can be used in a variety of highly regio- and stereocontrolled cyclizations [5, 29].

Ester tethers

In an early paper, Shea used a readily prepared ester tether to connect the diene and dienophile [29a, 30]. The effect of tether length on cyclization was investigated and was shown to be an important criterion for efficient cyclization. Just as in the Type I IMDA reaction, increasing the number of bridging atoms between the diene and dienophile provides a reaction which increasingly resembles the corresponding bimolecular process and the entropic gain from intramolecularization is expected to diminish. However, the strain energy contribution to the enthalpy term is expected to be much more important in the Type II IMDA reaction, with a longer tether being expected to reduce strain energy and facilitate cyclization.

A series of trienes were prepared [29a]. In the case of a four-atom tether, reaction was not possible, presumably as a result of highly unfavorable non-bonding interactions in the T. S. Increasing the tether length by one atom allowed reaction to proceed: in all cases, thermolysis in xylene at 185 °C afforded a single regioisomeric product in good yield (Scheme 10-25). Cyclization was even more facile in the case of a six-atom bridge, as illustrated by the reduced temperature required for cyclization: thus, thermolysis of triene **74** at 130 °C afforded a 92 : 8 ratio of cycloadducts **75** and **76**. This is presumably

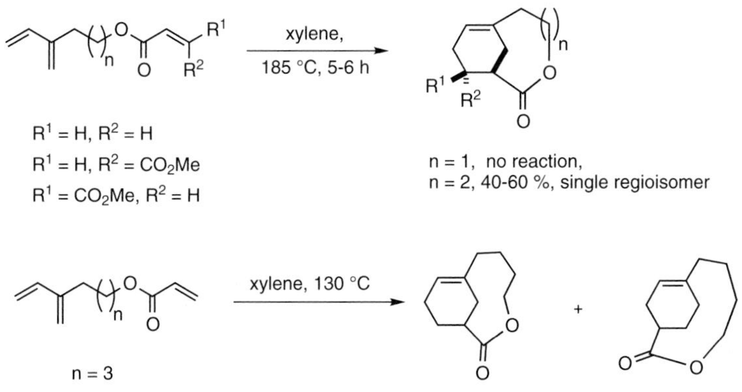

$R^1 = H, R^2 = H$
$R^1 = H, R^2 = CO_2Me$
$R^1 = CO_2Me, R^2 = H$

n = 1, no reaction,
n = 2, 40–60 %, single regioisomer

n = 3

74   **75**   92:8   **76**

Scheme 10-25   Ester groups can be used as the temporary connection in Type II IMDA reactions. Tether length affects the facility and regioselectivity of reaction.

a consequence of reduced strain and non-bonding interactions in the T. S. However, while reaction becomes more facile, the increased flexibility of a longer tether diminishes slightly the level of regiocontrol. Isolation of the minor cycloadduct **76** and saponification of the lactone bridge afforded a 1,4-disubstituted cyclohexene which corresponded to the major product of the analogous intermolecular Diels–Alder reaction. The regio-control of 2-alkyl-substituted 1,3-dienes in Diels–Alder cycloadditions is generally low although it usually favors the *"para"* cycloadduct [31]. Tethering the diene and dienophile therefore not only increases regioselectivity, but also overrides the inherent regiochemical bias of this reaction.

Silicon tethers

Shea and co-workers have also investigated the use of silyl ethers and silyl acetals as temporary connecting groups in the Type II IMDA reaction [29i, j, 1–n]. Moreover, they have demonstrated that efficient stereochemical transcription is possible by incorporating asymmetry into the linker [29l, m]. An elegant illustration of the power of this methodology for stereo- and regioselective bond construction is their recent synthesis of the adrenalcorticosteroid (+)-adrenosterone **77** [29m]. Their retrosynthesis is outlined in Scheme 10-26.

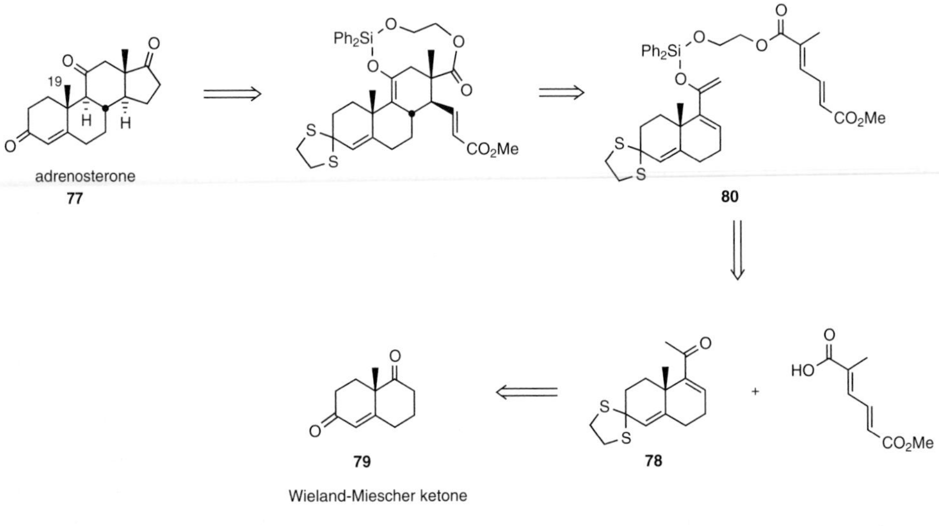

Scheme 10-26   Shea's retrosynthesis of adrenosterone incorporating a Type II IMDA reaction.

The diene precursor **78** is readily accessed in a few steps from the Wieland–Miescher ketone **79**, which also provides the A- and B-rings of the steroid skeleton. Furthermore, the presence of an axial methyl group at C(19) (steroid numbering) allows the possibility of controlling the absolute stereochemical outcome of the key cycloaddition used to install the C-ring. Preparation of the tethered Diels–Alder precursor **80** was achieved in a highly efficient manner (Scheme 10-27): kinetic deprotonation of ketone **78** with KHMDS and subsequent trapping of the enolate with Ph_2SiCl_2 generated the corresponding diphenylchlorosilyl dienol ether. Chloride substitution with glycol ester **81** afforded silyl acetal **80**. This was used directly, without purification, in the Diels–Alder reaction, from which a 1:10 mixture of diastereoisomers **82** and **83** was isolated in an excellent 90% yield from ketone **78**. Cyclization was not only completely regioselective, but both diastereoisomers resulted from an *endo* mode of addition. However, the major product proved to be that derived from β-attack of the dienophile, so the hoped-for α-directing effect of the C(19) methyl group was not realized.

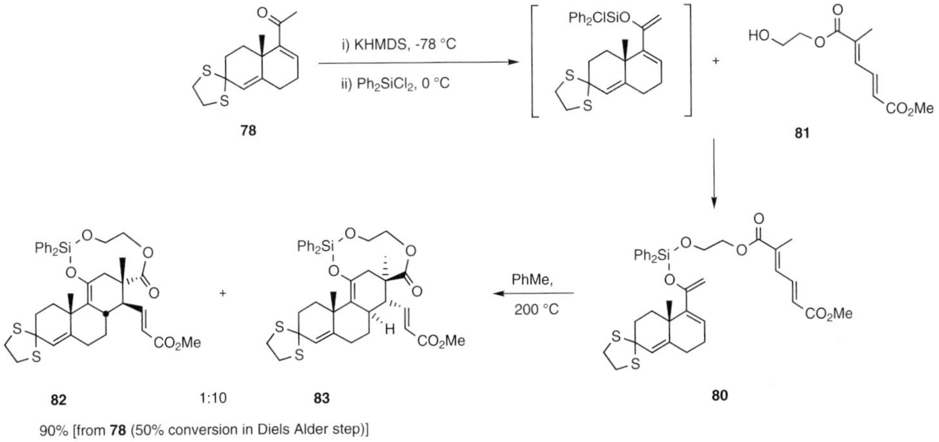

Scheme 10-27 Triene formation and subsequent IMDA reaction proceeds with high efficiency and selectivity.

A modified tether, incorporating (–)-hydrobenzoin instead of glycol, was readily prepared from ketone **78** as before, anticipating that steric interactions between substituents at these new stereogenic centers and the diphenylsilyl group would rectify the problem of π-facial selectivity in the cycloaddition. Thermolysis of triene **84** at 200 °C in toluene for 18 h afforded a 3:2 mixture of cyclized products **85** and **86** respectively, in which the major diastereoisomer resulted from the desired α-approach of the dienophile (Scheme 10-28). Subsequent manipulation generated (+)-adrenosterone in a total of only seven steps from enone **78**.

In Shea's approach to the polyhalogenated cyclohexane **87**, derived from a red marine alga, *Plocamium* sp., a Type II IMDA reaction utilizing a disposable (allyl)silyl tether was used as a key step [29n]. The triene cyclization precursor **88** was readily prepared and underwent Diels–Alder reaction in 74% yield with the expected, complete regiocontrol affording exclusively bicycle **89**. The rigid, bicyclic framework of the cycloadduct was then exploited to ensure a completely selective *exo–syn* chlorination of the bridge-head double bond with iodobenzene dichloride, which installed the desired *cis* relative configuration between chlorine and bromine. An interesting, reductive cleavage strategy was used to remove the tether: treatment of **90** with DIBALH not only effected ester reduction, but also installed the *exo* methylene group by elimination of the bridgehead chloride. Subsequent manipulations provided ready access to the monoterpene **87** (Scheme 10-29).

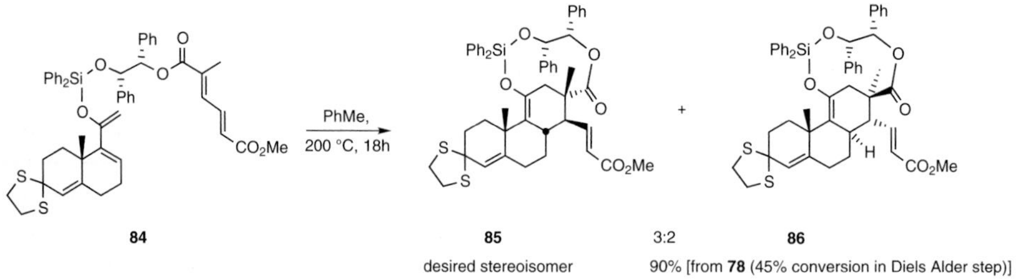

Scheme 10-28   A modified tether incorporating asymmetry provides the desired π-facial selectivity.

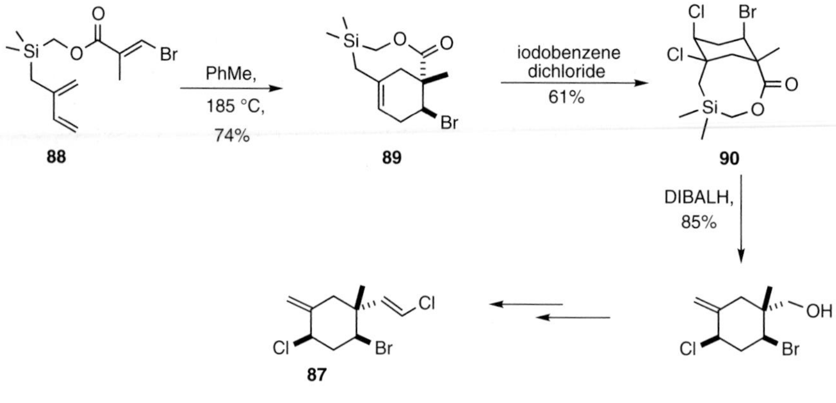

Scheme 10-29   A Type II IMDA strategy was used to prepare the terpenoid marine natural product **87**.

The work, by the Shea group in particular, exemplifies the power of the temporary connection in Type II IMDA reactions not only for controlling the regiochemical outcome of the cyclization but also for exerting absolute stereocontrol.

10.2.2   Other Cycloadditions

The problems of controlling the regio- and stereochemical outcome in cycloaddition reactions are obviously not only restricted to the Diels–Alder reaction; the strategy of utilizing a temporary connection between reacting π systems to dictate the outcome of the reaction has also been applied to a variety of cycloadditions, including [2+2], [3+2], and [5+2] reactions. Since most of the tethering strategies resemble those used in the Diels–Alder reaction, only a few examples will be discussed here – a more comprehensive list of work can be found in the references.

10.2.2.1   Use of temporary tethering strategies in [2+2] photocycloaddition

The [2+2] photocycloaddition between two olefin systems is the most efficient method for the preparation of cyclobutane rings [32–34]. In the first asymmetric synthesis of stoechospermol **91**, Koga and co-workers employed a tethering strategy to control the absolute stereochemical outcome of a key [2+2] photocycloaddition, which installed the correct stereochemistry at the two ring junctions [35]. By linking the two reacting olefins through an ester tether [32], it was anticipated that the cyclopentene derivative would be forced to react on the more hindered face of the enantiomerically pure butenolide, which is derived from (S)-glutamic acid. This would provide a route to the stereoisomeric cycloadducts obtained through the analogous intermolecular process in which cyclopentene derivatives had been previously shown to attack the butenolide *anti* to the hydroxymethyl substituent [36].

Formation of the ester cyclization precursor **92** was readily achieved as a 1:1 mixture of diastereoisomers and, upon irradiation, afforded a 1:1 mixture of cyclobutanes **93** differing only in the stereochemistry of the silyloxy substituent. The tether had therefore controlled both the regio- and stereoselectivity of the cycloaddition event. Subsequent manipulations afforded the desired natural product (Scheme 10-30).

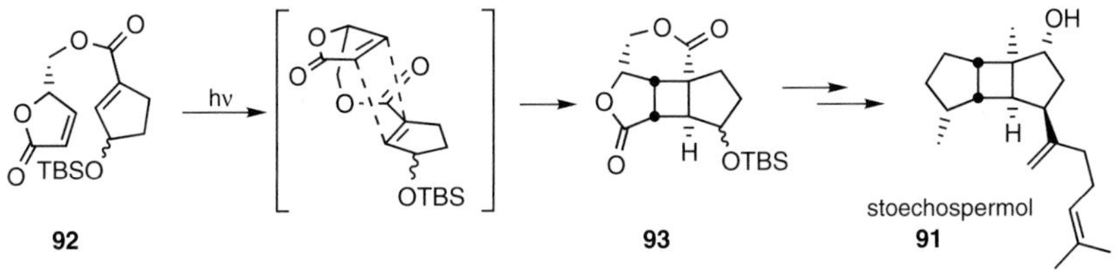

Scheme 10-30   An ester tether was used to control the outcome of a [2+2] cycloaddition in Koga's synthesis of stoechospermol.

A similar tethering strategy was more recently used by Inouye et al. in their formal, racemic synthesis of precapnelladiene **94** [37]. The reacting enone and cyclopentene were linked through an ether tether. Irradiation of **95** in hexane effected cyclization, generating a single product **96** in which the tether had governed the relative stereochemistry at the ring junction and the initial stereocenter in the cyclopentene. Retro aldol fission of the 6–4 ring system provided an elegant entry into the desired bicyclo[6.3.0]-undecane ring skeleton. Treatment with TMSI in benzene generated TMS enol ether **97**, which was converted directly into diketone **98** by radical deiodination. Further manipulation into ketone **99** provided an interesting formal synthesis of precapnelladiene (Scheme 10-31).

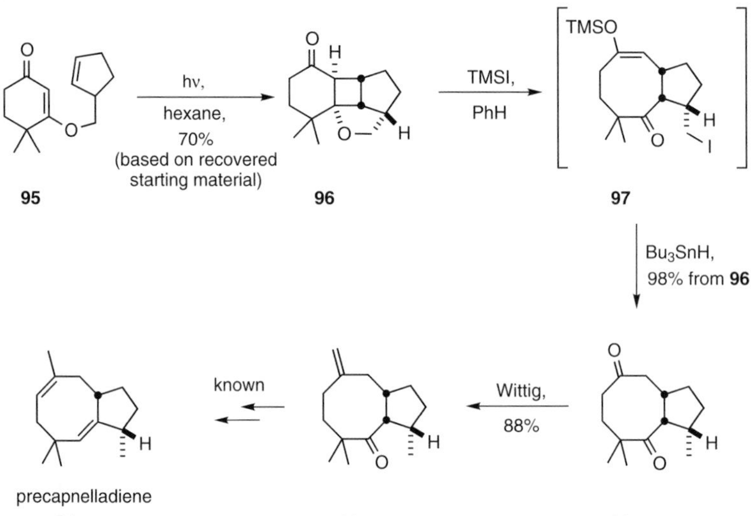

Scheme 10-31   Inouye et al. used a [2+2] cycloaddition/retro aldol cleavage strategy in their formal synthesis of precapnelladiene.

10.2.2.2   An efficient synthesis of (–)-detoxinine using a tandem [4+2]/[3+2] reaction

Recent reports from the Denmark group have demonstrated the high synthetic utility of tandem [4+2]/[3+2] nitroalkene cycloaddition reactions in the preparation of pyrrolizidine alkaloids [38]. This methodology was recently applied to the synthesis of the unusual amino acid detoxinine **100** [39], which provides the central core in the detoxins, a family of depsipeptides displaying detoxifying activity against the nucleoside antibiotic blasticidin S. It was envisaged that a two-atom silyl tether between the heterodiene (nitroalkene) and dipolarophile would be ideal for controlling the outcome of the intramolecular [3+2] cycloaddition in the tandem step. Furthermore, the silyl tether would not only serve as a protecting group for the secondary alcohol on the pyrrolidine but would also provide a masked alcohol functionality for the pendant side chain. Oxidative cleavage of the tether would release this diol functionality at a late stage in the synthesis (Scheme 10-32).

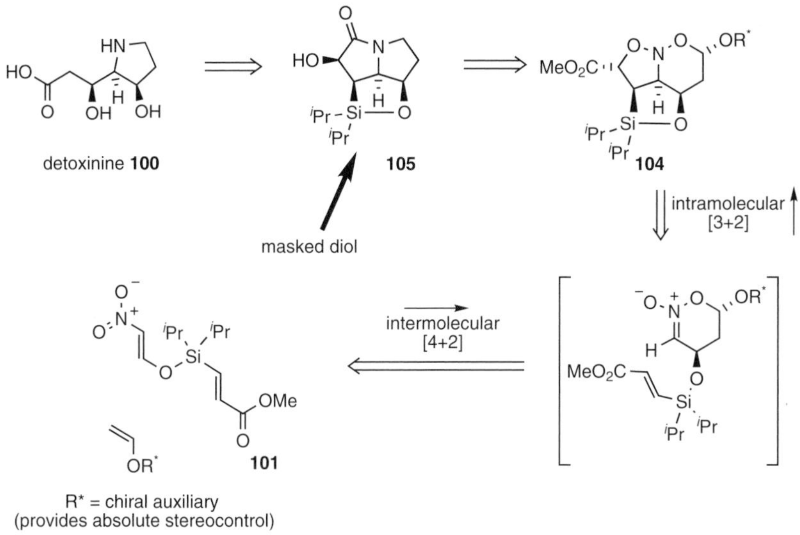

Scheme 10-32 A silyl tether can control the stereochemical outcome of the tandem [4+2]/[3+2] reaction for application to the synthesis of detoxinine.

Formation of the silyl tether **101** was achieved in a highly efficient manner (Scheme 10-33). A silyllithium reagent, prepared from ^{i}Pr$_2$PhSiCl, lithium, and CuCN, was reacted with methyl (*E*)-β-iodoacrylate, affording the coupled product in excellent yield and with complete retention of configuration. Subsequent dearylation was best achieved with dry HCl in CHCl$_3$ at 80 °C, providing the silyl chloride **102** in almost quantitative yield. Reaction with the potassium enolate of nitroacetaldehyde then afforded the tethered cycloaddition precursor **101**. In spite of bulky isopropyl substituents, **101** underwent hydrolysis during all purification attempts and was consequently used directly in the key tandem process.

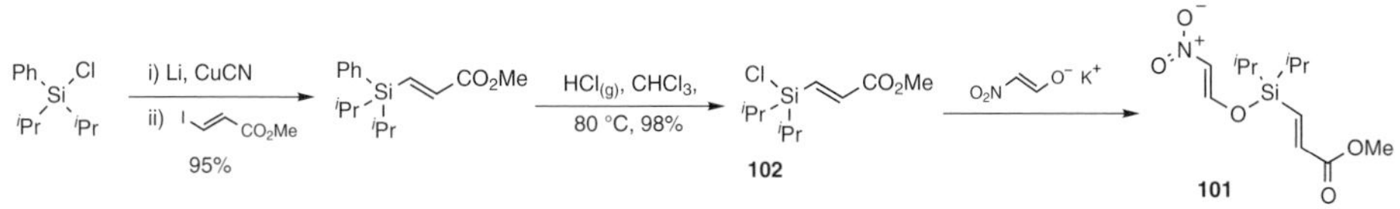

Scheme 10-33 Preparation of the silyl tether was achieved in excellent yield.

Optimum conditions in the crucial tandem cycloaddition reaction involved the addition of the bulky aluminum Lewis acid, MAPh, to a cooled (−78 °C) solution of the silyl species **101** and the enantiomerically pure vinyl ether **103** followed by slow warming to −15 °C over 14.5 h (Scheme 10-34). Nitroso acetal **104** was then obtained in good yield as an inseparable 27:1 mixture of diastereoisomers. The major product derives from initial *exo* addition of the s-*trans* conformer of the vinyl ether onto the *si* face of the nitro-alkene. The subsequent [3+2] cycloaddition is controlled completely by the silyl tether with addition of the nitronate to the same face to which the tether is attached [40]. An *endo*-disposed tether also forces an *exo* orientation of the ester functionality of the dipolarophile (Scheme 10-34). Hydrogenolysis of the nitrosoacetal provided hydroxylactam **105** in 51% yield. With the silyl tether protecting the requisite diol functionality, selective removal of the superfluous alcohol functionality was facile using a radical deoxygenation protocol. Finally oxidative cleavage of the silyl tether – which proved surprisingly easy in spite of the bulky isopropyl groups – and hydrolysis of the lactam provided (–)-detoxinine in ten steps and 13% overall yield from commercially available $^{i}Pr_2SiCl_2$ (Scheme 10-34).

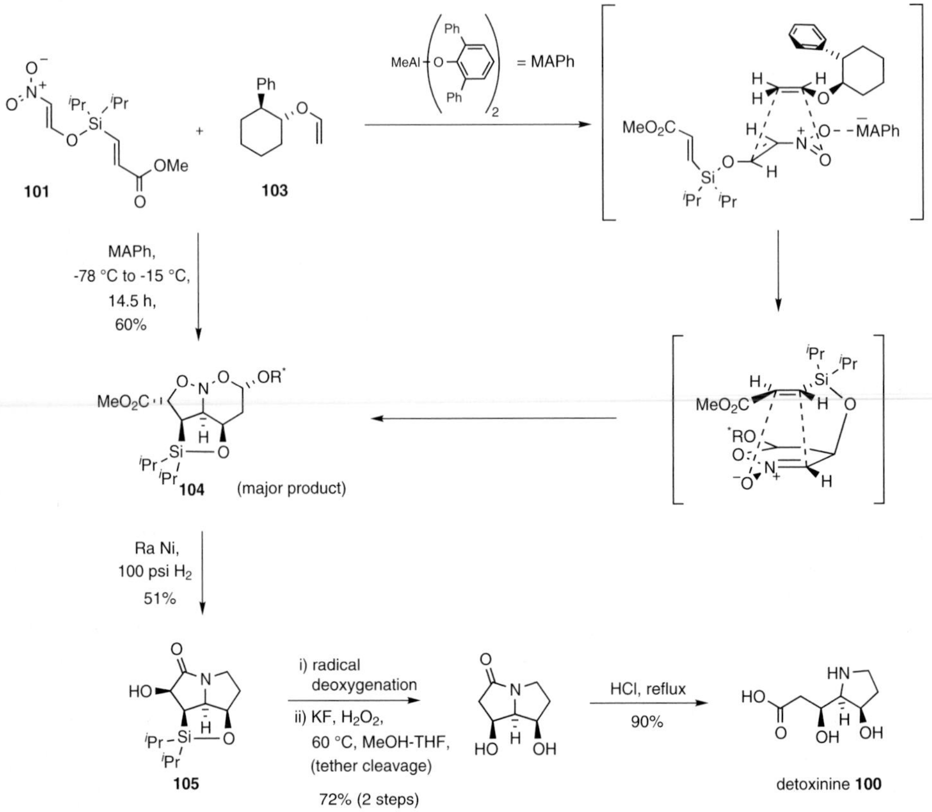

Scheme 10-34   The tandem [4+2]/[3+2] cycloaddition proceeded with excellent stereocontrol and allowed facile elaboration to (–)-detoxinine.

10.2.2.3   Sulfur tethers in [5 + 2] cycloaddition

The [5 + 2] cycloaddition reaction provides an interesting and potentially versatile entry into densely functionalized, seven-membered rings. However, it remains relatively unexplored. The reaction between 5-hydroxy-4-pyrones and alkenes has been shown to proceed only efficiently in the intramolecular cycloaddition mode [41]; cycloaddition in the analogous intermolecular variant either fails completely, or provides a mixture of stereoisomers. Mascareñas, Mouriño, and co-workers chose to investigate temporary tethering strategies to overcome the problems associated with this bimolecular reaction [42]. Commercially available kojic acid **106** provides a convenient [5π] precursor since it contains a primary alcohol functionality as a suitable site for attaching the [2π] reacting partner. A number of tethers were investigated, among them a sulfide linker. The starting thioether **107** was readily prepared by standard procedures [42a]. Thermolysis of **107** in toluene at 145 °C for 40 h provided a single *exo* cycloadduct **108** in good yield. Treatment of the cycloadduct with Raney nickel effected desulfurization and rearrangement to ketone **109** in 70% yield (Scheme 10-35). The net result is equivalent to a formal intermolecular cycloaddition between an unactivated alkene and α-deoxykojic acid, a transformation which cannot be achieved by the bimolecular process.

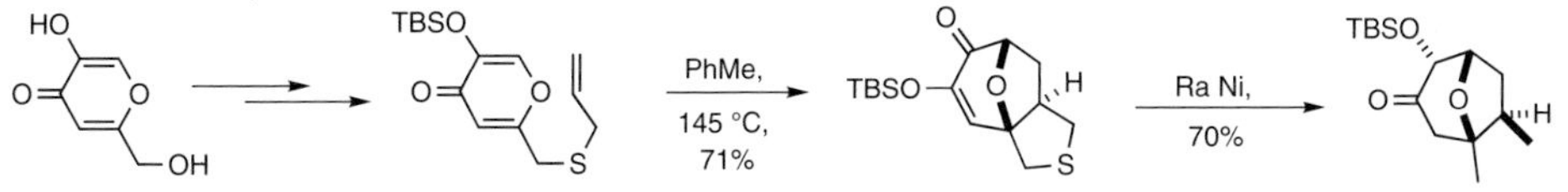

Scheme 10-35   A sulfide tether allows [5+2] cycloaddition between an unactivated alkene and a pyrone.

The success of the sulfide tether encouraged the investigation of related tethers [43]. A sulfoxide tether was readily accessed by *m*CPBA oxidation of sulfide **107**. Since the sulfoxide provides a stereogenic center in the tether, it was envisaged that this might induce a degree of diastereofacial selectivity in the cyclization. Indeed, thermolysis of **110** in 1,2-dichloroethane provided a separable 3:1 mixture of diastereoisomers in excellent yield (Scheme 10-36). Although the degree of asymmetric induction is modest, it once more illustrates the potential of using asymmetry in the disposable tether for imparting stereocontrol in the cyclization. It is also noteworthy that reaction with the sulfoxide tether proceeded at a more rapid rate at reduced temperature than with the corresponding sulfide. This result was attributed to a Thorpe–Ingold effect: increasing the substitution at the sulfur atom – by using a sulfone – would therefore be expected to facilitate the reaction further. This prediction was realized with cyclization of sulfone **111** proceeding readily at only 90 °C, again in excellent yield (Scheme 10-36).

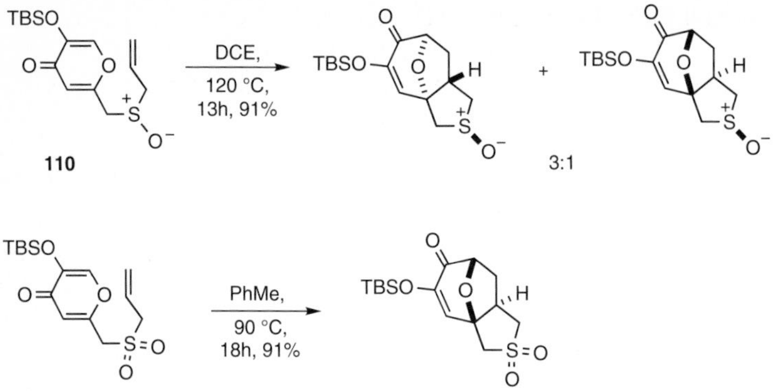

110

111

Scheme 10-36 Sulfoxide and sulfone tethers can also be used in [5 + 2] cycloadditions.

10.2.2.4 Azomethine ylides

Garner and co-workers have investigated the use of azomethine ylide cycloadditions [44] for the preparation of the 3,8-diazabicyclo[3.2.1]octane substructure of the bioxalomycins [45], a group of natural products which exhibit potent antimicrobial and antitumour activity (Scheme 10-37). To incorporate the desired absolute and relative stereochemistry into the bicyclic product, cycloaddition would necessitate an *endo* T. S. with *re*-face attack on the olefin dipolarophile. Since the intermolecular reaction provided the *exo* addition product with no diastereofacial selectivity, it was envisaged that linking the two reacting systems through a temporary tether would force the reaction to proceed through the desired *endo* T. S. and also provide some control over the facial selectivity.

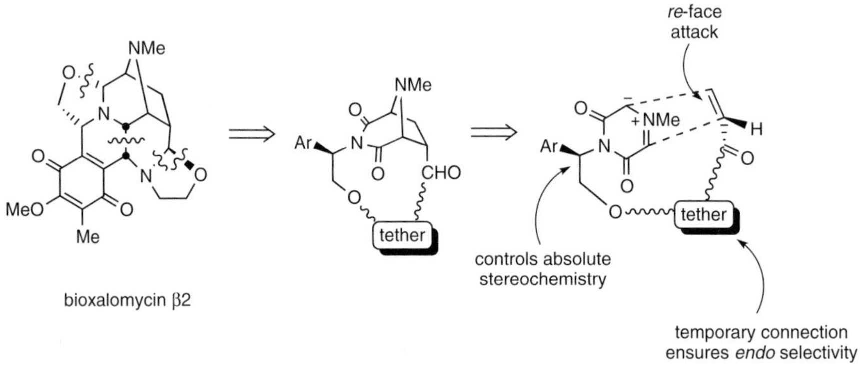

Scheme 10-37 An intramolecular azomethine ylide cycloaddition can access the 3,8-diazabicyclo[3.2.1]octane structure of the bioxalomycins.

The most simple method for connecting the two reacting fragments would be to combine the aldehyde and alcohol functionality into an ester. An aziridine acrylate precursor was readily prepared although all attempts to effect cyclization failed presumably owing to an unfavorable conformation adopted by the ester tether [45a]. A more flexible acetal functionality would possibly overcome these problems. Consequently, acetal **112** was synthesized as a 1:1 mixture of diastereoisomers in 77% yield by treating alcohol **113** with methoxyallene in the presence of mild acid. This time, photolysis of a dilute solution (0.001 M) of **112** in MeCN led to the isolation of a single stereoisomer **114** in 42% yield whose structure was confirmed by X-ray crystallography and was consistent with the desired *endo–re* mode of cyclization (Scheme 10-38) [45a, c]. Interestingly, only one of the acetal stereoisomers reacts; reaction of the other presumably requires a prohibitively high-energy T. S. to cyclize and instead is lost to non-productive side reactions. The outcome of this reaction was in accord with the group's T. S. model where the *endo–re* addition mode allowed the tether to adopt a conformation in which the controlling stereogenic center could place the smallest group (hydrogen) in the plane of the imide system, thereby minimizing $A^{1,3}$-strain and also allowing minimal distortion of the tether [45a, c].

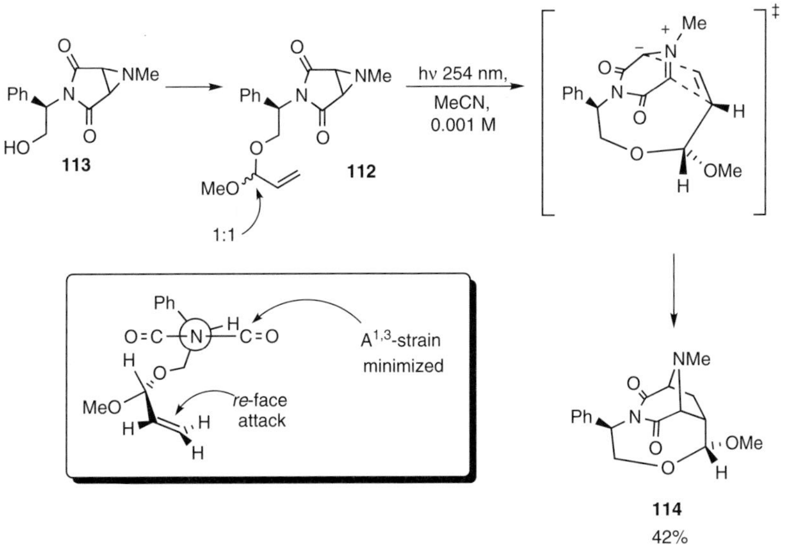

Scheme 10-38   An acetal tether allows the desired *endo–re* cyclization to occur.

Examination of the X-ray crystal structure of the cycloadduct, not surprisingly, revealed appreciable strain in the molecule since the generated eight-membered ring contains a *trans* amide functionality. It was therefore anticipated that a longer, achiral tether would provide a better cyclization precursor. A rational design approach was taken to discover the optimal tether length [45b]. Initially, different tether lengths comprising sp³-carbon atoms were evaluated using computational modeling of the two competing (*endo–si* and *endo–re*) diastereoisomeric T. S.s. These results suggested that short tether lengths favored the desired *endo–re* cycloaddition mode, while the *endo–si* product would be formed preferentially at increased tether length. More refined T. S. modeling using the elements in the real tether supported these results. A number of different silicon tethers of varying length were prepared and the computational results were validated. Thus, with the formation of a nine-membered ring (from **115**), the *re/si* ratio was 16:1 (Scheme 10-39), while an increased tether length producing a 13-membered ring gave a reversal in selectivity with the *endo–si* product being the predominant species (*re/si*, 1:12) (Scheme 10-39). Tether cleavage was readily accomplished using standard protocols.

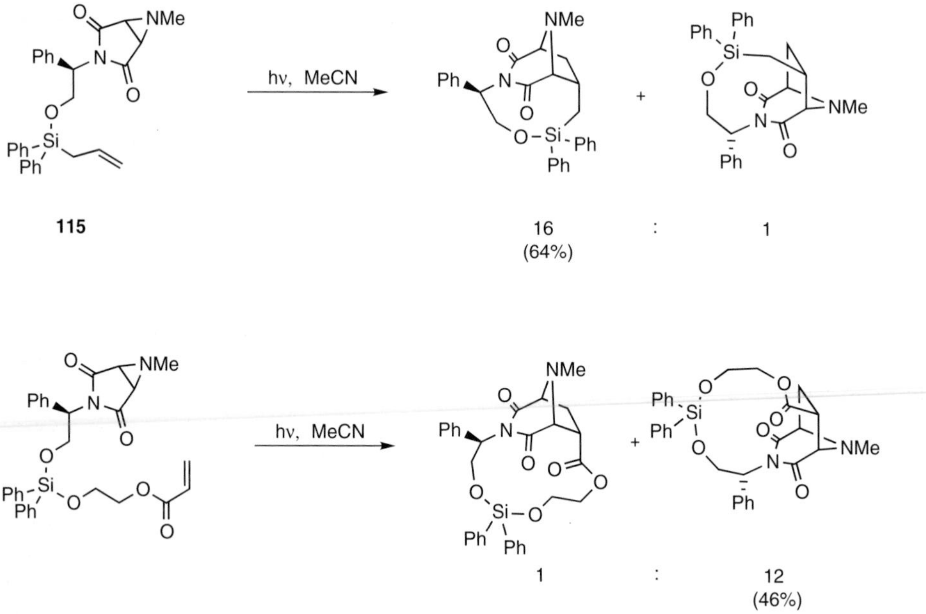

Scheme 10-39   Varying the length of the silyl tether allows access to either *endo–re* or *endo–si* products.

Garner and co-workers have once again illustrated the power of tether-mediated synthesis. In their example, the intermolecular reaction gave the undesired *exo* cycloadducts with no diastereofacial selectivity. Connecting the two reacting systems provided access to *endo* addition products. With further refinement of the tether length, the *re/si* diastereofacial selectivity was successfully addressed.

10.3 Temporary Tethering Strategies in Radical Cyclization Reactions

10.3.1 Introduction

The bimolecular reaction between a radical and radical acceptor such as an unsymmetrically substituted, non-activated double bond, is an inefficient process and only in special circumstances is it regio- or stereoselective. However, when the radical and radical acceptor are connected through a tether, the ensuing intramolecular reaction is much more efficient, and frequently exhibits high levels of regio- and stereoselectivity, especially when small rings (e.g., five- and six-membered rings) are generated [46]. Such radical cyclizations have been investigated in detail and now find widespread application in organic synthesis as a versatile route to carbocycles.

The high levels of control obtained in radical cyclization reactions may be exploited for acyclic regio- and stereocontrol by the use of a temporary connection, a strategy which was first investigated over a decade ago by Nishiyama and Stork, among others, and is now one of the most important applications of this method of intramolecularization. Of all the possible tethers which were discussed in the previous section on cycloadditions, silyl ethers, and to a lesser extent silyl acetals, have found most widespread use in radical cyclization reactions, primarily owing to their ease of formation and ready cleavage post-cyclization.

Before examples of this form of templated synthesis are discussed, it is noteworthy that the presence of a silicon atom in the connecting tether can have a profound effect on the outcome of the cyclization. Baldwin has provided a series of rules which may be used to predict the relative ease of ring formation when a carbon-centered radical adds to a double or triple bond which is connected to the starting radical through a carbon chain [47]. For example, with the 5-hexenyl radical, cyclization proceeds almost exclusively through the 5-*exo* mode providing a cyclopentane product (Scheme 10-40). Wilt and co-workers have shown that by replacing one of the methylene groups with a dimethylsilyl group, dramatic changes to both the 5-*exo*/6-*endo* ratio and the rate of reaction are observed [48]. In particular, they found that with 2- and 3-silahex-5-enyl radicals, the 6-*endo* cyclization mode was favored (Scheme 10-40). A number of factors are responsible for these changes in reactivity, including charge-stabilizing effects of the silyl group and the increased length of the C–Si bond (1.89 Å; cf. 1.54 Å for C–C), which brings the radical center into closer proximity with the terminus of the double bond.

The following discussion on the application of the temporary connection to radical cyclizations will be divided into five sections. In the first, a silyl ether is used as the tether in which one of the alkyl groups attached to the silicon possesses the radical precursor (usually a halogen). In the second section, it is the radical acceptor which is introduced on silyl ether formation. The third section concerns the use of silyl acetals as a temporary connection and in the fourth other templating strategies which do not fall into any of the aforementioned areas will be discussed. The final section is a discussion of the use of some of these strategies in C-glycosylation.

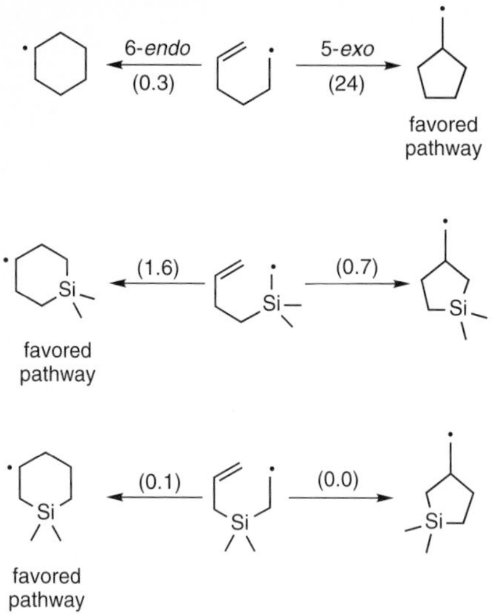

Scheme 10-40 2- and 3-silahex-5-enyl radicals preferentially undergo 6-*endo* cyclization (values in parentheses refer to the rate constants measured at 25 °C [10^{-4} s^{-1}]).

10.3.2   Silyl Ethers Containing the Radical Precursor

10.3.2.1   The α-(bromomethyl)dimethylsilyl ether in radical cyclizations

(Bromomethyl)dimethylsilyl chloride is the most widely investigated and important reagent for incorporating a radical precursor into a silyl tether. It is commercially available and readily introduced as a silyl ether under standard conditions. Moreover, the facile cleavage of the silicon tether under a variety of conditions after reaction yet further enhances its synthetic utility.

Radical cyclization onto a double bond

Nishiyama et al. first introduced the (bromomethyl)dimethylsilyl ether group in a new method for the synthesis of 1,3-diols from allylic alcohols [49]. Silyl ethers of a number of acyclic, allylic alcohols were readily prepared and, in a one-pot operation, heating a benzene solution with Bu$_3$SnH/AIBN at reflux effected radical formation, and cyclization onto the double bond provided the cyclic products. These were exposed directly to a Tamao oxidation [11], generating diols in good to excellent yields. The 5-*exo*-trig cyclization mode was found to predominate; only with terminally unsubstituted allylic systems were 1,4-diol products, derived from a 6-*endo* cyclization, observed (Scheme 10-41). The stereoselectivity of the process was also probed. For example, radical cyclization of (*E*)-allylic silyl ether **116** afforded an 84:16 mixture of *threo/erythro* diol products **117** and **118**, respectively, while the corresponding (*Z*)-isomer **119** generated the *threo* product **117** exclusively and in excellent yield (Scheme 10-41). These results are in accord with the Beckwith–Houk model for radical cyclization [50].

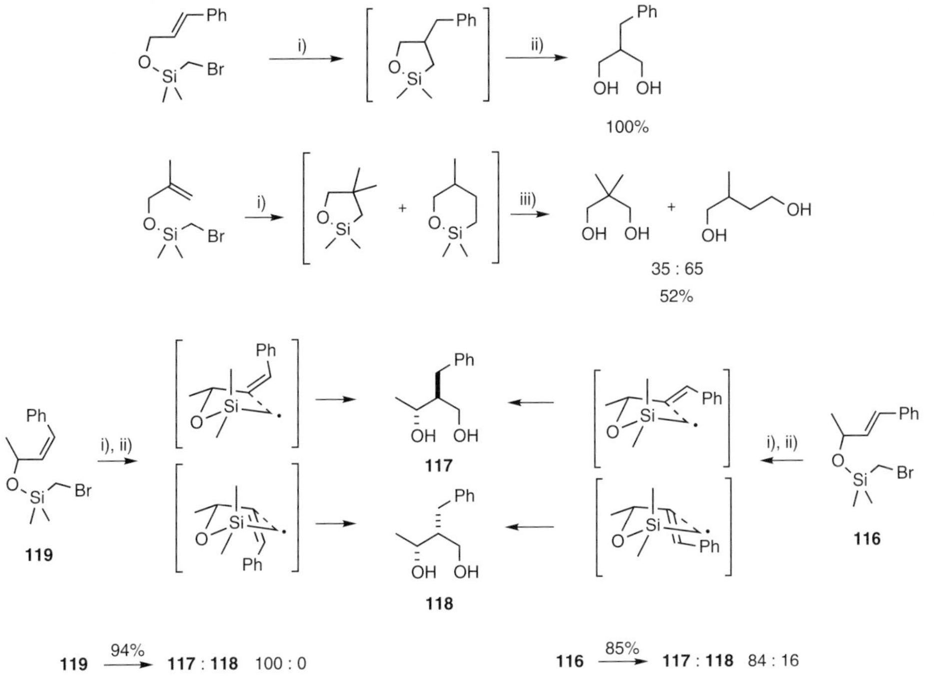

Scheme 10-41 Synthesis of 1,3- and 1,4-diols from (bromomethyl)dimethylsilyl ethers. Reagents and conditions: i) Bu_3SnH, AIBN, PhH, reflux; ii) 30% H_2O_2, DMF, KF, 60 °C; iii) 30% H_2O_2, MeOH, THF, Na_2CO_3, reflux.

Stork and co-workers extended this work with the application of the radical cyclization reaction to polycyclic systems, in particular steroids [51]. Nishiyama had already shown that cyclization onto a cyclohexene generated exclusively the *cis*-fused bicycle **120** and the corresponding *cis*-1,2-disubstituted cyclohexane **121** on oxidative tether cleavage (Scheme 10-42) [49]. In this example, the quenching of the cyclized radical by H-abstraction would not affect the outcome of the reaction. However, with cholestenol derivatives **122** and **123**, H-abstraction by the cyclized radical could provide either a *trans*- or a *cis*-fused A−B ring junction [51b]. This question was addressed by Stork: with the 3β-alcohol, silyl ether formation and subsequent radical cyclization afforded, after reductive desilylation, the *trans*-fused product **124** in 75% yield, while the epimeric precursor **123** provided the *cis*-fused A−B ring junction **125** in similarly excellent yield and stereoselectivity (Scheme 10-42). In both cases, methylation occurred *cis* to the alcohol and H-abstraction occurred *trans* to the starting alcohol. This therefore amounts to a *trans* hydromethylation across the allylic olefin with the tether controlling the initial facial selectivity of the radical cyclization. The cupped nature of the generated polycycle provides sufficient steric differentiation between the two faces such that H-abstraction proceeds from the less sterically demanding convex face.

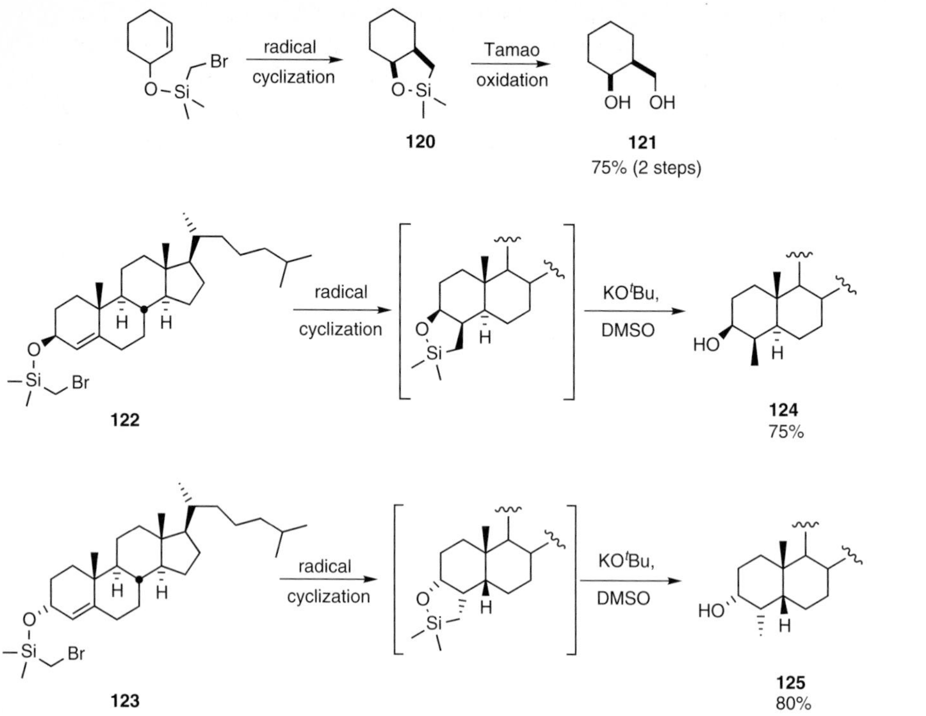

120

121
75% (2 steps)

122

124
75%

123

125
80%

Scheme 10-42   Radical cyclization and reductive desilylation allows stereospecific methylation of steroids and formation of a *cis* or a *trans* ring junction.

Formation of a *trans*-fused 5,6-ring junction can sometimes also be achieved selectively. In an early example from Stork, radical cyclization of silyl ether **126** proceeded in 36% yield but with complete stereoselectivity affording the *trans*-fused hydrindane system [51a]. In this case, the methyl and O^{t}Bu groups presumably reinforce the directing effect of the convex, newly formed silatricycle and H-abstraction is forced to occur from the sterically less hindered face (Scheme 10-43).

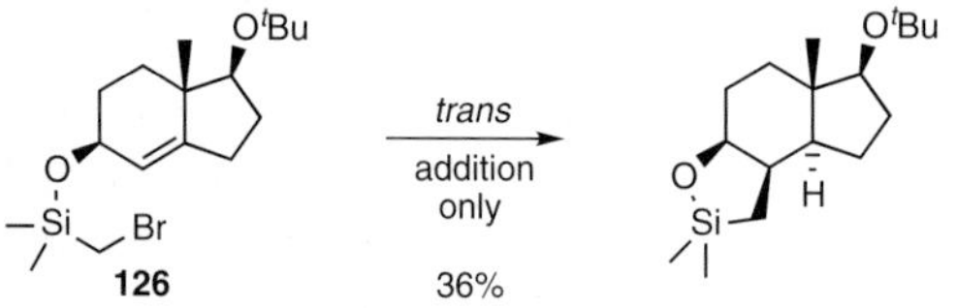

126   36%

Scheme 10-43   A *trans*-fused ring junction can be prepared selectively.

However, with other systems, the outcome of the final H-abstraction is less stereoselective. Jenkins and Wood have investigated the intramolecular radical cyclization onto cyclopentenols derived from glucose [52]. Two epimeric alcohols were prepared by reduction of enone **127** with either NaBH$_4$/CeCl$_3$ (α/β ratio 1:8) or L-selectride (α/β

ratio 8 : 1). Subsequent reaction with (bromomethyl)dimethylsilyl chloride afforded the corresponding silyl ether radical precursors **128** and **129**. Radical cyclization of the *β*-silyl ether **128** provided the *cis*-fused 5,6-ring junction where H-abstraction had proceeded, as before, *trans* to the alcohol functionality. However, in the case of *α*-silyl ether **129**, radical cyclization provided a mixture of products **130** and **131** in an unspecified ratio (Scheme 10-44). In this case, radical cyclization proceeds stereospecifically as normal, generating the *cis*-fused bicycle. If H-abstraction then occurs from the *α*-face (*trans* addition), a product possessing an energetically unfavorable *trans*-fused 5,6-ring junction is produced. As a result, the normally less favorable *cis* addition pathway leading to a *cis*-fused 5,6-ring junction can compete, leading to a more stable product.

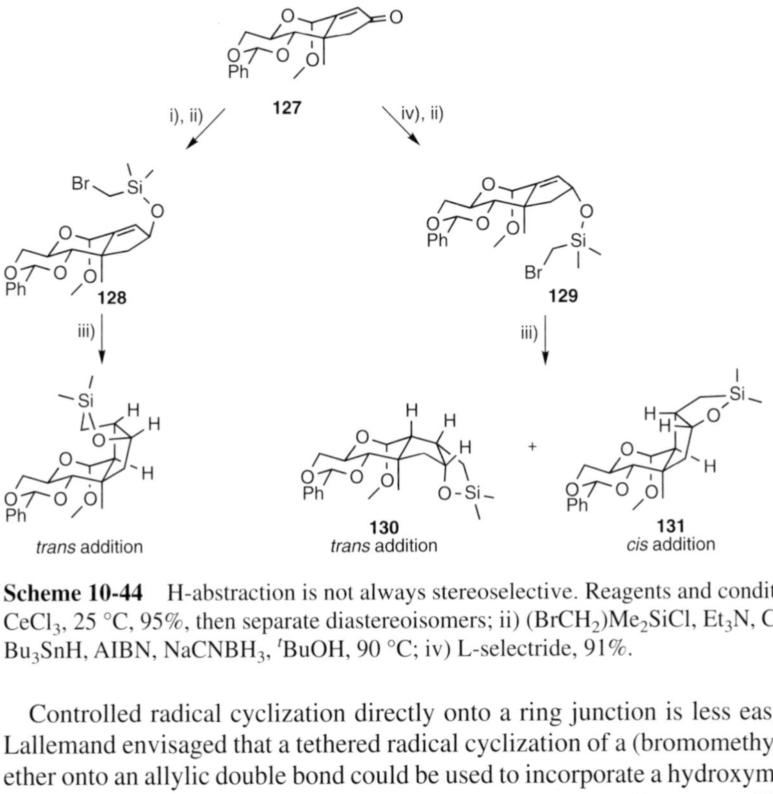

Scheme 10-44   H-abstraction is not always stereoselective. Reagents and conditions: i) NaBH$_4$, CeCl$_3$, 25 °C, 95%, then separate diastereoisomers; ii) (BrCH$_2$)Me$_2$SiCl, Et$_3$N, CH$_2$Cl$_2$, 89%; iii) Bu$_3$SnH, AIBN, NaCNBH$_3$, ^{t}BuOH, 90 °C; iv) L-selectride, 91%.

Controlled radical cyclization directly onto a ring junction is less easy. Lejeune and Lallemand envisaged that a tethered radical cyclization of a (bromomethyl)dimethylsilyl ether onto an allylic double bond could be used to incorporate a hydroxymethyl functionality into an angular position at the ring junction of a decalin system [53]. This would then provide an interesting entry into a variety of natural products containing this skeletal functionality, such as the insect antifeedant clerodin **132** (Scheme 10-45).

Two diastereoisomeric allylic silyl ethers **133** and **134** were first investigated and in both cases, the desired radical cyclization products were not observed. In the case of **133**, cyclization proceeded exclusively in a 6-*endo* mode, affording diol **135** after oxidative cleavage of the silyl linker. The preference for 6-ring formation was attributed to the conformational rigidity of the allylic system and the less hindered nature of the distal sp^2-

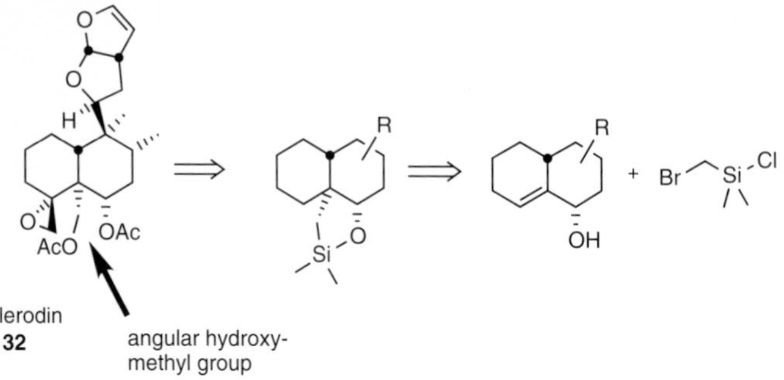

Scheme 10-45   A silyl-tethered radical cyclization could be used to incorporate angular hydroxy-methyl functionality at a ring junction.

carbon atom. Reaction was even less successful with silyl ether **134** and only the TMS-ether reduction product **136** was isolated, albeit in excellent yield. Thus, even with the advantages of intramolecularization, the desired radical addition cannot be guaranteed if the geometry is too rigid. The desired 5-*exo* cyclization mode could only be achieved by modifying the radical acceptor in **133**. Enone **137** underwent stereospecific cyclization onto the more hindered ring junction in quantitative yield, affording the thermodynami-cally less stable *cis* decalin product **138** after tether cleavage (Scheme 10-46). In the case of **133** and **137**, the axially oriented silyl ether allows the adoption of a pseudo-chair T. S. with axial attack on the olefin. This is not possible in the case of equatorially disposed silyl ether **134**, rendering cyclization much less efficient.

Temporary tethering of radical precursors has found other applications in natural prod-uct synthesis. Crimmins and O'Mahony utilized a silyl ether temporary connection to direct a hydro-hydroxymethylation of enol ether **139** in their synthesis of talaromycin A, **140** [54]. Since talaromycin A is susceptible to acid-catalyzed isomerization to the ther-modynamically more stable talaromycin B in which the hydroxymethyl substituent is equatorial, the use of the essentially neutral conditions of a radical cyclization to install the requisite axial hydroxymethyl group would avoid any potential isomerization prob-lems. Formation of enol ether **139** was achieved in five steps from (4*R*)-4-ethylvalerolac-tone **141**. Exposure of **139** to Bu$_3$SnH in benzene at reflux in the presence of AIBN as initiator effected radical cyclization with delivery of the radical to the same face to which the ether tether was attached. Tamao oxidation proceeded uneventfully, furnishing the desired natural product (Scheme 10-47).

In most cases the radical generated after cyclization is quenched by H-abstraction. However, another possibility is to utilize the cyclized radical in another C–C bond-form-ing event. Fraser-Reid and co-workers utilized a silyl-tethered radical cyclization of the (L)-rhamnal-derived silyl ether **142** to generate the anomerically stabilized radical **143**, which could be trapped in the presence of an excess of acrylonitrile to generate acetate **144** after tether cleavage and peracetylation (Scheme 10-48) [55a]. This reaction sequence occurred with complete regio- and stereoselectivity. The same group has also used an acetal tether (vide infra) to effect similar transformations [55b, 56].

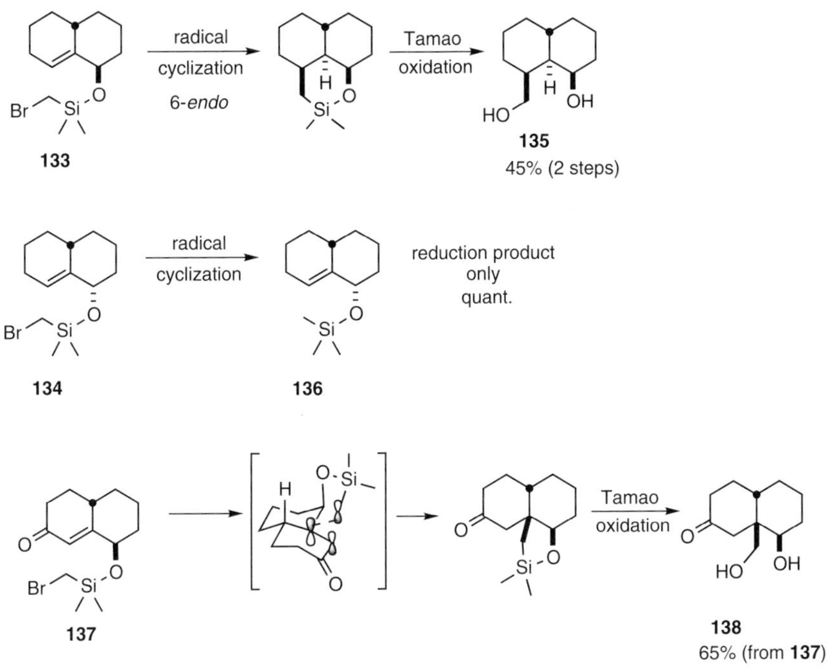

Scheme 10-46 Modification of the radical acceptor and appropriate choice of configuration of directing alcohol are necessary to ensure 5-*exo* cyclization.

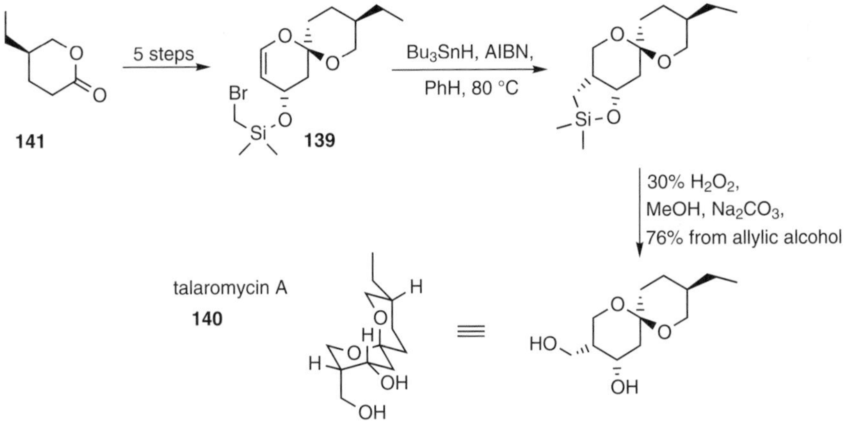

Scheme 10-47 Crimmins and O'Mahony used a silicon-tethered radical cyclization in a synthesis of talaromycin A.

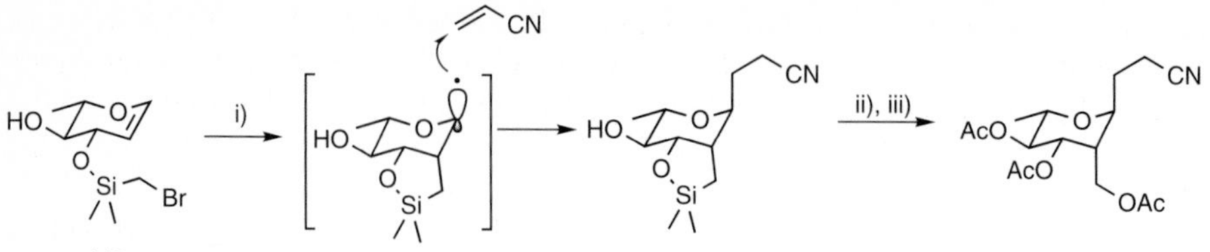

Scheme 10-48   Fraser-Reid showed that the cyclized radical could be trapped with acrylonitrile. Reagents and conditions: i) 0.1 eq. Bu$_3$SnCl, 2eq. NaCNBH$_3$, AIBN, CH$_2$=CHCN, ^{t}BuOH, reflux, 18h; ii) H$_2$O$_2$, KHCO$_3$, KF, THF, MeOH, reflux, 12h; iii) Ac$_2$O, py, rt, 8h, 51% from (L)-rhamnal.

The potential for sila-hexenyl radicals to undergo 6-*endo* cyclization has already been alluded to. Mayon and Chapleur have shown that conformationally restricted exo- and endocyclic allylic (bromomethyl)silyl ethers derived from carbohydrate precursors react exclusively through the 6-*endo* mode of cyclization [57]. Koreeda and co-workers have also exploited this mode of cyclization to provide stereocontrolled routes to 22-hydroxylated steroid side chains [58]. 16α-(Bromomethyl)silyl ether **145** underwent radical cyclization and subsequent conversion to diol **146** in 66% overall yield. Similarly, the epimeric 16β-radical precursor **147** reacted in good overall yield to give diol **148** (Scheme 10-49). Again, reaction through the 6-*endo* cyclization mode may be rationalized on the basis of reduced substitution at C(20) compared with C(17), and the conformational rigidity of the allylic system predisposing radical attack at the distal position of the olefin. It is also noteworthy that clean transcription of chirality is observed in the generation of two new stereocenters – net *cis* addition in the 16α-product and net *trans* addition in the 16β-isomer. Of further interest, the (bromomethyl)silyl ethers of the iso-

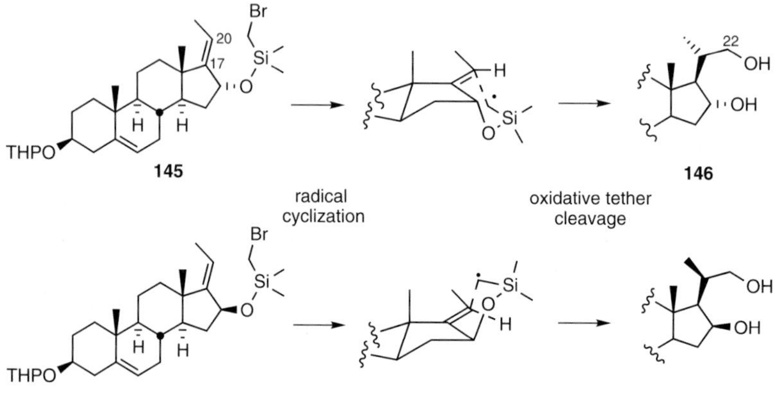

Scheme 10-49   6-*Endo* cyclization allows the stereocontrolled synthesis of 22-hydroxylated steroids.

meric (Z)-allylic alcohols afforded predominantly reduction products, suggesting the orientation of the methyl group at C(20) in these cases obstructs attack of the radical onto the olefin. Factors affecting the mode of cyclization have been investigated in more detail by the same group [59]. Wicha and co-workers have also demonstrated the utility of silyl-tethered radical cyclizations in functionalizing steroidal side chains [60].

Koreeda and Hamann have reported the use of silyl tethers in stereocontrolled syntheses of branched-chain 1,4-diols and 1,5-diols [61]. Exposure of (bromomethyl)silyl ethers prepared from the corresponding homoallylic alcohols with Bu_3SnH in the presence of AIBN allowed smooth conversion to the corresponding cyclic siloxanes, from which diol products were obtained using standard, oxidative cleavage protocols. While monosubstituted olefin **149** selectively underwent 7-*endo* cyclization, di- and trisubstituted olefins **150** and **151** preferentially reacted through the 6-*exo* mode with complete stereocontrol, affording the diol products **152** and **153**, respectively (Scheme 10-50).

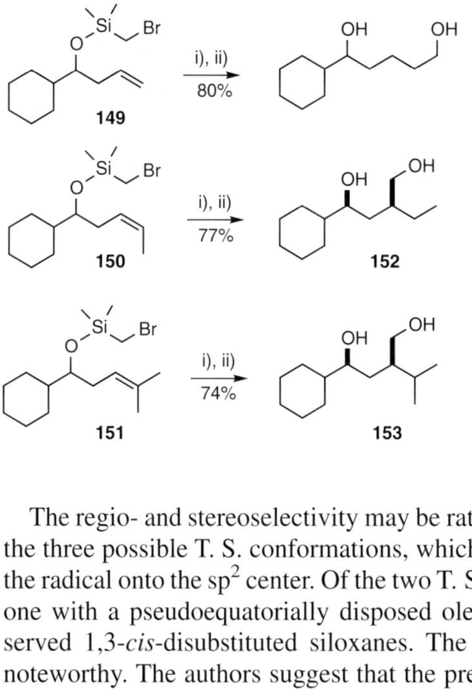

Scheme 10-50   Silyl-tethered radical cyclizations may be used to prepare branched-chain 1,4- and 1,5-diols. Reagents and conditions: i) Bu_3SnCl, $NaCNBH_3$, AIBN, ^tBuOH, reflux; ii) 30% H_2O_2, $KHCO_3$, THF, MeOH, reflux.

The regio- and stereoselectivity may be rationalized in terms of the relative stability of the three possible T. S. conformations, which permit a 107–109° approach trajectory of the radical onto the sp^2 center. Of the two T. S.s leading to 6-*exo* cyclization products, the one with a pseudoequatorially disposed olefinic group, is favored, leading to the observed 1,3-*cis*-disubstituted siloxanes. The regioselectivity of the cyclization is also noteworthy. The authors suggest that the preference of terminally unsubstituted olefins to undergo 7-*endo* cyclization reflects the increased stability of the secondary radical intermediate over the primary radical which would result from an *exo* cyclization.

Radical cyclization onto a triple bond

In a competition study, Malacria and co-workers have shown that a 5-*exo*-dig cyclization is appreciably faster than the 5-*exo*-trig process [62]. The same group has exploited this efficient cyclization using the (bromomethyl)silyl ether group as a radical precursor. In some early work, they demonstrated a simple use of this reaction in an efficient synthesis of 2-[(trimethylsilyl)methyl]prop-2-en-1-ol (**154**) [63], which has found use as a precursor of trimethylenemethane in the palladium-catalyzed [3 + 2] cycloadditions investigated principally by Trost [64]. Formation of silyl ether **155** from propargyl alcohol proceeded uneventfully. Subsequent radical cyclization provided exclusively the 5-*exo*-dig product **156**, which was not isolated owing to its instability; instead, treatment with MeLi generated the desired product **154** in 70% overall yield (Scheme 10-51).

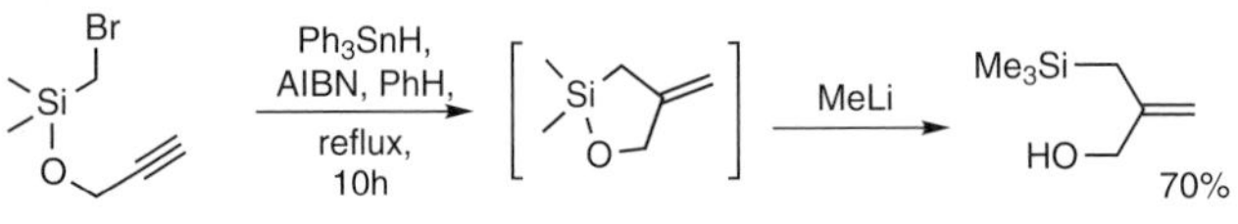

Scheme 10-51   Formation of 2-[(trimethylsilyl)methyl]prop-2-en-1-ol (**154**).

When the triple bond is terminally substituted, 5-*exo*-dig cyclization and subsequent H-abstraction generate a trisubstituted double bond. Malacria and co-workers have shown that the configuration of the double bond can be controlled by varying the nature of the terminal substituent, thus providing a novel entry into stereodefined trisubstituted double bonds [65], which are not always easy to access via more conventional protocols.

With alkyl substituents (e.g., **157**), the reaction is highly stereoselective, giving rise to the (*E*)-olefin product **158** (Scheme 10-52). H-abstraction proceeds *syn* to the sterically more demanding substitution site. Since inversion of the cyclized radical proceeds more rapidly than H-abstraction, quenching will occur on the vinyl intermediate in which $A^{1,3}$-interactions are minimized. Interactions between the H-donor and allylic substituents are not important in this case, owing to the angular nature of the radical.

When the triple bond is substituted with a trimethylsilyl or a phenyl group (e.g., **159**), stereoselectivity is reversed and now kinetically controlled, with H-abstraction providing predominantly the (*Z*)-olefinic product **160** (Scheme 10-52). This difference in reactivity may be attributed to a linear vinyl radical intermediate which is conjugatively stabilized by the terminal substituent removing $A^{1,3}$-interactions. Now, unfavorable steric interactions between the H-donor and the allylic substituents are product-determining.

Radical cyclization reactions have found widespread use in the preparation of polyquinanes. A highly functionalized diquinane framework was prepared by Malacria and co-workers using a radical cascade process [66]. It was anticipated that the (bromomethyl)-silyl ether **161** would serve as a suitable radical trigger. Thus, a solution of Ph3SnH and AIBN was added over 5 h to a benzene solution of **161** and 10 equiv of acrylonitrile at reflux. Further heating for 5 h followed by Tamao oxidation of the crude product mixture allowed isolation of diquinane **162** in 51% yield as a single stereoisomer (by ^{1}H-NMR) (Scheme 10-53).

More recently, the same group reported a route to linear triquinanes starting from acyclic precursors [67]. Once more, a (bromomethyl)silyl ether was used as the radical

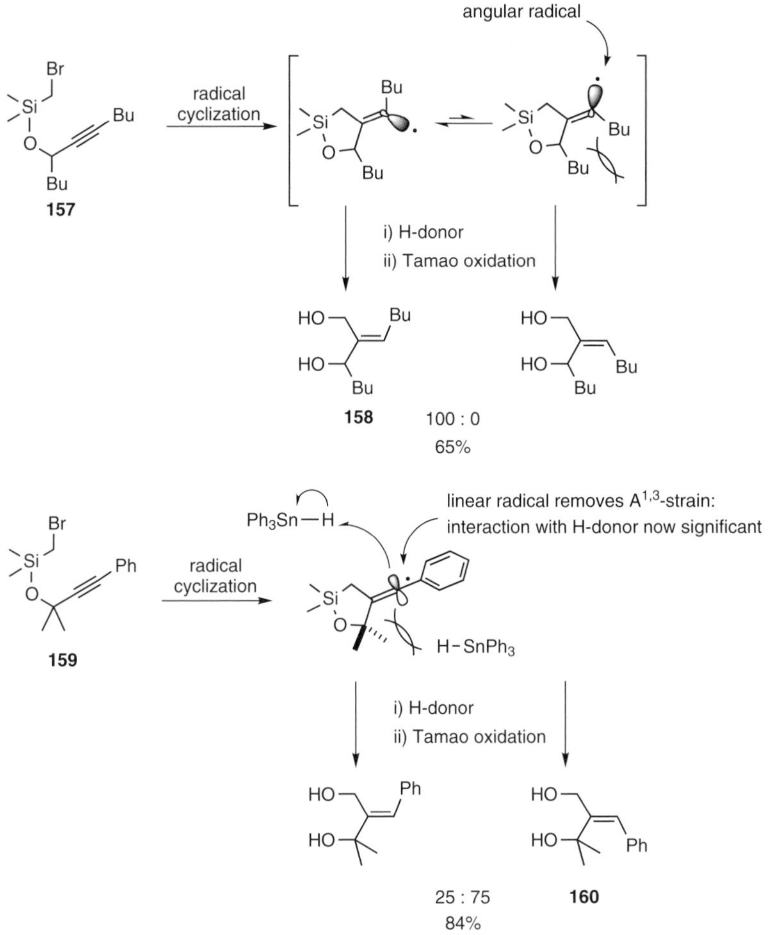

Scheme 10-52   Terminal substitution of the alkyne acceptor affects the stereochemical outcome of the product olefin.

trigger. Previous attempts at effecting such a cascade process had been thwarted by the propensity for an intermediate radical to undergo intramolecular H-abstraction. However, in a carefully developed sequence, cyclization of **163** afforded the triquinane **164** in 50% yield (Scheme 10-54).

10.3.2.2   Other (haloalkyl)dimethylsilyl ethers

α-(Bromomethyl)silyl ethers are useful precursors for incorporating a one-carbon branch via a tethered radical cyclization. Koreeda and George have shown that this meth-

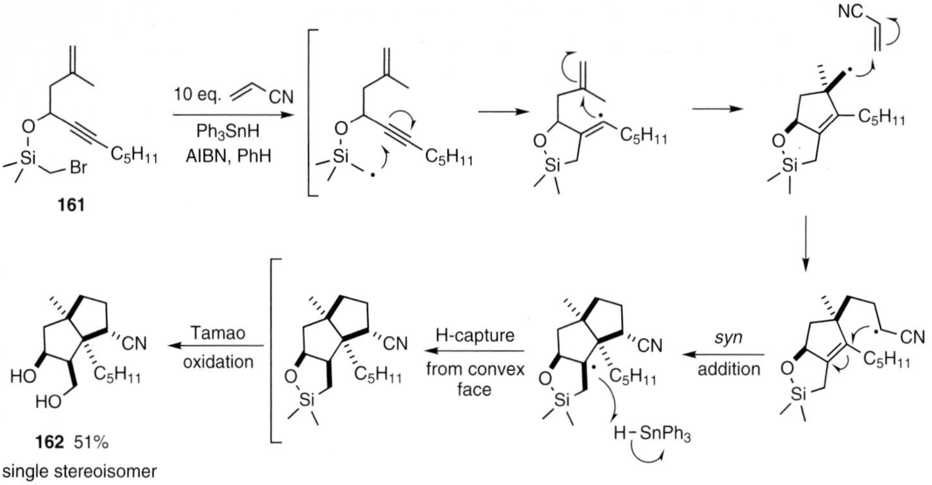

Scheme 10-53   Radical cyclization can be used to prepare diquinane **162** from an acyclic precursor.

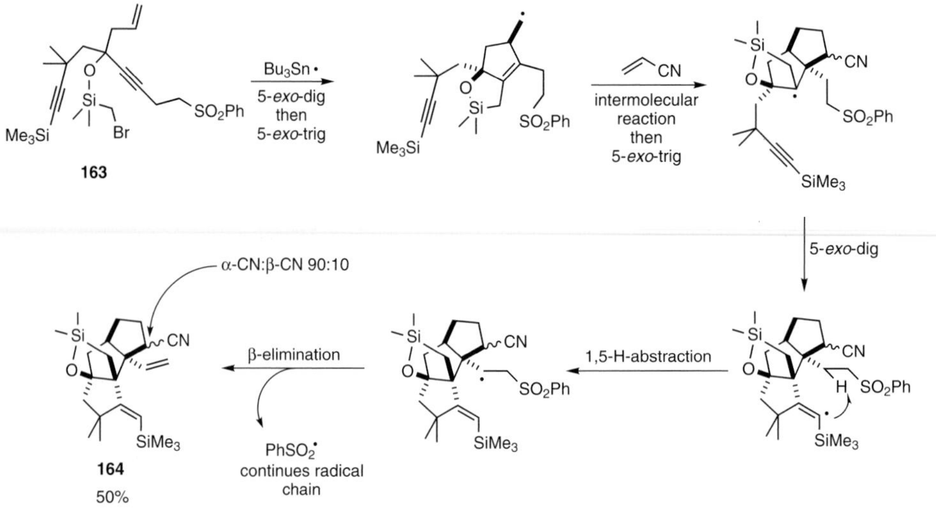

Scheme 10-54   Careful planning has allowed the preparation of triquinane **164** from an acyclic precursor.

od may be extended to incorporating longer alkyl chains with the preparation of (22R)-22-hydroxycholesterol **165** from the allylic alcohol **166** (Scheme 10-55) [68]. The requisite (α-bromoalkyl)silyl chloride **167** was prepared from 4-methylpentyne. Silylcupration with (PhMe$_2$Si)$_2$CuLi·LiCN afforded the (E)-vinylsilane **168** after aqueous workup. Regioselective hydroboration, oxidation, and bromination then furnished the α-bromo-silane **169** in excellent overall yield. Although aryl silanes are stable to a wide variety of reagents, they are susceptible to halodesilylation and may therefore be considered masked silyl halides. Treatment of **169** with ICl effected in situ generation of the desired silyl chloride **167**, which allowed preparation of silyl ether **170** (Scheme 10-55).

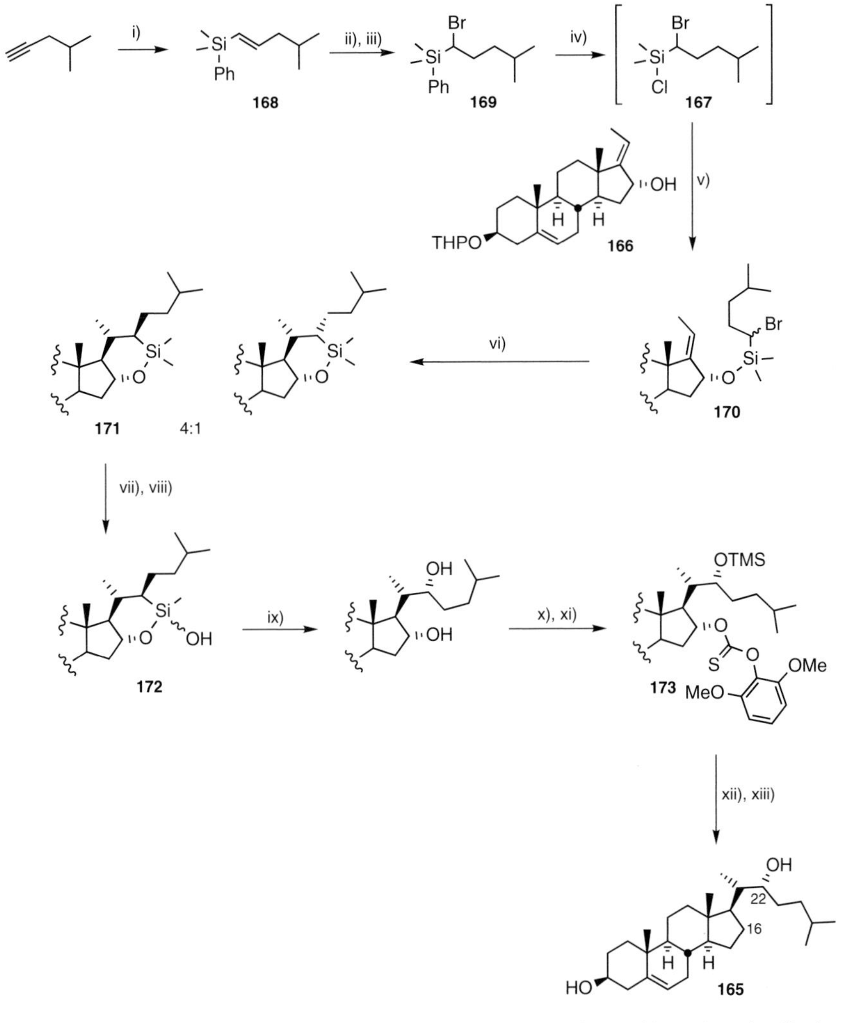

Scheme 10-55   Koreeda's synthesis of (22*R*)-22-hydroxycholesterol by tethered radical cyclization. Reagents and conditions: i) (PhMe$_2$Si)$_2$CuLi · LiCN, THF, 0 °C, 96%; ii) BH$_3$ · THF, THF, 0 °C to rt, then H$_2$O$_2$, 3M NaOH, 60 °C, 82%; iii) Ph$_3$PBr$_2$, MeCN, 0 °C to rt, 81%; iv,v) ICl, CH$_2$Cl$_2$, reflux, then **166**, Et$_3$N, CHCl$_3$, 0 °C to rt, 98%; vi) Bu$_3$SnH, AIBN, slow addition, PhH, reflux, 65%; vii) purify by recrystallization; viii) KOH, DMSO, 60 °C, 65%; ix) 70% H$_2$O$_2$, KHCO$_3$, MeOH, THF, reflux, 75%; x) 2,6-dimethoxyphenyl chlorothionocarbonate, py, DMAP, CH$_2$Cl$_2$, 0 °C, 88%; xi) TMSCl, py, 90%; xii) Bu$_3$SnH, AIBN, PhMe, reflux, 91%; xiii) *p*PTS, EtOH, THF, 60 °C, 90%.

Radical cyclization of (α-bromoalkyl)silyl ether **170** afforded a 4 : 1 mixture of 6-*endo* cyclization products in addition to a small quantity of directly reduced product. The major diastereoisomer **171** was readily purified by recrystallization. Oxidative cleavage of the tether proved to be much more difficult than was initially anticipated, probably owing to steric blocking by the isoamyl group. It was found that the electrophilicity of the siloxane could be increased by treatment of **171** with KOH/DMSO, affording **172** as a 1:1 mixture of stereoisomers in 65% yield; this then underwent smooth oxidative cleavage. Selective deoxygenation at C(16) exploited the slightly increased reactivity of this alcohol towards acylating reagents. Radical deoxygenation of the 22-TMS-ether derivative **173** proceeded smoothly in 91% yield providing (22*R*)-22-hydroxycholesterol **165** (Scheme 10-55).

Tamao et al. have investigated (dichloromethyl)dimethylsilyl ethers as radical cyclization precursors [69]. Silyl ether **174** was readily prepared from the commercially available silyl chloride and isophorol. 5-*Exo*-trig cyclization could be effected under high-dilution conditions, affording the bicycle **175** as a 6:4 mixture of stereoisomers. Subsequent oxidative tether cleavage afforded 2-formyl alcohols **176** and **177** where unfortunately, the presence of base caused partial epimerization of the center α to the formyl group (Scheme 10-56).

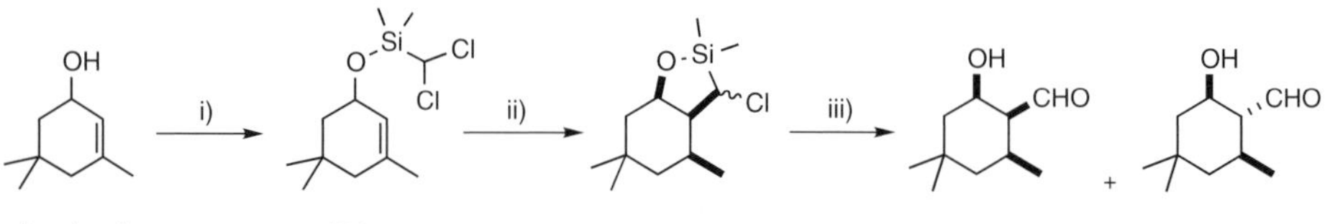

Scheme 10-56   A (dichloromethyl)dimethylsilyl ether can be used as a formyl radical equivalent. Reagents and conditions: i) (Cl$_2$CH)Me$_2$SiCl, Et$_3$N, Et$_2$O; ii) Bu$_3$SnH (1.2 eq.), AIBN, PhH, slow addition, 80 °C; iii) 30% H$_2$O$_2$, KF, KHCO$_3$, THF, MeOH, rt, 51% from isophorol.

The retention of a halo-substituent after radical cyclization means that this tether may also be used as a hydroxymethylidene equivalent. A *cis* hydrindane ring structure could be formed from silyl ether **178** by treatment with 2.4 equiv of Bu$_3$SnH under high-dilution conditions. Two successive 5-*exo*-trig cyclizations provided tricycle **179** and a 4:1 mixture of stereoisomeric diols after oxidative cleavage of the silyl tether (Scheme 10-57). The *cis* ring junction may be accounted for by the first H-abstraction proceeding from the convex face of the initially formed bicycle, and the major stereoisomer is a result of the second 5-*exo* cyclization occurring with attack on the terminal olefin from the concave face with an *exo*-oriented olefin (Scheme 10-57).

A *trans* hydrindane system was also accessible by trapping the radical intermediate resulting from the first cyclization with an allyl group with net *trans* addition across the double bond establishing the desired *trans* ring junction (Scheme 10-57).

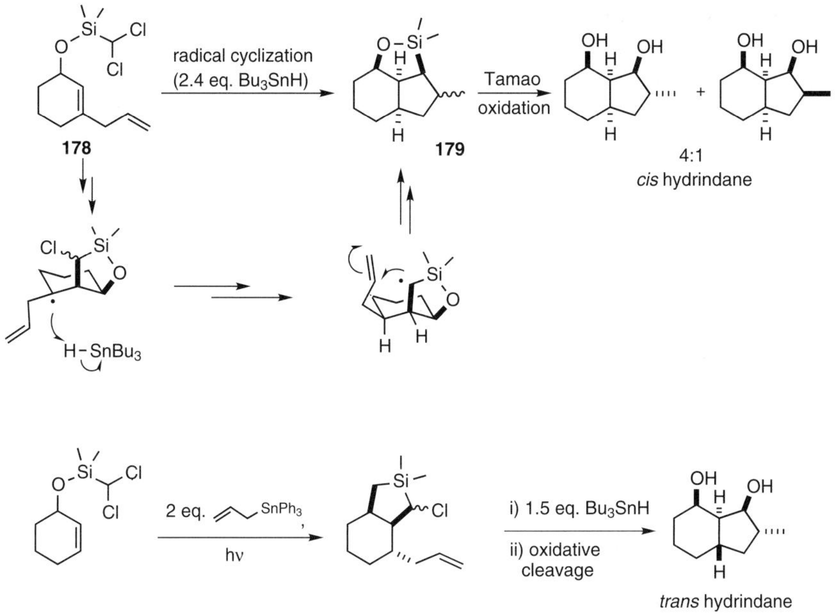

Scheme 10-57   Two radical cyclizations are possible with a (dichloromethyl)silyl ether tether.

Radical cyclization of (β-chloroethyl)silyl enol ethers has been investigated by Walkup et al. [70]. Cyclization in all cases proceeded through the 6-*endo* addition mode, affording γ-silyl alcohols in moderate yield after workup with MeLi. Net *trans* addition across the olefin double bond provides the major diastereoisomer.

10.3.2.3   Dimethylsilyl ethers possessing alkenyl and aryl radical precursors

Tamao, Ito, and co-workers have investigated the use of silyl ether tethers in the radical cyclization of vinyl radicals [71]. (1-Bromovinyl)dimethylsilyl chloride was used to prepare the corresponding silyl ether **180** of isophorol. Radical cyclization using Bu$_3$SnH/AIBN in benzene (0.4 M) provided the two possible cyclization products **181** and **182**, with the 5-*exo* product **181** predominating, along with the direct reduction product **183** (**181/182/183** 25 : 2 : 73). Formation of the reduction product could be suppressed using high-dilution conditions (0.02 M) (**181/182/183** 80 : 8 : 12). When tributylborane was used in place of AIBN as a radical initiator, a slight improvement in the 5-*exo*/6-*endo* ratio was apparent (**181/182/183** 76 : 5 : 19). The 5-*exo* product was formed stereoselectively in accord with previous examples with a net *trans* addition across the double bond. Tether cleavage was effected in a variety of manners, allowing elaboration of the major 5-*exo* product into the corresponding hydroxy ketone, homoallylic alcohol, and hydroxy vinyl bromide or silane (Scheme 10-58).

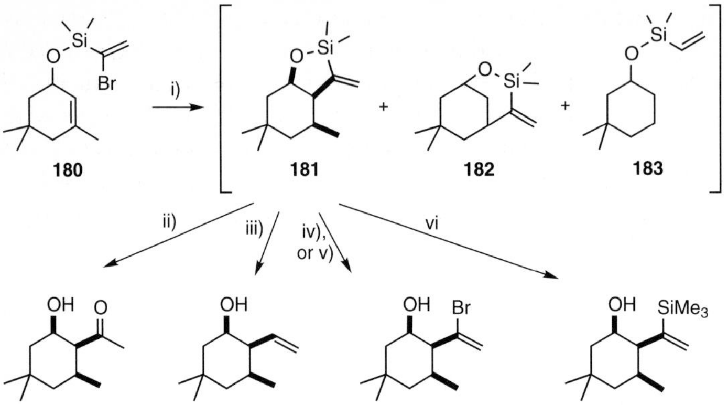

Scheme 10-58 Radical cyclization of (1-bromovinyl)dimethylsilyl ethers proceeds primarily through a stereoselective 5-*exo* pathway. Reagents and conditions: i) Bu₃SnH, Bu₃B, PhH, 25 °C, products not isolated; ii) H₂O₂, KF, KHCO₃, MeOH, THF, rt, 62 % from **180**; iii) ′BuOK, DMSO, H₂O, rt, 62% from **180**; iv) NBS, DMF, 0 °C to rt; v) Br₂, CCl₄, 0 °C then KHF₂, MeOH, rt, 30% from **180**; vi) MeLi, Et₂O, 52% from **180**.

When the (*Z*)-enriched 2-substituted vinylsilyl ether **184** was investigated in the same radical cyclization, only one cyclization product **185** was obtained with a (*Z*)-configured vinyl silane [71]. This may be accounted for by a rapid equilibration of the vinyl radical prior to cyclization. While reaction of the (*Z*)-isomer is favorable, in the corresponding (*E*)-isomer, interactions between the terminal hexyl substituent and the cyclohexane ring render cyclization much less efficient (Scheme 10-59).

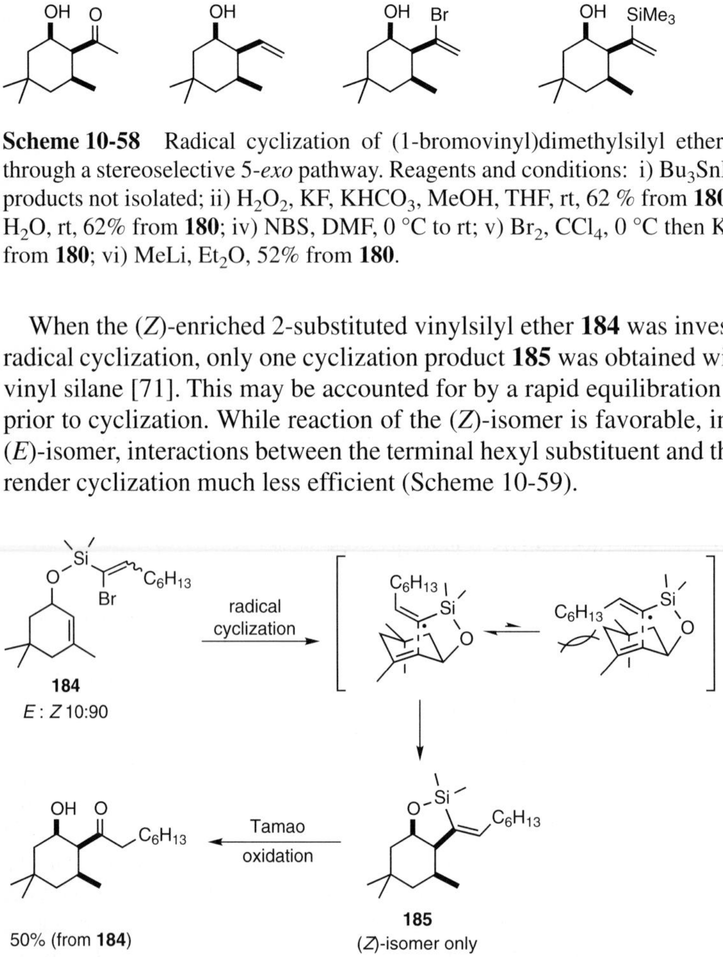

Scheme 10-59 Substituted vinylsilyl ether **184** reacts stereoselectively in a radical cyclization.

In an example from the Curran group, a silyl tether provides a method of incorporating a radical precursor into a molecule at a site remote from where it is desired [72]. 1,5-

H-abstraction is generally a facile process, which often competes with desired transformations. However, in this case it can be utilized favorably to translocate the radical to the desired position for further reaction. The radical precursor in **186** is just a modified phenyldimethylsilyl ether and is much more readily incorporated than the alternative product, which would possess the radical precursor at the site of cyclization. Exposure of (*o*-bromophenyl)dimethylsilyl ether **186** to Corey's catalytic tin hydride conditions [73] formed cyclopentane **187** as a 1 : 1.1 mixture of *cis/trans* isomers, in addition to some directly reduced product **188** (Scheme 10-60).

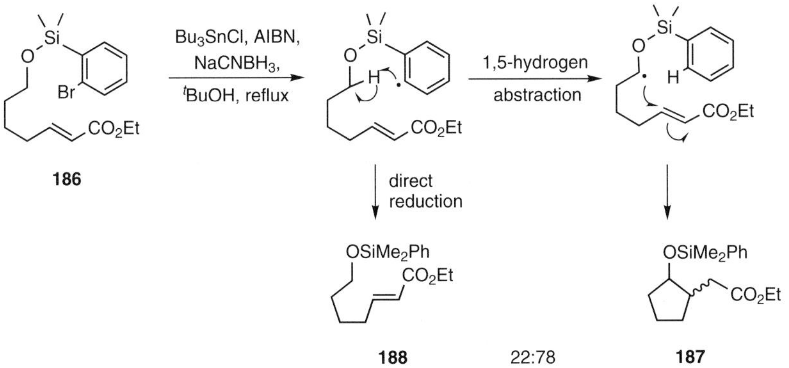

Scheme 10-60   A silyl tether provides a more readily prepared radical precursor for effecting cyclization.

10.3.3   Silyl Ethers Containing the Radical Acceptor

Previous examples of silyl-tethered radical cyclizations have involved the incorporation of the radical precursor during the formation of the silyl ether tether. An alternative would be to incorporate the radical acceptor. A number of suitable silyl chlorides are commercially available which facilitate the preparation of such cyclization precursors.

In an early example, Hiemstra, Speckamp, and co-workers demonstrated the synthetic utility of this form of tether-directed radical cyclization in a formal synthesis of the *β*-hydroxy-*γ*-amino acid, statine **189** [74]. Diacetate **190** was readily prepared in three steps from (*S*)-malic acid. Under normal circumstances, treatment of **190** with a Lewis acid effects formation of the corresponding *N*-acyliminium intermediate which may be trapped intermolecularly with a nucleophile, preferentially from the less hindered face, affording a *trans*-substituted lactam (Scheme 10-61). For the purposes of a synthesis of statine however, a *cis*-substitution pattern was desired. It was envisaged that this might be achieved using a tether-directed radical cyclization. To that end, incorporation of a thiophenyl radical precursor was achieved by transacetalization. Subsequent deacetylation and silyl ether formation using aminosilane **191** generated the radical cyclization precursor **192**. Radical cyclization proceeded exclusively in the 5-*exo* mode affording the *cis*-fused bicycle as an inconsequential 3:2 mixture of stereoisomers. Subsequent reductive cleavage of the silyl ether allowed conversion into lactam **193** which had been previously converted to statine **189** by Johnson and co-workers (Scheme 10-61) [75].

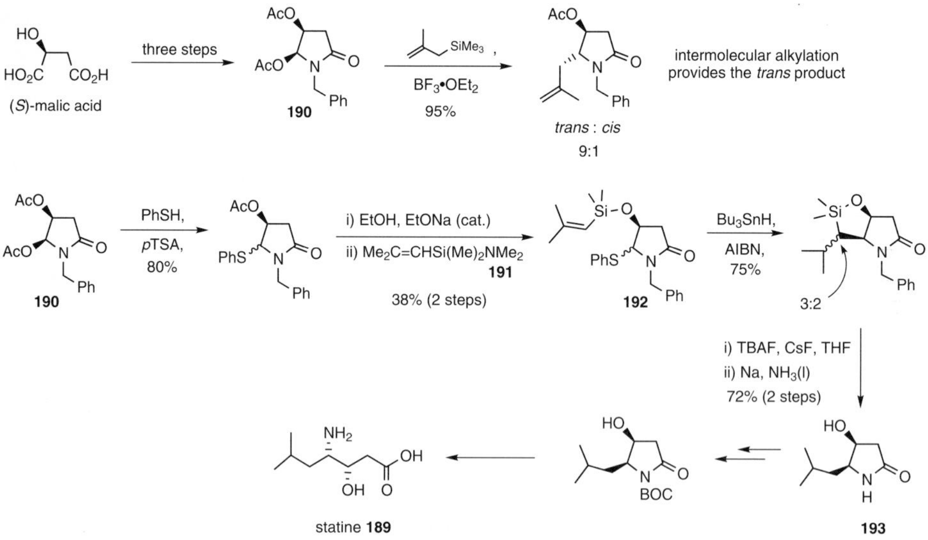

Scheme 10-61   A tether-directed radical cyclization provided the key step in the formal synthesis of statine **189** by Hiemstra, Speckamp, and co-workers.

C-branched nucleosides have been found to possess a number of biological properties, including antibacterial, antitumour, and antiviral activity. This has fueled a drive for preparing novel analogs of such molecules. Chattopadhyaya and co-workers have published extensively on the manipulation of nucleosides using intramolecular radical cyclizations [76], which are carried out under mild and, importantly, neutral conditions, making them compatible with a range of functionality.

In an early example, the incorporation of phenylseleno functionality into the 3′-position of a number of thymidine nucleosides allowed the stereoselective preparation of 3′-C-branched derivatives (Scheme 10-62) [76a]. An ester tether was used to connect the radical acceptor olefin functionality to the 5′-hydroxy group. Radical cyclization of **194** under high-dilution conditions using $Bu_3SnH/AIBN$ in benzene at reflux afforded the corresponding δ-lactone bicyclic products **195**. In all cases, retention of configuration was observed at the site of radical generation with the formation of *cis*-fused bicycles. Furthermore, only products resulting from a 6-*exo* cyclization mode were isolated. Cyclization through a chair-like T. S. with a pseudo-equatorially disposed substituent accounts for the observed stereoisomer. Hydrolytic cleavage of the tether proceeded efficiently with aqueous ammonia, generating the corresponding hydroxy acids **196** in excellent yield (Scheme 10-62).

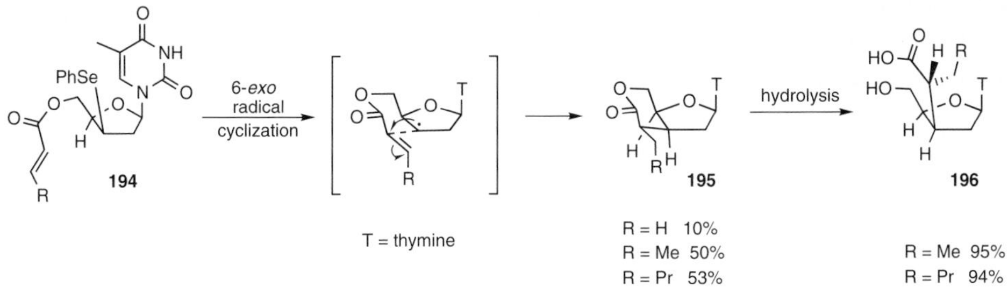

Scheme 10-62 Tethering the radical acceptor and precursor through an ester linkage allows stereoselective radical cyclization of 3′-phenylseleno thymidine nucleosides.

The same group have also investigated the incorporation of the radical acceptor moiety via a silyl ether tether [76b]. A phenylseleno substituent was stereospecifically incorporated into either the 2′- or 3′-position of four thymidine nucleosides, followed by incorporation of an olefinic radical acceptor into the vicinal hydroxyl group by reaction with (allyl)dimethylsilyl chloride. In all cases, only products resulting from a 7-*endo* cyclization mode were isolated after treatment with Bu$_3$SnH/AIBN.

The stereochemical outcome of the reaction depended on the configuration of the (allyl)silyl ether group. In the cases of **197** and **198**, regardless of whether the radical was generated at the 2′- or 3′-position, cyclization onto the allyl group on the α-face provided single diastereoisomeric *cis*-fused bicyclic products **199** and **200** in good yields. However, when the radical was trapped on the β-face, reaction either afforded a 1:1 mixture of *cis*- and *trans*-fused bicycles in the case of **201**, or the unusual tetracyclic product **202** resulting from further reaction with the proximal thymine base (Scheme 10-63). Oxidative cleavage of the silyl tether afforded the corresponding 1,5-diol products in good yields.

Chattopadhyaya and co-workers have also demonstrated that alkynyl groups can be used as radical acceptors when linked through a silyl ether [76c]. Radical cyclization and subsequent oxidative cleavage of the tether provided access to 2′- and 3′-C-branched α-keto-β-D-ribonucleosides, usually in good to excellent yields.

In a more recent study on the use of silyl-tethered radical cyclizations for the preparation of 4′-α-C-branched-2′-deoxyadenosines, Shuto, Matsuda, and co-workers observed an interesting dependence of the apparent mode of cyclization on the reaction conditions [77]. The known 4′-phenylseleno derivative **203** of 2′-deoxyadenosine was readily converted to the 3′-vinylsilyl ether **204** using standard conditions. Treatment of **204** with 3 equiv of Bu$_3$SnH at 80 °C in benzene, using AIBN as radical initiator followed by oxidative tether cleavage, provided a 2:1 diastereoisomeric mixture of diols, **205** being derived from a 5-*exo* cyclization mode (Scheme 10-64). However when the Bu$_3$SnH/AIBN was added slowly over 4 h to a solution of **204** at 110 °C, maintaining a low concentration of Bu$_3$SnH, a reversal in regioselectivity was observed with the diol **206** being obtained after Tamao oxidation (Scheme 10-64). The results suggested that the

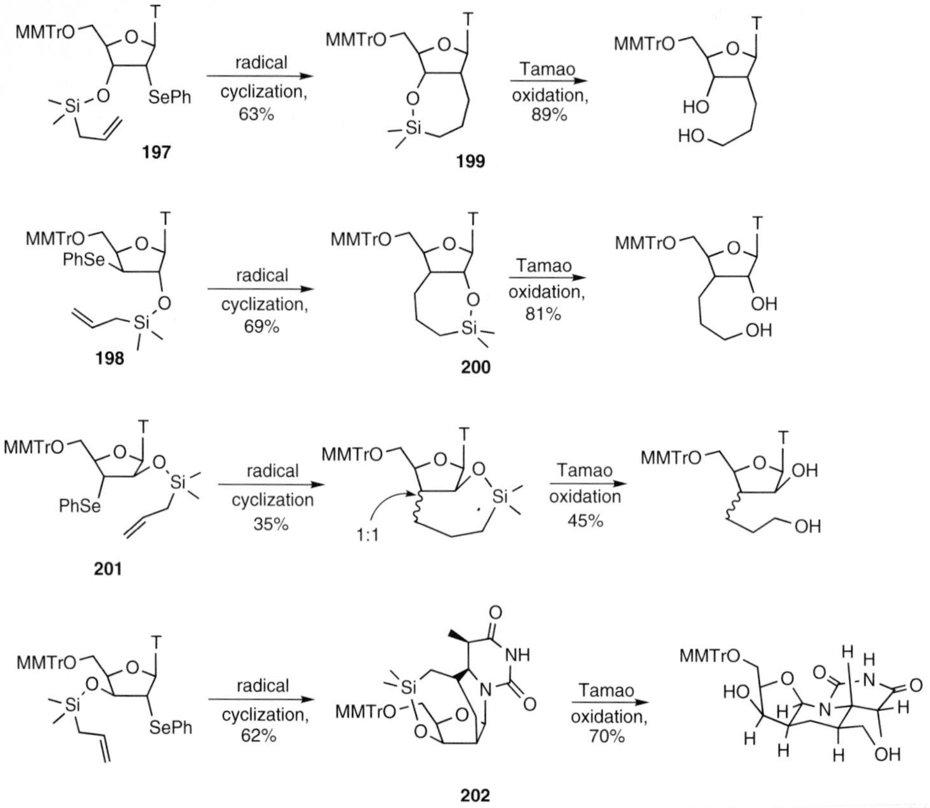

Scheme 10-63 The stereochemical outcome in the silyl-tethered radical cyclization of nucleosides is dictated by the configuration of the silyl ether group.

formation of the 6-*endo* product **207** was under thermodynamic and not kinetic control, since the *endo/exo* ratio would not be expected to change upon varying the concentration of Bu$_3$SnH if the reaction was kinetically controlled.

A mechanism was proposed which may account for these observations (Scheme 10-65). In both cases, radical cyclization proceeds in the normal 5-*exo*-trig mode affording the intermediate radical **208**. At low concentrations of Bu$_3$SnH, rearrangement to the more stable secondary radical **209** is possible and this species is then quenched by H-abstraction. The precise mechanism of the rearrangement is not known although it may proceed via a fragmentation–recombination process or through a silicon-bridging intermediate.

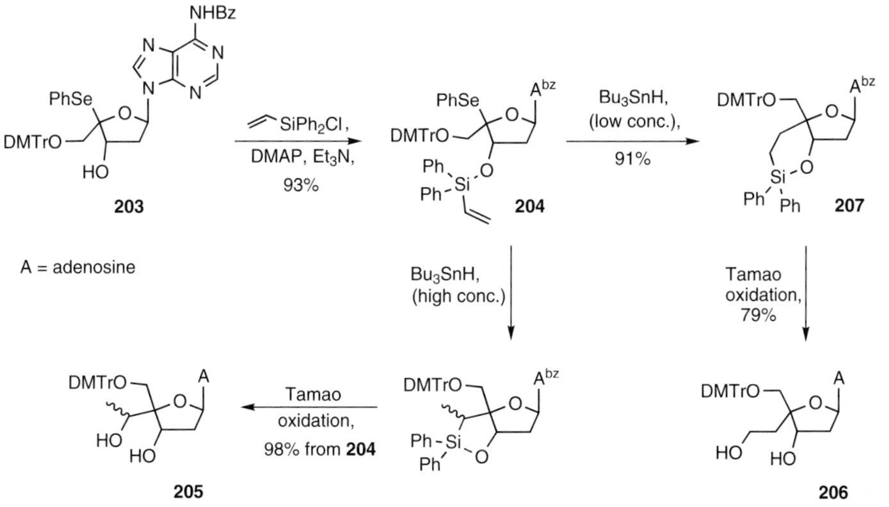

Scheme 10-64 The concentration of Bu$_3$SnH can be crucial in determining the product outcome.

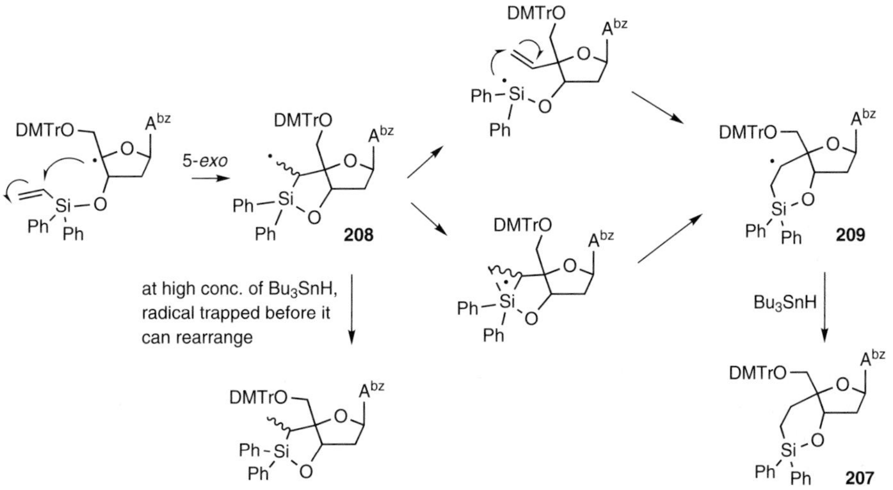

Scheme 10-65 A possible mechanism to account for isolation of a 6-*endo* addition product.

10.3.4 Silyl Acetals

Silyl acetals have been investigated by Hutchinson and co-workers as tethers in radical cyclization reactions [78]. A number of commercially available dialkyldichlorosilanes were investigated as tether precursors although ^{i}Pr$_2$SiCl$_2$ was found to be the most suitable, providing a compromise between hydrolytic stability and ease of introducing the second alcohol group. 2-Bromoethanol was used as the radical precursor throughout the study while the radical acceptor was varied. The unsymmetrical silyl acetal unit was best prepared in a stepwise procedure: treating ^{i}Pr$_2$SiCl$_2$ with a three-fold excess of 2-bromoethanol allowed isolation of the monochlorosilane **210** by distillation in 60% yield. Subsequent reaction with a variety of radical acceptors (hydroxyalkenes and lithium enolates) generated the silyl acetals in good to excellent yield (Scheme 10-66).

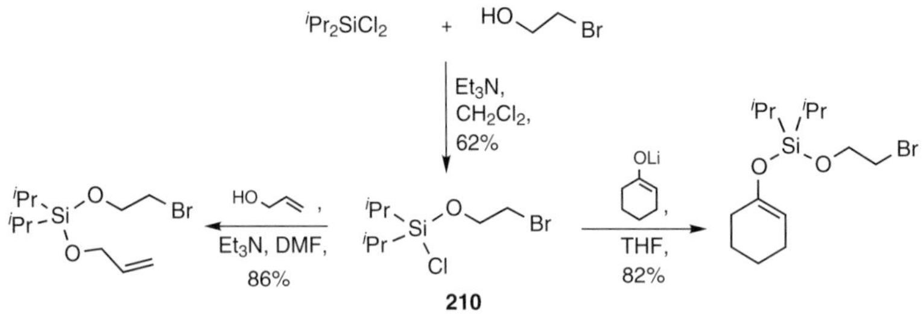

Scheme 10-66   Stepwise preparation of silyl acetals.

Radical cyclization using standard Bu$_3$SnH/AIBN conditions revealed a preference for the formation of the larger of the two possible rings via an *endo*-trig cyclization, although in some cases it was possible to induce the *exo*-trig cyclization by polarizing the radical acceptor (Scheme 10-67). The preferential 7-*endo*-trig cyclization is in contrast with the 6-*exo* trig cyclization observed with silyl ethers (vide supra). In some cases, a reduction product was also obtained. Deuteration studies using Bu$_3$SnD indicated that this product arose from a competing intramolecular H-abstraction. Cleavage of the tether was possible using standard acidic hydrolysis or by treatment with TBAF (Scheme 10-67).

The preparation of medium-sized rings can be problematic; however, in this case there are a number of factors which may account for the observed facility of ring formation and regioselectivity. These include the relatively long Si–O bond, large O–Si–O bond angle (Si–O bond 1.64 Å; O–Si–O angle 149°) and silicon anomeric effects. Furthermore, the presence of two oxygen atoms in the ring reduces unfavorable transannular interactions.

A highly convergent approach was taken by Myers and co-workers in their elegant synthesis of tunicamycin V, **211** [79]. In a key step, a silyl acetal tether-directed radical cyclization was used to unite the uridine nucleoside with the disaccharide fragment, which contains an unusual "trehalose" glycosidic linkage. The silyl acetal radical precursor was prepared in a single operation from the allylic alcohol **212** and uridine 5′-aldehyde **213**. Thus 3 equiv of pyridine were added dropwise to a toluene solution of the aldehyde (2 equiv) and PhSeH (3 equiv) and this mixture was then added to a solution of Me$_2$SiCl$_2$ (20 equiv) in pyridine. After 4.5 h at 23 °C, removal of the volatiles provided

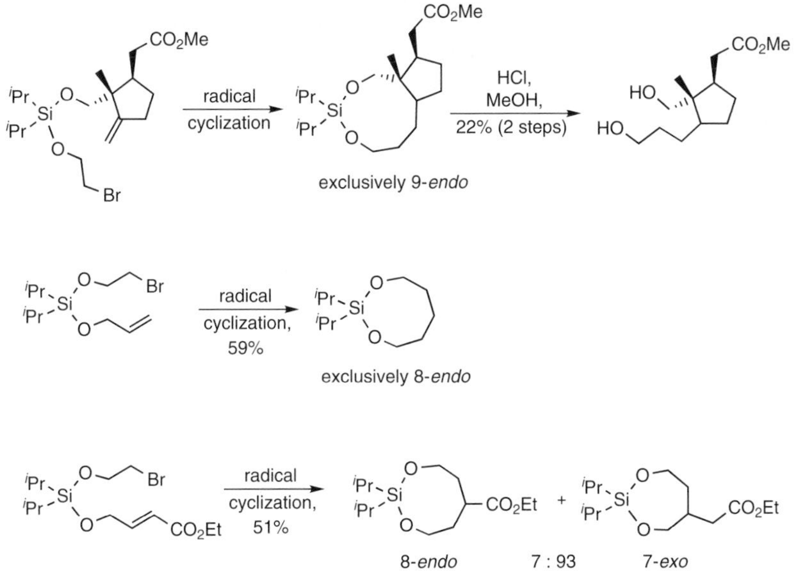

Scheme 10-67   Radical cyclization of silyl acetals proceeds preferentially via the *endo*-trig mode.

the crude monochloride. Addition of the allylic alcohol **212** in pyridine to a solution of the intermediate monochloride in toluene effected rapid conversion to the silyl acetal **214**, which could be isolated in a very impressive 81% yield after purification by flash column chromatography (Scheme 10-68).

The protecting groups on the 2′- and 3′-hydroxyl groups of the ribonucleoside were found to have a marked influence on the configurational outcome at the C(5′) stereocenter in the radical cyclization. In all cases investigated, use of Bu_3SnH with Et_3B as a radical initiator allowed efficient conversion to the desired cyclic siloxane; however, the use of sterically bulky TBS ethers on C(2′) and C(3′) gave rise to the undesired stereochemistry at C(5′). This was attributed to destabilizing steric interactions between the glucosamine residue and the 3′-silyl ether in the T. S. leading to the desired product, which were not present in the T. S. leading to the epimeric product. Consequently, the radical cyclization was ultimately carried out on the free diol **215** (Scheme 10-69). It was anticipated that the diminished size at C(2′) and C(3′) would not only reduce unfavorable steric interactions but the desired T. S. would even be preferred owing to the presence of a hydrogen bond between the C(3′)-hydroxyl group and the glucosyl amide oxygen atom. Cyclization in toluene at 0 °C provided the desired product **216** with the correct configuration at C(5′) (7.5:1 ratio). Use of the silyl acetal not only ensured highly efficient coupling of the two fragments but subsequent cleavage of the siloxane tether could also be achieved under very mild conditions using $KF \cdot 2 H_2O$ in MeOH, affording the sensitive tetraol **217** in an excellent yield of 60% from **215** (Scheme 10-69). Subsequent manipulations allowed the preparation of tunicamycin V [79b].

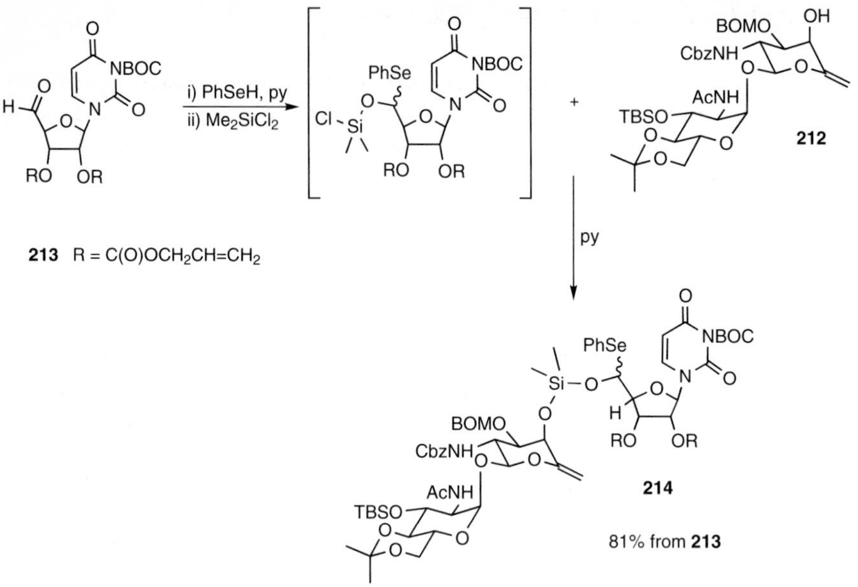

Scheme 10-68   Preparation of the key silyl acetal tether in one step, during the synthesis of tuni-camycin V.

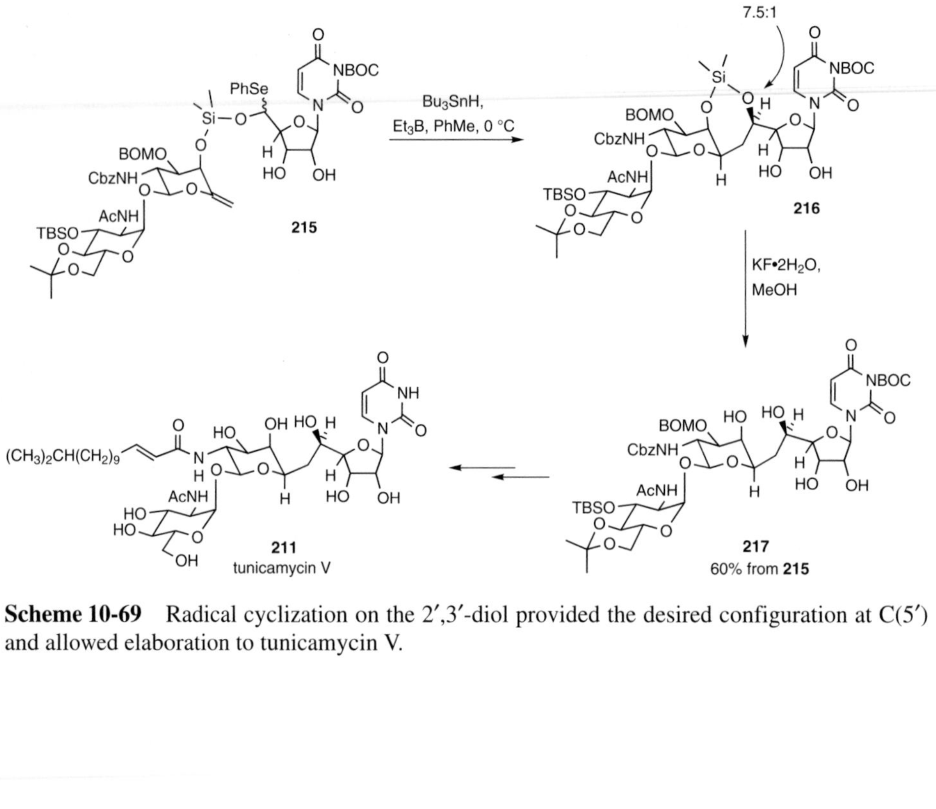

Scheme 10-69   Radical cyclization on the 2′,3′-diol provided the desired configuration at C(5′) and allowed elaboration to tunicamycin V.

10.3.5 Acetal Tethers in Radical Cyclizations

Although the (bromomethyl)silyl ether connection has been more extensively utilized in radical cyclizations, Stork also introduced, at a similar time, the use of a mixed acetal function [51a, 73b, 80]. This tether differs in that a two-carbon unit is introduced on the proximal carbon atom of the olefin, whereas the silyl tether allows the incorporation of only one carbon atom. The chemistry of this tether is dominated by 5-*exo*-trig cyclizations onto allylic double bonds which proceed with the usual degree of high stereoselectivity.

Two examples in natural product synthesis will serve to illustrate the regio- and stereodirecting ability of this tether system. In the first example, Stork and co-workers used an acetal tether to direct a radical cyclization for incorporating the *trans* relationship between the pendant chains in prostaglandin $F_{2\alpha}$ [81]. Treatment of enantiomerically pure monosilyl ether **218** with ethyl vinyl ether in the presence of NIS provided the mixed iodoacetal radical precursor **219** in excellent yield. Introduction of the "lower" side chain was then approached in two ways. In the first, radical cyclization of **219** was effected using the Bu$_3$SnCl/NaCNBH$_3$/AIBN system and the cyclized radical was trapped with *tert*-butyl isocyanide, providing nitrile **220** in 71% yield as a single stereoisomer (Scheme 10-70). In accord with the radical cyclization of α-(bromomethyl)silyl ethers, 5-*exo*-trig cyclization produced the bicyclic intermediate radical which was then trapped exclusively from the convex face with *tert*-butyl isocyanide. The presence of the α-silyl ether group enhances the *trans* selectivity by further blocking the concave face of the bicycle. Subsequent introduction of the rest of the side chain proceeded uneventfully with reduction of the nitrile to the corresponding aldehyde followed by Wadsworth–Emmons homologation using the Masamune–Roush conditions [82]. In a second approach, 2-(trimethylsilyl)oct-1-en-3-one was used as the radical trap, enabling incor-

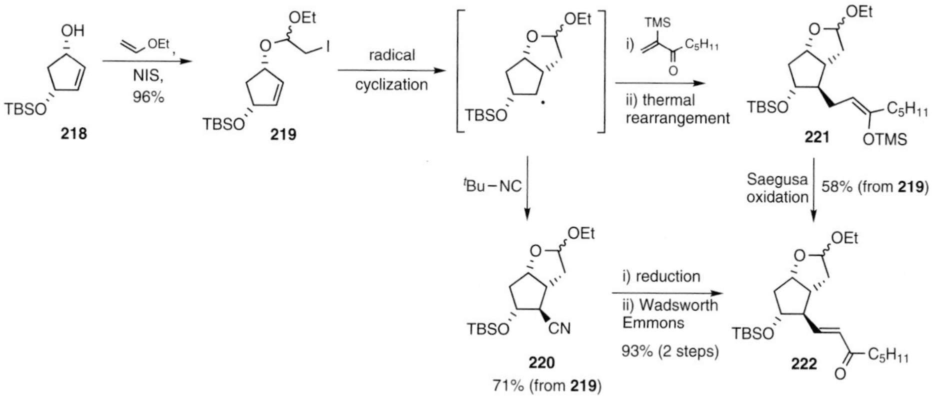

Scheme 10-70   Use of two radical traps to incorporate the "lower" side chain of prostaglandin $F_{2\alpha}$.

poration of the whole side chain in one step. The presence of the α-(trimethylsilyl) group was crucial for ensuring regioselective re-incorporation of the enone functionality. Radical cyclization proceeded efficiently as before with the expected excellent regio- and stereoselectivity to give, after thermal rearrangement, TMS-enol ether **221**, which allowed regioselective incorporation of the olefin functionality with a Saegusa oxidation (Scheme 10-70) [83]. Enone **222** was obtained in 58% yield over the three steps.

Completion of the synthesis was readily achieved. Diastereoselective reduction of the ketone was accomplished with (S)-BINOL/LiAlH$_4$. Then, removal of the tether by acidic hydrolysis, which also effected silyl ether deprotection, afforded dihydroxy lactol **223** allowing introduction of the final side chain using a Wittig reaction (Scheme 10-71).

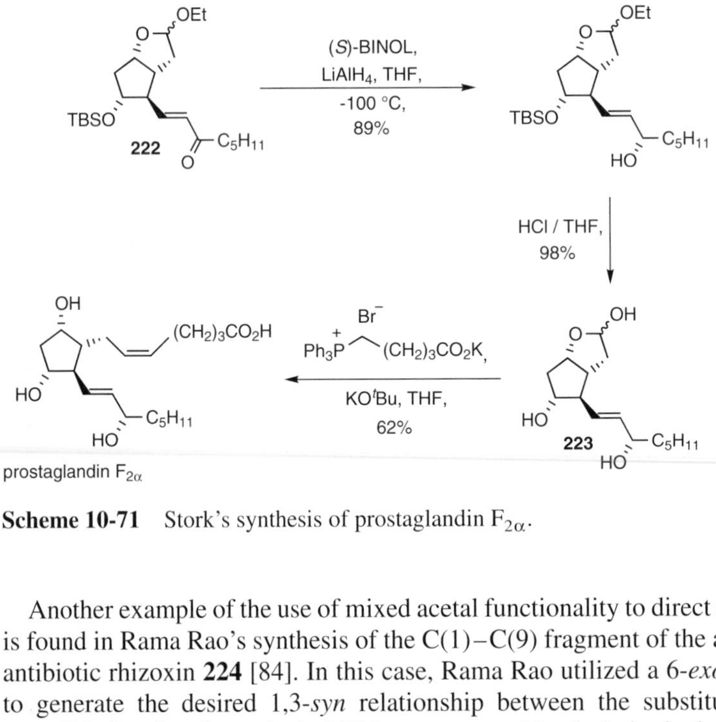

Scheme 10-71   Stork's synthesis of prostaglandin F$_{2\alpha}$.

Another example of the use of mixed acetal functionality to direct a radical cyclization is found in Rama Rao's synthesis of the C(1)–C(9) fragment of the antitumor macrolide antibiotic rhizoxin **224** [84]. In this case, Rama Rao utilized a 6-*exo* radical cyclization to generate the desired 1,3-*syn* relationship between the substituents on the δ-lactone **225**. Starting from alcohol **226**, treatment with ethyl vinyl ether in the presence of NBS produced the radical precursor. Stereoselective radical cyclization produced exclusively the 6-*exo*-trig addition product **227**. The stereochemical outcome of the reaction may be accounted for by cyclization proceeding through a chair-like T. S. with pseudo-equatorially disposed substituents generating the desired *syn* relation. Subsequent manipulations including hydrolysis of the tether and oxidation of the resultant lactol provided the desired lactone product **225** (Scheme 10-72).

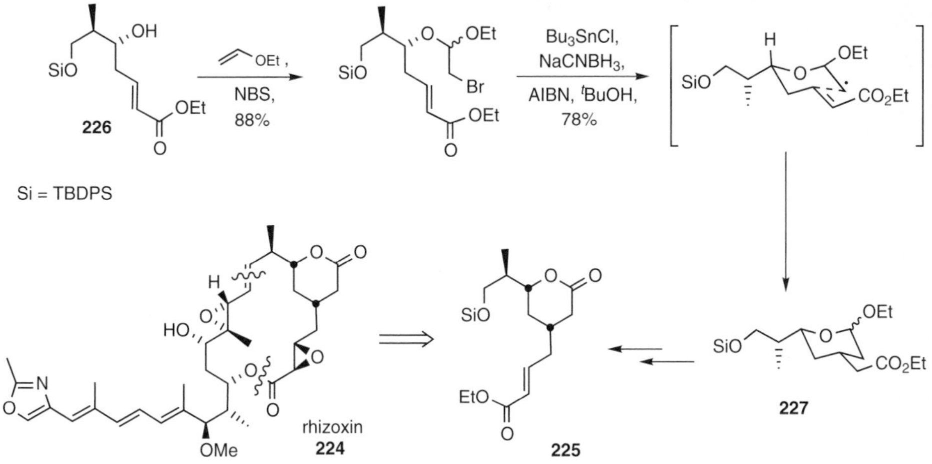

Scheme 10-72   Synthesis of the C(1)–C(9) fragment of rhizoxin in which Rama Rao used an acetal tether to direct a radical cyclization.

10.3.6 Tether-directed Radical Cyclization Approaches to the Synthesis of *C*-Glycosides

Stork first demonstrated the use of the temporary silicon connection in stereospecific *C*-glycoside synthesis [2]. (Phenylethynyl)silyl ethers were prepared from a number of phenylseleno glycosides. Generating the anomeric radical by treatment with Bu₃SnH/AIBN in benzene at reflux effected efficient radical cyclization onto the tethered alkynyl group in a highly stereocontrolled manner. Thus, a *β*-*C*-glycoside could be obtained by tethering the radical acceptor to a *β*-oriented hydroxyl group, while tethering to an *α*-oriented hydroxyl group allowed access to the corresponding *α*-*C*-glycoside (Scheme 10-73).

In all three cases outlined in Scheme 10-73, cyclization proceeds in an *exo* mode. For the 2-substituted glucoside **228**, 5-*exo*-dig cyclization was extremely efficient, providing the desired *α*-*C*-glucoside **229** in an excellent 83% yield after silyl deprotection with TBAF. Attaching the radical acceptor to the 3-hydroxyl group allowed an efficient 6-*exo*-dig cyclization to proceed, generating alcohol **230** in 73% yield. The high yield of this process is particularly noteworthy since the sugar must undergo a conformational change in which the 3-silyloxy group adopts an axial orientation to ensure close proximity to the radical generated at the anomeric center. When the radical acceptor was attached to the 6-hydroxyl group, slow addition of the Bu₃SnH was required to maintain a good yield of the less favorable 7-*exo*-dig cyclization product. Once again the reaction was completely stereospecific, generating the *β*-*C*-glucoside **231**. In all cases examined, an (*E*)-styryl group was produced as the major product, often with excellent stereocontrol.

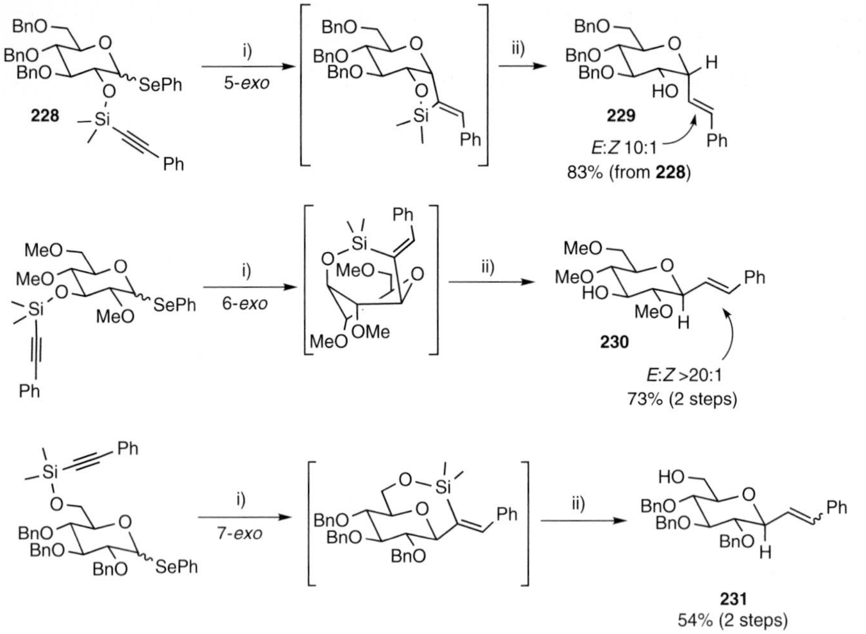

Scheme 10-73 Intramolecular radical cyclization provides a stereospecific route to α- and β-C-glycosides. Reagents and conditions: i) Bu₃SnH, AIBN, PhH, reflux, 0.01 M, (syringe pump used in 6-*exo* cyclization); ii) TBAF, THF.

The introduction of an olefin at the anomeric position is synthetically interesting as it permits, via ozonolytic cleavage, access to the anomeric aldehyde, useful for further manipulations. Furthermore, the cleavage of the tether after cyclization releases one hydroxyl group selectively, which may be used chemoselectively in subsequent reactions. This reaction was also shown to be applicable to furanose sugar derivatives [2].

Sinaÿ and co-workers have extended this type of reaction to the preparation of *C*-disaccharides [85]. These molecules are conformationally similar to normal disaccharides but since the interglycosidic oxygen is replaced with a methylene group, hydrolysis of the glycosidic linkage is prevented. A 9-*endo*-trig radical cyclization of silyl acetal **232**, which had previously been shown to be viable by Hutchinson et al. [78], was used to prepare the methyl α-*C*-disaccharide **233** (Scheme 10-74) [85a]. A stepwise procedure was used to prepare the radical cyclization precursor: an excess of Me₂SiCl₂ was first added to the lithium salt of phenylseleno glycoside **234** at low temperature. After warming to room temperature and removing the volatiles (including the excess Me₂SiCl₂) in vacuo, a solution of the primary alcohol **235** and imidazole in THF was added, allowing formation of the second silyl ether linkage. This procedure allowed isolation of silyl acetal **232** in an impressive 95% yield. Slow addition of Bu₃SnH and AIBN to a solution of **232** maintained a low concentration of tin hydride and provided the cyclization product

233 as a single diastereoisomer in 40% yield after cleavage of the silyl tether (Scheme 10-74). (Spectroscopic analysis of **233** suggested that the *C*-glucosyl moiety deviates from the $^4C_1$ chair form – hence its depiction in the Haworth representation in the scheme.)

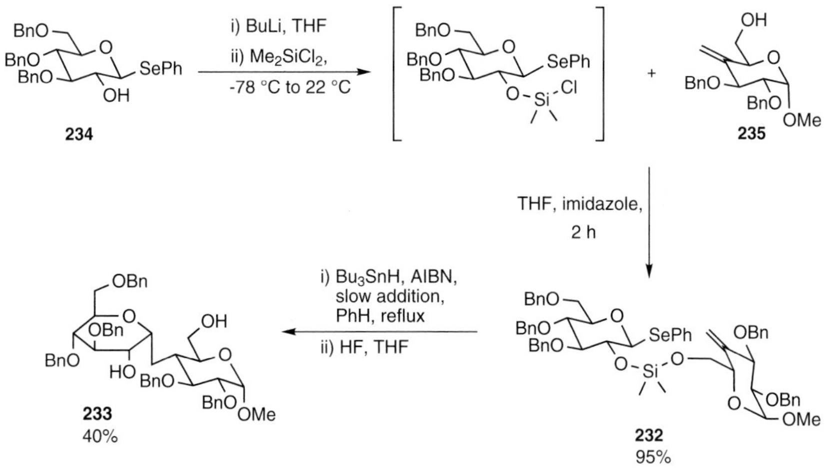

Scheme 10-74   9-*endo*-trig radical cyclization used by Sinaÿ in the synthesis of methyl α-*C*-maltoside **233**.

The same group used a ketal tether as an alternative connecting group in the synthesis of the 1,4-linked *C*-disaccharide **236** [85b]. Tebbe methylenation of acetate **237** provided the corresponding enol ether **238**, which upon treatment with alcohol **235** in the presence of CSA at –40 °C in acetonitrile, furnished linked disaccharide **239** in 81% yield. Subsequent radical cyclization, acidic hydrolysis of the tether and peracetylation provided the D-mannose-containing *C*-disaccharide **236** as the major product in 35% yield from **239** (Scheme 10-75). Cyclization was not completely stereoselective and a small amount of the β-*C*-manno isomer was also isolated (α/β 10:1). This result is in contrast to similar studies on tether-directed β-*O*-mannoside syntheses (vide infra) where a much shorter tether attached to the axial 2-hydroxyl group forces obtention of the desired β-configuration.

The C-glycosylation strategy described above is, in principal, quite general and has been alikened to an enzymatic reaction in which the tether provides the pre-organizing role of an enzyme (a *C*-glycosyl transferase), the seleno glycoside is the glycosyl donor, and the *exo*-methylene sugar the glycosyl acceptor. The Bu₃SnH plays the role of co-enzyme and initiates the reaction.

Although phenylseleno sugars have been most frequently used as radical precursors, their preparation requires the use of odorous phenylselenol. Furthermore, formation of the anomeric radical usually requires Bu₃SnH, which is also unpleasant to work with. An

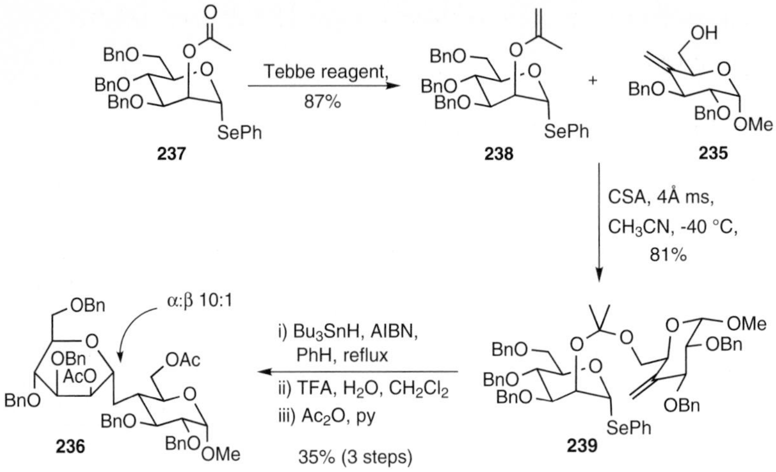

Scheme 10-75   A ketal tether was used in a 9-*endo*-trig radical cyclization to prepare *C*-disaccharide **236**.

alternative and more environmentally friendly approach to the same goal, of generating a radical at the anomeric position of a sugar, is to use single-electron transfer from samarium diiodide to anomeric sulfones. Anomeric phenyl sulfones have been investigated as radical precursors by Sinaÿ and co-workers and shown to participate in efficient radical cyclizations similar to those described above [85c]. One disadvantage of using phenylsulfones however, is the need for the carcinogenic cosolvent HMPA to increase the reducing potential of the SmI$_2$. Fortunately Skrydstrup, Beau, and co-workers have shown that the more readily reduced 2-pyridyl sulfones circumvent the need for adding HMPA – reaction proceeds in THF alone [86]. Using this methodology, they prepared methyl α-*C*-isomaltoside **240** in 48% yield from the tethered precursor **241** (Scheme 10-76). In this example, a short tether length ensured exclusive formation of the α-*C*-glyco-side.

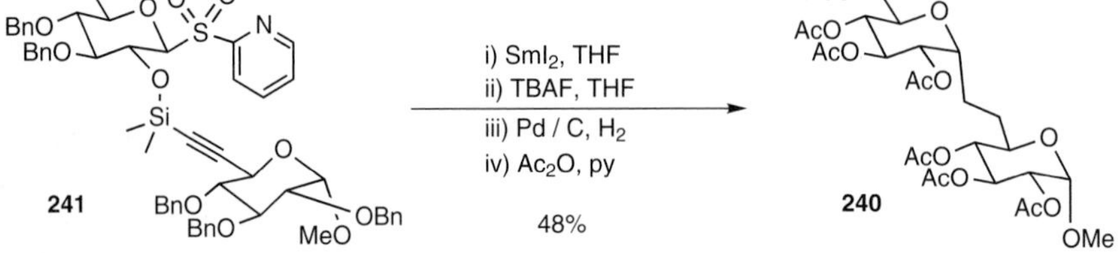

Scheme 10-76   A 5-*exo*-dig cyclization initiated by the action of SmI$_2$ on 2-pyridyl sulfone **241** allowed the preparation of α-*C*-isomaltoside **240**.

10.4 Use of the Temporary Connection in O-Glycosylation and Nucleosidation

10.4.1 Tether-Directed O-Glycosylation

The last decade has seen a resurgence in oligosaccharide chemistry. This has been fueled by an increasing realization of the important role oligosaccharides play in a wide variety of biological processes ranging from viral and bacterial adhesion to cells, protein trafficking, and intercellular recognition [87]. Access to complex oligosaccharides from natural sources is possible, although usually only in small quantities. One method of providing larger quantities of these important molecules is through chemical synthesis. Although the field of carbohydrate chemistry has been intensively studied, regio- and stereoselective glycosylation remains a difficult problem which usually relies on elaborate protecting group strategies for controlling regioselectivity (i.e., reaction at the appropriate hydroxyl group) and the use of an array of activation strategies for performing the actual glycosylation reaction, which all too frequently give a mixture of α- and β-anomers requiring tedious separation. Stereocontrolled glycosylation where either α- or β-anomer linkage is obtained in a predictable manner remains an important area of research [88].

Synthesis of the β-mannopyranoside linkage is notoriously difficult since the axial hydroxyl group in the 2-position favors the formation of the α-anomer on both steric and electronic grounds. This problem is compounded by the fact that it is an important glycosyl linkage found in the pentasaccharide core **242** of all asparagine-linked glycoproteins (Figure 10-6); efficient and stereoselective routes to the synthesis of this linkage are therefore highly desirable.

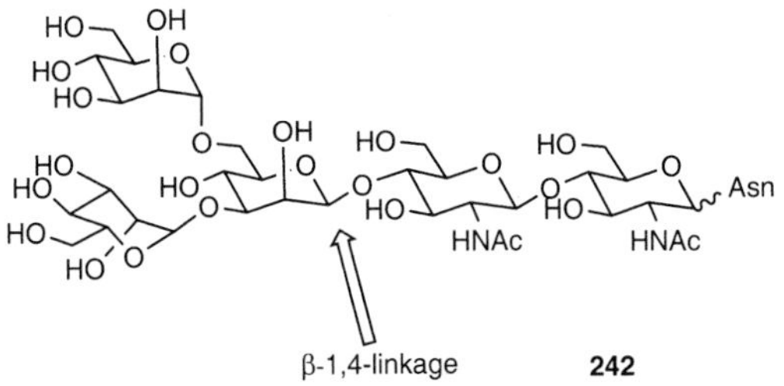

β-1,4-linkage   **242**

Figure 10-6   The pentasaccharide core of all asparagine glycoproteins contains a β-mannoside linkage.

The groups of Hindsgaul and Stork addressed this problem using a temporary connecting group to force intramolecular delivery of the glycosyl acceptor to the β-face of the mannosyl donor. Barresi and Hindsgaul chose to use an acetal tether to connect the reacting partners [89a]. For example, formation of the mixed acetal **243** was achieved by reacting enol ether **244** with glucosyl acceptor **245** in the presence of catalytic pTSA. Subsequent glycosylation was initiated by NIS, affording exclusively the desired β-

linked disaccharide **246** in 42% yield. It was later found that the addition of 4-methyl-2,6-di-*tert*-butylpyridine (4-Me-DTBP) improved the yield further, especially in the case of less reactive secondary alcohols, presumably by preventing decomposition of the acetal tether prior to the glycosylation step [89b]. With these modified conditions, disaccharide **246** could be obtained in a very respectable 77% yield, again as a single stereoisomer (Scheme 10-77).

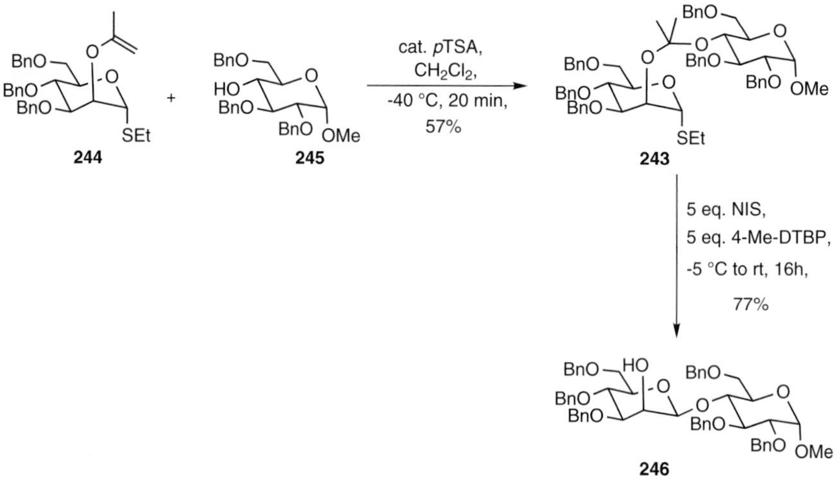

Scheme 10-77   Tethering the glucosyl acceptor through an acetal to the 2-OH of the mannosyl donor affords exclusively the β-1,4-linked mannoglucoside.

The proposed mechanism of reaction is outlined in Scheme 10-78. Activation of the anomeric group is followed by intramolecular delivery of the glycosyl acceptor from the β-face through a five-membered ring affording intermediate **247**, which upon aqueous workup provides the desired product. The intramolecularity of the reaction combined with delivery through a small ring ensures the excellent levels of stereocontrol.

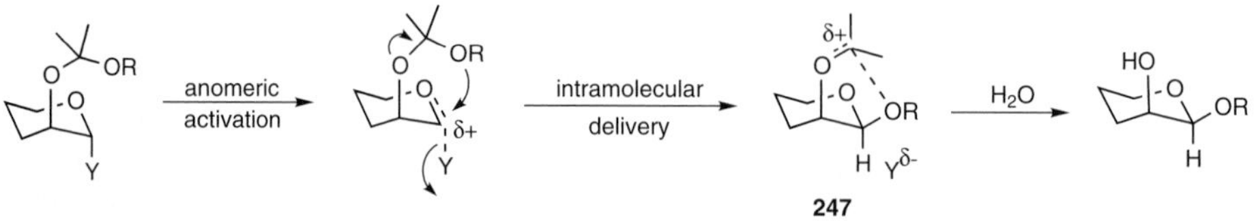

Scheme 10-78   Proposed mechanism for intramolecular glycosylation from a tethered acceptor.

The concerted nature of the reaction was further probed by performing the glycosylation in the presence of 1 equiv of methanol [89a, c]. In the case of mixed acetal **248**, the yield of the β-1,6-linked disaccharide **249** was found to be unchanged by the presence of MeOH. However, with acetal **243**, the addition of 1 equiv of MeOH resulted in a decrease in the yield of disaccharide **246** which reflects the reduced reactivity of secondary alcohols in sugars. Transfer of MeOH was responsible for the only other product isolated. However, the presence of only the β-methyl mannopyranoside **250** suggested that this too was formed by intramolecular delivery. Transacetalization with MeOH to form **251** would account for the observance of exclusively this product (Scheme 10-79).

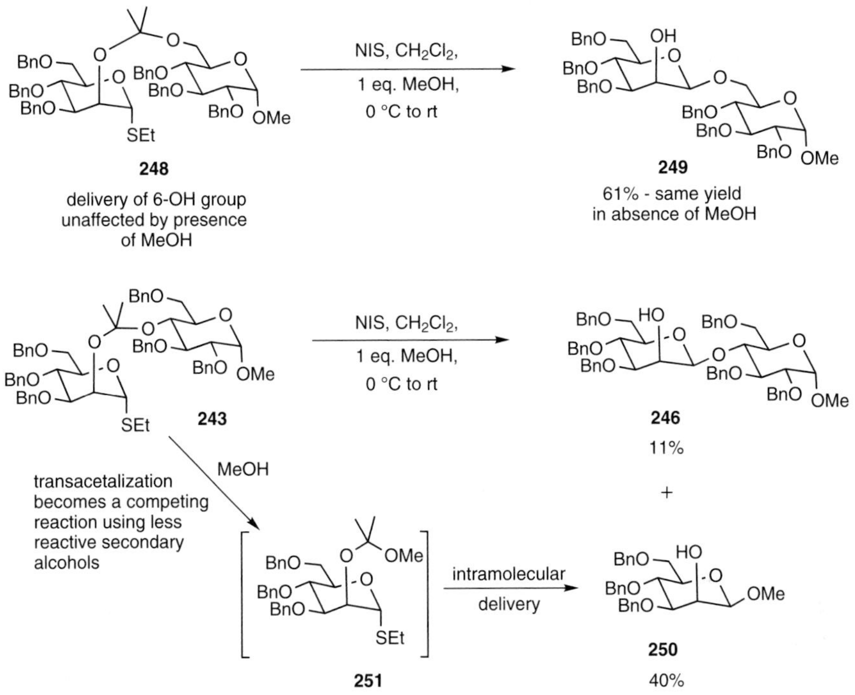

Scheme 10-79   The addition of MeOH did not affect the stereospecificity of the glycosylation.

Although the reaction clearly provides a solution to the synthesis of β-mannopyranosides, its execution is non-trivial and appreciable optimization is required to achieve good results [89b, c]. This is especially true when preparing more complex oligosaccharides or when low-reactivity secondary hydroxyl sugars are used as glycosyl acceptors. These reactions then require much longer reaction times, allowing product-consuming side reactions to compete with the desired process.

Stork and co-workers investigated silyl acetals as the tether-directing group for preparing the difficult β-mannopyranoside linkage [90]. The desired mixed silyl acetal precursor was initially prepared in a stepwise fashion analogous to that described earlier, by first reacting the glycosyl donor with Me$_2$SiCl$_2$ to form the chlorodimethylsilyl ether and then adding the glycosyl acceptor to complete formation of the silyl acetal. Oxidation of the phenylthio group with *m*CPBA provided the sulfoxide which was used in the glycosylation reaction according to the procedure of Kahne et al. (Scheme 10-80) [91]. It was later found that if coupling with a secondary hydroxyl group was desired, the silyl acetal tether could be formed directly by simply mixing an equimolar mixture of the preformed sulfoxide donor and glycosyl acceptor with Me$_2$SiCl$_2$, thus circumventing the need for manipulating the sensitive silyl chloride intermediate [90b].

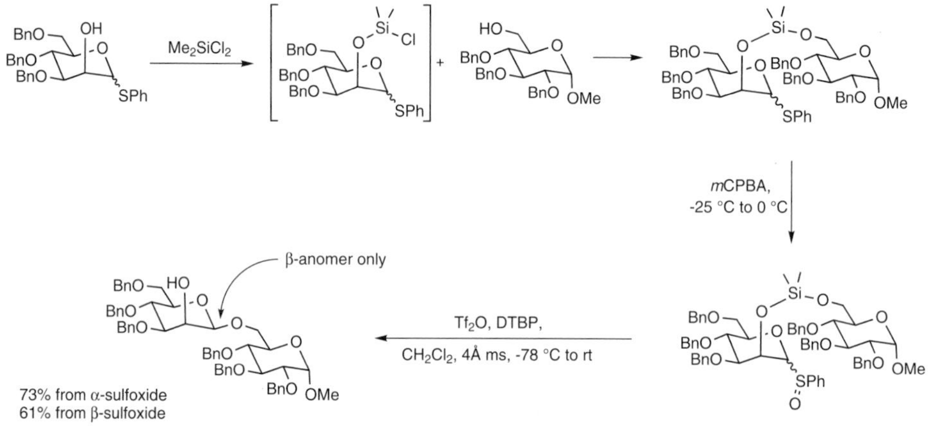

Scheme 10-80   Linking the mannosyl donor with the glucosyl acceptor via a silyl acetal allows stereospecific β-mannoside formation.

The outcome of the glycosylation is in accordance with the results observed by Barresi and Hindsgaul and is suggestive of intramolecular delivery of the glycosyl donor from the β-face. Both α- and β-sulfoxides give the same result – the stereochemical outcome is controlled only by the axial orientation of the 2-hydroxyl tether. Using a silyl acetal tether, it was possible to form *O*-glucosyl β-mannopyranosides at the 2-, 3-, and 6-positions in good to excellent overall yields [90]. Formation of the 4-*O*-glucosyl β-mannopyranoside **246** by intramolecular delivery, however, proved to be much more problematic and proceeded in a low yield of 12% (this is in contrast to the analogous acetal tether; vide supra) [90b]. The major product from this reaction was found to be the 1,6-β-disaccharide **252** in which the silyl acetal bridge remained intact and debenzylation of the 6-hydroxyl group had occurred allowing formation of the observed 1,6-linkage (4% of the α-anomer was also isolated) (Scheme 10-81). A similar observation has been noted by Bols [92d].

Bols has also investigated the use of silyl acetals in intramolecular O-glycosylation [92]. In early work, he showed that α-glucosides could be readily prepared stereospecifically by attaching the glycosyl acceptor through a dimethylsilyl acetal to the 2-position of phenyl 3,4,6-tri-*O*-acetyl-1-thio-α-glucopyranoside [92a, c]. Activation of the anomeric group using NIS in the presence of TfOH effected rapid conversion to the corresponding α-glucoside (Scheme 10-82). Several observations were suggestive of intramolecular delivery of the glycosyl acceptor. Stereocontrol was in all cases excellent, with exclusive formation of the α-anomer – analogous intermolecular reactions gave the expected mixture of α- and β-anomers. The yields and rates of reaction were also not dependent on the steric hindrance of the aglycon; thus octanol and *tert*-butanol were both transferred with similar efficiency (Scheme 10-82) [92a, c]. Finally, in a competition experiment between intramolecular transfer of *tert*-butanol and intermolecular reaction with the TMS ether of octanol, the major product was the α-glucoside resulting from addition of the more sterically demanding alcohol.

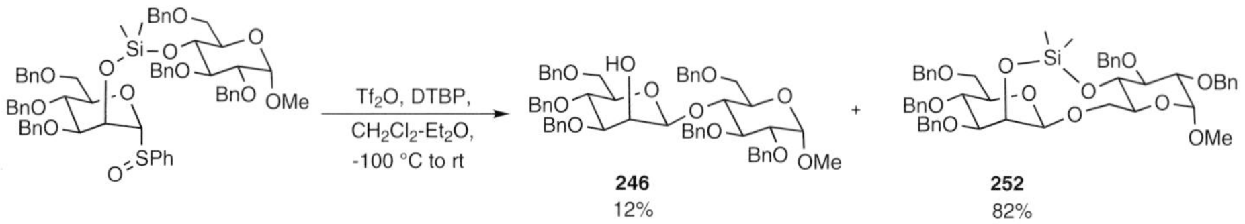

Scheme 10-81   Formation of the β-1,4-linkage using a silyl acetal tether proved problematic.

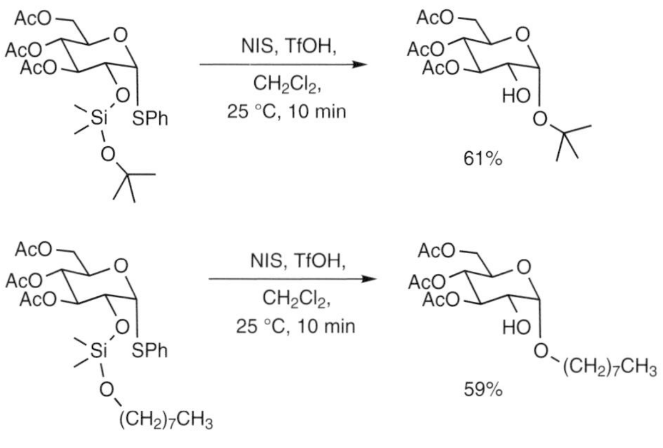

Scheme 10-82   α-Glucosidation is possible through intramolecular transfer.

Silyl acetals were also employed in the preparation of disaccharides containing α-gluco and α-galacto linkages [92b, d]. The conditions used in the glycosylation were found to be important in achieving the desired coupling products (Scheme 10-83). When silyl acetal **253** was treated with NIS in the presence of TfOH, only 19% of the desired α-glucoside **254** was isolated; the major product resulted from cleavage of the tether. Clearly the presence of TfOH, required to effect reaction on such a deactivated glycosyl donor, was cleaving the tether more rapidly. Fortunately it was found that by changing the solvent from CH_2Cl_2 to the more polar nitromethane, heating at 100 °C, allowed glycosylation to proceed without the need for acid and provided **254** in very good yield and as a single anomer [92b, d].

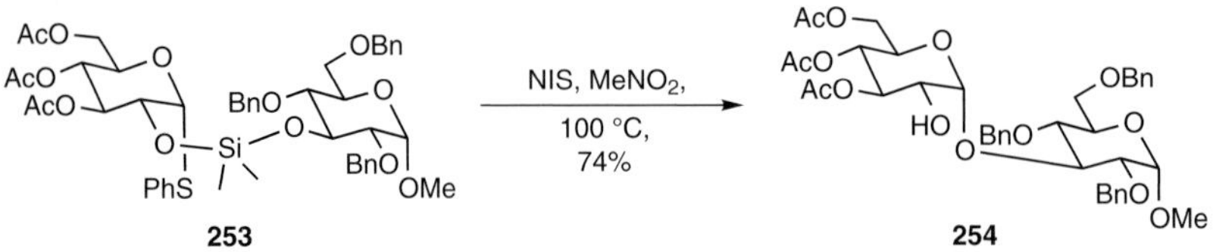

Scheme 10-83   The use of nitromethane circumvents the need for TfOH and allows the formation of disaccharide **254**.

In previous studies, including those of Stork and Hindsgaul [89, 90], the glycosyl acceptor was tethered to the 2-hydroxyl group of the donor. The short tether length (three atoms between reacting centers) and intramolecularity of the reaction ensures complete stereocontrol. Bols and Hansen have also investigated the use of more remote hydroxyl centers for tethering a simple aglycon, octan-1-ol, to the glycosyl donor [93]. Thioglucosides **255**, **256**, and **257** were prepared uneventfully and subjected to glycosylation using NIS as activator in nitromethane. The results are outlined in Scheme 10-84. When the aglycon was attached to the 3-hydroxyl group, reaction was not stereoselective, affording a 1:4 mixture of α/β anomers in a low yield of 22%. Tethering to the 4-position proved more successful, providing the α-glucoside **258** exclusively, in 45% yield. The stereoselectivity of this reaction is interesting since it requires a conformational change in which the silyl linker assumes an axial position to allow delivery of the aglycon to the anomeric position. Finally, attaching the aglycon to the 6-position provided the β-glucoside **259** with complete stereocontrol although the major product from this reaction was tribenzyl-levoglucosan **260** (Scheme 10-84). This very brief study demonstrates that more remote tethers can be used for stereoselective glycosylation although yields are usually poorer and levels of stereocontrol not always good.

The use of the so-called aglycon delivery method provides the most reliable method of forming the β-mannoside linkage. However, the tethers discussed so far are fairly limited in their compatibility with other standard manipulations carried out during oligosaccharide synthesis. Ogawa and co-workers have recently introduced an alternative which possesses a number of advantages over the previously investigated silyl acetal and dimethylacetal tethers [94]. They realized that a *p*-methoxybenzyl (PMB) ether, a commonly used protecting group in oligosaccharide chemistry, could be used as a latent tethering unit. This protecting group is readily cleaved under oxidative conditions by treatment with DDQ or CAN. If a glycosyl acceptor containing a free hydroxyl group is present during the "deprotection" reaction, then the intermediate oxonium species may be intercepted, resulting in the formation of an acetal tether (Scheme 10-85). Thus, treat-

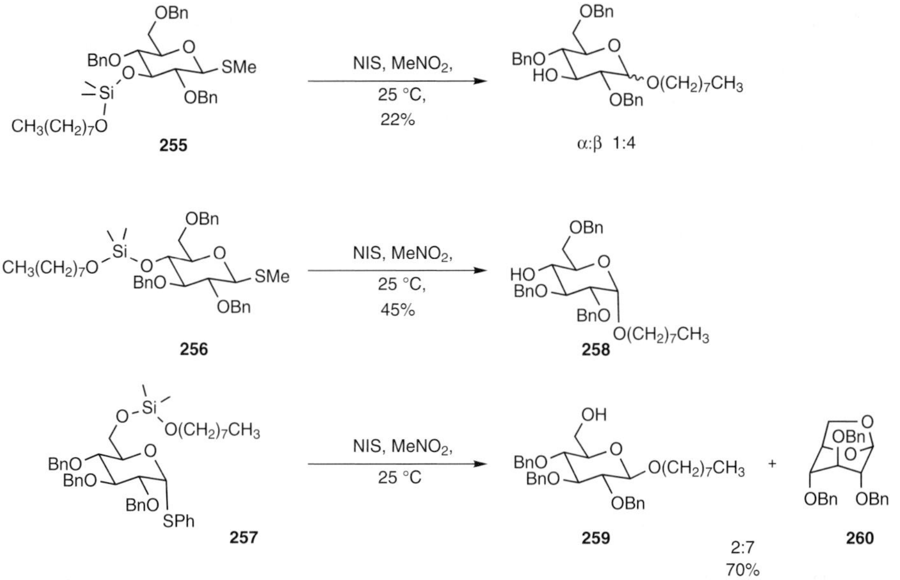

Scheme 10-84 Remote tethers are less useful for stereoselective intramolecular glycosylation.

ing the 2-PMB-protected mannosyl donor **261** with DDQ in the presence of molecular sieves and the glucosamine acceptor **262**, tethered intermediate **263** could be formed. Without the need for isolation, activation of the anomeric linkage with silver tri-flate/tin(II) chloride provided exclusively the β-1,4-mannopyranoside **264** in 40% yield (Scheme 10-85). A variety of solvents were investigated, and of those surveyed only toluene proved to be unsuitable; reaction in this case was sluggish, probably owing to poor stabilization of the cationic intermediate, and gave a mixture of products.

This methodology has been applied to a number of important oligosaccharides, in-cluding the core pentasaccharide structure of asparagine-linked glycoproteins shown in Figure 10-6 [94b, f]. Glycosylation is not restricted to the use of anomeric fluorides as glycosyl donors; thioglycosides have also been used [94b, c]. The β-mannoside reaction has recently been optimized and, with a suitable choice of protecting groups, can be per-formed in yields around 80% [94e].

The application of solid supports in oligosaccharide synthesis has received much attention in recent years [95], one of the main reasons for this being the facilitated pur-ification of the products. Ito and Ogawa have shown that their intramolecular aglycon delivery strategy can be applied to a polymer-supported system in a novel approach to "clean" glycosylation [94d]. A modified PMB ether protecting group was used to attach a mannosyl thioglycoside to a poly(ethylene glycol) (PEG) support. Treatment of **265** with DDQ in the presence of an excess of the glycosyl acceptor **266** allowed tether for-mation as before. The tethered intermediate now attached to PEG was then precipitated with *tert*-butyl methyl ether, and washing allowed facile purification and recovery of the

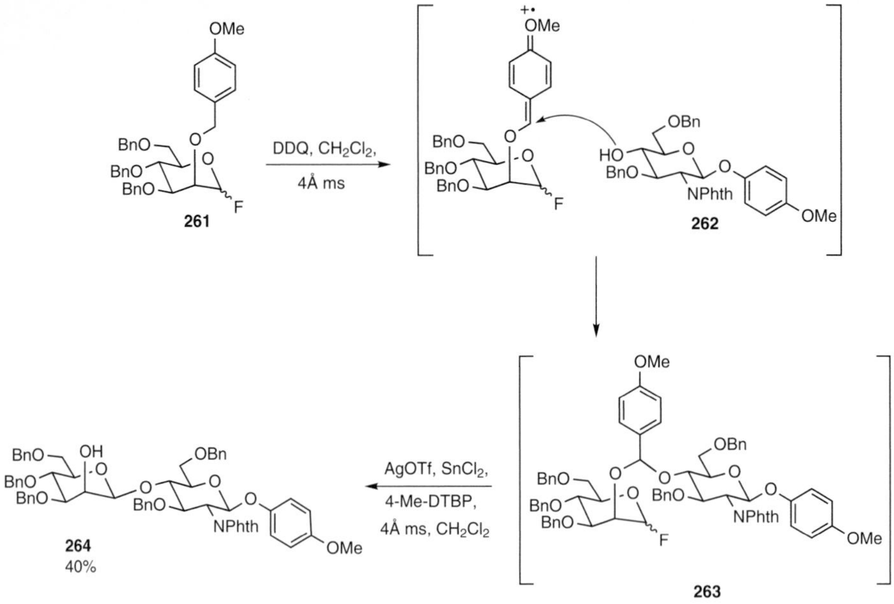

Scheme 10-85 β-Mannosides may be produced by intramolecular aglycon delivery using a PMB acetal tether.

excess glycosyl acceptor. Redissolution in 1,2-dichloroethane and activation of the thioglycoside with MeOTf/MeSSMe in the presence of 4-Me-DTBP effected glycosylation. By using a polymer-bound tether, only the product resulting from tether cleavage – i.e., the desired disaccharide **267** – would enter the solution phase, leaving by-products resulting from hydrolysis or elimination of the thio group still attached to the PEG support. Simple aqueous workup and facile chromatographic purification provided the desired disaccharide **267** in good yield (Scheme 10-86).

Fructofuranosides are widespread in nature and, with few exceptions, are β-linked. The use of the PMB intramolecular aglycon delivery strategy developed by Ogawa and Ito has also been successfully applied to the stereospecific synthesis of β-D-fructofuranosides by Krog-Jensen and Oscarson [96]. They showed that preparation of the PMB acetal was possible with the PMB ether attached to either the glycosyl donor **268** or acceptor **269**. Further improvements to this step were made by adding the DDQ solution dropwise to a solution of the PMB ether and free alcohol over 2 h and by quenching the reaction by the addition of Et₃N. Of a range of different activating agents investigated, dimethyl(methylthio)sulfonium trifluoromethanesulfonate (DMTST), iodonium collidine perchlorate (IDCP) and iodonium collidine triflate (IDCT) all proved superior to the conditions originally reported by Ogawa. NIS was also useful, although in this case the relatively nucleophilic succinimide anion intercepted the intermediate cationic intermediate and provided the corresponding *N,O*-acetal product **270**, albeit in good yield and with complete β-selectivity at the new anomeric linkage (Scheme 10-87).

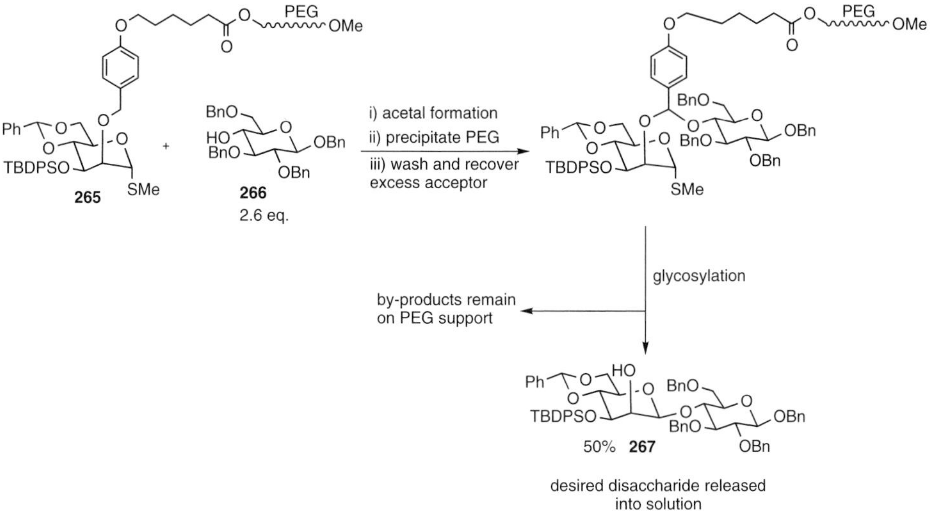

Scheme 10-86   Use of a polymer-supported glycosyl donor provides a "clean" approach to β-mannosylation.

The use of bis-ester tethers in remote intramolecular glycosylation has been investigated by a number of groups. In contrast to the short tethers used by Stork, Hindsgaul, and Ogawa, these more remote tethers provide a much less predictable outcome and as a result they remain less synthetically useful. Another difference between the short tethers and remote bis-ester tethers is that the reacting alcohol functionality is not masked in the tether but is already free to react as soon as the anomeric center is activated.

Valverde and co-workers have investigated the use of bis-ester tethers derived from phthalic and succinic acids in remote glycosylation reactions [97]. Formation of the tether was achieved in a stepwise fashion from phthalic or succinic anhydride. For example, reaction of phenyl 2,3,4-*O*-acetyl-1-thio-β-D-glucopyranoside with phthalic anhydride, followed by treatment of the intermediate carboxylic acid with SOCl$_2$, generated the acid chloride **271**. This was then reacted with glycosyl acceptor **272** with complete regioselectivity at the 2-OH position using a known procedure [98], providing the tethered glycosylation precursor **273**. The presence of two free hydroxyl groups in the acceptor sugar now presents the further complication of regiocontrol in addition to that of stereocontrol at the anomeric linkage. Reaction at the 3-OH position would provide a 13-membered ring while a 14-membered macrocycle would result from glycosylation using the 4-OH group as the nucleophile. It was hoped that geometrical constraints imposed in the different reacting cyclic T. S.s would control both the regio- and the stereochemical outcome of the reaction. Activation of the thioglycoside **273** with NIS/TfOH at room temperature effected rapid reaction, providing a single stereo- and regioisomer **274** in excellent yield (Scheme 10-88). The use of a more flexible succinyl tether gave the same result, albeit in slightly diminished yield; reaction occurred exclusively at the 3-hydrox-

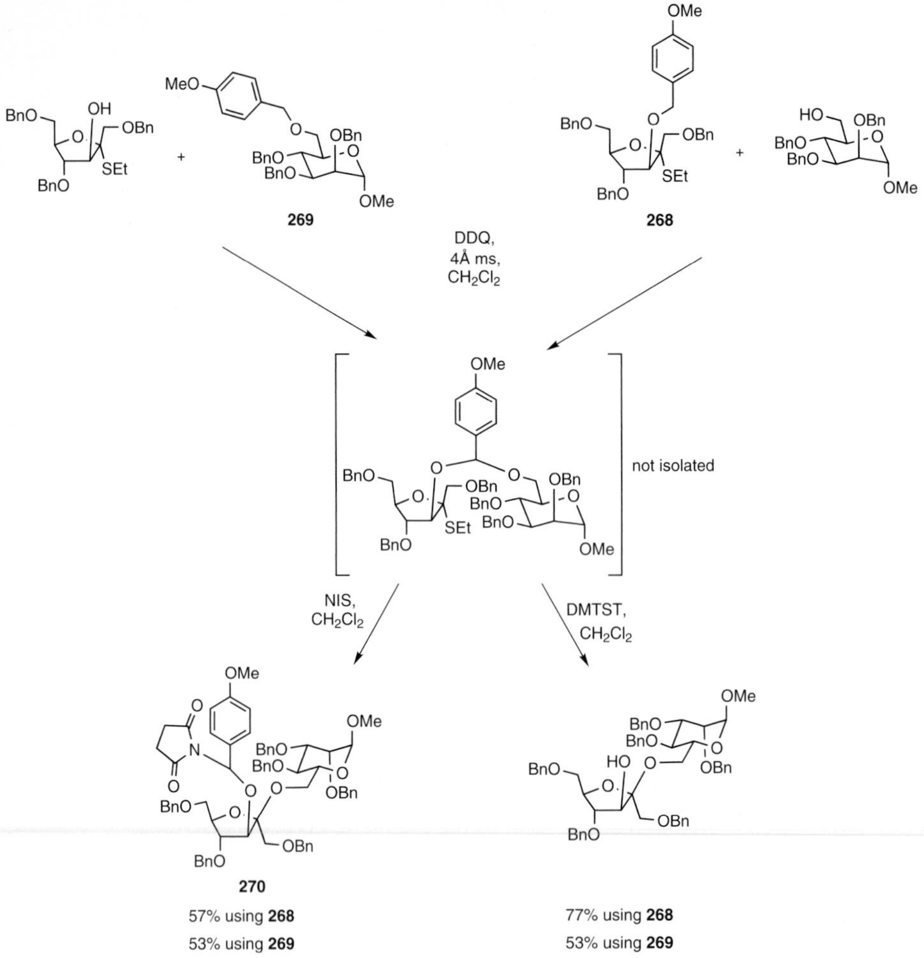

Scheme 10-87   A PMB acetal tether can be used to prepare β-fructofuranosides.

yl position providing the β-anomer **275** (Scheme 10-88). Changing the protecting groups on the glycosyl donor from acetates to ethers, which cannot participate in the reaction (methyl or benzyl), provided the same regiochemical outcome but stereocontrol was reduced, with up to 15% of the α-anomer also being isolated. Nevertheless, the fact that the major product was the 3-β-glycoside suggests that the tether remains instrumental in governing the outcome of the reaction.

By changing the anchoring sites of the bis-ester template, it is possible to reverse completely the regioselectivity of the coupling [97b]. For example, when a phthalyl tether was used to link the 2-OH of a glucosyl acceptor to the 6-OH of a mannosyl donor to

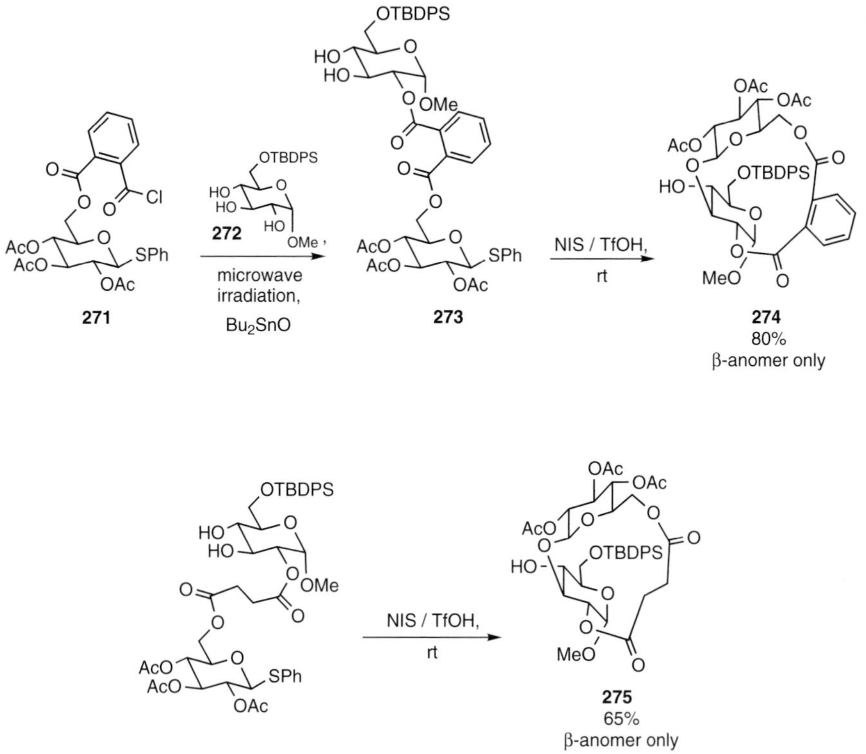

Scheme 10-88   Use of phthalyl and succinyl ester tethers provides another approach to regio- and stereocontrolled glycosylation.

form **276**, glycosylation proceeded with complete regioselectivity (and stereoselective-ly), with reaction occurring exclusively at the 3-OH. Alternatively, if the glucosyl accep-tor was tethered through the 6-OH affording precursor **277**, treatment with NIS/TfOH provided exclusively the regioisomeric 4'-α glycoside **278** (Scheme 10-89).

Ziegler and co-workers have also investigated the use of phthalyl, succinyl, and mal-onyl ester tethers in a wide variety of intramolecular glycosylation reactions although they have only concentrated on controlling the stereochemical outcome of the glycosy-lation reaction [99]. They used this approach to prepare β-L-rhamnoside linkages [99a], which, like β-D-mannosides, are notoriously difficult to prepare owing to the presence of an axially oriented oxygen at C-(2). A number of ethylthio and phenylthio α-rhamnosyl donors were linked through the 2-OH to the 3-OH of a glucosyl acceptor by a phthalyl, succinyl, or malonyl spacer group. The prearranged saccharides were then treated with a variety of activators, in a range of solvents and at different temperatures, affording good to excellent yields of the glycosylation product as a mixture of α- and β-anomers. The best results were obtained with the rhamnosyl derivative **279**, which uses a flexible suc-

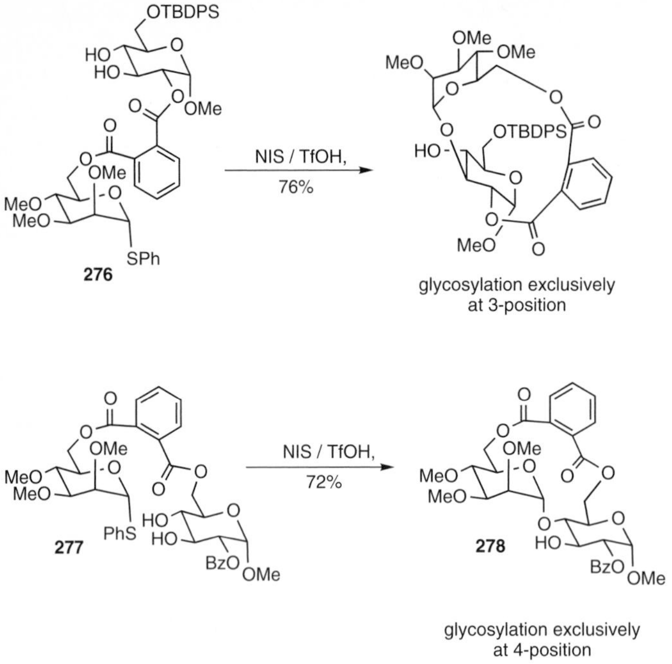

Scheme 10-89   Different anchoring sites can be used to vary the regioselective outcome in cyclo-glycosylation.

cinyl spacer: activation of the anomeric position with NIS/TMSOTf in acetonitrile at −30 °C for 10 min provided a 16 : 84 mixture of α/β anomers **280** and **281** in excellent overall yield (Scheme 10-90).

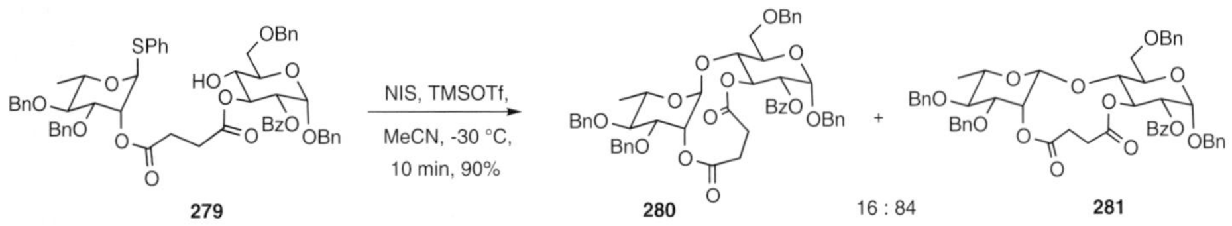

Scheme 10-90   Careful choice of tether and reaction conditions allows the formation of a β-L-rhamnoside.

The presence of a 2-acyl group in the rhamnosyl donor would be expected to give rise to high α-selectivity in the product. The solvent is also very important in the reaction outcome – only when MeCN is used does the reaction show any signs of useful stereo-control in favor of the desired product. Force-field calculations (using Insight II) on the two products revealed that the desired β-anomer was around 10 kcal mol^{-1} more stable than the α-anomer analog. This is, at least in part, a result of a more favorable staggered conformation of the methylene groups in the succinyl tether found in the β-anomer.

This and other investigations into the synthesis of difficult glycosidic linkages illustrate that the use of a bis-ester tether offers distinct advantages over the corresponding intermolecular reaction in terms of improved levels of stereocontrol. However, to obtain the best results requires appreciable optimization involving variation of the length and type of tether, in addition to the anchoring sites to the reacting sugars – very small conformational changes in a large cyclic T. S. can have a profound effect on the stereochemical outcome of the reaction. Although the reaction outcome is clearly influenced by the nature of the bis-ester bridging group, other effects such as solvent, method of activation, and temperature are crucial in controlling the reaction.

Takahashi and co-workers have used molecular mechanics calculations in designing a suitable tether for a synthesis of the branched trisaccharide **282** [100]. It was envisaged that appropriate tethering of the glycosyl fluoride donor to the disaccharide acceptor, which has three sites for reaction, would impart a high degree of regio- and stereocontrol in the key glycosylation reaction. Succinyl and phthalyl tethers attached in both cases to the 6-position of the disaccharide acceptor and to the 2- and 3-positions of the fluoride glycosyl donor provided four possible glycosylation precursors. Monte Carlo (MC) conformational searches were performed on all isomers generating a number of low-energy conformations for each isomer. These were further minimized using the AMBER forcefield implemented in MacroModel. Subsequent analysis of a Boltzmann distribution of all these conformers at 25 °C revealed that the anomeric carbon of the glycosyl donor was close to the 4-OH group of the glycosyl acceptor in 90% of the total conformer population. In particular, isomer **283** seemed to possess the most favorable arrangement (in the ground state) with a distance of 3.59 Å between the C(4)-oxygen atom and the reacting anomeric carbon atom. Similar conformational analysis and energy minimization of the cyclized products revealed that this precursor linked by a phthalyl ester between the 2-OH position of the donor and the 6-position of the acceptor also provided the most stable product. With these results in hand, the isomer predicted to be most favorable for cyclization was prepared and subjected to glycosylation. The expected trisaccharide **284**, where glycosylation had indeed proceeded on the 4-OH group affording a β-linkage, was isolated in 37% yield. The analogous intermolecular reaction was also investigated to provide a measure of the utility of the tether. This produced a mixture of trisaccharides in poor yield, where reaction had occurred at the 2-OH and 3-OH positions only. This clearly demonstrates that intramolecularization using a tether not only improves the yield, but also ensures that reaction occurs exclusively at the least reactive 4-OH group (Scheme 10-91).

Huchel and Schmidt have used a remote glycosylation strategy to prepare the cellobioside **285** [101]. In this case, they utilized a bis-ether linkage prepared from α,α'-dibromo-*m*-xylene to pre-organize the glycosyl donor and acceptor. It was found that a linkage through the 6-position of the donor and the 3-position of the acceptor gave the best result. Treatment of **286** with NIS/TMSOTf provided the 14-membered macrocycle disaccharide **287** in excellent yield as a single anomer (Scheme 10-92). Global debenzylation followed by peracetylation provided the desired cellobioside **285**.

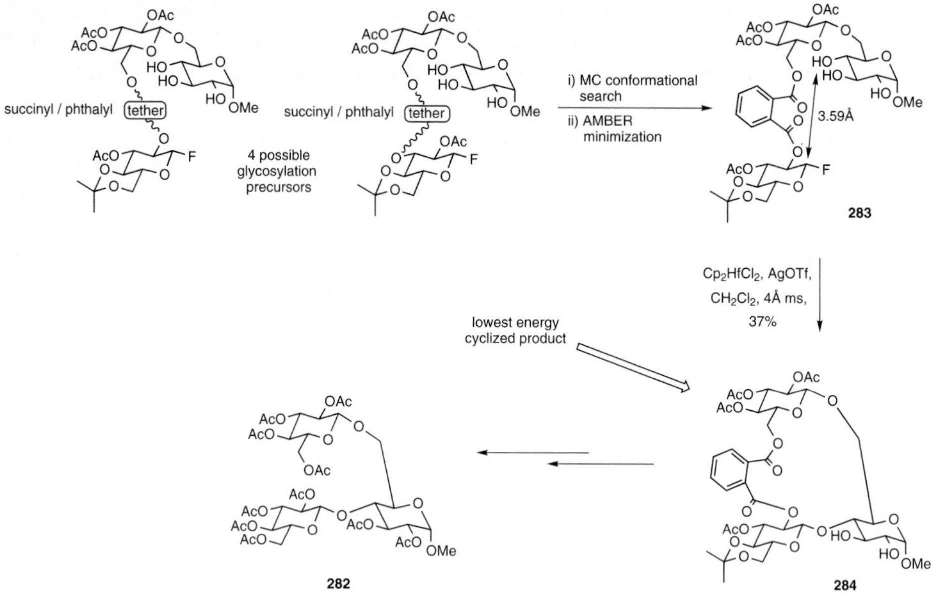

Scheme 10-91   Molecular mechanics can be used to design tethers for remote glycosylation.

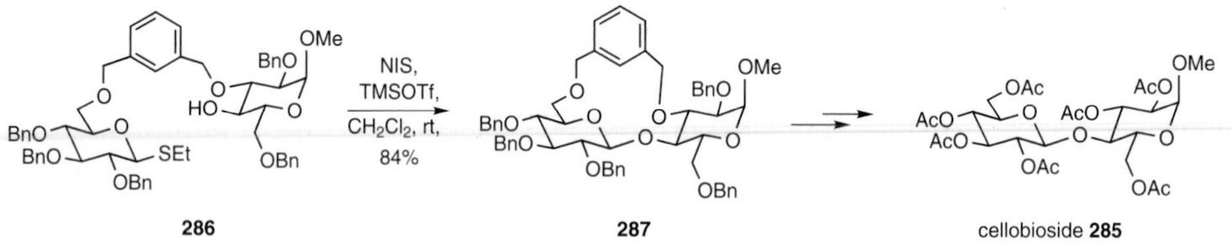

Scheme 10-92   A bis-ether tether was used in a stereoselective synthesis of cellobioside **285**.

Remote glycosylation strategies using bis-ester and ether tethering strategies can provide useful levels of regio- and stereocontrol which are not always possible using analogous intermolecular reactions. Levels of stereocontrol are not always as high as when the shorter tethers discussed earlier are used. Nevertheless, the stereoselective synthesis of β-L-rhamnoside and β-D-mannoside linkages is illustrative of the potential of this approach. The tethered precursors are relatively easy to prepare, stable to the glycosylation conditions, and readily varied, allowing fine tuning of the reactive conformation for improving or varying selectivity. However, the major drawback with this approach is the difficulty in predicting the outcome of the reaction – the much larger cyclic T. S.s require fairly complicated molecular mechanics computational studies for understanding favored reactive conformations.

All the examples of tether-directed intramolecular glycosylation discussed so far have involved connecting the two reacting partners through the hydroxyl substituents around the sugar rings. An alternative site of attachment for the glycosyl acceptor is through the anomeric leaving group of the donor [102–104]. This method of glycosylation has received less attention but is potentially more versatile. Since the glycosyl donor leaving group is also the glycosyl acceptor, the need for preparing potentially unstable donor precursors is circumvented. Furthermore, the protecting group strategy is less affected by the need to form a tether.

Ideally, the reaction would proceed stereospecifically through an S_Ni-type mechanism, where the configuration of the starting material would dictate that in the product glycoside. The concept is outlined in Scheme 10-93.

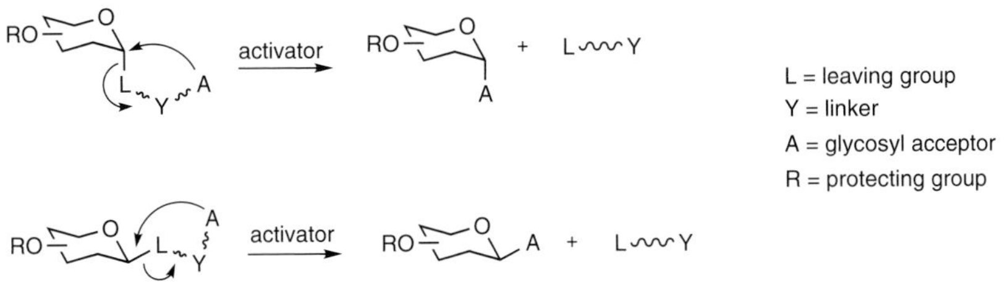

L = leaving group
Y = linker
A = glycosyl acceptor
R = protecting group

Scheme 10-93   Leaving group-based intramolecular glycoside bond formation is an attractive route to stereoselective glycosylation.

Behrendt and Schmidt provided an early example of this approach [102]. Reaction of *O*-glycosyl trichloroacetimidates **288** and **289** with cyclohexanecarboxylic acid proceeded with inversion of configuration, providing the corresponding anomeric carboxylic esters **290** and **291**, respectively. Subsequent aldol reaction with benzaldehyde provided the β-hydroxy carboxylates **292** and **293** as a mixture of diastereoisomers. When β-configured **292** was reacted with the 6-*O*-triflate of methyl 2,3,4-tri-*O*-benzyl-α-D-glucopyranoside in the presence of NaH and 15-crown-5 at 0 °C, the disaccharide **294** was isolated in 68% yield as a 10:1 mixture of β/α anomers. Reaction of the α-configured precursor **293** under similar conditions but at −30 °C, provided **295** in 81% yield as a 1:2 mixture of β/α anomers in addition to small quantities of products arising from retroaldolization and O-alkylation (Scheme 10-94).

The isolation of a β-lactone product **296** suggests a possible reaction mechanism (Scheme 10-95). Reaction of NaH/15-crown-5 with the aldol product generates the corresponding alkoxide **297**. As a result of steric crowding, this is not directly alkylated – instead, cyclization onto the ester functionality and trapping of this alkoxide provides an intermediate orthoester **298**. Relief of strain by 1,3-glycosyl transfer then furnishes the observed products (Scheme 10-95). It is also interesting to note that the stereoselectivity in the reaction depends on the configuration of the starting material, as is desirable in this type of transformation: thus the β-carboxylate precursor provides the β-anomer prefe-

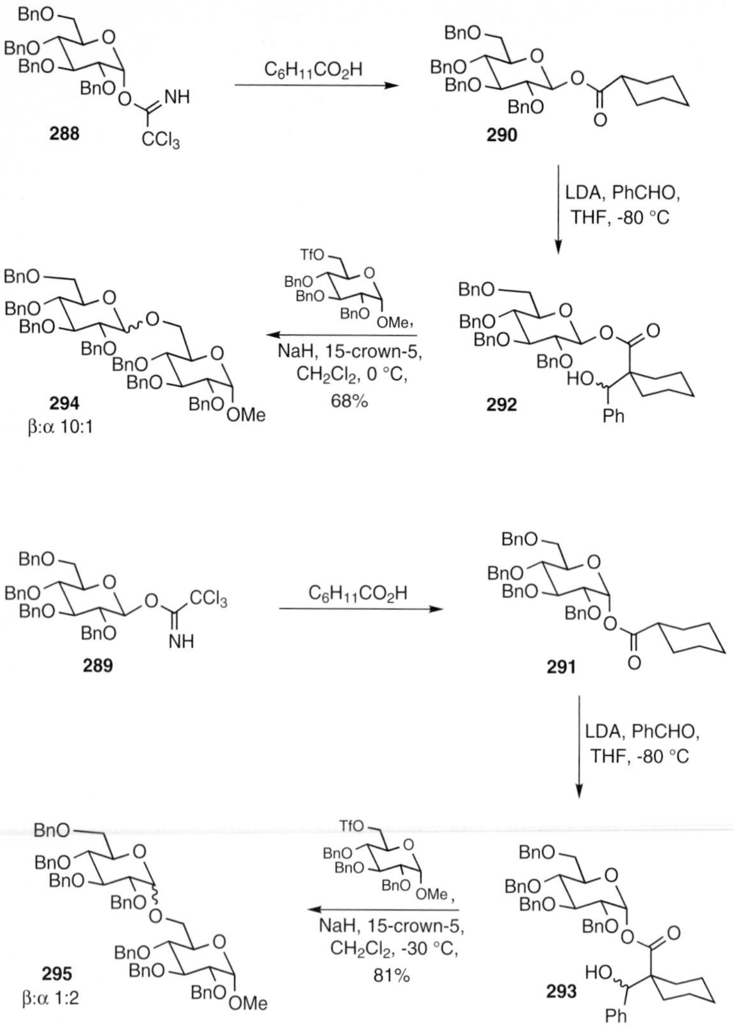

Scheme 10-94   Leaving group-based intramolecular glycosylation.

rentially, while the α-isomer generates the α-glycoside preferentially. These ratios exhibit a temperature dependency; they are improved at low temperature.

An alternative linker which can serve as both a leaving group *and* a latent glycosyl acceptor is a mixed carbonate [103]. In this case, suitable activation with a Lewis acid causes extrusion of carbon dioxide and formation of the reactive glycosyl acceptor, which can then be trapped by the glycosyl donor. The mixed carbonate precursor **299** for this decarboxylative glycosylation strategy was prepared as a mixture of stereoisomers

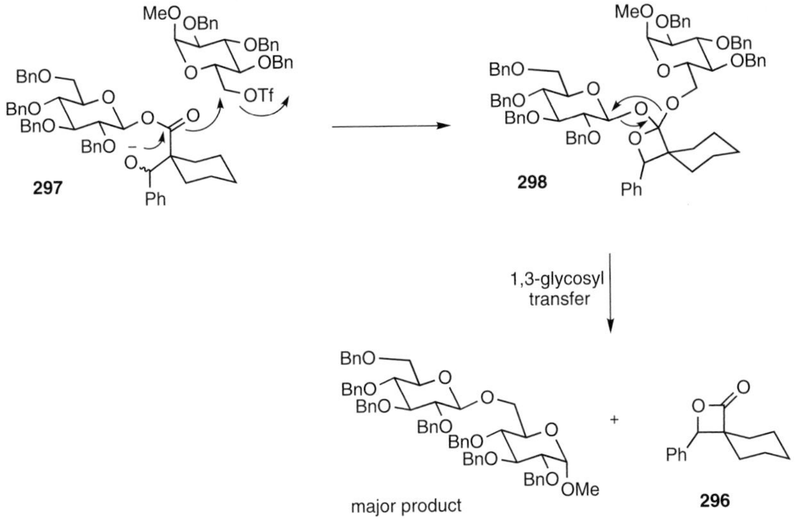

Scheme 10-95 Formation of an orthoester followed by 1,3-glycosyl transfer are possible steps in the mechanism.

in a stepwise fashion by treating the glycosyl donor **300** with the activated carbonate **301** (Scheme 10-96). Of the Lewis acid activation systems which were examined, TMSOTf, SnCl$_4$/AgClO$_4$, and Cp$_2$HfCl$_2$/AgClO$_4$ proved to be most useful. However, the results suggest that the tether has relatively little effect on the stereochemical outcome of the reaction, which is more dependent on the structure of the glycosyl donor, the solvent, and the temperature.

This apparent lack of tether-directing stereocontrol in the decarboxylative glycosylation strategy highlights an important caveat in all such reactions where the nucleophile is generated as the tether is being broken. The tether is only likely to provide good levels of stereocontrol if the bond-breaking and bond-forming events proceed at a very similar time. This seems to be the case in the short acetal and silyl acetal tethers of Stork, Hindsgaul, and Bols (vide supra). If the reaction is non-concerted, it more closely resembles an

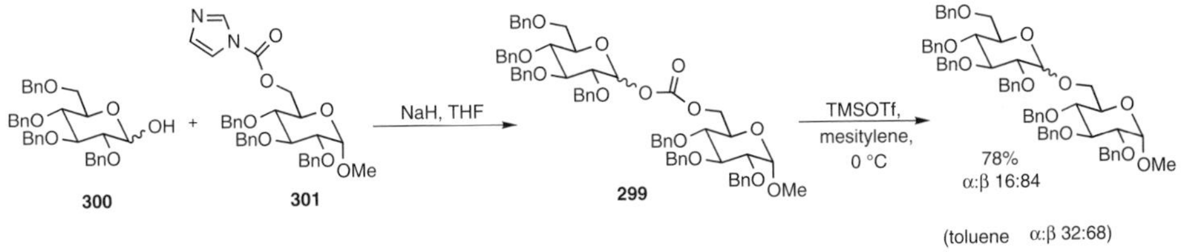

Scheme 10-96 Formation of a mixed carbonate allows a decarboxylative glycosylation pathway.

intermolecular process and as such will be affected by the usual external variables such as solvent and mode of activation. Indeed, Scheffler and Schmidt have performed the appropriate competition experiments, suggesting that the decarboxylative glycosylation procedure, at least in the case of silyl triflate activation, proceeds through an intermolecular, non-concerted pathway [105], and as such, this strategy shows little or no advantage over more conventional intermolecular glycosylation protocols.

In summary, the increasing interest in oligosaccharides necessitates improved methods for their synthesis in diastereoisomerically pure form. One strategy which has been investigated for controlling the outcome of the glycosylation reaction is the use of a temporary connection. This has provided some of the best synthetic routes to date for the synthesis of difficult linkages such as β-L-rhamnosides and β-D-mannosides. However, there remains appreciable scope for improvement. The silyl acetal and acetal tethering methodology developed primarily by Stork, Hindsgaul, and Bols become much less efficient when they are applied to the synthesis of anything larger than a disaccharide and then require appreciable optimization. Ogawa's PMB-acetals are perhaps the most promising tethers in this family. The remote glycosylation strategies using bis-esters and ethers are more robust but the lack of simple rules for predicting the stereochemical outcome of the glycosylation event hinder their more widespread use. Leaving group-based glycosylation is an attractive alternative to the substituent tethers. However, ensuring a concerted reaction is paramount to obtaining high levels of stereocontrol in these reactions – for the most part, this problem has not been solved.

10.4.2   Directed Nucleosidations

One of the most frequent ways of preparing nucleosides is to use the Vorbrüggen modification of the Hilbert–Johnson reaction in which a silylated base reacts with an activated sugar to form the nucleoside linkage [106]. The stereoselectivity of this process is highly dependent on the substitution pattern at the 2′-position of the sugar. In the case of ribosides containing a 2′-α-acyloxy group, neighboring group participation ensures that the β-anomer is formed almost exclusively. However, in the absence of this functionality, nucleosidation is much less stereoselective and a mixture of α- and β-anomers is usually obtained.

Modified nucleosides have been shown to exhibit a range of biological properties and are attracting increasing attention as therapeutic agents. Efficient methods for the stereoselective synthesis of these compounds, in particular controlling the anomeric linkage, are therefore desirable. The groups of Jung [107] and Sugimura [108] have both addressed the problem of anomeric selectivity in the preparation of 2′-deoxyribonucleosides by intramolecular delivery of the base through a temporary attachment to the 5′-position of the riboside. The geometry of the tetrahydrofuran ring then ensures that the base is delivered exclusively to the β-face, providing the desired β-anomer after hydrolysis of the tether (Scheme 10-97).

Jung and Castro prepared 2′-deoxyuridine **302** using this strategy [107]. Reaction of the potassium salt of ribal **303** with 2-methylthiopyrimidone **304** incorporated the base.

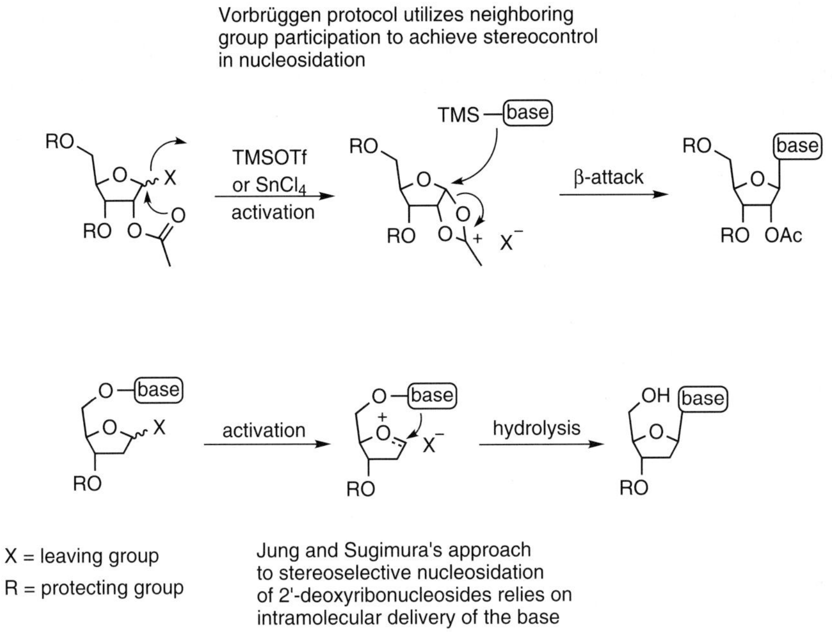

Scheme 10-97 shows:

Vorbrüggen protocol utilizes neighboring group participation to achieve stereocontrol in nucleosidation

TMSOTf or SnCl₄ activation → TMS—base → β-attack

X = leaving group
R = protecting group

Jung and Sugimura's approach to stereoselective nucleosidation of 2'-deoxyribonucleosides relies on intramolecular delivery of the base

Scheme 10-97 Intramolecular delivery of the nucleoside base provides a solution to the problem of stereocontrol in 2'-deoxyribonucleosides.

Subsequent hydration of the enol ether followed by BOM-deprotection and formation of the anomeric acetate provided the nucleosidation precursor **305**. In situ silylation of the pyrimidone and activation of the anomeric group with TMSOTf gave the desired 2'-deoxy-3'-*O*-methyluridine **302** in moderate yield after basic hydrolysis of the tether with no contaminating α-anomer (Scheme 10-98). A similar strategy was used by Sujino and Sugimura to prepare similar compounds [108].

10.5 Tether-mediated Nucleophile Delivery

The tether directing strategies for C- and O-glycosylation are one subset of a much wider area of using the temporary connection for controlling the regio- and stereochemical outcome in the reaction of nucleophiles with an electrophilic substrate. Some examples of this form of tether-mediated synthesis will be illustrated with some applications to natural product synthesis [109].

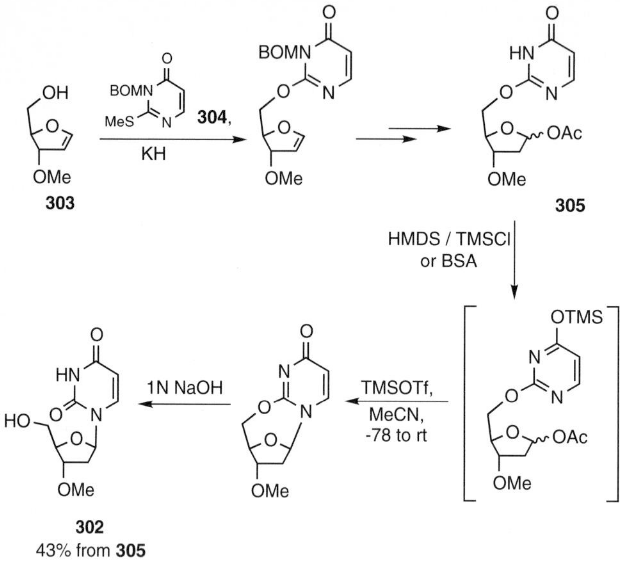

Scheme 10-98   Intramolecular Vorbrüggen coupling allows stereoselective synthesis of β-2'-deoxyribonucleosides.

10.5.1   Synthesis of (+)-Hydantocidin

The spironucleoside hydantocidin **306** exhibits a number of interesting properties, including herbicidal and plant growth regulatory activities. All possible diastereoisomers have been synthesized and only the naturally occurring isomer was shown to display significant biological activity [110]. The most challenging task in a synthesis of **306** was correct installation of the stereochemistry at the anomeric center. Furthermore, since the α-nitrogen is thermodynamically more stable than the β-isomer, it was not possible to rely on thermodynamic equilibration to install this key stereocenter.

In Chemla's approach, it was decided to tether a suitable nitrogen nucleophile to the 6-position of the sugar moiety to ensure β-facial selectivity in the key nucleosidation-type reaction [111]. Hydroxylamine **307**, readily prepared from D-fructose, was treated with *p*-methoxybenzyl isocyanide to afford the corresponding urea derivative **308** in excellent yield (Scheme 10-99). Treatment with the Lewis acid TMSOTf effected smooth cyclization to isoxazolidine **309**, installing the correct stereochemistry at the anomeric position. Oxidation and removal of the PMB protecting group generated the tricyclic isoxazolidine hydantocidin **310**, which just required cleavage of the hydroxylamine tether and removal of the acetonide protecting group to reveal the desired compound. A number of reducing agents were examined for cleaving the N–O bond, of which Mo(CO)$_6$ proved to be the reagent of choice. Acid-catalyzed deprotection of the acetonide then provided hydantocidin **306**.

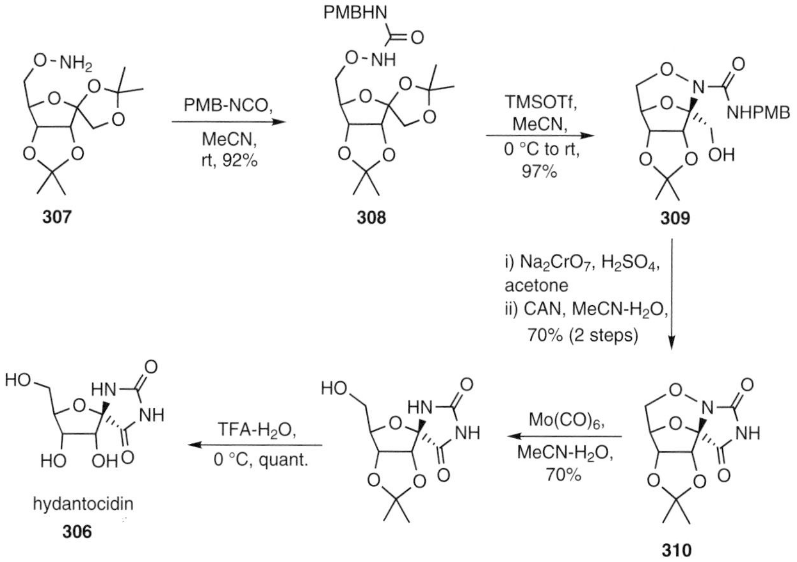

Scheme 10-99   Intramolecular delivery of a nitrogen nucleophile provides the correct stereo-chemistry at the anomeric position for a synthesis of hydantocidin.

10.5.2   Synthesis of Lincomycin

Knapp and Kukkola have reported a synthesis of the antibiotic lincomycin **311** (Scheme 10-100) [112]. Earlier synthetic approaches to this molecule had revealed that the C(6) position was very hindered precluding efficient incorporation of the amino functionality via conventional intermolecular protocols. To circumvent this problem they chose to deliver a nitrogen nucleophile intramolecularly from a tether at C(4). It was anticipated that epoxide opening would proceed regioselectively in a 6-*exo* manner, incorporating the desired stereochemistry at both C(6) and C(7) positions. To this end, epoxy alcohol **312** was synthesized in good overall yield from commercially available methyl α-D-galactopyranoside. After some experimentation, treatment of the sodium alkoxide of **312** with dimethylcyanamide created the isourea tether which underwent spontaneous, ster-eospecific epoxide ring opening furnishing, after rearrangement, oxazoline **313** in 95% yield. Benzylation of the 4-OH and subsequent hydrolysis of the tether generated amino alcohol **314**, which was subsequently converted into lincomycin **311**.

10.5.3   Intramolecular Allylation –
Use in the Synthesis of Tricyclic β-Lactam Antibiotics

The stereoselective allylation of aldehydes is one of the most important, and intensively investigated, reactions in synthetic chemistry [113]. Nowadays, one of the most efficient

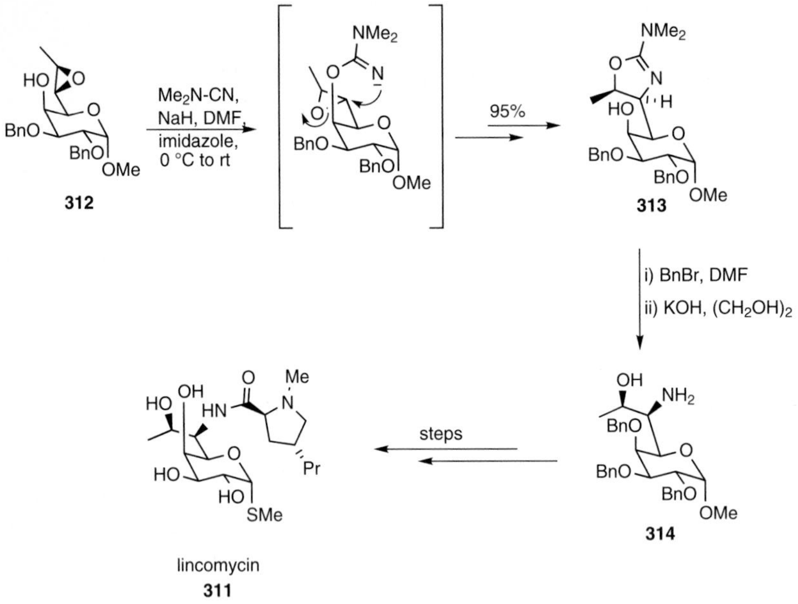

Scheme 10-100   Intramolecular delivery of a nitrogen nucleophile allows stereospecific incorporation of the amino functionality at the sterically hindered C(6) position.

ways of carrying out this transformation is to use a chiral Lewis acid in the presence of an allylstannane [114], or to use chiral ligands on the allyl metal reagent [115]. However, prior to the development of these methodologies, substrate control was widely used to achieve asymmetric induction in the reaction. Reetz and co-workers have made a number of important studies into the effect of α- and β-stereocenters on the addition of nucleophiles to carbonyl functionality. In one such investigation, the sense of 1,3-asymmetric induction using a tethered allylsilane was shown to be highly dependent on the choice of Lewis acid [116]. They found that reaction with $TiCl_4$ proceeded in excellent diastereoselectivity, providing predominantly the *syn* product **315**. This observation is in agreement with an intramolecular transfer of the allyl nucleophile through an eight-membered T. S. in which the β-silyl ether and aldehyde form a chelated structure outlined in Scheme 10-101. The intramolecularity of the reaction was proved by a series of crossover experiments. The result is complementary to that obtained using $SnCl_4$ as the Lewis acid, where the *anti* diol product predominates. This reaction was shown to be an intermolecular process.

Allylation using tethered nucleophiles has been investigated by other groups and has been shown to typically exhibit excellent levels of stereocontrol [117], which in some cases are complementary to those obtained in the corresponding intermolecular process. A Glaxo group used such an approach in the synthesis of a key intermediate for the preparation of tricyclic β-lactam antibiotics [118].

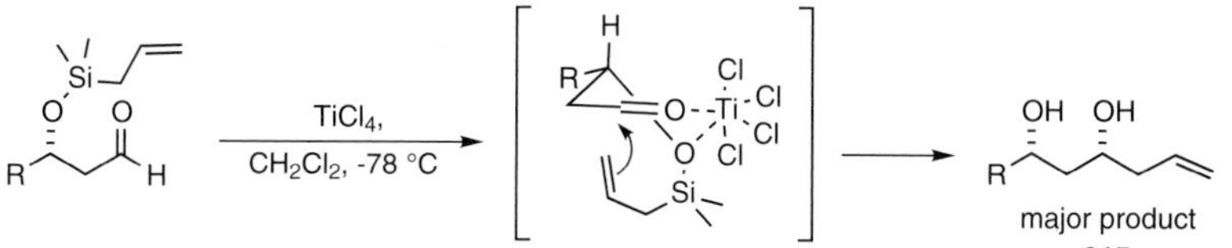

R = Me
R = Bu

major product
315
R = Me, *syn:anti* 92:8, (70%)
R = Bu, *syn:anti* 90:10, (80%)

Scheme 10-101   Intramolecular allylation provides the complementary stereochemical result to the analogous intermolecular process.

Preparation of the temporary connection was achieved by reaction of acetoxy lactam **316** with allyl(chloro)silane **317**, generating the *N*-silyl species **318**. Exposure to TMSOTf at 0 °C effected in situ formation of the *N*-acyliminium species, which was trapped by the pendant allyl nucleophile, providing two out of the possible four diastereoisomeric cyclized products, **319** and **320** in a 4:1 ratio (Scheme 10-102). Formation of the two products may be accounted for by cyclic T. S.s where addition occurs onto the *re* face of the azetinone, *anti* to the sterically demanding silyl ether group through a chair-like or boat-like conformation.

10.5.4   Synthesis of Corticosteroids

A variety of different covalent tethering strategies have been used in stereo- and regioselective manipulations of the side chains of steroids. Livingston et al. utilized a temporary connection for nucleophilic delivery in a facile synthesis of corticosteroids from readily available androst-4-ene-3,17-dione **321** [119]. Reaction of **321** with KCN proceeded chemoselectively at the 17-ketone, providing a mixture of cyanohydrins, from which the 17β-cyanohydrin **322** could be obtained after equilibration and recrystallization. Conversion to the (chloromethyl)silyl ether **323** provided the additional carbon required for homologation. Treatment of **323** with LDA effected deprotonation of the chloromethyl group and cyclization onto the nitrile providing 21-chloroketone **324** in 93% yield from **323** after acidic workup (Scheme 10-103). Reductive dehalogenation or nucleophilic displacement of the chloride with KOAc afforded 17α-hydroxyprogesterone **325** and cortexolone acetate **326** respectively, both in quantitative yield. Although the homologation may be achieved using standard intermolecular reactions, the described sequence is noteworthy in that there is no need to protect either alcohol or 3-keto functionality.

10.5.5   Synthesis of (–)-α-Kainic Acid

Kainic acid **327** exhibits an array of biological properties, including insecticidal, antithelmic, and neuroexcitatory activity. One of the challenging aspects to the synthesis of

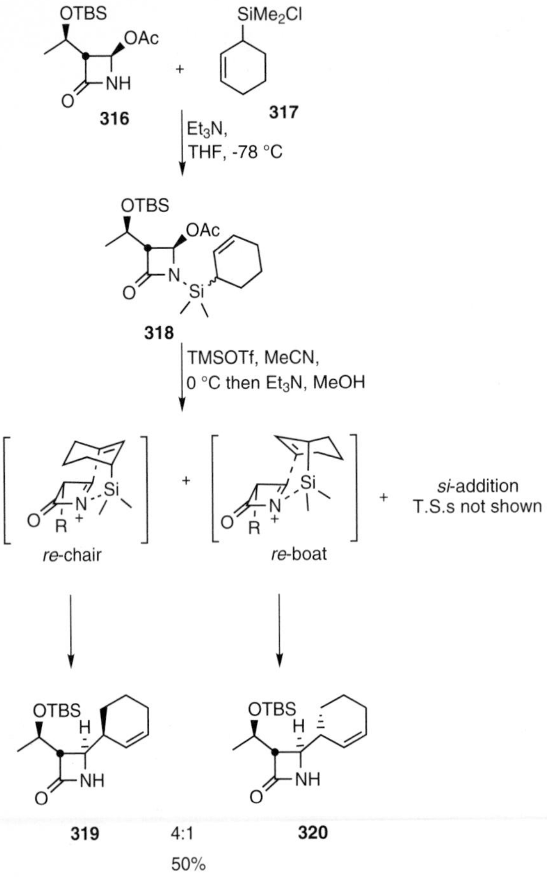

Scheme 10-102 Tether-directed allylation provides a diastereoselective route to precursors in the synthesis of tricyclic β-lactam antibiotics.

such a trisubstituted pyrrolidine is the formation of the *cis* relationship between the C(3) and C(4) positions. Bachi and Melman showed that a potential precursor **328** could be prepared in enantioselective form [120]. To complete the synthesis of kainic acid, it just remained to carry out a formal S$_N$2 substitution of the tosylate with an acetic acid residue. Attempts to do this using an intermolecular reaction were hampered by the steric hindrance at the reacting site and facile elimination involving the particularly acidic hydrogen at C(2). To overcome these problems, an intramolecular nucleophile delivery strategy was devised (Scheme 10-104) [120]. Addition of the sulfenyl chloride **329** derived from methyl mercaptoacetate to the isopropenyl group proceeded with complete regio- and stereocontrol affording **330**, and subsequent oxidation with *m*CPBA provided the corresponding sulfone **331**. The use of a sulfone tether instead of the simple sulfide

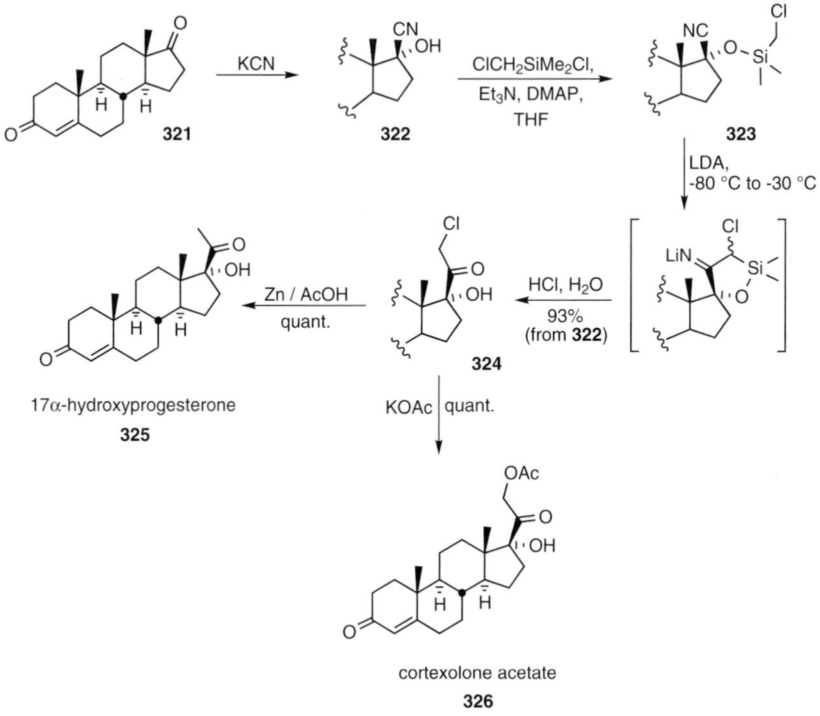

Scheme 10-103 Intramolecular nucleophile delivery provides an efficient route to biologically important corticosteroids.

was found to be necessary since, under the basic conditions used to prepare the enolate nucleophile, elimination of the sulfenyl chloride to the isopropenyl precursor **328** was observed. Additional advantages of the sulfone tether include its increased stability and a more acidic α-proton which allows the use of milder reaction conditions to generate the enolate nucleophile. The desired stereospecific intramolecular substitution was achieved in good yield upon treatment of **331** with potassium methoxide in THF. Samarium diiodide proved to be the reagent of choice for cleaving the temporary connection, and after global deprotection provided kainic acid **327**.

10.6 Silicon-tethered Ene Cyclization

Robertson and co-workers have investigated a Type II intramolecular version of the ene reaction in which the reacting partners, an olefin possessing an allylic hydrogen and an aldehyde, are tethered at an internal site [121]. In analogy with intramolecular variants in which the reactants are linked through an all-carbon tether [122], this reaction was antic-

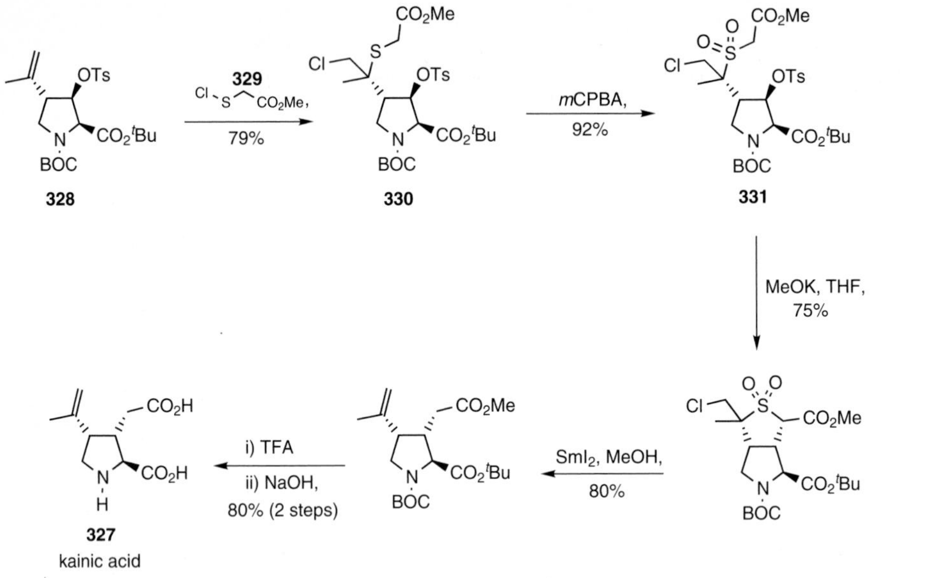

Scheme 10-104   Synthesis of kainic acid in which Bachi and Melman used intramolecular nucle-ophile delivery.

ipated to proceed with high levels of stereocontrol but with the added advantage of pos-sible cleavage of the silicon bridge at a later stage providing acyclic products.

A number of routes were investigated for synthesizing the tethered precursors. The first involved the sequential addition of organometallic nucleophiles to Me$_2$SiCl(NMe$_2$) [121a]. However, this suffered from a number of drawbacks, including the limited avail-ability of substituted propionaldehyde homoenolates, and the need to work with the highly reactive aminosilyl chloride. Two new routes have recently been reported [121b]. Both utilize the reaction of propenyllithium with oxasilacyclopentanes which are readi-ly prepared in two ways, namely, a 5-*exo*-trig radical cyclization of (bromomethyl)silyl ethers or intramolecular hydrosilylation of an allylic double bond (Scheme 10-105).

Treatment of **332** or **333** with MeAlCl$_2$ at −78 °C effected reaction providing the cyclized products with excellent levels of stereocontrol. The observed products can be rationalized by invoking the T. S.s outlined in Scheme 10-106 in which the alkyl sub-stituent occupies an equatorial site in the chair conformation and the aldehyde is axial. Reaction was found to be significantly slower than the all-carbon analogs; this may be attributed to the longer C−Si bond length (1.89 Å; cf. 1.54 Å for C−C) and the reduced ability for silicon to stabilize any buildup of positive charge on the α-carbon atom.

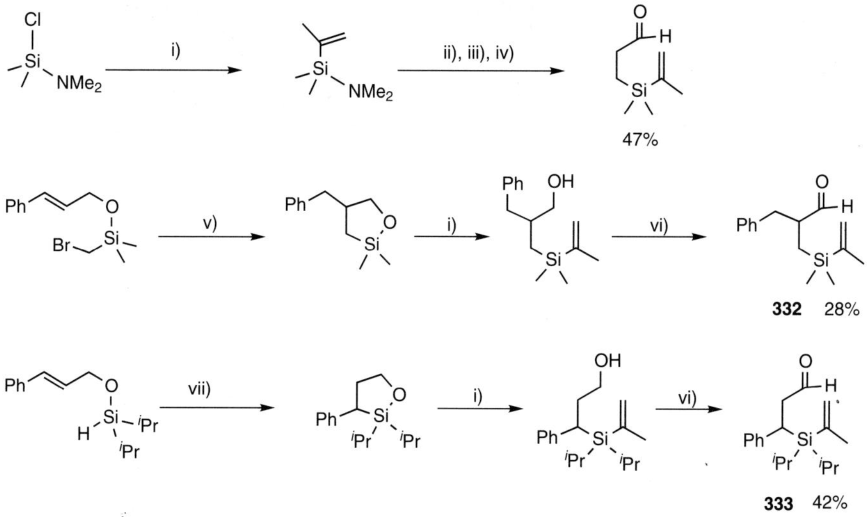

Scheme 10-105 Silicon-tethered ene precursors can be prepared in a number of ways. Reagents and conditions: i) 2-propenyllithium, THF, -78 °C; ii) AcCl, rt; iii) (MeO)₂CH(CH₂)₂MgBr, THF, rt; iv) *p*TsOH, aq. *ᵢ*PrOH, THF, heat; v) Bu₃SnH, AIBN, PhH; vi) PDC, 4Å ms, CH₂Cl₂; vii) [(PPh₃)₃RhCl], 4Å ms, THF.

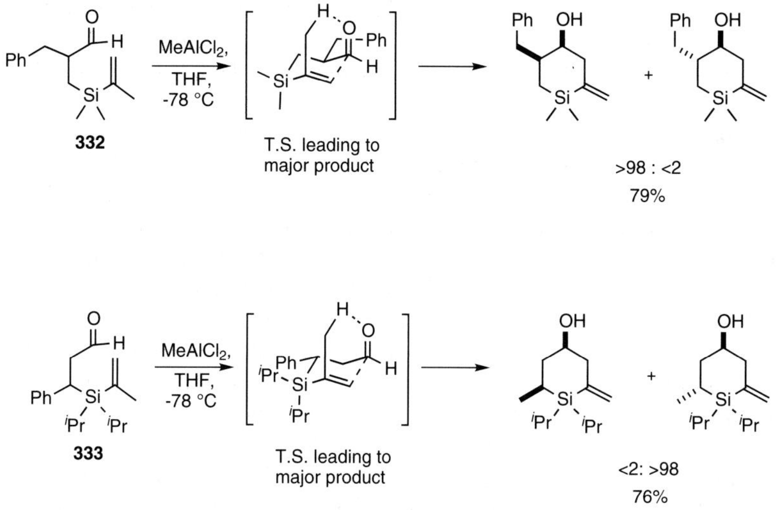

Scheme 10-106 Silicon-tethered ene cyclization is highly stereoselective.

10.7   Intramolecular Hydrosilylation and Related Reactions

10.7.1   Hydrosilylation of Olefins

The intramolecular hydrosilylation of terminal olefins and acetylenes is a highly efficient reaction when activated silanes (e.g., $HSiCl_3$) are used in the presence of a metal catalyst (e.g., $[H_2PtCl_6 \cdot 6H_2O]$) [123]. In this case, the reaction is normally highly regioselective with silyl addition proceeding onto the less hindered terminal position. Reaction of trialkylsilanes is appreciably less efficient and often does not proceed at all. Hydrosilylation of internal olefins is also problematic – they exhibit reduced reactivity, allowing side reactions such as catalyst-mediated double bond isomerization to become significant. Furthermore, regioselectivity is also difficult to address.

One solution to these problems is to temporarily tether the hydrosilylating reagent to the substrate. The ensuing intramolecular reaction has been shown to be highly efficient and, in certain cases, proceeds with excellent selectivity (intramolecular hydrogermylation [124] and hydroboration [125] reactions have also been investigated). Tamao, Ito, and co-workers provided some of the earliest studies [126] and have also investigated the mechanism of the reaction [127]. Treatment of a number of allylic and homoallylic alcohols with $(Me_2HSi)_2NH$ provided the corresponding silyl ethers, which on exposure to 0.1 mol% of Wilkinson's catalyst $[(PPh_3)_3RhCl]$ or hexachloroplatinic acid $[H_2PtCl_6 \cdot 6H_2O]$, underwent hydrosilylation. Subsequent oxidative cleavage of the silyl tether afforded the corresponding diol products in good overall yield (Scheme 10-107). In the case of cyclic systems, e.g., **334**, reaction proceeds with excellent stereo- and regiocontrol, providing the *syn*-diol product. Acyclic systems provide more variable results; the outcome is highly dependent on the presence of additional stereocenters and the position and substitution pattern of the reacting olefin. The alkyl substituents on the silicon tether can also influence the stereochemical outcome: use of cyclohexylsilyl hydrides generally provides greater stereoselectivity than in the case of dimethylsilyl analogs [128]. α-Hydroxy enol ethers can be used to provide stereoselective routes to 1,2,3-triols, although in these cases the neutral Karstedt catalyst, $[Pt\{[(CH_2=CH)Me_2Si]_2O\}_2]$ has to be used to prevent decomposition of the acid-labile substrate [126c]. Bosnich and co-workers have shown that dihydrosilanes in the presence of an $[Rh(Ph_2PCH_2CH_2PPh_2)]^+$ catalyst form the silyl ether tether in situ, providing the cyclized siloxanes directly from the corresponding allylic alcohol [129].

A variety of oxidative tether cleavage protocols have also been evaluated which allow regioselective monoprotection of the resulting 1,3-diol products, facilitating differentia-

Scheme 10-107   Homoallylic and allylic silyl ethers undergo efficient hydrosilylation. Reagents and conditions: i) $(HMe_2Si)_2NH$, NH_4Cl cat.; ii) $[(PPh_3)_3RhCl]$ (0.1 mol%), 100 °C; iii) 30% H_2O_2, $NaHCO_3$, MeOH, THF, 60 °C; iv) $[H_2PtCl_6 \cdot 6H_2O]$ (0.1mol%), rt to 60 °C; v) $[Pt\{[(CH_2=CH)Me_2Si]_2O\}_2]$ (0.1 to 0.5 mol%), 60 °C; vi) 30% H_2O_2, 15% KOH, MeOH, THF, rt; vii) (cyclohexyl)chlorosilane, Et_3N, DMAP; viii) $\{[Rh(Ph_2PCH_2CH_2PPh_2)]^+ClO_4^-\}$, Ph_2SiH_2.

[a] Yields are based on the starting alcohols.

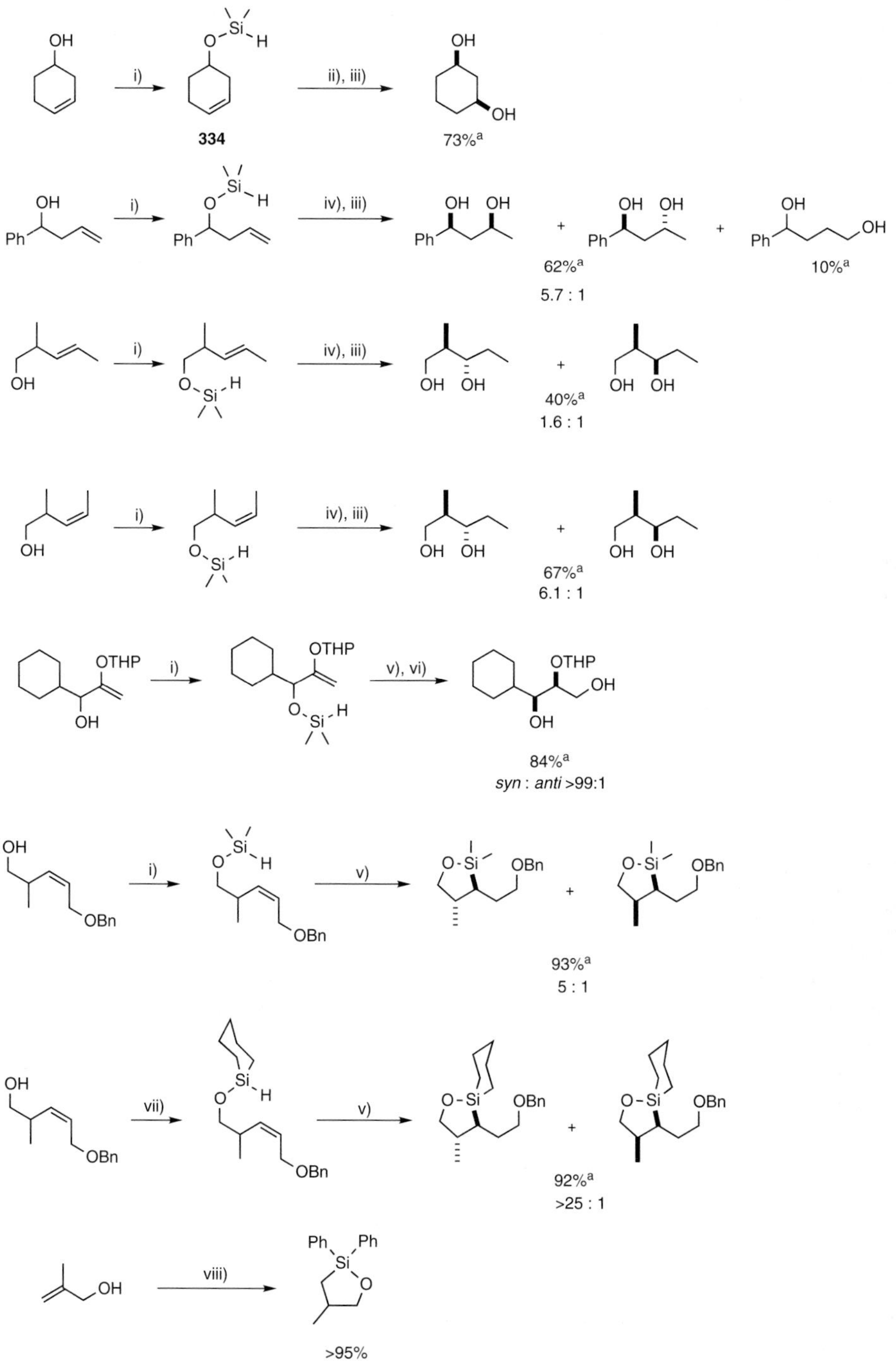
i)
ii), iii)
334
73%[a]
iv), iii)
62%[a]
5.7 : 1
10%[a]
40%[a]
1.6 : 1
67%[a]
6.1 : 1
v), vi)
84%[a]
syn : anti >99:1
v)
93%[a]
5 : 1
vii)
92%[a]
>25 : 1
viii)
>95%

tion of the two alcohol functionalities, and further enhancing the synthetic utility of this hydroxylation protocol (Scheme 10-108) [126b].

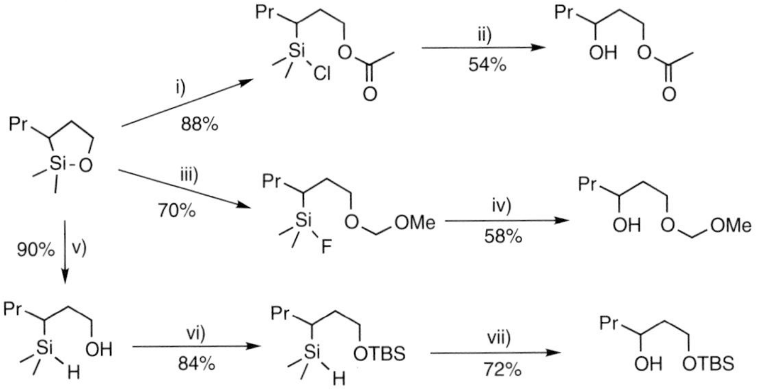

Scheme 10-108 Different tether cleavage strategies provide access to monoprotected diols. Reagents and conditions: i) MeCOCl, ZnCl₂ cat., rt; ii) KF, KHCO₃, MeOH, THF then 30% H₂O₂, rt; iii) MOMCl, CsF, MeCN, rt; iv) 30% H₂O₂, KHCO₃, MeOH, rt to 50 °C; v) DIBALH, Et₂O, 0 °C to 60 °C; vi) TBSCl, Et₃N, DMAP, CH₂Cl₂; vii) 30% H₂O₂, NaHCO₃, MeOH, THF, 60 °C.

Curtis and Holmes provided an early display of the synthetic utility of the intramolecular hydrosilylation in their synthesis of the *trans*-diol **335**, an intermediate in their approach to obtusenyne **336** [130]. The allylic silyl ether **337** was prepared from the corresponding alcohol using an excess of 1,1,3,3-tetramethyldisilazane in the presence of a catalytic amount of ammonium chloride, and then used without purification. A wide variety of catalysts were investigated with the rhodium complex [Rh(acac)(norbornadiene)] providing the best results in terms of stereoselectivity (*trans/cis* >95:5) (Scheme 10-109). Unfortunately, partial loss of the silyl group under the reaction conditions resulted in a moderate yield of the diol product which was not improved when the corresponding diisopropylsilyl ether was used.

Hale and Hoveyda have also produced an interesting synthesis of the C(27)–C(33) fragment of the potent immunosuppressant, rapamycin, using the siloxane **338**, generated from intramolecular hydrosilylation, in remote acyclic stereocontrol (Scheme 10-110) [131]. Reaction of homoallylic silyl ether **339** gave a 4:1 mixture of siloxane products and further manipulation of the major stereoisomer **338** provided Weinreb amide **340**. Osmylation of the adjacent double bond not only proceeded with excellent diastereoselectivity, but also with complete rearrangement of the siloxane to an internal position, serving as an efficient method for in situ differentiation of the two newly generated secondary hydroxyl groups. Further reactions provided the correct stereochemical relation of stereocenters found in the C(27)–C(33) fragment of rapamycin.

An enantioselective version of the hydrosilylation reaction would greatly extend its synthetic utility. The reaction mechanism of this catalytic asymmetric process has been investigated in great detail and shown to be extremely complicated [132]. Nevertheless,

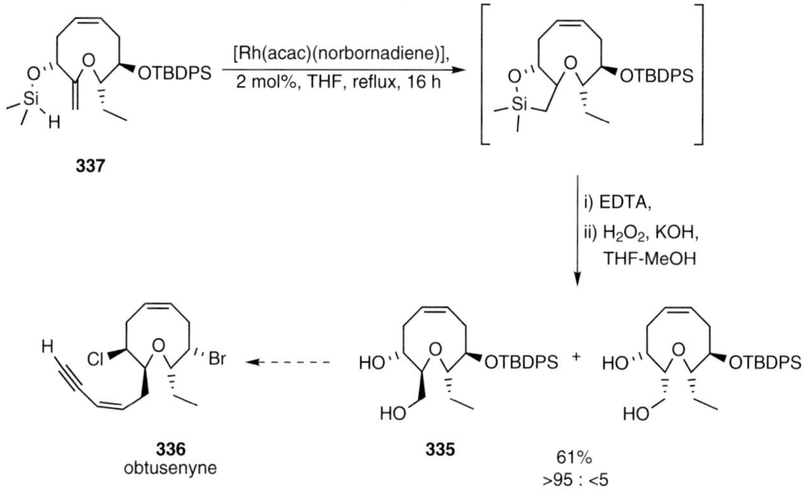

Scheme 10-109 Intramolecular hydrosilylation used by Curtis and Holmes in the synthesis of a key intermediate towards the preparation of obtusenyne.

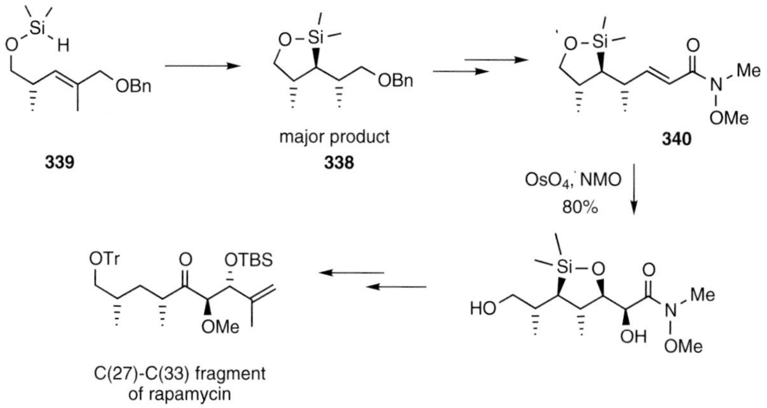

Scheme 10-110 Use of a siloxane template in the preparation of the C(27)–C(33) fragment of rapamycin.

a number of catalyst systems have been developed based on rhodium(I) complexes with chiral diphosphines, and excellent levels of enantiocontrol have been achieved.

Hydrosilylation of allylic amines proceeds under platinum catalysis with the opposite regioselectivity to their oxygen analogs [133]. The intermediate 1-aza-2-silacyclobutanes resulting from a 4-*exo*-trig addition are thermally stable and may be purified by distillation. Alternatively, direct oxidation using the Tamao conditions provides the corresponding 1,2-amino alcohols with excellent *syn* selectivity (Scheme 10-111).

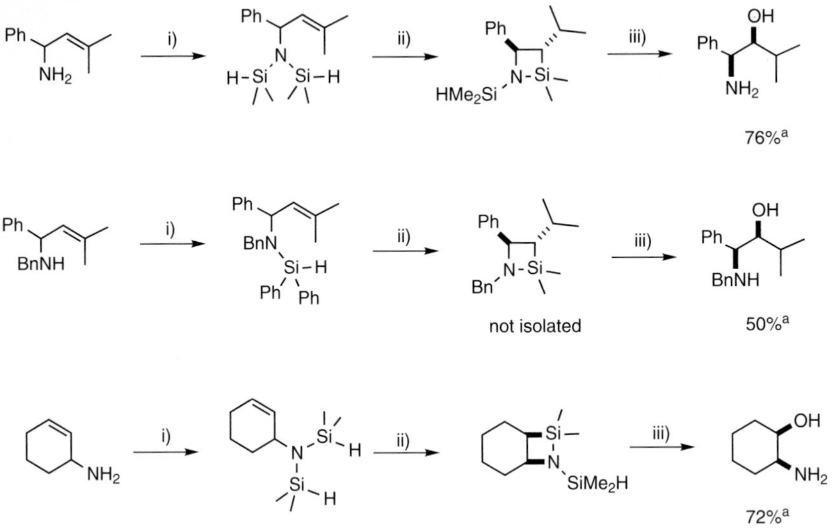

Scheme 10-111 Pt-catalyzed intramolecular hydrosilylation of allylic amines provides 1,2-amino alcohols. Reagents and conditions: i) BuLi, Me₂SiHCl, Et₂O; ii) [Pt{[(CH₂=CH)Me₂Si]₂O}₂],
rt; iii) EDTA · 2Na, hexane, rt then 30% H₂O₂, KF, KHCO₃, MeOH, THF, rt.

10.7.2 Hydrosilylation of Acetylenes

The attachment of a silyl hydride reagent to homopropargylic alcohols provides a good method for hydrosilylating triple bonds. The temporary attachment again allows regioselective *cis* hydrosilylation of internal acetylenes with products resulting from an *exo*-dig-cyclization mode being observed (Scheme 10-112) [134]. A number of variations have also been reported, including silylformylation [135] and cyanosilylation [136], which also exploit the regio- and stereocontrol from the intramolecular reaction.

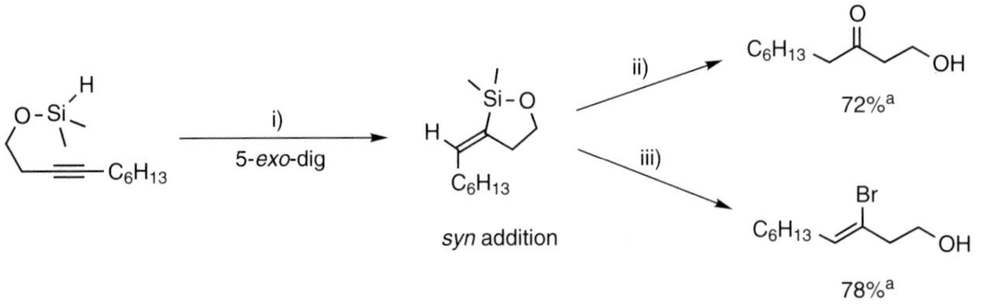

Scheme 10-112 Intramolecular hydrosilylation of homopropargylic acetylenes proceeds with excellent stereo- and regiocontrol. Reagents and conditions: i) [H₂PtCl₆ · 6H₂O] (0.1 mol%), CH₂Cl₂, rt; ii) 30% H₂O₂, KF, KHCO₃, MeOH, THF, 30 °C; iii) Br₂, 0 °C then KHF₂, MeOH, rt.

___
[a] Yield based on allylic amine. *syn : anti* ratio >99:1 in all cases.

10.7.3 Directed Bis-Silylation Reactions

The bis-silylation of olefins and acetylenes is a potentially desirable reaction on account of the fact that two C−Si bonds are generated in one step. Intramolecular versions would also be attractive. The reaction is analogous to the tethered hydrosilylation described above but has received less attention. One of the major problems which had to be over-

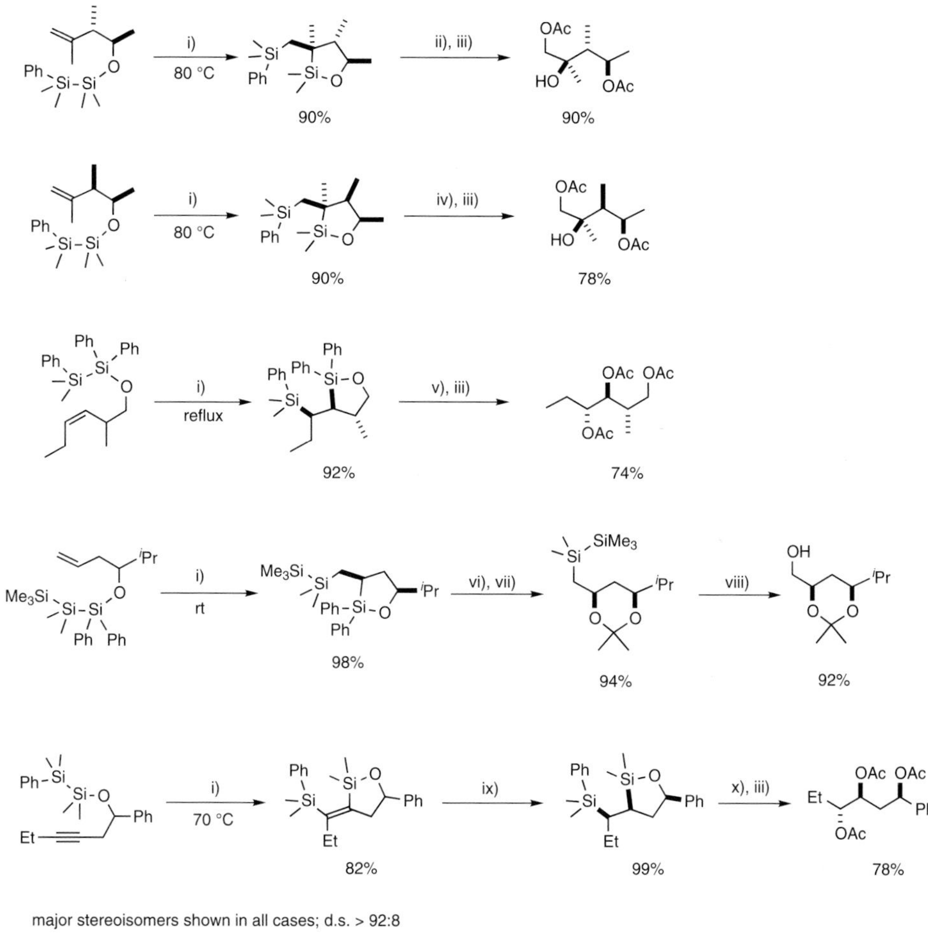

major stereoisomers shown in all cases; d.s. > 92:8

Scheme 10-113 Intramolecular bis-silylation followed by oxidative tether cleavage provides a stereoselective route to 1,2,4-triols. Reagents and conditions: i) Pd(OAc)$_2$, 1,1,3,3-tetramethylbutyl isocyanide, toluene; ii) TFA then H$_2$O$_2$, KF, KHCO$_3$, KHF$_2$; iii) Ac$_2$O, Et$_3$N, DMAP; iv) KO^{t}Bu, DMSO then H$_2$O$_2$, KF, KHCO$_3$, KHF$_2$; v) TFA then H$_2$O$_2$, TBAF, KHCO$_3$; vi) H$_2$O$_2$, KF, KHCO$_3$, MeOH, THF; vii) 2-methoxypropene, CSA, acetone; viii) TBAF, THF then H$_2$O$_2$, KHCO$_3$, MeOH; ix) diimide, EtOH; x) KO^{t}Bu, DMSO then TBAF, KHCO$_3$, H$_2$O$_2$.

come was finding a suitable catalyst system. Ito and co-workers made the major break-through by finding that Pd(OAc)$_2$ in the presence of an excess (with respect to Pd) of 1,1,3,3-tetramethylbutyl isocyanide was a highly efficient catalyst system for the bis-silylation of homoallylic olefins and acetylenes [137]. Although the precise nature of the active species is not known, it is most probably a palladium(0) species with coordinated isocyanide groups since no reaction was observed in the absence of the isocyanide. A variety of disilanyl ethers were readily prepared under standard conditions for silylation and reacted with high efficiency with terminal olefins and many types of acetylenes. Internal olefins proved more difficult although the use of phenyl substituents on the silicon atom proximal to the ether tether provided a solution to this problem [137e]. The reaction is stereospecific with *syn* addition across the reacting group, and also stereoselective. The stereoselectivity of the addition is usually excellent and may be explained by invoking a chair-like T. S. Oxidative cleavage of the silyl ether tether provides an interesting route to 1,2,4-triols – the reaction may therefore be viewed as a directed stereoselective glycolation (Scheme 10-113) [137c]. When a trisilanyl ether is used in the bis-silylation of double bonds, the difference in reactivity towards oxidation allows a stepwise oxidation of each C–Si bond (Scheme 10-113) [138].

The bis-silylation reaction has found application in the synthesis of the antifungal metabolite (–)-avenaciolide [139] and in the enantioselective preparation of (*E*)-allylsilanes [140]. The use of chiral isocyanides prepared from the terpenoid (+)-ketopinic acid has also been investigated in an enantioselective bis-silylation of olefins, although the levels of enantioselectivity are, in most cases, modest [141].

10.7.4 Intramolecular Hydrosilylation of Ketones

The *anti*-1,3-diol motif is widespread among biologically important natural products, and efficient routes towards this unit are desirable. Davis and co-workers have investigated the possibility of using a silyl ether tether to connect a hydridic nucleophile to a β-hydroxy ketone substrate [142]. Reaction of commercially available ^{i}Pr$_2$SiHCl with a β-hydroxy ketone provided the stable silyl ether product **341** in good yield. Subsequent exposure to a variety of Lewis acids effected a highly stereoselective, intramolecular reduction of the ketone group providing the *anti*-1,3-diol protected as a silyl acetal **342**, which was sufficiently stable to be used as a protecting group for the diol functionality, yet could also be readily removed with aqueous HF in MeCN (Scheme 10-114). SnCl$_4$ gave the best results although a wide variety of Lewis acids and Brønsted acids were effective (the use of TBAF and zinc Lewis acids gave poor results). The observed stereochemical outcome can be accounted for by invoking a chair-like T. S. in which diaxial interactions with the bulky silyl substituents are minimized.

The reaction is not chelation-controlled and does not involve hydride transfer to the Lewis acid since BF$_3$ · OEt$_2$, which can only be four-coordinate, also gives excellent *anti* stereoselectivity. Tethering the reagent and intramolecularizing the process is also important for the rate of reaction – the analogous intermolecular process was found to be sluggish, even at 0 °C. Rhodium(I) complexes also catalyze the reduction reaction. Burk

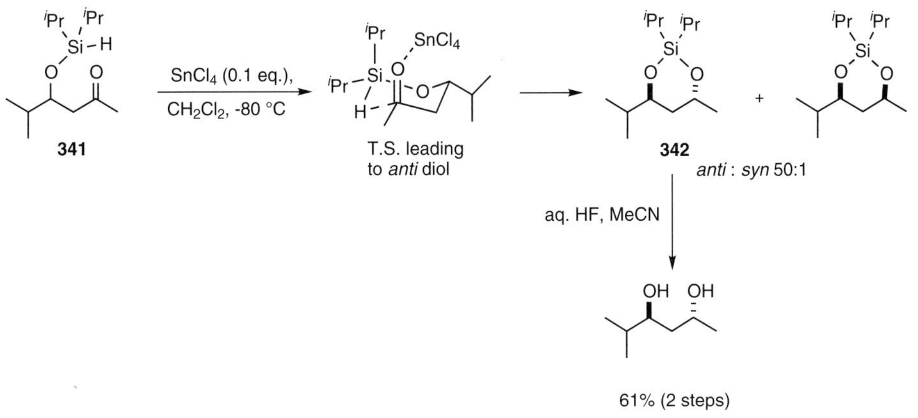

Scheme 10-114 Intramolecular hydrosilylation of β-hydroxy ketones provides 1,3-*anti*-diols.

and Feaster have used chiral diphosphine ligands to provide an enantioselective version of the reaction using α-hydroxy ketones as precursors [143].

The reaction could be extended to β-hydroxy esters (Scheme 10-115) [144]. In this case, TBAF proved the most effective reagent for providing an almost quantitative conversion to the corresponding alkoxysiladioxanes. Activation of the acetal functionality was best achieved with the superacid $[TfOH_2^+B(OTf)_4^-]$, and trapping with allyltrimethylsilane provided the *anti* diol products in excellent yield and stereoselectivity. The outcome is in accord with chelation-controlled intermolecular addition with axial attack of the nucleophile onto the cationic oxonium intermediate.

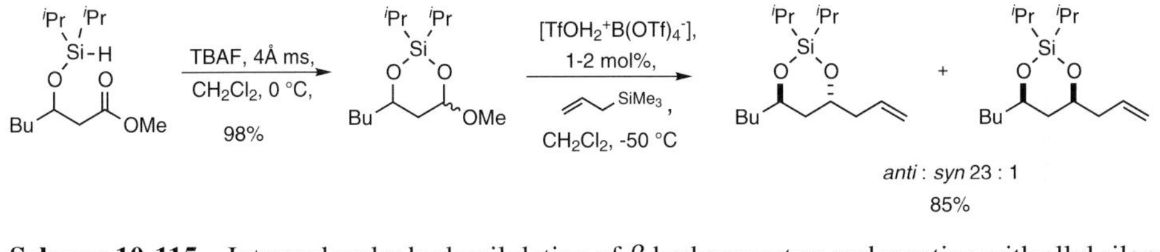

Scheme 10-115 Intramolecular hydrosilylation of β-hydroxy esters and reaction with allyl silane provides *anti*-1,3-diol products.

10.8 Olefination

The complex structure of the macrolide bryostatin 1, **343**, in addition to its activity against various forms of cancer, has made it the subject of a number synthetic studies [145]. One of the more difficult functionalities to incorporate in a stereoselective fashion into the molecule are the two trisubstituted exocyclic unsaturated esters. Evans and Carreira addressed this problem in a novel way by using a tethered phosphonate in a macro-olefination reaction (Scheme 10-116) [146].

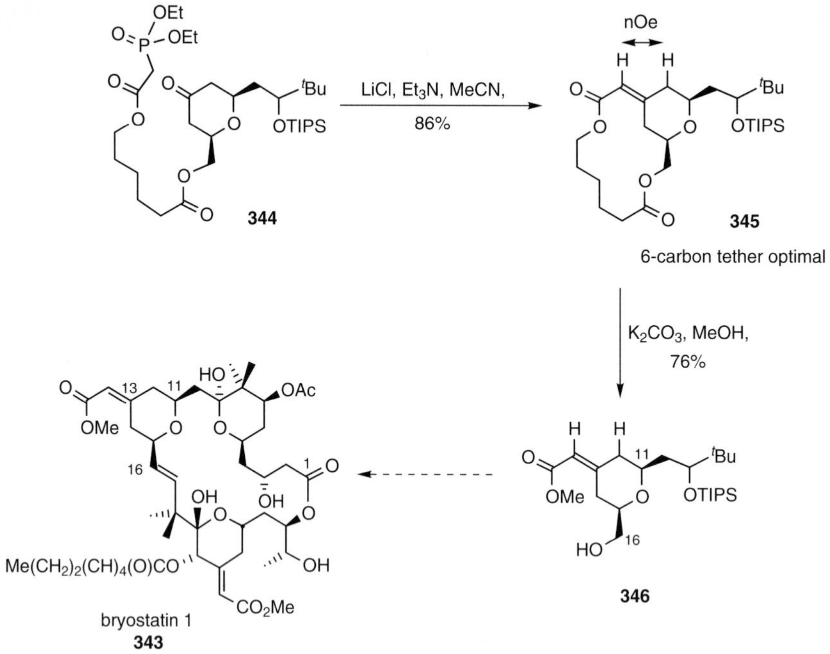

Scheme 10-116   The correct choice of tether length ensures stereoselective macroolefination.

Preliminary studies on a model system suggested that a six-carbon tether, which would produce a 14-membered ring, would be optimal in minimizing transannular interactions and satisfying the stereoelectronic requirements of the ester groups (Scheme 10-116). Additional computer modeling enabled low-energy conformations of the two diastereoisomeric enoates obtained using such a tether to be evaluated more quantitatively. It was pleasing to see that the desired enoate product was around 10 kcal mol^{-1} more stable than the alternative geometry and the two ester functionalities adopted the preferred s-*cis* conformation. The calculated enoate dihedral angle (O=C–C=C) of 165° also implied additional stabilization through conjugation. Horner–Wadsworth–Emmons olefination of the model compound **344** provided the desired macrocycle **345** in 86% yield and as a single diastereoisomer. Subsequent basic methanolysis resulted in facile tether cleavage, providing the corresponding hydroxy ester **346**.

A similarly tethered phosphonate was used to prepare the exocyclic olefin in Danishefsky's synthesis of calicheamicinone **347** [147]. Chemoselective acylation of the advanced intermediate **348** provided the olefination precursor **349**, and subsequent cyclization gave δ-lactone **350**, enforcing the desired stereochemistry of the double bond. The additional conjugative stabilization of the "exocyclic" olefin with the ester also facilitated subsequent transformations of the azide to the methyl carbamate functionality, and further manipulations ultimately led to the first synthesis of calicheamicinone **347** (Scheme 10-117).

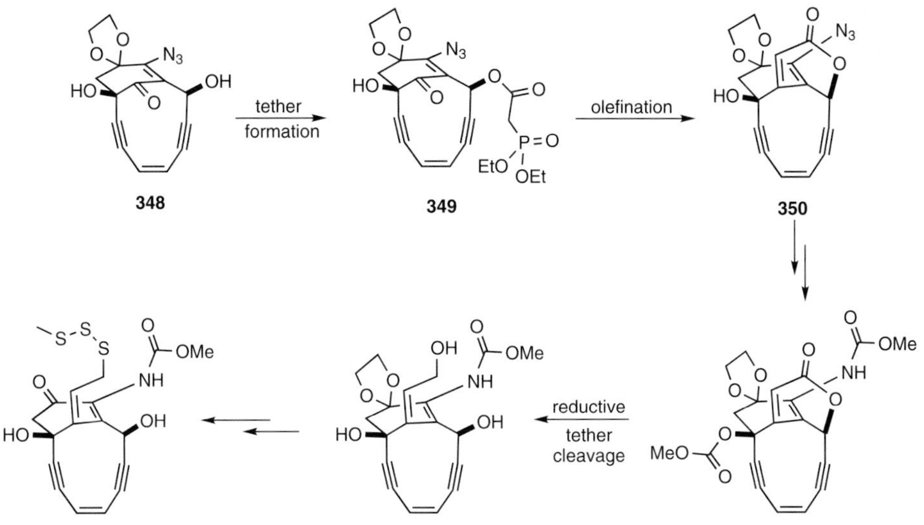

Scheme 10-117 A short tether ensures stereoselective incorporation of the exocyclic olefin functionality in calicheamicinone.

10.9 Ullmann Coupling

The Ullmann coupling and related reactions provide a useful route to biaryl ring systems [148], which occur in a number of biologically important natural products. Restricted rotation about the C–C bond connecting the two aryl rings leads to the formation of atropisomers, which are frequently configurationally stable even at elevated temperatures. As a result, biaryl systems have also become an extremely important source of chiral ligands for use in asymmetric catalysis. Methods for selectively preparing one atropisomer directly would circumvent the need for a tedious resolution of the racemate, which is the typical method for obtaining enantiomerically pure compounds. A number of research groups have tackled this problem by using a chiral, covalent tether to connect the two reacting aryl groups, thereby taking advantage of the intramolecularization of the process to transmit the asymmetry in the tether to the key C–C bond construction [149–153]. Clearly, if this approach is to compete with resolution protocols, tether formation and subsequent cleavage must be facile and high yielding, and the stereochemical transcription excellent.

Lipshutz and co-workers used a variety of readily available chiral diols to tether bromonaphthyl precursors through a bis-ether linkage [149a]. The nature of the tether was found to be crucial for achieving high asymmetric induction. BINOL **351** was prepared using three different tethers with varying degrees of success (Scheme 10-118). Using a lactic acid-derived tether (**352**), the binaphthyl product **353** was formed in 66% e.e., while a tether derived from mandelic acid (**354**) provided **355** in an improved 80–90% e.e. It was anticipated that increasing the *gauche* interactions in the tether, which were assumed to account for the stereochemical transcription, would increase the level of asymmetric induction. Diol **356**, derived from tartaric acid, was therefore twice coupled to 1-bromo-2-naphthol using Mitsunobu reactions, to form the tethered precursor **357**. Dilithiation with ^{t}BuLi and treatment with CuCN presumably led to the formation of the higher-order cyanocuprate **358**, which on exposure to O_2 at $0\,°C$ underwent oxidative coupling, providing a single diastereoisomeric product **359** in 78% yield. Benzylic oxidation with NBS followed by the addition of KOH solution accomplished tether cleavage, providing enantiomerically pure BINOL **351** (Scheme 10-118).

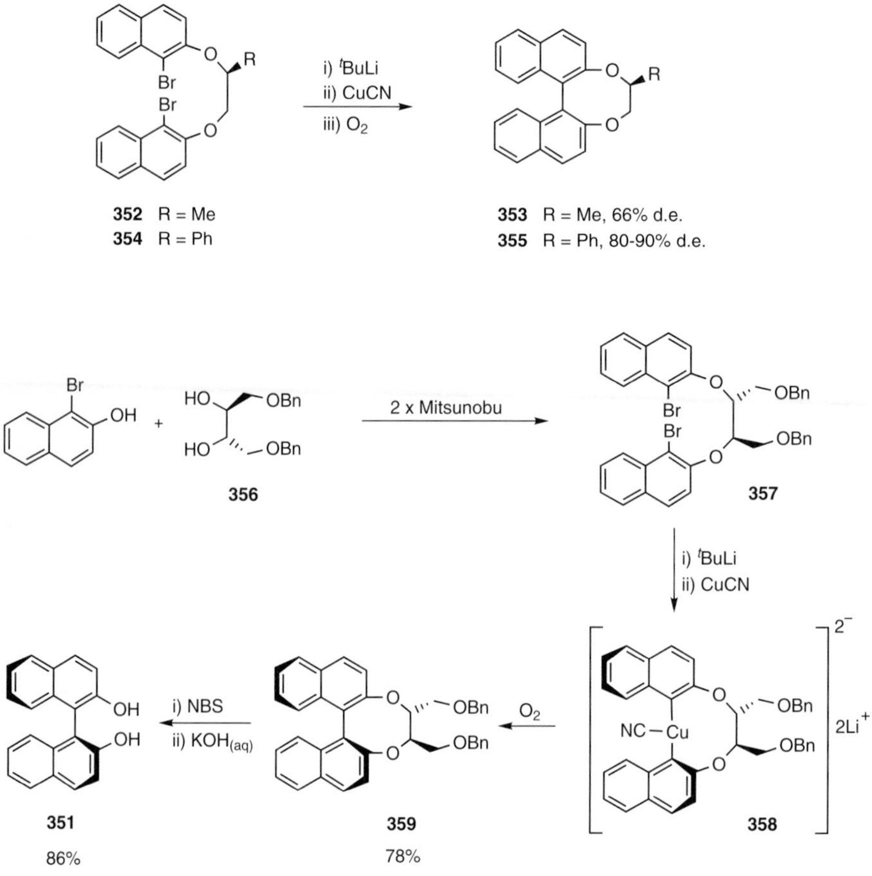

352 R = Me	**353** R = Me, 66% d.e.
354 R = Ph	**355** R = Ph, 80-90% d.e.

356

2 x Mitsunobu

357

i) ^{t}BuLi
ii) CuCN

351
86%

i) NBS
ii) KOH$_{(aq)}$

359
78%

O_2

358

Scheme 10-118 Appropriate choice of tether provided enantiomerically pure BINOL.

Iwasaki and co-workers have used a similar approach to prepare unsymmetrical biaryls [150]. Although they did not address the issue of enantioselectivity, their use of salicyl alcohol **360** as the temporary connection illustrates a number of desirable features in good tether design. The different nucleophilicity of a phenolic, compared with a benzylic, OH group allows the possibility for sequential, regioselective acylation of differently substituted 2-iodobenzoyl chlorides. For example, acylation of **360** with the benzoyl chloride **361** at low temperature proceeded regioselectively on the phenolic position but underwent complete acyl transfer on warming to room temperature to afford ester **362**. Without isolation, addition of the benzoyl chloride **363** to the reaction mixture at −20 to −30 °C provided the diester **364** in 84% yield (Scheme 10-119). Intramolecular Ullmann coupling also proceeded in excellent yield when a dilute solution of the diester **364** in DMF was added to a suspension of copper powder in DMF at reflux. Regioselective and sequential cleavage of the ester groups might be a desirable process if separate elaboration of each released acid functionality was necessary. Hydrogenolysis proved to be most effective in selectively cleaving the benzylic ester, providing monoester **365** in 82% yield (Scheme 10-119). The salicyl alcohol template has proved to be highly efficient for the synthesis of a wide variety of unsymmetrical diphenic acid derivatives in addition to heterobiaryls [150].

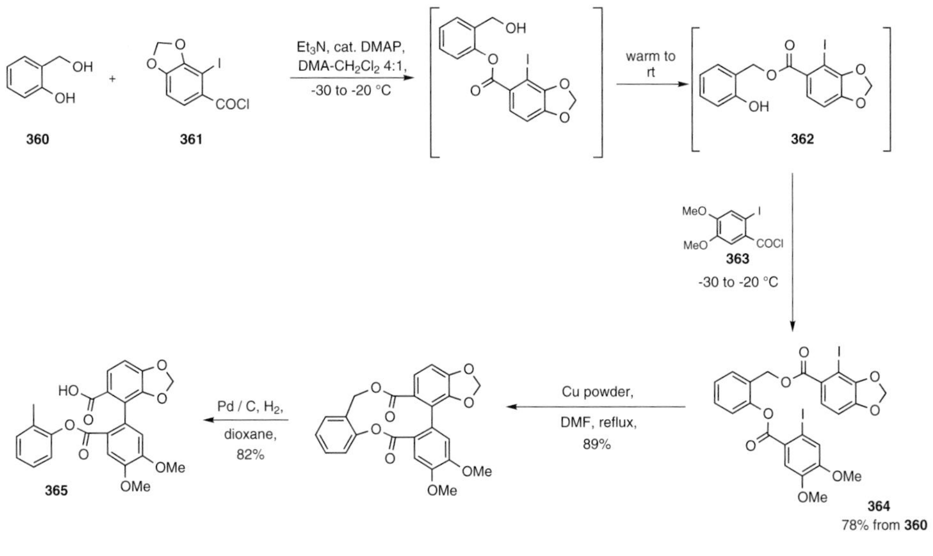

Scheme 10-119   Salicyl alcohol is a useful template for the preparation of unsymmetrical biaryls.

10.10   Use of the Temporary Connection in Carbenoid Chemistry

Marsden and co-workers have recently used α-diazo esters tethered through a silyl ether in stereoselective C–H insertion reactions to provide a novel route to 1,2,4-triols [154]. For example, treatment of α-diazo silyl ether **366** with Rh^{II} octanoate generated the corresponding carbenoid species, which underwent 1,5 C–H insertion exclusively into the equatorial C–H bond, providing oxasilacyclopentane **367** after reduction of the ester functionality with DIBALH (Scheme 10-120). It is noteworthy that the initially formed bicycle possesses a *trans* ring junction, whereas entry to similar systems using radical cyclization almost exclusively provides the *cis* product. This may be attributed to the avoidance of steric interactions between the cyclohexyl ring and the bulky rhodium catalyst in the insertion reaction. The silicon tether which had served to direct the insertion reaction then had to be unmasked to reveal the desired alcohol functionality. Standard protocols for achieving oxidative tether cleavage led to a competing Peterson-type olefination. Fortunately, the use of hot DMF as solvent alleviated this problematic side reaction, allowing triol **368** to be isolated in 35% yield from silyl ether **366** (Scheme 10-120).

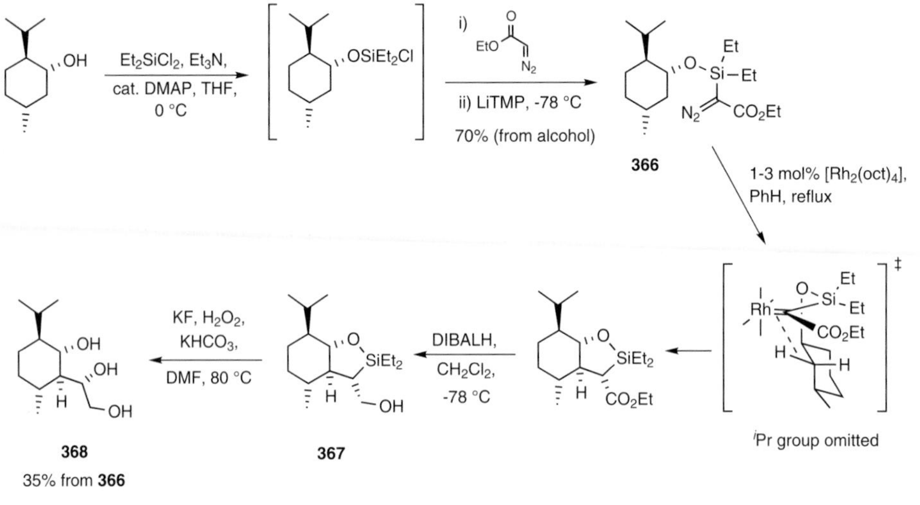

Scheme 10-120 α-Diazo silyl ethers can be used in stereoselective 1,5 C-H insertion reactions.

A similar tethering system was used to functionalize the A-ring of a number of steroidal derivatives regio- and stereoselectively. However, attempts to extend the reaction to acyclic systems were less successful, providing mixtures of diastereoisomers.

Transition metal-catalyzed carbenoid insertion into the π system of aryl compounds followed by ring expansion of the resultant [4.1.0]bicyclohepta-2,4-diene provides the best route to tropylidenes. Sugimura and co-workers have shown that a chiral pentane-2,4-diol linker can be used to connect a diazo carbenoid precursor with a phenolic substrate to control the regio- and stereochemical outcome of the initial cyclopropanation reaction [155]. Phenol-derived diazo ester **369** reacted in the presence of [Rh$_2$(OAc)$_4$] with complete regio- and stereocontrol, providing cyclopropane **370**. This rearranged stereospecifically to a single, diastereoisomeric cycloheptatriene product **371**, which was isolated in 80% yield after column chromatography (Scheme 10-121). The same group have used the same chiral linker to control the regio- and stereochemical outcome of other reactions including a *meta* arene–alkene photocycloaddition and a diastereodifferentiating *E*–*Z* photoisomerization of cyclooctene [156].

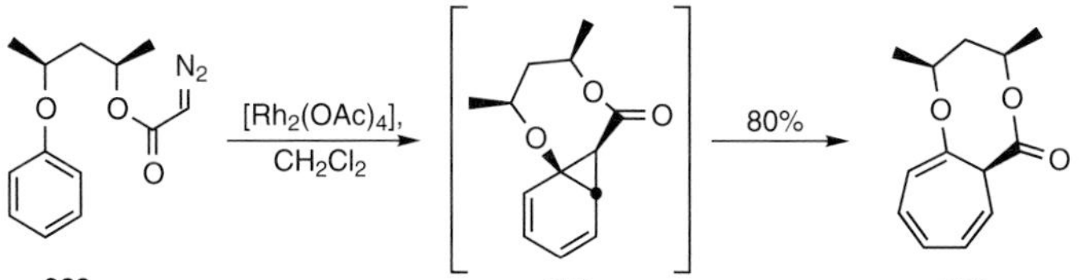

369 **370** **371**

Scheme 10-121   A chiral pentane-2,4-diol tether provides a diastereoselective route to tropylidenes.

10.11   The Temporary Sulfur Connection

It would be impossible to write a chapter on the use of the temporary connection in organic synthesis without discussing the Eschenmoser sulfide contraction developed by the Eschenmoser group during their pioneering studies towards the total synthesis of vitamin B$_{12}$ [157, 158]. Disconnecting the corrin nucleus of the vitamin B$_{12}$ precursor, cobyric acid **372** revealed two similarly sized fragments **373** and **374** (Scheme 10-122). The most convergent route to the preparation of the thioamide fragment **374** entailed cleavage of the vinylogous amidine. The low reactivity of possible reaction partners rendered an intermolecular approach to this bond formation difficult and unviable. Eschenmoser elegantly by-passed this problem by constructing a sulfide bridge between the two reacting centers, bringing them into close proximity and enabling this crucial bond formation [157].

   There are two versions of the so-called sulfide contraction although both use a thioamide as the source of sulfur (Scheme 10-123). In the alkylative version [159d, e], the thioamide behaves as a nucleophile, displacing a suitable leaving group α to a ketone and providing thioimino ester **375**. In the presence of a suitable base, enolization can occur, allowing nucleophilic attack on the proximal electrophilic imine and generating an episulfide intermediate **376**. Treatment with a thiophilic phosphine or phosphite causes extrusion of the sulfur tether, providing vinylogous amide **377**, which is readily con-

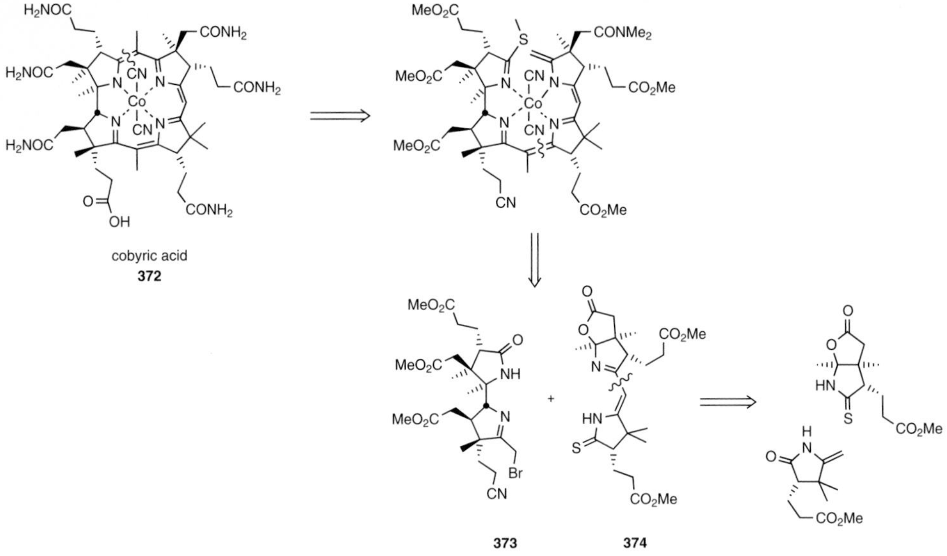

Scheme 10-122   Partial retrosynthesis for vitamin B$_{12}$.

verted to the corresponding vinylogous amidine **378**. The alternative procedure employs an oxidative coupling protocol to form the sulfide tether [159a, c]. Treatment of thioamide **379** with benzoyl peroxide provides either the disulfide or thiolactam-*S*-oxide. Whichever is produced, the net result is to make the sulfur sufficiently electrophilic for it to be attacked by the enamide **380**, providing the key sulfide intermediate **381**. Heating in the presence of a phosphine effects C–C bond formation and sulfur extrusion in one step (Scheme 10-123). In the actual synthesis of amidine **374**, the oxidative coupling protocol was used, providing a highly efficient and convergent approach to this key intermediate in the synthesis of the corrin nucleus of cobyric acid **372** [159a].

The ease with which C–S bonds are reductively cleaved makes a sulfide connection an attractive single-atom tether. Moyano, Pericàs, and co-workers have used such a linker in their synthesis of (+)-β-cuparenone **382** [160]. It was envisaged that the cyclopentanone framework could be accessed using a Pauson–Khand reaction in which the acetylene and alkene components would be tethered through a sulfide bridge. A chiral auxiliary attached to the olefin would impart diastereocontrol on the cyclization, forming a bicyclic product, and the sulfide tether would intramolecularize the process, facilitating the cyclization. The cupped shape of the resulting bicycle would then be exploited in a conjugate addition of a nucleophile from the less hindered convex face, installing the key stereocenter. Reductive cleavage of the sulfur bridge would unmask two methyl groups, with the third being incorporated using standard manipulations. Reaction of enyne **383** with [Co$_2$(CO)$_8$] using NMO as a promoter gave an 8:1 diastereoisomeric mixture of bicycles with the desired (4*R*,5*S*)-stereoisomer **384** predominating (Scheme 10-124).

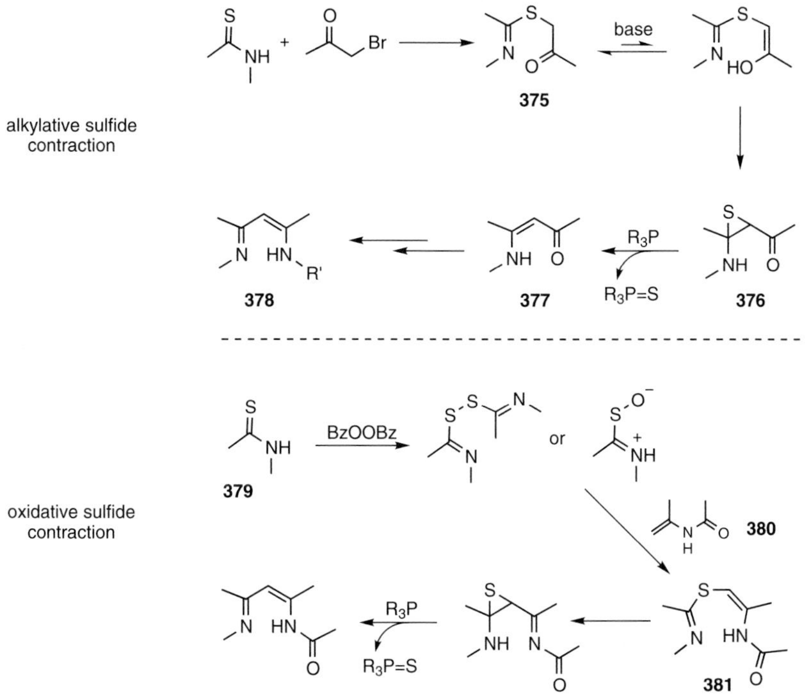

alkylative sulfide contraction

oxidative sulfide contraction

Scheme 10-123 The two versions of the Eschenmoser sulfide contraction.

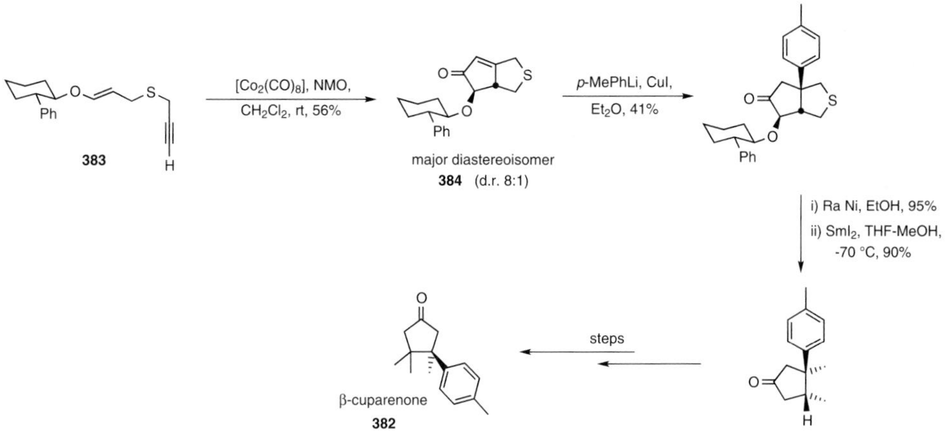

Scheme 10-124 A sulfide tether was used in the synthesis of β-cuparenone.

Conjugate addition of a cuprate generated from *p*-MePhLi and CuI also proceeded with complete stereocontrol as predicted. Subsequent cleavage of the sulfide bridge with Raney nickel (Ra-Ni) revealed two of the three methyl groups, and further manipulations led to the synthesis of *β*-cuparenone **382**. The same group used a similar approach in the preparation of *cis*-bicyclo[5.3.0]decan-8-ones and *cis*-bicyclo[6.3.0]decan-9-ones, providing an entry into the guaiane and homoguaiane classes of natural products [161].

10.12   Metathesis

Ring closing metathesis (RCM) has emerged as an important addition to the arsenal of methodologies used in modern organic synthesis and the introduction of readily handled ruthenium catalysts, e.g., **385**, by Grubbs and co-workers, has further increased the synthetic utility of this reaction [162]. Not surprisingly, the use of temporary tethering protocols in combination with the high efficiency of forming large rings (>12) by RCM, has been explored (see Chapter 9).

In a recent example, Grubbs wished to create an all-carbon analog of the cysteine dimer **386**, which contains a disulfide bridge [163]. Intermolecular metathesis on (*S*)-*C*-allylglycine proved ineffective, even under forcing conditions. Once again, the use of a temporary connection proved to be an efficient solution to the problem. By connecting the amino acid monomers to a catechol template **387**, RCM proceeded with good efficiency providing the 12-membered macrocycle **388** as a mixture of olefin isomers. Reduction of the olefin mixture and ester hydrolysis led to the desired amino acid analog **389** (Scheme 10-125).

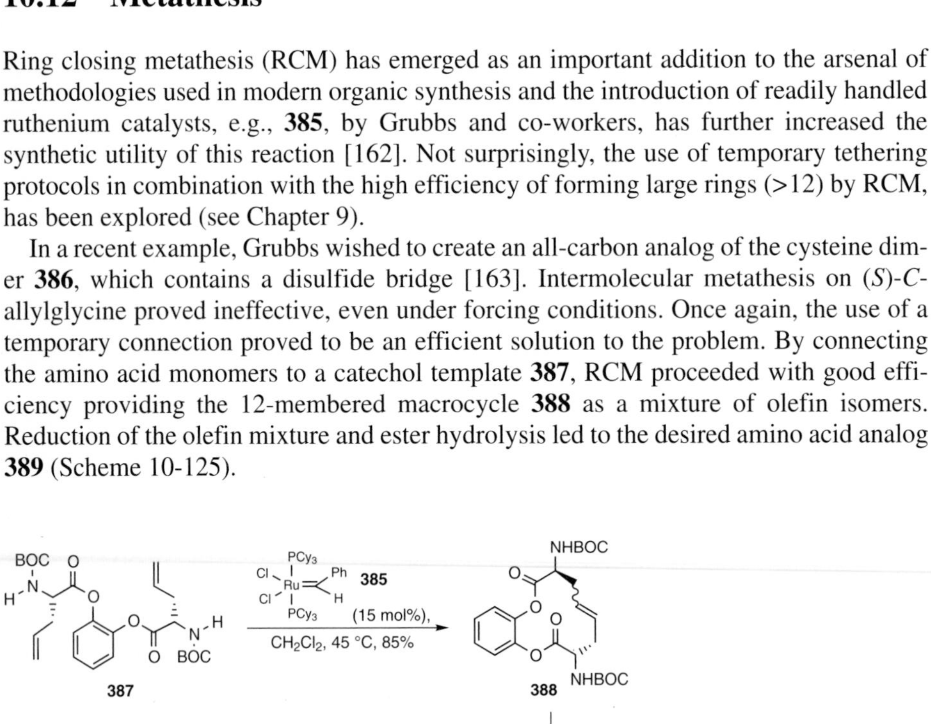

Scheme 10-125   A catechol template makes possible the synthesis of a cysteine-dimer analog using RCM.

Evans and Murthy used RCM of enantiomerically enriched allylic alcohols tethered through a silicon atom to prepare the reduced carbohydrate D-altritol **390** [164]. Treatment of diphenylsilyl acetal **391** with Grubbs' catalyst **385** in CH₂Cl₂, heating at reflux, provided the C_2-symmetrical silacycle **392** in excellent yield. Dihydroxylation of the *cis*-olefin and simple manipulations led to a facile, expeditious route to D-altritol (Scheme 10-126).

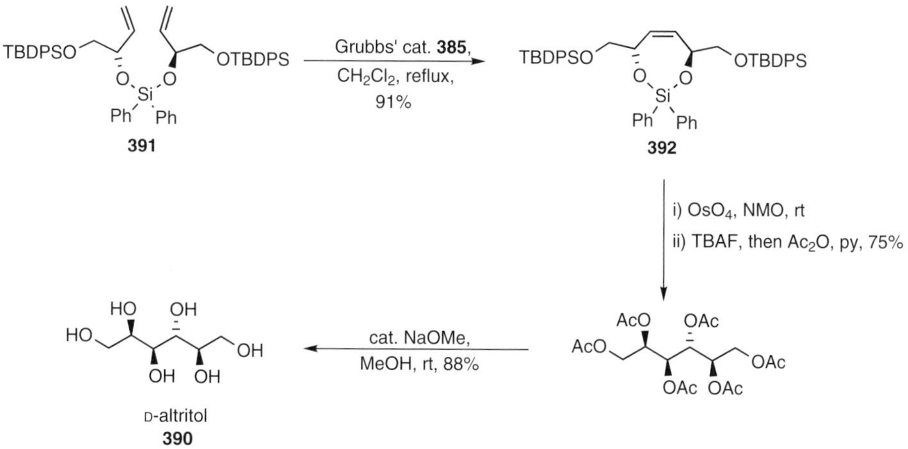

Scheme 10-126   RCM provides a rapid synthesis of D-altritol.

The mild conditions and high chemoselectivity of the metathesis reaction ensures that a wide variety of temporary connection strategies are compatible, and it is envisaged that this strategy will find increasing use in the future.

10.13   Dötz Benzannulation

The benzannulation reaction is a versatile method for the formation of polysubstituted aromatic compounds such as naphthoquinones. This three-component coupling involves the reaction between an α,β-unsaturated Fischer carbene, an acetylene, and a CO ligand, and initially proceeds by cycloaddition of the alkyne with the carbene complex. The regioselectivity of this step is highly dependent on the substituents on the acetylene moiety and is usually low in the case of internal acetylenes.

To overcome potential problems, Semmelhack et al. employed a tethered disubstituted acetylene to control the regioselectivity of the synthesis of the internal six-membered ring in his synthesis of deoxyfrenolicin **393** [165]. Fischer carbene complex **394**, bearing a pendant acetylene, was reacted in Et₂O at 35 to 37 °C for 64 h, and after oxidative workup with DDQ, provided the quinone **395** in 51% yield as a single regioisomer with the correct substitution pattern for subsequent conversion to deoxyfrenolicin (Scheme 10-127).

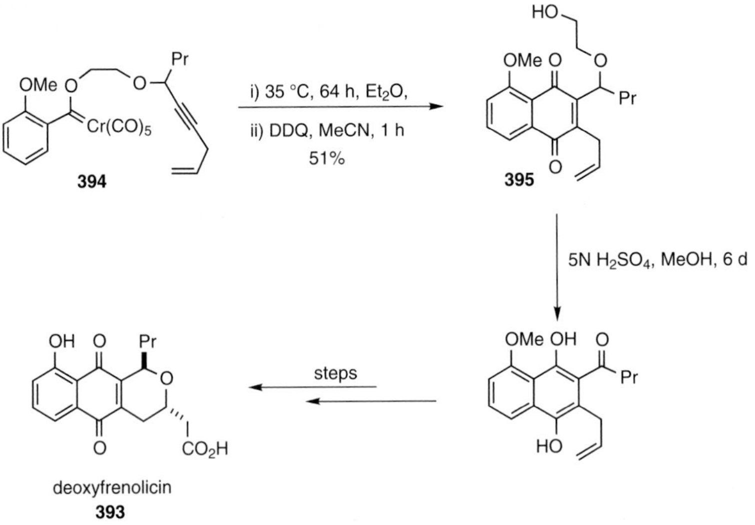

Scheme 10-127   A tethered acetylene ensures the correct regioselectivity in the benzannulation reaction.

10.14   Miscellaneous

Moeller and co-workers have shown that electrochemically initiated oxidative cyclization between enol ethers and between enol ethers and vinyl or allyl silanes proceeds with high stereoselectivity (Scheme 10-128) [166]. They envisaged that the use of a temporary tether to connect the reacting partners would provide a stereoselective route to quaternary centers in an acyclic system. A silicon tether was deemed most appropriate since it could be incorporated into one of the reacting partners as a vinylsilane. Accordingly, oxidation of precursor **396** provided a single isomeric product **397** in good yield, demonstrating the compatibility of the silicon tether to the electrolysis conditions (Scheme 10-128).

In a related study, intramolecular oxidative coupling of enolate derivatives has been investigated by Schmittel and co-workers [167]. The intermolecular version of this reaction provides a useful route to 1,4-dicarbonyl compounds but typically suffers from low levels of stereoselectivity. Furthermore, in mixed systems, the desired heterocoupling products are often accompanied by appreciable amounts of homocoupling products. It was hoped that the use of a single metal for both enolate precursors with concomitant intramolecularization of the bond-forming event might overcome some of these problems.

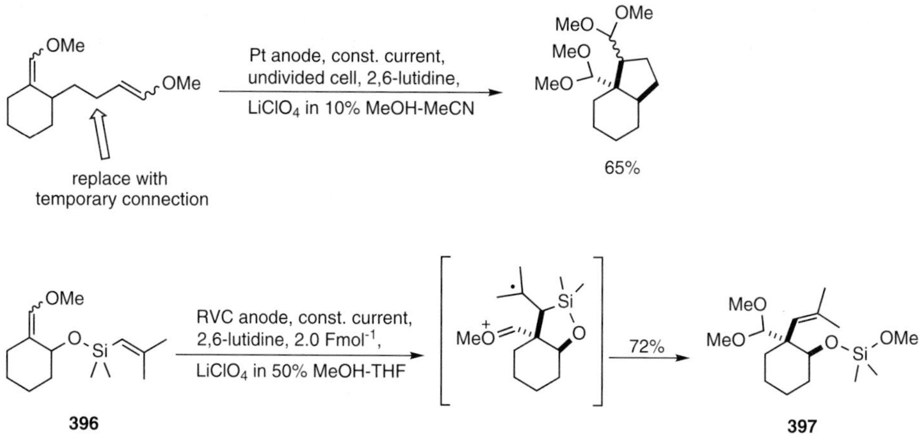

Scheme 10-128 Electrochemical oxidative cyclization provides a route to stereodefined quaternary centers.

The silyl and titanium bis-enolates **398** and **399** were prepared and treated with a variety of single-electron transfer reagents. In the case of silyl bis-enolate **398**, oxidation with $[Fe(phen)_3]^{3+}$ provided diketone **400** in moderate yield and good d.e. (Scheme 10-129). The titanium analog **399** reacted much more rapidly (1 min; cf. 12 h for **398**) under similar conditions – which is in accord with its lower redox potential – but provided a 1:1 mixture of *meso/dl* products. This result once more emphasizes the importance of the nature of the tether when C–C bond formation and tether cleavage both occur during the reaction. C–C bond formation occurs prior to tether cleavage only in the case of silyl enol ether **398**, illustrating the increased levels of diastereocontrol typical of an intramolecular process.

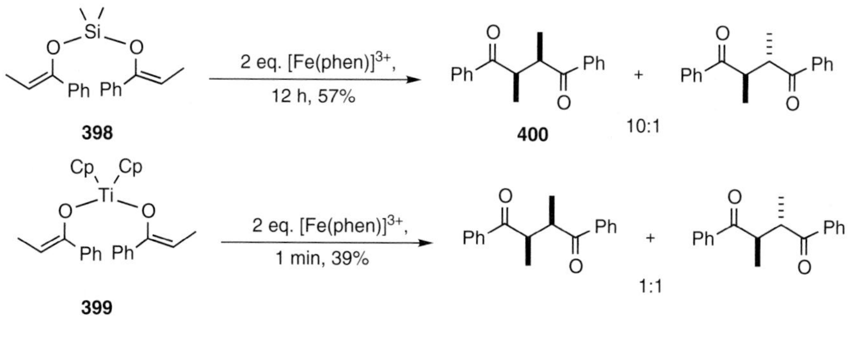

Scheme 10-129 Intramolecular coupling of silyl bis-enolates occurs with good diastereocontrol.

The rigid, well defined structures of functionalized steroids provide a cheap and readily available source of chiral templates for use in a variety of molecular recognition processes. Maitra and Bandyopadhyay have used methyl 7-deoxycholate as a template in a stereoselective synthesis of the 1,1'-binaphthyl derivative **401** [168]. The two hydroxy groups at C-3 and C-12 are suitably sited for attaching naphthyl coupling precursors via ester linkages **402**. Oxidative coupling between the naphthol units in **402** was best achieved using [Mn(acac)₃] in MeCN, providing the binaphthyl product **403** in 65% yield and 65% d.e. Reductive cleavage and in situ acetylation then gave the enantiomerically enriched binaphthyl compound **401** (Scheme 10-130). Thus the steroidal template imparted a moderate degree of diastereoselection on the coupling reaction although further modification of the reaction conditions and the use of different steroid templates could improve the degree of asymmetric induction. The same group have also reported an enantioselective synthesis of Tröger's base using a similar templating strategy [169].

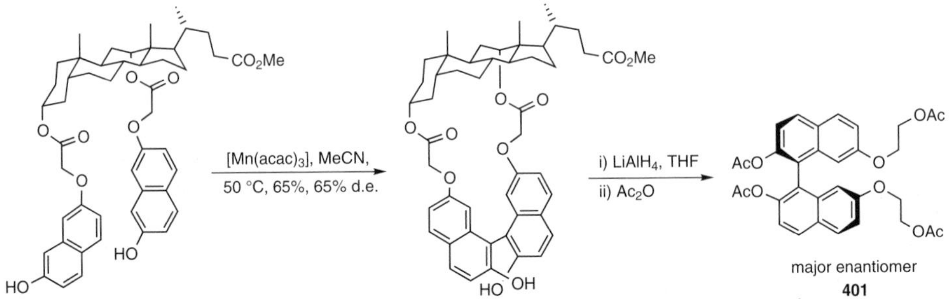

Scheme 10-130   Methyl 7-deoxycholate provides a chiral template for the enantioselective synthesis of the binaphthyl compound **401**.

10.15   Concluding Remarks

In this review we have tried to give some idea of the importance of the concept of using a temporary connection between two compatible reacting components for organic synthesis. This process of intramolecularizing an event often gives considerable advantages over its intermolecular counterpart and is a fairly established tactic for synthesis. However, although a wide variety of methods are now available for linking the two reactive centers together and, similarly, for eventual removal of the temporary connection, we are not yet using the full power of this method as a *primary* design feature in synthesis; rather it is more commonly used to overcome problems encountered during the execution of the initially designed pathway. Nevertheless, as our knowledge increases and more methods become available, we can consider using more imaginative routes based upon the use of the temporary connection. The whole area continues to grow and will become a crucial part of the synthetic planning process in the future.

10.16 Experimental Procedures

10.16.1 Use of an Acetal Tether in Diels–Alder Reactions: IMDA Reaction of Dimethyl Acetal 28 (Scheme 10-9) [15]

10.16.1.1 Tether formation

Tebbe's reagent (1.35 mL of a 0.50 M solution in PhMe, 0.68 mmol) was added dropwise over 5 min to a solution of acetate **29** (87 mg, 0.62 mmol) in PhMe (0.93 mL), THF (0.31 mL), and pyridine (10 μL) at −40 °C. After 15 min, the solution was warmed to rt over 2 h and then stirred for a further 2 h. The reaction mixture was cooled to −10 °C and NaOH (0.19 mL of a 15% aqueous solution) was carefully added. The mixture was warmed to rt and stirring continued until no effervescence was observed. Et_2O (8 mL) was then added and the reaction mixture dried (Na_2SO_4) and filtered through Celite®. Removal of the Et_2O in vacuo provided an orange solution, which was further diluted with pentane (30 mL). Filtration through Celite® and removal of the pentane in vacuo provided a yellow solution of the vinyl ether **30** (ca. 0.3 M solution in PhMe) which was used directly in the next step. A solution of methyl 4-hydroxybut-2-enoate **3** (60 mg, 0.52 mmol) in PhMe (1 mL) and [Pd(COD)Cl$_2$] (35 mg, 0.12 mmol) were added sequentially to the solution of **30**, and the resulting mixture was stirred at rt for 24 h. Pyridine (20 μL) was then added. The mixture was diluted with Et_2O and filtered through Celite®. Removal of the solvent in vacuo provided a yellow oil which was purified by flash column chromatography [Et_2O/petroleum ether (40–60 °C boiling point fraction), 1:4] to yield the triene **28** as a colorless oil (112 mg, 71%).

10.16.1.2 IMDA reaction and tether cleavage

A solution of triene **28** (100 mg, 0.39 mmol) in Et_2O was filtered through an alumina column, dried by azeotroping with PhMe (3 × 10 mL) and then redissolved in dry PhMe (15 mL). The resulting solution was degassed by the freeze–pump–thaw technique and then heated at 165 °C for 2.5 h in a sealed tube. Removal of the solvent in vacuo provided a mixture of diastereoisomeric cycloadducts (ratio 2.7:1, 94 mg, 94% mass recovery). Concentrated HCl (two drops) was added to a solution of the cycloadducts in MeOH (10 mL) and the reaction mixture stirred for 2 h. Solid $NaHCO_3$ was then added until effervescence ceased. The solid was removed by filtration and the residue washed with CH_2Cl_2. The combined organic fractions were concentrated in vacuo and purification of the residue by flash column chromatography [EtOAc/petroleum ether (40–60 °C boiling point fraction), 3:17] provided the hydroxy lactones **31** and **32** (2.7:1, 76 mg, 72%).

10.16.2 Use of a Boronate Tether in Diels–Alder Reactions: IMDA Reaction between Anthrone 50 and Methyl 4-Hydroxybut-2-enoate 3 (Scheme 10-19) [20a]

10.16.2.1 Tether formation and IMDA reaction

A solution of methyl 4-hydroxybut-2-enoate **3** (20 mg, 0.17 mmol) in pyridine (1.5 mL) was added to a solution of anthrone **50** (33 mg, 0.17 mmol) and PhB(OH)$_2$ (22 mg, 0.18 mmol) in pyridine (1.5 mL). The resulting solution was then heated at reflux for 5 h with azeotropic removal of H$_2$O. Removal of the solvent in vacuo and purification of the residue by preparative thin layer chromatography (hexane/EtOAc, 3:1) provided the Diels–Alder adduct **51** (55 mg, 81%), m.p. 218 °C.

10.16.2.2 Tether cleavage

A solution of the boronate **51** (38 mg, 0.10 mmol) in THF (3 mL) was added to hydrogen peroxide (30% solution, 5 mL) and the mixture was stirred at rt for 12 h. NaHSO$_4$ (1 g) was then added and the product was extracted with CH$_2$Cl$_2$ (2×25 mL). The combined organic extracts were washed with brine and dried (Na$_2$SO$_4$). Removal of the solvent in vacuo and purification of the residue by preparative thin layer chromatography (hexane/EtOAc, 3:1) provided the diol **52** (27 mg, 90%), m.p. 136 °C.

10.16.3 Use of the (Bromomethyl)dimethylsilyl Ether Group in a Radical Cyclization in the Synthesis of Talaromycin A, 140 [54]

10.16.3.1 Tether formation

Et$_3$N (40 µL, 0.29 mmol) and imidazole (20 mg, 0.29 mmol) were added to a solution of (4*R*,8*R*)-8-ethyl-4-hydroxy-1,7-dioxaspiro[5.5]undec-2-ene (38 mg, 0.29 mmol) in DMF (2 mL) and the resulting solution was stirred for 5 min. (Bromomethyl)dimethylsilyl chloride (40 µL, 0.29 mmol) was then added and the reaction was stirred for 3 h. The reaction was quenched by the addition of NaHCO$_3$ solution (satd) and diluted with Et$_2$O. The organic layer was washed with brine, dried (MgSO$_4$) and then concentrated in vacuo. Purification of the residue by flash column chromatography (EtOAc/hexane, 1:1) provided the silyl ether **139** as a labile, colorless oil (62 mg, 93%).

10.16.3.2 Radical cyclization and tether cleavage (Scheme 10-47)

A solution of Bu$_3$SnH (60 µL, 0.21 mmol) and AIBN (9 mg, 0.05 mmol) in benzene (1 mL) was added via syringe pump over 2 h to a solution of silyl ether **139** (62 mg, 0.18 mmol) in benzene (4 mL, degassed) heated at reflux. The reaction mixture was heated for a further 5 h and then concentrated in vacuo to provide the crude silacycle, which was used directly: the crude product was added to a mixture of H$_2$O$_2$ (30% solution, 0.2 mL), NaHCO$_3$ (30 mg), MeOH (2 mL) and THF (2 mL) and the mixture heated at reflux for 12 h. After cooling to rt, the mixture was diluted with H_2O, $NaHSO_3$ (10% solution) and $NaHCO_3$ (10% solution) and then placed in a continuous extractor and extracted with Et_2O for 4 h. Concentration of the Et_2O layer and purification by flash column chromatography (EtOAc/hexane, 1:9 to neat EtOAc) afforded talaromycin A, **140** (34 mg, 84%).

10.16.4   Use of Acetals in Intramolecular O-glycosylation: Synthesis of ethyl 2,3,6-Tri-*O*-benzyl-4-*O*-(3,4,6-tri-*O*-benzyl-*β*-D-mannopyranosyl)-*α*-D-glucopyranoside 246 (Scheme 10-77) [89c]

10.16.4.1   Tether formation

*p*TSA · H_2O (5 mg, 0.028 mmol) was added to a cooled (−40 °C) solution of vinyl ether **244** (374 mg, 0.699 mmol) and alcohol **245** (325 mg, 0.699 mmol) in CH_2Cl_2 (15 mL). The progress of the reaction was carefully monitored by thin layer chromatography and quenched by the addition of Et_3N (0.1 mL) when the product spot was major. Concentration of the reaction mixture and purification of the residue by flash column chromatography (EtOAc/hexane, 18:82 + 0.1% Et_3N) provided the acetal **243** as a colorless syrup (397 mg, 57%). The product decomposed after extended periods (>24 h) unless stabilized in an organic solvent (CH_2Cl_2 or EtOAc/hexane) with added Et_3N (0.5%) (shelf-life >2 months).

10.16.4.2   Glycosylation and tether cleavage

4-Me-DTBP (56 mg, 0.028 mmol) was added to a solution of acetal **243** (55 mg, 0.055 mmol) in CH_2Cl_2 (5 mL) at 0 °C. NIS (62 mg, 0.280 mmol) was then added in one portion and the reaction mixture was allowed to warm to rt. After 18 h, $Na_2S_2O_3$ (3 mL of a 0.5 M solution) was added. CH_2Cl_2 (10 mL) was then added and the resulting solution was washed with H_2O (2×10 mL). The phases were separated and the organic fraction dried ($MgSO_4$), filtered, and concentrated. Purification of the residue by flash column chromatography (EtOAc/hexane, 2:3) provided the disaccharide **246** as a clear syrup (38 mg, 77%).

References

[1] "Intramolecularization" is used to describe the process of making an intermolecular transformation intramolecular, hence the verb "intramolecularize"; for a previous use see, for example, T.-L. Ho in *Tactics of Organic Synthesis*, Wiley, New York, **1994**, Chapter 6.

[2] G. Stork, H. S. Suh, G. Kim, *J. Am. Chem. Soc.* **1991**, *113*, 7054–7056.

[3] (a) M. Bols, T. Skrydstrup, *Chem. Rev.* **1995**, *95*, 1253–1277; (b) L. Fensterbank, M. Malacria, S. McN. Sieburth, *Synthesis* **1997**, 813–854; (c) D. R. Gauthier, Jr., K. S. Zandi, K. J. Shea, *Tetrahedron* **1998**, *54*, 2289–2238.

[4] (a) W. Oppolzer, *Angew. Chem., Int. Ed. Engl.* **1977**, *16*, 10–23; (b) D. Craig, *Chem. Soc. Rev.* **1987**, *16*, 187–238.

[5] K. J. Shea, S. Wise, *J. Am. Chem. Soc.* **1978**, *100*, 6519–6521.

[6] (a) D. Craig, J. C. Reader, *Tetrahedron Lett.* **1990**, *31*, 6585–6588; (b) D. Craig, J. C. Reader, *Tetrahedron Lett.* **1992**, *33*, 4073–4076; (c) D. Craig, J. C. Reader, *Tetrahedron Lett.* **1992**, *33*, 6165–6168; (d) P. J. Ainsworth, D. Craig, J. C. Reader, A. M. Z. Slawin, A. J. P. White, D. J. Williams, *Tetrahedron* **1995**, *51*, 11601–11622.

[7] J. W. Gillard, R. Fortin, E. L. Grimm, M. Maillard, M. Tjepkema, M. A. Bernstein, R. Glaser, *Tetrahedron Lett.* **1991**, *32*, 1145–1148.

[8] G. H. Posner, C.-G. Cho, T. E. N. Anjeh, N. Johnson, R. L. Horst, T. Kobayashi, T. Okano, N. Tsugawa, *J. Org. Chem.* **1995**, *60*, 4617–4628.

[9] For a recent example of the use of a vinyl silyl *amine* in the synthesis of (+)-aloperine: A. D. Brosius, L. E. Overman, L. Schwink, *J. Am. Chem. Soc.* **1999**, *121*, 700–709.

[10] G. Stork, T. Y. Chan, G. A. Breault, *J. Am. Chem. Soc.* **1992**, *114*, 7578–7579.

[11] (a) K. Tamao, T. Kakui, M. Akita, T. Iwahara, R. Kanatani, Y. Yoshida, M. Kumada, *Tetrahedron* **1983**, *39*, 983–990; (b) I. Fleming, *Chemtracts (Org. Synth.).* **1996**, *9*, 1–64.

[12] (a) S. McN. Sieburth, L. Fensterbank, *J. Org. Chem.* **1992**, *57*, 5279-5281; (b) S. McN. Sieburth, J. Lang, *J. Org. Chem.* **1999**, *64*, 1780–1781.

[13] K. Tamao, K. Kobayashi, Y. Ito, *J. Am. Chem. Soc.* **1989**, *111*, 6478–6480.

[14] For a more recent example: K. Kahle, P. J. Murphy, J. Scott, R. Tamagni, *J. Chem. Soc., Perkin Trans. 1* **1997**, 997–999.

[15] P. J. Ainsworth, D. Craig, A. J. P. White, D. J. Williams, *Tetrahedron* **1996**, *52*, 8937–8946.

[16] R. K. Boeckman Jr., K. G. Estep, S. G. Nelson, M. A. Walters, *Tetrahedron Lett.* **1991**, *32*, 4095–4098.

[17] (a) K. Okumura, K. Okazaki, K. Takeda, E. Yoshii, *Tetrahedron Lett.* **1989**, *30*, 2233–2236; (b) W. R. Roush, B. B. Brown, *Tetrahedron Lett.* **1989**, *30*, 7309–7312.

[18] R. L. Funk, C. J. Mossman, W. E. Zeller, *Tetrahedron Lett.* **1984**, *25*, 1655–1658.

[19] (a) D. Craig, J. C. Reader, *Synlett* **1992**, 757–758; (b) P. J. Ainsworth, D. Craig, J. C. Reader, A. M. Z. Slawin, A. J. P. White, D. J. Williams, *Tetrahedron* **1996**, *52*, 695–724.

[20] (a) K. Narasaka, S. Shimada, K. Osoda, N. Iwasawa, *Synthesis* **1991**, 1171–1172; (b) S. Shimada, K. Osoda, K. Narasaka, *Bull. Chem. Soc. Jpn.* **1993**, *66*, 1254–1257.

[21] (a) K. C. Nicolaou, J. J. Liu, C.-K. Hwang, W.-M. Dai, R. K. Guy, *J. Chem. Soc., Chem. Commun.* **1992**, 1118–1120; (b) K. C. Nicolaou, J.-J. Liu, Z. Yang, H. Ueno, E. J. Sorensen, C. F. Claiborne, R. K. Guy, C.-K. Hwang, M. Nakada, P. G. Nantermet, *J. Am. Chem. Soc.* **1995**, *117*, 634–644; (c) K. C. Nicolaou, Z. Yang, J. J. Liu, H. Ueno, P. G. Nantermet, R. K. Guy, C. F. Claiborne, J. Renaud, E. A. Couladouros, K. Paulvannan, E. J. Sorensen, *Nature (London)* **1994**, *367*, 630–634.

[22] (a) R. A. Batey, A. N. Thadani, A. J. Lough, *J. Am. Chem. Soc.* **1999**, *121*, 450–451; (b) R. A. Batey, A. N. Thadani, A. J. Lough, *Chem. Commun.* **1999**, 475–476.

[23] M. Koerner, B. Rickborn, *J. Org. Chem.* **1990**, *55*, 2662–2672.

[24] For an alternative tethered IMDA approach to taxol: (a) D. S. Millan, T. T. Pham, J. A. Lavers, A. G. Fallis, *Tetrahedron Lett.* **1997**, *38*, 795–798; (b) T. Wong, P. D. Wilson, S. Woo, A. G. Fallis, *Tetrahedron Lett.* **1997**, *38*, 7045–7048; (c) A. G. Fallis, *Pure Appl. Chem.* **1997**, *69*, 495–500.

[25] (a) E. Bovenschulte, P. Metz, G. Henkel, *Angew. Chem., Int. Ed. Engl.* **1989**, *28*, 202–203; (b) P. Metz, M. Fleischer, R. Fröhlich, *Synlett* **1992**, 985–987; (c) P. Metz, M. Fleischer, *Synlett* **1993**, 399–400; (d) P. Metz, E. Cramer, *Tetrahedron Lett.* **1993**, *34*, 6371–6374; (e) P. Metz, U. Meiners, R. Fröhlich, M. Grehl, *J. Org. Chem.* **1994**, *59*, 3687–3689; (f) P.

Metz, J. Stölting, M. Läge, B. Krebs, *Angew. Chem., Int. Ed. Engl.* **1994**, *33*, 2195–2197; (g) P. Metz, M. Fleischer, R. Fröhlich, *Tetrahedron* **1995**, *51*, 711–732.

[26] E. F. Landau, *J. Am. Chem. Soc.* **1947**, *69*, 1219.

[27] S. Doye, T. Hotopp, R. Wartchow, E. Winterfeldt, *Chem. Eur. J.* **1998**, *4*, 1480–1488.

[28] Note that caution must be exercised on adding the ^{S}BuLi as the reaction mixture is highly flammable.

[29] (a) K. J. Shea, P. S. Beauchamp, R. S. Lind, *J. Am. Chem. Soc.* **1980**, *102*, 4544–4546; (b) K. J. Shea, S. Wise, L. D. Burke, P. D. Davis, J. W. Gilman, A. C. Greeley, *J. Am. Chem. Soc.* **1982**, *104*, 5708–5715; (c) K. J. Shea, J. W. Gilman, *Tetrahedron Lett.* **1983**, *24*, 657–660; (d) K. J. Shea, L. D. Burke, *J. Org. Chem.* **1985**, *50*, 725–727; (e) K. J. Shea, W. M. Fruscella, R. C. Carr, L. D. Burke, D. K. Cooper, *J. Am. Chem. Soc.* **1987**, *109*, 447–452; (f) K. J. Shea, W. M. Fruscella, W. P. England, *Tetrahedron Lett.* **1987**, *28*, 5623–5626; (g) K. J. Shea, L. D. Burke, *J. Org. Chem.* **1988**, *53*, 318–327; (h) K. J. Shea, L. D. Burke, W. P. England, *J. Am. Chem. Soc.* **1988**, *110*, 860–864; (i) K. J. Shea, K. S. Zandi, A. J. Staab, R. Carr, *Tetrahedron Lett.* **1990**, *31*, 5885–5888; (j) K. J. Shea, A. J. Staab, K. S. Zandi, *Tetrahedron Lett.* **1991**, *32*, 2715–2718; (k) T. G. Lease, K. J. Shea, *J. Am. Chem. Soc.* **1993**, *115*, 2248–2260; (l) K. J. Shea, D. R. Gauthier Jr., *Tetrahedron Lett.* **1994**, *35*, 7311–7314; (m) C. D. Dzierba, K. S. Zandi, T. Möllers, K. J. Shea, *J. Am. Chem. Soc.* **1996**, *118*, 4711–4712; (n) J. M. Whitney, J. S. Parnes, K. J. Shea, *J. Org. Chem.* **1997**, *62*, 8962–8963.

[30] For the application of ester and related tethers in Type I IMDA: (a) M. Yoshioka, H. Nakai, M. Ohno, *J. Am. Chem. Soc.* **1984**, *106*, 1133–1135; (b) J. D. White, E. G. Nolen Jr., C. H. Miller, *J. Org. Chem.* **1986**, *51*, 1150–1152; (c) F. E. Ziegler, B. H. Jaynes, M. T. Saindane, *J. Am. Chem. Soc.* **1987**, *109*, 8115–8116; (d) D. H. Birtwistle, J. M. Brown, M. W. Foxton, *Tetrahedron* **1988**, *44*, 7309–7318; (e) G. A. Kraus, D. Bougie, R. A. Jacobson, Y. Su, *J. Org. Chem.* **1989**, *54*, 2425–2428; (f) D. Craig, M. J. Ford, J. A. Stones, *Tetrahedron Lett.* **1996**, *37*, 535–538.

[31] H. E. Hennis, *J. Org. Chem.* **1963**, *28*, 2570–2572.

[32] For examples of intramolecular [2+2] cycloadditions using ester and related tethers: (a) B. S. Green, Y. Rabinsohn, M. Rejtö, *J. Chem. Soc., Chem. Commun.* **1975**, 313–314; (b) B. S. Green, Y. Rabinsohn, M. Rejtö, *Carbohydr. Res.* **1975**, *45*, 115–126; (c) B. S. Green, A. T. Hagler, Y. Rabinsohn, M. Rejtö, *Isr. J. Chem.* **1976/77**, *15*, 124–130; (d) M. Sato, Y. Abe, H. Ohuchi, C. Kaneko, *Heterocycles* **1990**, *31*, 2115–2119; (e) B. König, S. Leue, C. Horn, A. Caudan, J.-P. Desvergne, H. Bouas-Laurent, *Liebigs Ann.* **1996**, 1231–1233; (f) S. Faure, S. Piva-Le Blanc, O. Piva, J.-P. Pete, *Tetrahedron Lett.* **1997**, *38*, 1045–1048; (g) C. L. Muller, J. R. Bever, M. S. Dordel, M. M. Kitabwalla, T. M. Reineke, J. B. Sausker, T. R. Seehafer, Y. Li, J. P. Jasinski, *Tetrahedron Lett.* **1997**, *38*, 8663–8666.

[33] For examples of intramolecular [2+2] cycloadditions using silicon tethers: (a) S. A. Fleming, S. C. Ward, *Tetrahedron Lett.* **1992**, *33*, 1013–1016; (b) S. C. Ward, S. A. Fleming, *J. Org. Chem.* **1994**, *59*, 6476–6479; (c) M. T. Crimmins, L. E. Guise, *Tetrahedron Lett.* **1994**, *35*, 1657–1660; (d) C. L. Bradford, S. A. Fleming, S. C. Ward, *Tetrahedron Lett.* **1995**, *36*, 4189–4192; (e) K. I. Booker-Milburn, S. Gulten, A. Sharpe, *Chem. Commun.* **1997**, 1385–1386.

[34] For other examples of intramolecular [2+2] cycloadditions: (a) B. A. Pearlman, *J. Am. Chem. Soc.* **1979**, *101*, 6398–6404; (b) B. A. Pearlman, *J. Am. Chem. Soc.* **1979**, *101*, 6404–6408; (c) D. Burdi, S. Hoyt, T. P. Begley, *Tetrahedron Lett.* **1992**, *33*, 2133–2136.

[35] M. Tanaka, K. Tomioka, K. Koga, *Tetrahedron Lett.* **1985**, *26*, 3035–3038.

[36] K. Tomioka, M. Tanaka, K. Koga, *Tetrahedron Lett.* **1982**, *23*, 3401–3404.

[37] Y. Inouye, M. Shirai, T. Michino, H. Kakisawa, *Bull. Chem. Soc. Jpn.* **1993**, *66*, 324–326.

[38] S. E. Denmark, A. Thorarensen, *Chem. Rev.* **1996**, *96*, 137–165.

[39] S. E. Denmark, A. R. Hurd, H. J. Sacha, *J. Org. Chem.* **1997**, *62*, 1668–1674.

[40] For other examples of tethered [3+2] cycloadditions: (a) J. Rong, P. Roselt, J. Plavec, J. Chattopadhyaya, *Tetrahedron* **1994**, *50*, 4921–4936; (b) P. Righi, E. Marotta, A. Landuzzi, G. Rosini, *J. Am. Chem. Soc.* **1996**, *118*, 9446–9447; (c) R. Dogbéavou, L. Breau, *Synlett* **1997**, 1208–1210; (d) P. Righi, E. Marotta, G. Rosini, *Chem. Eur. J.* **1998**, *4*, 2501–2512.

[41] M. E. Garst, B. J. McBride, J. G. Douglass III, *Tetrahedron Lett.* **1983**, *24*, 1675–1678.

[42] (a) A. Rumbo, L. Castedo, A. Mouriño, J. L. Mascareñas, *J. Org. Chem.* **1993**, *58*, 5585–5586; (b) J. L. Mascareñas, A. Rumbo, L. Castedo, *J. Org. Chem.* **1997**, *62*, 8620–8621.

[43] A. Rumbo, L. Castedo, J. L. Mascareñas, *Tetrahedron Lett.* **1997**, *38*, 5885–5886.

[44] For a similar tether-mediated intramolecular dipolar cycloaddition approach to the synthesis of (–)-kainic acid: S. Takano, Y. Iwabuchi, K. Ogasawara, *J. Chem. Soc., Chem. Commun.* **1988**, 1204–1206.

[45] (a) P. Garner, K. Sunitha, W.-B. Ho, W. J. Youngs, V. O. Kennedy, A. Djebli, *J. Org. Chem.* **1989**, *54*, 2041–2042; (b) P. P. Garner, P. B. Cox, S. J. Klippenstein, W. J. Youngs, D. B. McConville, *J. Org. Chem.* **1994**, *59*, 6510–6511; (c) P. Garner, P. B. Cox, J. T. Anderson, J. Protasiewicz, R. Zaniewski, *J. Org. Chem.* **1997**, *62*, 493–498.

[46] D. P. Curran, N. A. Porter, B. Giese (Eds.), *Stereochemistry of Radical Reactions*, VCH, Weinheim, **1996**, 23–115.

[47] J. E. Baldwin, *J. Chem. Soc., Chem. Commun.* **1976**, 734–736.

[48] (a) J. W. Wilt, *Tetrahedron* **1985**, *41*, 3979–4000; (b) J. W. Wilt, J. Lusztyk, M. Peeran, K. U. Ingold, *J. Am. Chem. Soc.* **1988**, *110*, 281–287.

[49] H. Nishiyama, T. Kitajima, M. Matsumoto, K. Itoh, *J. Org. Chem.* **1984**, *49*, 2298–2230.

[50] (a) A. L. J. Beckwith, C. J. Easton, A. K. Serelis, *J. Chem. Soc., Chem. Commun.* **1980**, 482–483; (b) A. L. J. Beckwith, T. Lawrence, A. K. Serelis, *J. Chem. Soc., Chem. Commun.* **1980**, 484–485; (c) A. L. J. Beckwith, C. J. Easton, T. Lawrence, A. K. Serelis, *Aust. J. Chem.* **1983**, *36*, 545–556; (d) A. L. J. Beckwith, C. H. Schiesser, *Tetrahedron Lett.* **1985**, *26*, 373–376; (e) D. C. Spellmeyer, K. N. Houk, *J. Org. Chem.* **1987**, *52*, 959–974.

[51] (a) G. Stork, M. Kahn, *J. Am. Chem. Soc.* **1985**, *107*, 500-501; (b) G. Stork, M. J. Sofia, *J. Am. Chem. Soc.* **1986**, *108*, 6826–6828; (c) G. Stork, R. Mah, *Tetrahedron Lett.* **1989**, *30*, 3609–3612.

[52] P. R. Jenkins, A. J. Wood, *Tetrahedron Lett.* **1997**, *38*, 1853–1856.

[53] J. Lejeune, J. Y. Lallemand, *Tetrahedron Lett.* **1992**, *33*, 2977–2980.

[54] M. T. Crimmins, R. O'Mahony, *J. Org. Chem.* **1989**, *54*, 1157–1161.

[55] (a) J. C. López, A. M. Gómez, B. Fraser-Reid, *J. Chem. Soc., Chem. Commun.* **1993**, 762–764; (b) J. C. López, A. M. Gómez, B. Fraser-Reid, *J. Chem. Soc., Chem. Commun.* **1994**, 1533-1534.

[56] J. C. López, B. Fraser-Reid, *J. Am. Chem. Soc.* **1989**, *111*, 3450–3452.

[57] P. Mayon, Y. Chapleur, *Tetrahedron Lett.* **1994**, *35*, 3703–3706.

[58] M. Koreeda, I. A. George, *J. Am. Chem. Soc.* **1986**, *108*, 8098–8100.

[59] M. Koreeda, D. C. Visger, *Tetrahedron Lett.* **1992**, *33*, 6603–6606.

[60] (a) A. Kurek-Tyrlik, J. Wicha, G. Snatzke, *Tetrahedron Lett.* **1988**, *29*, 4001–4004; (b) A. Kurek-Tyrlik, J. Wicha, A. Zarecki, G. Snatzke, *J. Org. Chem.* **1990**, *55*, 3484–3492.

[61] M. Koreeda, L. G. Hamann, *J. Am. Chem. Soc.* **1990**, *112*, 8175–8177.

[62] G. Agnel, M. Malacria, *Tetrahedron Lett.* **1990**, *31*, 3555–3558.

[63] G. Agnel, M. Malacria, *Synthesis* **1989**, 687–688.

[64] B. M. Trost, *Angew. Chem., Int. Ed. Engl.* **1986**, *25*, 1–20.

[65] (a) E. Magnol, M. Malacria, *Tetrahedron Lett.* **1986**, *27*, 2255–2256; (b) M. Journet, E. Magnol, W. Smadja, M. Malacria, *Synlett* **1991**, 58–60; (c) M. Journet, M. Malacria, *J. Org. Chem.* **1992**, *57*, 3085–3093.

[66] M. Journet, W. Smadja, M. Malacria, *Synlett* **1990**, 320–321.

[67] P. Devin, L. Fensterbank, M. Malacria, *J. Org. Chem.* **1998**, *63*, 6764–6765.

[68] M. Koreeda, I. A. George, *Chem. Lett.* **1990**, 83–86.

[69] K. Tamao, K. Nagata, Y. Ito, K. Maeda, M. Shiro, *Synlett* **1994**, 257–259.

[70] R. D. Walkup, R. R. Kane, N. U. Obeyesekere, *Tetrahedron Lett.* **1990**, *31*, 1531–1534.

[71] K. Tamao, K. Maeda, T. Yamaguchi, Y. Ito, *J. Am. Chem. Soc.* **1989**, *111*, 4984–4985.

[72] D. P. Curran, D. Kim, H. T. Liu, W. Shen, *J. Am. Chem. Soc.* **1988**, *110*, 5900–5902.

[73] (a) E. J. Corey, J. W. Suggs, *J. Org. Chem.* **1975**, *40*, 2554–2555; (b) G. Stork, P. M. Sher, *J. Am. Chem. Soc.* **1986**, *108*, 303–304.

[74] W.-J. Koot, R. van Ginkel, M. Kranenburg, H. Hiemstra, S. Louwrier, M. J. Moolenaar, W. N. Speckamp, *Tetrahedron Lett.* **1991**, *32*, 401–404.

[75] R. G. Andrew, R. E. Conrow, J. D. Elliot, W. S. Johnson, S. Ramezani, *Tetrahedron Lett.* **1987**, *28*, 6535–6538.

[76] (a) Z. Xi, P. Agback, A. Sandström, J. Chattopadhyaya, *Tetrahedron* **1991**, *46*, 9675–9690; (b) Z. Xi, P. Agback, J. Plavec, A. Sandström, J. Chattopadhyaya, *Tetrahedron* **1992**, *48*, 349-370; (c) Z. Xi, J. Rong, J. Chattopadhyaya, *Tetrahedron* **1994**, *50*, 5255–5272.

[77] (a) S. Shuto, M. Kanazaki, S. Ichikawa, A. Matsuda, *J. Org. Chem.* **1997**, *62*, 5676–5677; (b) S. Shuto, M. Kanazaki, S. Ichikawa, N. Minakawa, A. Matsuda, *J. Org. Chem.* **1998**, *63*, 746–754.

[78] J. H. Hutchinson, T. S. Daynard, J. W. Gillard, *Tetrahedron Lett.* **1991**, *32*, 573–576.

[79] (a) A. G. Myers, D. Y. Gin, K. L. Widdowson, *J. Am. Chem. Soc.* **1991**, *113*, 9661–9663; (b) A. G. Myers, D. Y. Gin, D. H. Rogers, *J. Am. Chem. Soc.* **1993**, *115*, 2036–2038.

[80] (a) G. Stork, P. M. Sher, *J. Am. Chem. Soc.* **1983**, *105*, 6765–6766; (b) G. Stork, *Bull. Chem. Soc. Jpn.* **1988**, *61*, 149–154.

[81] G. Stork, P. M. Sher, H.-L. Chen, *J. Am. Chem. Soc.* **1986**, *108*, 6384–6385.

[82] M. A. Blanchette, W. Choy, J. T. Davis, A. P. Essenfeld, S. Masamune, W. R. Roush, T. Sakai, *Tetrahedron Lett.* **1984**, *25*, 2183–2186.

[83] Y. Ito, T. Hirao, T. Saegusa, *J. Org. Chem.* **1978**, *43*, 1011–1013.

[84] A. V. Rama Rao, G. V. M. Sharma, M. N. Bhanu, *Tetrahedron Lett.* **1992**, *33*, 3907–3910.

[85] (a) Y. C. Xin, J.-M. Mallet, P. Sinaÿ, *J. Chem. Soc., Chem. Commun.* **1993**, 864–865; (b) B. Vauzeilles, D. Cravo, J.-M. Mallet, P. Sinaÿ, *Synlett* **1993**, 522–524; (c) A. Chénedé, E. Perrin, E. D. Rekaï, P. Sinaÿ, *Synlett* **1994**, 420–422.

[86] D. Mazéas, T. Skrydstrup, O. Doumeix, J.-M. Beau, *Angew. Chem., Int. Ed. Engl.* **1994**, *33*, 1383–1386.

[87] For an overview on the role of carbohydrates in biology: J. Lehmann, *Carbohydrates – Structure and Biology*, Thieme, Stuttgart, **1998**.

[88] For an overview on the preparation of oligosaccharides: S. Hanessian (Ed.), *Preparative Carbohydrate Chemistry*, Marcel Dekker, New York, **1996**.

[89] (a) F. Barresi, O. Hindsgaul, *J. Am. Chem. Soc.* **1991**, *113*, 9376–9377; (b) F. Barresi, O. Hindsgaul, *Synlett* **1992**, 759–761; (c) F. Barresi, O. Hindsgaul, *Can. J. Chem.* **1994**, *72*, 1447–1465.

[90] (a) G. Stork, G. Kim, *J. Am. Chem. Soc.* **1992**, *114*, 1087–1088; (b) G. Stork, J. J. La Clair, *J. Am. Chem. Soc.* **1996**, *118*, 247–248.

[91] D. Kahne, S. Walker, Y. Cheng, D. Van Engen, *J. Am. Chem. Soc.* **1989**, *111*, 6881–6882.

[92] (a) M. Bols, *J. Chem. Soc., Chem. Commun.* **1992**, 913–914; (b) M. Bols, *J. Chem. Soc., Chem. Commun.* **1993**, 791–792; (c) M. Bols, *Acta Chem. Scand.* **1993**, *47*, 829–834; (d) M. Bols, *Tetrahedron* **1993**, *49*, 10049–10060; (e) M. Bols, *Acta Chem. Scand.* **1996**, *50*, 931–937.

[93] M. Bols, H. C. Hansen, *Chem. Lett.* **1994**, 1049–1052.

[94] (a) Y. Ito, T. Ogawa, *Angew. Chem., Int. Ed. Engl.* **1994**, *33*, 1765–1767; (b) A. Dan, Y. Ito, T. Ogawa, *J. Org. Chem.* **1995**, *60*, 4680–4681; (c) A. Dan, Y. Ito, T. Ogawa, *Tetrahedron Lett.* **1995**, *36*, 7487–7490; (d) Y. Ito, T. Ogawa, *J. Am. Chem. Soc.* **1997**, *119*, 5562–5566; (e) Y. Ito, Y. Ohnishi, T. Ogawa, Y. Nakahara, *Synlett* **1998**, 1102–1104; (f) A. Dan, M. Lergenmüller, M. Amano, Y. Nakahara, T. Ogawa, Y. Ito, *Chem. Eur. J.* **1998**, *4*, 2182–2190.

[95] For a recent review: P. H. Seeberger, S. J. Danishefsky, *Acc. Chem. Res.* **1998**, *31*, 685–695.

[96] (a) C. Krog-Jensen, S. Oscarson, *J. Org. Chem.* **1996**, *61*, 4512–4513; (b) C. Krog-Jensen, S. Oscarson, *J. Org. Chem.* **1998**, *63*, 1780–1784.

[97] (a) S. Valverde, A. M. Gómez, A. Hernández, B. Herradón, J. C. López, *J. Chem. Soc., Chem. Commun.* **1995**, 2005–2006; (b) S. Valverde, A. M. Gómez, J. C. López, B. Herradón, *Tetrahedron Lett.* **1996**, *37*, 1105–1108.

[98] A. Morcuende, S. Valverde, B. Herradón, *Synlett* **1994**, 89–91.

[99] (a) R. Lau, G. Schüle, U. Schwaneberg, T. Ziegler, *Liebigs. Ann.* **1995**, 1745–1754; (b) T. Ziegler, R. Lau, *Tetrahedron Lett.* **1995**, *36*, 1417–1420; (c) T. Ziegler, G. Lemanski, A. Rakoczy, *Tetrahedron Lett.* **1995**, *36*, 8973–8976; (d) T. Ziegler, G. Lemanski, *Eur. J. Org. Chem.* **1998**, 163–170; (e) T. Ziegler, G. Lemanski, *Angew. Chem., Int. Ed. Engl.* **1998**, *37*, 3129–3132.

[100] H. Yamada, K. Imamura, T. Takahashi, *Tetrahedron Lett.* **1997**, *38*, 391–394.

[101] U. Huchel, R. R. Schmidt, *Tetrahedron Lett.* **1998**, *39*, 7693–7694.

[102] M. E. Behrendt, R. R. Schmidt, *Tetrahedron Lett.* **1993**, *34*, 6733–6736.

[103] (a) T. Iimori, T. Shibazaki, S. Ikegami, *Tetrahedron Lett.* **1996**, *37*, 2267–2270; (b) T. Iimori, I. Azumaya, T. Shibazaki, S. Ikegami, *Heterocycles* **1997**, *46*, 221–224.

[104] C. Mukai, T. Itoh, M. Hanaoka, *Tetrahedron Lett.* **1997**, *38*, 4595–4598.

[105] G. Scheffler, R. R. Schmidt, *Tetrahedron Lett.* **1997**, *38*, 2943–2946.

[106] H. Vorbrüggen, B. Bennua, *Tetrahedron Lett.* **1978**, 1339–1342.

[107] M. E. Jung, C. Castro, *J. Org. Chem.* **1993**, *58*, 807–808.

[108] K. Sujino, H. Sugimura, *Chem. Lett.* **1993**, 1187–1190.

[109] For some other examples of tether-directed nucleophile delivery: (a) J. B. Hendrickson, P. S. Palumbo, *J. Org. Chem.* **1985**, *50*, 2110–2112; (b) D. M. Hedstrand, S. R. Byrn, A. T. McKenzie, P. L. Fuchs, *J. Org. Chem.* **1987**, *52*, 592–598; (c) H. Ishibashi, T. S. So, H. Nakatani, K. Minami, M. Ikeda, *J. Chem. Soc., Chem. Commun.* **1988**, 827–828; (d) H. S. Park, I. S. Lee, D. W. Kwon, Y. H. Kim, *Chem. Commun.* **1998**, 2745–2746; (e) S. Knapp, *Chem. Soc. Rev.* **1999**, *28*, 61–72.

[110] (a) S. Mio, M. Ueda, M. Hamura, J. Kitagawa, S. Sugai, *Tetrahedron* **1991**, *47*, 2145–2154; (b) S. Mio, M. Shiraishi, S. Sugai, H. Haruyama, S. Sato, *Tetrahedron* **1991**, *47*, 2121–2132.

[111] P. Chemla, *Tetrahedron Lett.* **1993**, *34*, 7391–7394.

[112] S. Knapp, P. J. Kukkola, *J. Org. Chem.* **1990**, *55*, 1632–1636.

[113] W. R. Roush, *Comprehensive Organic Synthesis, Vol. 2* (Eds.: I. Fleming, B. M. Trost), Pergamon, Oxford, **1990**, pp. 1–53.

[114] (a) G. E. Keck, K. H. Tarbet, L. S. Geraci, *J. Am. Chem. Soc.* **1993**, *115*, 8467–8468; (b) A. Yanagisawa, H. Nakashima, A. Ishiba, H. Yamamoto, *J. Am. Chem. Soc.* **1996**, *118*, 4723–4724.

[115] (a) H. C. Brown, K. S. Bhat, *J. Am. Chem. Soc.* **1986**, *108*, 5919–5923; (b) R. O. Duthaler, P. Herold, W. Lottenbach, K. Oertle, M. Riediker, *Angew. Chem., Int. Ed. Engl.* **1989**, *28*, 494–495.

[116] M. T. Reetz, A. Jung, C. Bolm, *Tetrahedron* **1988**, *44*, 3889–3898.

[117] (a) O. R. Martin, S. P. Rao, K. G. Kurz, H. A. El-Shenawy, *J. Am. Chem. Soc.* **1988**, *110*, 8698–8700; (b) S. Uyeo, H. Itani, *Tetrahedron Lett.* **1991**, *32*, 2143–2144; (c) H. Hioki, M. Okuda, W. Miyagi, S. Itô, *Tetrahedron Lett.* **1993**, *34*, 6131–6134.

[118] C. Bismara, R. Di Fabio, D. Donati, T. Rossi, R. J. Thomas, *Tetrahedron Lett.* **1995**, *36*, 4283–4286.

[119] D. A. Livingston, J. E. Petre, C. L. Bergh, *J. Am. Chem. Soc.* **1990**, *112*, 6449–6450.

[120] M. D. Bachi, A. Melman, *J. Org. Chem.* **1997**, *62*, 1896–1898.

[121] (a) J. Robertson, G. O'Connor, D. S. Middleton, *Tetrahedron Lett.* **1996**, *37*, 3411–3414; (b) J. Robertson, D. S. Middleton, G. O'Connor, T. Sardharwala, *Tetrahedron Lett.* **1998**, *39*, 669–672.

[122] M. I. Johnston, J. A. Kwass, R. B. Beal, B. B. Snider, *J. Org. Chem.* **1987**, *52*, 5419–5424.

[123] (a) J. L. Speier, *Adv. Organomet. Chem.* **1979**, *17*, 407–447; (b) I. Ojima, *The Chemistry of Organic Silicon Compounds, Part 2* (Eds.: S. Patai, Z. Rappoport), Wiley, Chichester, **1989**, pp. 1479–1526.

[124] M. Taniguchi, K. Oshima, K. Utimoto, *Chem. Lett.* **1993**, 1751–1754.

[125] T. Harada, Y. Matsuda, J. Uchimura, A. Oku, *J. Chem. Soc., Chem. Commun.* **1989**, 1429–1430.

[126] (a) K. Tamao, T. Nakajima, R. Sumiya, H. Arai, N. Higuchi, Y. Ito, *J. Am. Chem. Soc.* **1986**, *108*, 6090–6093; (b) K. Tamao, T. Yamauchi, Y. Ito, *Chem. Lett.* **1987**, 171–174; (c) K. Tamao, Y. Nakagawa, H. Arai, N. Higuchi, Y. Ito, *J. Am. Chem. Soc.* **1988**, *110*, 3712–3714.

[127] (a) S. E. Denmark, D. C. Forbes, *Tetrahedron Lett.* **1992**, *33*, 5037–5040; (b) K. Tamao, Y. Nakagawa, Y. Ito, *Organometallics* **1993**, *12*, 2297–2308.

[128] D. G. J. Young, M. R. Hale, A. H. Hoveyda. *Tetrahedron Lett.* **1996**, *37*, 827–830.

[129] X. Wang, W. W. Ellis, B. Bosnich, *Chem. Commun.* **1996**, 2561–2562.

[130] N. R. Curtis, A. B. Holmes, *Tetrahedron Lett.* **1992**, *33*, 675–678.

[131] M. R. Hale, A. H. Hoveyda, *J. Org. Chem.* **1992**, *57*, 1643–1645.

[132] (a) K. Tamao, T. Tohma, N. Inui, O. Nakayama, Y. Ito, *Tetrahedron Lett.* **1990**, *31*, 7333–7336; (b) S. H. Bergens, P. Noheda, J. Whelan, B. Bosnich, *J. Am. Chem. Soc.* **1992**, *114*, 2121–2128; (c) S. H. Bergens, P. Noheda, J. Whelan, B. Bosnich, *J. Am. Chem. Soc.* **1992**, *114*, 2128–2135; (d) R. W. Barnhart, X. Wang, P. Noheda, S. H. Bergens, J. Whelan, B. Bosnich, *Tetrahedron* **1994**, *50*, 4335–4346; (e) X. Wang, B. Bosnich, *Organometallics* **1994**, *13*, 4131–4133.

[133] K. Tamao, Y. Nakagawa, Y. Ito, *J. Org. Chem.* **1990**, *55*, 3438–3439.

[134] K. Tamao, K. Maeda, T. Tanaka, Y. Ito, *Tetrahedron Lett.* **1988**, *29*, 6955–6956.

[135] I. Ojima, E. Vidal, M. Tzamarioudaki, I. Matsuda, *J. Am. Chem. Soc.* **1995**, *117*, 6797–6798.

[136] M. Suginome, H. Kinugasa, Y. Ito, *Tetrahedron Lett.* **1994**, *35*, 8635–8638.

[137] (a) M. Murakami, P. G. Andersson, M. Suginome, Y. Ito, *J. Am. Chem. Soc.* **1991**, *113*, 3987–3988; (b) Y. Ito, M. Suginome, M. Murakami, *J. Org. Chem.* **1991**, *56*, 1948–1951; (c) M. Murakami, H. Oike, M. Sugawara, M. Suginome, Y. Ito, *Tetrahedron* **1993**, *49*, 3933–3946; (d) M. Murakami, M. Suginome, K. Fujimoto, H. Nakamura, P. G. Andersson, Y. Ito, *J. Am. Chem. Soc.* **1993**, *115*, 6487–6498; (e) M. Suginome, A. Matsumoto, K. Nagata, Y. Ito, *J. Organomet. Chem.* **1995**, *499*, C1–C3.

[138] M. Suginome, S. Matsunaga, T. Iwanami, A. Matsumoto, Y. Ito, *Tetrahedron Lett.* **1996**, *37*, 8887–8890.
[139] M. Suginome, Y. Yamamoto, K. Fujii, Y. Ito, *J. Am. Chem. Soc.* **1995**, *117*, 9608–9609.
[140] M. Suginome, A. Matsumoto, Y. Ito, *J. Am. Chem. Soc.* **1996**, *118*, 3061–3062.
[141] M. Suginome, H. Nakamura, Y. Ito, *Tetrahedron Lett.* **1997**, *38*, 555–558.
[142] (a) S. Anwar, A. P. Davis, *J. Chem. Soc., Chem. Commun.* **1986**, 831–832; (b) S. Anwar, A. P. Davis, *Tetrahedron* **1988**, *44*, 3761–3770; (c) S. Anwar, G. Bradley, A. P. Davis, *J. Chem. Soc., Perkin Trans. 1* **1991**, 1383–1389.
[143] M. J. Burk, J. E. Feaster, *Tetrahedron Lett.* **1992**, *33*, 2099–2102.
[144] A. P. Davis, S. C. Hegarty, *J. Am. Chem. Soc.* **1992**, *114*, 2745–2746.
[145] D. A. Evans, P. H. Carter, E. M. Carreira, J. A. Prunet, A. B. Charette, M. Lautens, *Angew. Chem., Int. Ed.* **1998**, *37*, 2354–2359, and references cited therein.
[146] D. A. Evans, E. M. Carreira, *Tetrahedron Lett.* **1990**, *31*, 4703–4706.
[147] M. P. Cabal, R. S. Coleman, S. J. Danishefsky, *J. Am. Chem. Soc.* **1990**, *112*, 3253–3255.
[148] D. W. Knight, *Comprehensive Organic Synthesis, Vol. 3* (Eds.: I. Fleming, B. M. Trost), Pergamon, Oxford, **1990**, pp. 481–520.
[149] (a) B. H. Lipshutz, F. Kayser, Z.-P. Liu, *Angew. Chem., Int. Ed. Engl.* **1994**, *33*, 1842–1844; (b) B. H. Lipshutz, Z.-P. Liu, F. Kayser, *Tetrahedron Lett.* **1994**, *35*, 5567–5570; (c) B. H. Lipshutz, B. James, S. Vance, I. Carrico, *Tetrahedron Lett.* **1997**, *38*, 753–756.
[150] (a) M. Takahashi, T. Kuroda, T. Ogiku, H. Ohmizu, K. Kondo, T. Iwasaki, *Tetrahedron Lett.* **1991**, *32*, 6919–6922; (b) M. Takahashi, Y. Moritani, T. Ogiku, H. Ohmizu, K. Kondo, T. Iwasaki, *Tetrahedron Lett.* **1992**, *33*, 5103–5104; (c) M. Takahashi, T. Kuroda, T. Ogiku, H. Ohmizu, K. Kondo, T. Iwasaki, *Heterocycles* **1992**, *34*, 2061–2064; (d) M. Takahashi, T. Kuroda, T. Ogiku, H. Ohmizu, K. Kondo, T. Iwasaki, *Heterocycles* **1993**, *36*, 1867–1882.
[151] (a) S. Miyano, M. Tobita, H. Hashimoto, *Bull. Chem. Soc. Jpn.* **1981**, *54*, 3522–3526; (b) S. Miyano, S. Handa, M. Tobita, H. Hashimoto, *Bull. Chem. Soc. Jpn.* **1986**, *59*, 235–238; (c) S. Miyano, H. Fukushima, S. Handa, H. Ito, H. Hashimoto, *Bull. Chem. Soc. Jpn.* **1988**, *61*, 3249–3254.
[152] G. Bringmann, J. R. Jansen, H.-P. Rink, *Angew. Chem., Int. Ed. Engl.* **1986**, *25*, 913–915.
[153] T. Takada, M. Arisawa, M. Gyoten, R. Hamada, H. Tohma, Y. Kita, *J. Org. Chem.* **1998**, *63*, 7698–7706.
[154] S. N. Kablean, S. P. Marsden, A. M. Craig, *Tetrahedron Lett.* **1998**, *39*, 5109–5112.
[155] T. Sugimura, S. Nagano, A. Tai, *Chem. Lett.* **1998**, 45–46.
[156] (a) T. Sugimura, N. Nishiyama, A. Tai, T. Hakushi, *Tetrahedron: Asymmetry* **1994**, *5*, 1163–1166; (b) T. Sugimura, H. Shimizu, S. Umemoto, H. Tsuneishi, T. Hakushi, Y. Inoue, A. Tai, *Chem. Lett.* **1998**, 323–324.
[157] For an excellent overview of the Woodward/Eschenmoser synthesis of vitamin B_{12}: K. C. Nicolaou, E. J. Sorensen, *Classics in Total Synthesis*, VCH, Weinheim, **1996**, pp. 99–136.
[158] For a more recent use of the Eschenmoser sulfide contraction: (a) T. G. Minehan, Y. Kishi, *Tetrahedron Lett.* **1997**, *38*, 6811–6814; (b) T. G. Minehan, Y. Kishi, *Angew. Chem., Int. Ed.* **1999**, *38*, 923–925.
[159] (a) Y. Yamada, D. Miljkovic, P. Wehrli, B. Golding, P. Löliger, R. Keese, K. Müller, A. Eschenmoser, *Angew. Chem., Int. Ed. Engl.* **1969**, *8*, 343–348; (b) A. Eschenmoser, *Pure Appl. Chem.* **1969**, *20*, 1–23; (c) P. Dubs, E. Götschi, M. Roth, A. Eschenmoser, *Chimia* **1970**, *24*, 34–35; (d) M. Roth, P. Dubs, E. Götschi, A. Eschenmoser, *Helv. Chim. Acta* **1971**, *54*, 710–734; (e) E. Götschi, W. Hunkeler, H.-J. Wild, P. Schneider, W. Fuhrer, J. Gleason, A. Eschenmoser, *Angew. Chem., Int. Ed. Engl.* **1973**, *12*, 910–912.

[160] J. Castro, A. Moyano, M. A. Pericàs, A. Riera, A. E. Greene, A. Alvarez-Larena, J. F. Piniella, *J. Org. Chem.* **1996**, *61*, 9016–9020.

[161] (a) J. Castro, A. Moyano, M. A. Pericàs, A. Riera, *J. Org. Chem.* **1998**, *63*, 3346–3351; (b) see also: A. Stumpf, N. Jeong, H. Sunghee, *Synlett* **1997**, 205–207.

[162] For recent reviews: (a) R. H. Grubbs, S. J. Miller, G. C. Fu, *Acc. Chem. Res.* **1995**, *28*, 446–452; (b) R. H. Grubbs, S. Chang, *Tetrahedron* **1998**, *54*, 4413–4450.

[163] D. J. O'Leary, S. J. Miller, R. H. Grubbs, *Tetrahedron Lett.* **1998**, *39*, 1689–1690.

[164] P. A. Evans, V. S. Murthy, *J. Org. Chem.* **1998**, *63*, 6768–6769.

[165] M. F. Semmelhack, J. J. Bozell, T. Sato, W. Wulff, E. Spiess, A. Zask, *J. Am. Chem. Soc.* **1982**, *104*, 5850–5852.

[166] (a) K. D. Moeller, C. M. Hudson, L. V. Tinao-Wooldridge, *J. Org. Chem.* **1993**, *58*, 3478–3479; (b) C. M. Hudson, K. D. Moeller, *J. Am. Chem. Soc.* **1994**, *116*, 3347–3356.

[167] M. Schmittel, A. Burghart, W. Malisch, J. Reising, R. Söllner, *J. Org. Chem.* **1998**, *63*, 396–400.

[168] U. Maitra, A. K. Bandyopadhyay, *Tetrahedron Lett.* **1995**, *36*, 3749–3750.

[169] U. Maitra, B. G. Bag, *J. Org. Chem.* **1992**, *57*, 6979–6981.

Subject Index

absolute configurations   203
1,5-H-abstraction   322
H-abstraction   311, 316
acetal-bridged carceplex   107, 108
acetals in intramolecular O-glycosylation   387
acetal tether in Diels–Alder reactions   385
acetal tethers   284, 285, 305
acetal tethers in radical cyclizations   331
acrylate oligomerization   221
methyl acrylate polymers   227
methyl acrylate telomers   224, 229
acrylic acid telomers   224
acrylonitrile   238
activation entropy   275
ACT reactions   225, 227, 245
ACT sequence   224, 228, 229
ACT strategy   221, 227
acyloin condensation   253
adamantanecarboxylate   7
4-*exo*-trig addition   367
C_{60} *cis*-1 addition pattern   210
C_{60} bis-addition patterns   199
C_{60} *cis*-1 bis-adduct   210
C_{60} *e,e,e*-tris-adduct   190
C_{60} *trans*-1 bis-adduct   206
C_{60} *trans*-4 bis-adduct   208
C_{60} tris-adduct   194
C_{60} bis-adducts   194, 198
C_{60} *cis*-2 bis-adducts   200
C_{60} *cis*-3 bis-adducts   202
adenine self-pairing   144
(+)-adrenosterone   296, 297
aglycon delivery method   342
alanyl-PNA   143
aldolase mimics   184
alkaline hydrolysis of ester   61
allenes   252
π-allylpalladium species   252, 265
(*E*)-allylsilanes   370
allyltributylstannane   222
D-altritol   381
aluminum tris(2,6-diphenylphenoxide)   125
Alzheimer's disease   144
AMBER forcefield   349

amidine binding site   56
amidine groups   61
amidinium benzoate   64
5-amidopyrazole   56
amidopyrazoles   56
amino acid derivatives   49
amino acids   48
1,2-amino alcohols   368
aminoethyl linkage   139, 141, 151
amphiphilic helices   8
amplification   151
amyloid diseases   144, 145
anchor–tether–reactive group conjugate   194, 213
androstan-17β-ol   162
androstan-3,17-dione   60
androstane-3,17-diol   180, 181
androst-4-ene-3,17-dione   359
anomerically stabilized radical   312
anomeric phenyl sulfones   336
anomeric selectivity   354
methyl 9-anthracenecarboxylate   232
anthracene endcaps   232
anthracenes   189
anthracene-transfer reaction   192
antibodies   45, 61
1,3-*anti*-diols   371
antisense molecules   133
apparent association constants   120
aromatic stacking   137, 146
aromatic substitution   172
aromatic templates   84, 90
artificial enzymes   159
artificial proteins   8
asparagine glycoproteins   337
asymmetric cyclocopolymerizations   41, 58
1,3-asymmetric induction   358
asymmetric induction   202
asymmetric Sharpless bis-osmylation   202
asymmetric syntheses   60
autocatalysis   18, 147
(–)-avenaciolide   370
aza-bridged fullerenes   210
1-aza-2-silacyclobutanes   367
azomethine ylide cycloadditions   304

„back-biting" reactions   230
backbone flexibility   137
backbone-modified nucleic acids   133
bacteria   49, 52, 65
Baldwin rules   307
barbiturate drugs   56
base pairing   152
B-DNA   147
B-DNA conformation   140
Beckwith–Houk model for radical cyclization
   308
benzannulation reaction   382
p-benzenediazonium sulfonate   173
benzhydryl radical   163
benzo[18]crown-6   10, 33
benzophenone-4-carboxylic acid   174
benzophenone-4-hexanoic acid   162
benzophenone photochemistry   160, 163
benzophenone-4-propionic acid   161
benzophenones   175
N-benzyl-D,L-valine   55
benzylic ether tether   289
benzylthia-bridged carceplex   111
cis-bicyclo[5.3.0]decan-8-ones   380
cis-bicyclo[6.3.0]decan-9-ones   380
bicyclo-DNAs   137
[4.1.0]bicyclohepta-2,4-diene   376
bicyclo[6.3.0]undecane ring skeleton   300
„billiard-ball" effect   111
bimolecular rate constant   23
bimolecular termination   239
bimolecular terminators   241
binary RCM catalyst   264
binding constant   25
binding-site interactions   52
Bingel addition   190
Bingel cyclopropanation   189
Bingel macrocyclization   196, 198, 201, 213
BINOL   373, 374
biological catalysts   152
biological templates   2
biomimetic catalysis   184
biomimetic chemistry   159
biomimetic oxidizing catalysts   180
biopolymers   65
biosynthesis of cholesterol   159
bioxalomycins   304
4,4′-bipyridine   11
4,4′-bipyridinium unit   11
2,6-bis(acrylamido)pyridine   56
bis-adduct of C_{60} 193
bis-adducts of C_{60} 190
bis(alkynyl)methanofullerene   190

bis-aza-fullerenes   210
bis(boronic acid)-substituted fullerene   207
bis(carceplex)   110
bis-ester tethers   345
bis-ether tether   350
bisimidazoles   55
bis-lactonization   235
bis(*o*-quinodimethanes)   205
bis(1,10-phenanthroline)copper(II) core   201
boronate template   290
boronate tether in Diels–Alder reactions   386
boronic acid   54, 207
boron templates   6
boron-tethered cycloaddition   292
boron tethers   290
branched-chain 1,4-diols   315
branched-chain 1,5-diols   315
4′-*α*-C-branched-2′-deoxyadenosines   325
C-branched nucleosides   324
branched trisaccharide   349
bridgehead double bond   277, 298
α-(bromomethyl)dimethylsilyl ether   308
(bromomethyl)dimethylsilyl ether group in a
   radical cyclization   386
(bromomethyl)silyl ether connection   331
bromosteroid   170
bryostatin 1   371
tert-butyl acrylate   238
di-*tert*-butylsilyl acetals   279

C_{60} 189
cage compound   7
calicheamicinone   372, 373
calix[4]arene–cavitand hybrid carceplexes   112
calix[4]arene/resorc[4]arene carceplex   112
calixarenes   112
calixene heterodimers   127
camphanic acid   122
capsules   105, 118
7-carbaadenine   144
carbenoid chemistry   376
carbohydrate recognition   206
carbon-acetal tether   285
carbonyl insertion   161
3-carboxybenzisoxazoles   61
carceplexes   10, 105, 106
carcerands   10
carceroisomerism   113
catalytic antibodies   61
catalytic DNA   136
catalytic epoxidations   178
catalytic ring expansion   251
catalytic RNA   136

catalytic template directed ligation   134
catalytic turnover   5, 27
catechol template   380
[2]catenane   76, 78, 81, 83, 84
[3]catenane   86
[4]catenane   86
[5]catenane   86
[6]catenane   86
[7]catenane   86
catenane   18
catenanes   11, 18, 75
catenate   18
cavitands   109
cavity effect   55
cellobioside   349, 350
„Central Dogma"   133
cetyl trimethylammonium bromide (CTAB)   175
chain transfer   221, 224
chain transfer agent   220, 226
chain transfer process   229
chalcogen-based terminators   236
charge transfer interactions   11
chelation-controlled intermolecular addition   371
chemical template   4
chemosensors   50, 64
chiral auxiliaries   219
chiral cavities   45, 57
C_{60} chiral *cis*-3 bis-adducts   211
C_{60} chiral functionalization pattern   203
chiral oxazolidine auxiliaries   227
chiral „softballs"   122
chiral stationary phase (CHIRALCEL OD®)   206
chiral template molecules   40, 42
chiroptical properties   57, 203
chlorinase   174
exo–syn chlorination   298
chlorination of anisole   172
3α-cholestanol   161, 163
cholestanyl acetate   163
$\Delta^{9(11)}$-cholestenol   163
Δ^{14}-3α-cholestenol   162
chromatographic resolution   47
chromatographic supports   52
chromatography   44, 58
CH···X (X=O) hydrogen bonding   108
circular dichroism (CD)   143
circular dichroism (CD) spectra   203
circular DNA   150
circumrotation   78, 85
clerodin   311
clipping reaction   201

coated silicas   52
cobyric acid   377
co-conformation   76
co-conformers   88
coenzyme B-12   184
coiled-coil peptide templates   148
co-monomers   41
compactin   287
conformations of flexible molecules   184
conjugated π-chromophore   189
C_{60}-containing catenane   204
controlled oligomerizations   230
coordinated template catalysts   183
coordination-induced proximity   252
coordinative bonds   55
copolymers   41
Corey–Nicolaou method for macrolide
   synthesis   249
corrin   377
cortexolone acetate   360
corticosteroids   167, 359, 361
cortisone   167, 168
Cotton effects   203, 206
C–C coupling reactions   249
covalent template   189
covalent tether   276
p-cresol   173
cross-catalytic self-replication   147
crosslinked polymers   45, 48, 57
crosslinking agents   45, 50
crosslinking polymerization   52
crown ether formation   249
crown ethers   3
crownophanes   254
CTV heterodimers   127
cubic cyclophane   195
cucurbituril   14, 18, 27
cucurbituril capsules   127
β-cuparenone   379
(+)-β-cuparenone   378
cyanosilylation   368
cyclic porphyrin dimer   12
cyclic porphyrin oligomers   11
cyclic porphyrin tetramer   13
cyclic porphyrin trimer   12
5-*exo* cyclization   312, 313, 325
5-*exo*-dig cyclization   315, 316, 333, 336
5-*exo*-trig cyclization   308, 320, 331
6-*exo*-dig cyclization   333
6-*exo* trig cyclization   328
7-*endo-trig* cyclization   328
7-*exo*-dig cyclization   333
endo–re cyclization   305

endo-trig cyclization   328
cyclization effective molarities   33
6-*exo* cyclization mode   324
cyclizations   21
cyclization templates   29
cyclized-DNAs   137
[5+2] cycloaddition   303
endo–re cycloaddition mode   306
cycloaddition reactions   276
[5+2] cycloadditions   304
cycloadditions   60
cyclobarbital   56
cyclobutanedicarboxylic acids   60
cyclocholate dimers   127
cyclodextrin   172
α-cyclodextrin   172
β-cyclodextrin   172
cyclodextrin-containing [2]catenane   81
cyclodextrin-containing rotaxanes   82
cyclodextrin dimers   127
cyclodextrin inclusion   183
cyclodextrins   81, 178
cycloglycosylation   348
cyclohexyl iodide   222
cyclononylprodigiosin   263
cyclophane   11, 191
cyclopropanedicarboxylic acids   60
cyclotelomerization   221, 226
cyclotrimerization of butadiene   249
cylinder-shaped molecular capsules   125
cysteine-dimer analogue   380
cytochrome P-450   183
cytochrome P-450s   177

dactylol   263
deboronation   42
cis decalin   312
decarboxylations   61
decarboxylative glycosylation strategy   353
deep-cavity resorcinarene   125
dehydrofluorination   61
demetallation   78
2′-deoxyadenosine   325
deoxyadenosine oligomers   146
methyl 7-deoxycholate template   384
deoxyfrenolicin   382
α-deoxykojic acid   303
2′-deoxy-3′-*O*-methyluridine   355
2′-deoxyribonucleosides   354, 355
β-2′-deoxyribonucleosides   356
2′-deoxyuridine   355
Dess–Martin reagent   228
detoxinine   301

(–)-detoxinine   300, 302
dialkylammonium-containing rotaxanes   91
dialkylammonium guests   91
dialkylsilyl tether   283
dialkylvinylsilyl halides   281
diaryl sulfides   167
diastereofacial selectivity   303, 306
re/*si* diastereofacial selectivity   306
diastereomeric dyads   42
diastereoselective Bingel macrocyclization   202
diastereoselective bis-additions   211
diastereoselective tether-directed bis-cyclo-
    propanation of C_{60} 203
diastereoselectivity   199
3,8-diazabicyclo[3.2.1]octane   304
α-diazo silyl ethers   376
dicarbonyl compounds   49
(dichloromethyl)dimethylsilyl ethers   320
Dieckmann condensation of diesters   253
Diels–Alder additions   190, 193, 205, 206, 208
Diels–Alder adducts   126, 233
Diels–Alder cycloaddition   136, 232, 285
Diels–Alder reactions   15, 121, 231, 277, 290
Diels–Alder strategy   285
dienyl silyl ethers   281, 284
di(ethoxycarbonyl)methano addend   190
diglycoluril isomers   123
1,25-dihydroxyvitamin D_3 280
dimeric hydrogen-bonded capsules   120
9,10-dimethylanthracene (DMA)   190
dimethylene sulfone RNA   141
dimethyl(methylthio)sulfonium trifluoromethane-
    sulfonate (DMTST)   344
dinuclear macrocycle   83
1,5-diol products   325
1,3-diols   309
1,4-diols   309
diols   48
dioxoporphyrins   14
dipeptide–amidopyrazole complex   57
dipeptide template   56
diphenylmethane stoppers   79
diphenylsilyl tether   283
15,16-diphenyl-1,4,7,10-tetraoxa(10.2)[20]para-
    cyclophan-15-ene   269
1,3-dipolar cycloaddition   14
diquinane   318
diquinane framework   316
directed bis-silylation reactions   369
directed nucleosidations   354
directed stereoselective glycolation   370
C-disaccharides   334, 336
distance selectivity   44

disulfides 49
4-methyl-2,6-di-*tert*-butylpyridine (4-Me-DTBP) 338
divinylbenzene 51
dizinc(II) bisporphyrins 126
DMAP catalysis 235
DNA 2, 7, 143
DNA-catalyzed reductive ligation 150
DNA-directed imine formation 148
DNA double helix 133
DNA duplex 29
DNA ligase 136
DNA polymerases 135
DNA replication 14, 105
DNA template 151
A-DNA 147
DNOE NMR 238
dodecanedioic acid 176
π-donor character 85
Dötz benzannulation 381
double [3+2] cycloadditions 209, 210
double Diels–Alder addition 205
double-helical template 150
double helix 18
doubly ion-paired template 176
drugs 49
duplex stability 141
dyes 50

effective molarity 18, 25, 32, 134, 151
electrochemically initiated oxidative cyclization 382
electrochemical reducibility 193
electrochemical reduction 205
electrolysis 170
electronic structure of the fullerene 205
electron microscopy 145
π-electron ring-current effects 195
electron transfer 205
enantiomeric resolution 48, 50, 53
enantioselective catalysis 219
enantioselective synthesis 202
enantioselective synthesis of myltaylenol 293
encoded enzymes 149
6-*endo* addition 327
6-*endo* cyclization 308, 314, 315, 320, 325
endohedral ^{3}He compounds 195
endohedral metallofullerenes 205
endopeptidase inhibitor 267
$^1O_2$–ene reaction (the Schenk reaction) 195
energy transfer 205
enthalpy–entropy compensation 119
enzymes 5, 159

enzyme catalysis 5, 46
enzyme–substrate complex 159
epilachnene 263
equilibrium association constants 121
Eschenmoser sulfide contraction 377, 379
ester tethers 295
ether-based templates 228
ether tethers 287
ethylene dimethacrylate 51
ethylthio donor 347
(4*R*)-4-ethylvalerolactone 312
N-ethyl-4-vinylbenzamide 67
N-Ethyl-4-vinylbenzocarboximide acid ethyl ester 67
exo/endo selectivity 283
exponential amplification 147

β-facial selectivity 356
ferrochelatase 136
ferrulactone 258
ferrulactone I 259
fiber–fiber association 144
fibril 145, 152
fibril arrays 152
fibrillogenesis 151
filipin III 219
Fischer carbene complex 381
Fischer indole synthesis 8
fluoride catalysis 233
fluoromethacrylate-bearing template 239
4-fluoro-4-(4-nitrophenyl)butanone 61
folding topology 8
footprint catalysis 61
force-field calculations 348
B-form DNA 140
N-formylpiperidine 106
formyl radical equivalent 320
fragmentation–recombination process 326
free-radical addition 219
free-radical halogenations 163
free-radical halogenations of steroids 164
free-radical oligomerizations 219
free-radical polymerizations 236
free-radical reactions 164
fructofuranosides 344
β-fructofuranosides 346
β-D-fructofuranosides 344
fullerene–calixarene conjugate 201
fullerene chirality 193
fullerene–crown ether conjugates 205
fullerene dendrimer 201
fullerene–glycodendron conjugate 200
fullerene π-chromophore 195

fullerene–porphyrin conjugates   201, 205
fullerene properties   189
fullerene radical anion   205
fullerenes   189
C_{60} bis-functionalization patterns   204
trans-fused hydrindane   310
cis-fused 5,6-ring junction   311
trans-fused 5,6-ring junction   309, 311

gel-like phases   120
genomes   135, 153
Glaser coupling   32
Glaser–Hay coupling   11
(–)-gloeosporone   262, 263, 264
α-glucosidation   341
glycoluril   120
α-*C*-glycoside   333
β-*C*-glycoside   333
glycosyl acceptor   340, 353
O-glucosyl *β*-mannopyranosides   340
glycosylation   340, 349
C-glycosylation   307
O-glycosylation   340
glycosyl donor   340
1,3-glycosyl transfer   353
C-glycosyl transferase   335
O-glycosyl trichloroacetimidates   351
graft polymerizations   233
group transfer polymerization   233, 234
Grubbs ruthenium carbene catalyst   80
Grubbs-type catalysts   261
guanidinium group   61
guest-determining step (GDS)   107
guest selectivity   107

(haloalkyl)dimethylsilyl ethers   318
halodesilylation   319
„head to tail" isomer   279
heat of formation   193
hemicarceplex   111, 114, 115
hemicarceplexes   105, 114
^{3}He-NMR   195
HETCOR   227
heteroduplex   141
hexadeoxynucleotide template   28
C_{60} hexakis-adducts   190, 191, 192, 195, 212
hexanucleotide template   6
(E)-hex-3-ene-1,5-diyne   199
hexose-oligonucleotides   137
high dilution   266
higher-ordered aggregates   144
HOMO   232
β-homoalanyl-PNA   143

homo-DNA (h-DNA)   137
homopropargylic acetylenes   368
Hoogsteen pairing   149
Horner–Wadsworth–Emmons olefination   372
hydantocidin   357
(+)-hydantocidin   356
cis hydrindane   320
trans hydrindane   320
(–)-hydrobenzoin   297
hydrodesilylation   282
hydrogen abstraction   165
hydrogen atom abstraction   226
hydrogen bonded capsule   108
hydrogen bonding   11, 137
hydrogen bonding interactions   78
hydrogen bonds   94
[C–H···O] hydrogen bonds   84, 87, 91, 95
[N$^+$–H···O] hydrogen bonds   91
hydrogenolysis   110
hydrolysis of diphenyl carbonate   64
hydropalladation of an allene   267
hydrophobically driven syntheses   81
hydrophobically driven syntheses of
    [2]catenanes   83
hydrophobic side chain interdigitation   146
hydroquinone   11
hydrosilylation   365
hydrosilylation of acetylenes   368
hydrosilylation of allylic amines   367
hydrosilylation of olefins   364
hydroxycarboxylic acids   48
(22*R*)-22-hydroxycholesterol   319, 320
β-hydroxy esters   371
22-hydroxylated steroid side chains   314
17*α*-hydroxyprogesterone   360

IMDA reaction   278, 287
imidazole derivatives   49
immobilized template   174
immunoassay   49
immunosorbents   50
immunosuppressive alkaloids streptorubin B
    251
imprinted cavity   46, 60
imprinting template   206, 208
induced fit   53
inherently chiral fullerenes   203
initial rates   29
initiator   231
initiator–terminator functionalized template   234
initiator–terminator gap   235, 238
initiator–terminator pair   233
A^{1,3} interactions   277, 316

[C–H⋯π] interactions   84, 87, 108
π-π interactions   11
inter-bowl hydrogen bonds   115
intercapsular communication process   110
interchromophoric spatial relationship   205
intermolecular associations   152
intermolecular chain-growth polymerization   230
inter-strand hydrogen bonding   146
interweaving templates   18
intramolecular aglycon delivery   343, 344
intramolecular allylation   358, 359
intramolecular Bingel addition   202
intramolecular bis-silylation   369
intramolecular carbopalladation reactions of allenes and alkenes   265
intramolecular carbopalladation reactions of ω-haloallenes   251
intramolecular chlorination   163
intramolecular coupling of silyl bis-enolates   383
intramolecular [3 + 2] cycloaddition   300
intramolecular delivery of a nitrogen nucleophile   358
intramolecular delivery of the nucleoside base   355
intramolecular Diels–Alder (IMDA) reaction   277
intramolecular epoxidations   171
intramolecular glycosylation   338
intramolecular hydroboration   364
intramolecular hydrogen atom abstraction   229
intramolecular hydrogermylation   364
intramolecular hydrosilylation   364, 367, 368
intramolecular hydrosilylation of ketones   370, 371
intramolecularization   275, 276, 307, 349, 373, 384
intramolecular McMurry coupling   269
intramolecular nucleophile delivery   361, 362
intramolecular oxidative coupling of enolate derivatives   382
intramolecular radical cyclization   310
intramolecular transformations   275
intramolecular Ullmann coupling   375
in-vitro selections   136
in-vitro selection strategies   134
iodine atom templates   167
iodoaryl template   164
iodonium collidine perchlorate (IDCP)   344
iodonium collidine triflate (IDCT)   344
ion-exchange mechanism   59
ion-paired templates   174

ion pairing   177, 266
iron(II)phthalocyanine   3
iron porphyrin   178
iron tetraphenylporphyrin   183
α-C-isomaltoside   336
methyl α-C-isomaltoside   336
isophorol   321
isophthaloyl chloride   80
isotactic dyad   227
isotactic telomer   227
ivangulin   293

jasmin ketolactone   263
„jelly donut“   124
„jelly donut“ dimer   123

α-kainic acid   361
Karstedt catalyst   364
Kemp's triacid   18
ketal tether   336
16-keto-3α,5α-androstanol   174
12-keto-3α-cholestanol   161
α-keto-β-D-ribonucleosides   325
(+)-ketopinic acid   370
ketyl radical anions   256
key/lock principle   46
kinetic control   95
kinetic deprotonation   297
kinetic templates   5, 21, 22
knots   18
kojic acid   303

lamination   146
Langmuir monolayers   200
(−)-lardolure   219
large carceplexes   109
(+)-lasiodiplodin   263
L-DOPA methyl ester   60
leaving group-based glycosylation   354
leaving group-based intramolecular glycoside bond formation   351
leaving group-based intramolecular glycosylation   352
leucine-zipper motif   148
Lewis acid complexes   136
Lewis-acid/Lewis-base interaction   261
Lewis acid-promoted, IMDA cycloaddition   280
ligand exchange chromatography   55
ligation   135, 146
ligation catalyst   151
light initiated polymerization   223
lincomycin   357
linear coupling reactions   21

linear oligomers   19
linear porphyrin octamer   20
linear porphyrin oligomers   32
linear templates   14, 22, 23
linear triquinanes   317
linear vinyl polymers   40
β-1,4-linkage   341
β-1,6-linked disaccharide   339
β-1,4-linked mannoglucoside   338
liquid secondary ionization mass spectrometry
   86
locked-nucleic acids (LNA's)   137
lowest unoccupied molecular orbitals   190
low-molecular-weight template   39
low-order hydrogen-bonded aggregates   120
low-valent titanium reagents   253

macrocycle syntheses   259
macrocyclic lactam   79
macrocyclic polyether   85, 90, 91, 94
macrocyclization   220
– metal cation templated   5
macrocyclization rates   228, 236
macrocyclization reactions   2
macrocyclizations   228, 238
macro model   349
macro-olefination reaction   371
macroporous imprinted polmyers   52
macroporous polymers   45, 50
main chain chirality   40
methyl α-C-maltoside   335
β-mannopyranoside linkage   337
β-mannopyranosides   339
β-mannoside formation   340
β-mannoside linkage   342
β-D-mannoside linkage   350
β-mannosides   344
β-D-mannosides   354
β-mannosylation   345
Masamune–Roush conditions   331
mass transfer   53, 55, 65
materials science   4
McMurry reactions   256
McMurry-type macrocyclizations   254
mechanical bonds   75
mechanically interlocked molecule   75
medium-sized rings   251
mesoporous solids   4
meta arene–alkene photocycloaddition   377
metal-bridged carceplexes   113
metal cations   2
metal cation templates   10, 33
metal-induced self-assembly   113

metal ion catalysis   183
metallomacrocycle   90
metalloporphyrin templates   177
metallosalen templates   177
metal nanocylinders   4
metal templates   3, 76
metathesis reaction   261
methyl methacrylate   40
methacrylate polymerizations   233
methacryloyl radical   240
methano[60]fullerenecarboxylic acid   193, 212
2-methoxy-4,4,5,5-tetramethyl-1,3,2-dioxa-
   borolane   292
3-*O*-methyl-D-glucofuranose   208
micelles   175
Michaelis–Menten analysis   22
Michaelis–Menten kinetics   63, 134
microparticles   65
microreactors   48, 50, 60
miniature reaction chamber   126
Mitsunobu reactions   374
MMA oligomerization   240
MM2 calculations   110
Mn^{3+}-porphyrin   180, 181
5-*exo*-trig mode   326
modified silicas   44
molar optical rotation   57, 58
molecular capsules   105
molecular container   110
molecular dyads   205
molecular dynamics   88
molecular dynamics simulations   140, 142
molecular imprinting   45, 47, 208
molecularly imprinted beads   52
molecularly imprinted cavities   64
molecularly imprinted materials   39
molecularly imprinted polymers   48
– catalysis   60
– chromatography   58
molecular mechanics   233, 350
molecular mechanics calculations   166, 349
molecular nanoscaffolding   195
molecular recognition   135, 137
molecular replication   147
molecular scaffolding   189
molecular square   191
molecular switches   93, 113
molecular switching devices   110
molecular template
– definition   4, 33
monolayers   8
– self assembled   9
monomethanofullerene   190

monoprotected diols   366
Monte Carlo (MC) conformational searches   349
Mosher esters   232
multiple adducts of C_{60} 189
multiple functionalization   189
m-xylyl-bridged hemicarceplexes   117
myltaylenol   293, 294

(NBD)$_6$-based template   233
(NBD)$_3$-derived template   233, 235, 237
(NBD)$_4$-derived template   238
NBD hexameric spacer   233
(NBD)*n*-based templates   231
(NBD)$_4$ templates   232
nearest-neighbor approximation   140
negative cooperativity   27
„negative" template   116
negative templates   20
neutron scattering experiments   145
Ni^0-catalyzed hydrosilylation   284
nickel(0)-mediated polymerization
   of norbornadiene   231
NiCOD$_2$ 231
nicotinic acid   171
nitromethane   342
N-methyl-2-pyrrolidinone (NMP)   107
N,N'-Diethyl-4-vinylbenzamidine   67
N,N-dimethylacetamide (DMA)   106
non-bonding interactions   284
noncovalent bonds   75
non-covalent imprinting   53
non-covalent interactions   47
non-covalent template   189
non-enzymatic oligonucleotide ligations   133
norbornadiene   231
norbornadiene oligomers   232
[*n*]pseudorotaxanes   92
nucleic acid backbone modifications   153
nucleic acid replicators   18
nucleic acids   135
nucleic acid template   151
nucleobase modification   136
nucleosides   49
nucleotides   49

obtusenyne   367
C_{60} octakis(ethyl ester)   195
octalactin A21   258
(−)-octalactin A   259
octanedioic acid   176
olefination   371
oligomeric natural products   219

oligomerization of methyl methacrylate   236
oligomerization of methyl methacrylate with
   template   246
oligomethylene-tethered bis(cyclopropenone
   acetals)   209
oligonucleotide biopolymers   153
oligonucleotide catalysts   152
oligonucleotide templates   134, 149
oligopyridine templates   14
oligosaccharide chemistry   337
oligoselective free-radical polymerization
   of MMA   237
oligoselective free-radical polymerizations   243
oligoselective polymerization   238
oligoselectivity   233
Olympiadane   86
one-dimensional property transfer   40
OPNA   143
optically active copolymers   41
optically active fullerene derivatives   203
optical rotation   58
organopalladium reagents   252
oxacyclohexadec-11-en-2-one   269
oxazolidine acrylamides   228
oxidative tether cleavage   366
oxophilicity of titanium   256
oxy-PNA (OPNA)   143

palladium-catalyzed alkenyl epoxide
   cyclization   269
palladium-catalyzed [3+2] cycloadditions   316
paracrystalline materials   144
5-*exo* pathway   322
Pauson–Khand reaction   378
PEG block copolymer   152
C_{60} pentakis-adducts   195
pentamidine   56
pentasaccharide   337
peptide-nucleic acids (PNAs)   142
peptide–peptide association   144
peptides   49
A-*β*-peptides   144
pesticides   50
π-facial selectivity   298
phenanthroline-containing rotaxanes   77
phenanthroline ligands   76
phenyl-*α*-D-mannopyranoside   46
L-phenylalanine anilide   48
L-phenylalanine-anilide template   54
phenyliodine dichloride   163, 164
phenyliodochloride radical   167
phenylseleno glycoside   334
phenylseleno sugars   335

3'-phenylseleno thymidine nucleosides   325
phenylthio α-rhamnosyl donor   347
phenyl 3,4,6-tri-*O*-acetyl-1-thio-α-gluco-
   pyranoside   341
pheromone   219
phosphate backbone   153
phosphate esters   184
phosphate substitution   138
phosphodiester formation   151
methylphosphonate derivatives   138
phosphonate esters   50
phosphorothiolate derivatives   138
photochemical [4+4] cycloaddition   205
photochemical functionalizations   174
photochemical insertion   167
photochemically initiated polymerization   66
photochemical remote functionalization of long-
   chain alkyl esters   160
[2+2] photocycloaddition   299
photodimerization of a norbornene precursor
   233
photoinduced intramolecular pinacolization
   254
E–Z photoisomerization   377
photolysis   165
photonic crystals   4
phthalocyanine   6
phthalyl   347
PMB acetal tether   344, 346
p-methylanisole   173
PNA   143
PNA–DNA–PNA triple helix   142
polyene macrolide antibotics   219
polyether template   229
polymerase chain reaction   149
polymeric materials   45
polymeric template   39
polymerization   51
polymers   51
polymer-supported glycosyl donor   345
polynorbornane-based template   230
polyols   48
polyquinanes   316
poly(triacetylene) oligomers   199
porphyrin–cyclodextrin conjugates   179
porphyrin-fullerene conjugates   205
porphyrin luminescence   205
porphyrin metallation   136
porphyrins   126
„positive" template   116
precapnelladiene   299
precatenanes   76
preorganization   126, 134

prerotaxanes   76
5-*exo*-trig process   315
product dissociation   149
product distribution analysis   22
product histogram   225
product histogram for ACT   229
product–template duplex   135
property transfer   40
propionaldehyde homoenolates   362
prostaglandin $F_{2\alpha}$ 331
proteins   49
pseudo-dilution effect   266
pseudo-octahedral addition pattern   194
[3]pseudorotaxane   91
[4]pseudorotaxane   91
[5]pseudorotaxane   91
[2]pseudorotaxanes   85, 95
Pt-catalyzed intramolecular hydrosilylation   368
purine derivatives   49
pyranosyl-RNA (p-RNA)   137
pyrazine   107
pyridine-based tethered templates   169
pyridoxal   184
pyridoxal phosphate   184
pyridoxamine   184
pyridoxamine phosphate   184
2-pyridyl sulfones   336
pyrrolizidine alkaloids   300

o-quinodimethanes   291

racemic dyads   42
racemic resolution   46
radical acceptor   323
6-*exo* radical cyclization   332
9-*endo-trig* radical cyclization   335, 336
radical cyclization   42, 309, 310, 312, 318
radical cyclization of (1-bromovinyl)dimethyl-
   silyl ethers   322
radical cyclization of silyl acetals   329
radical cyclization on the 2',3'-diol   330
radical cyclization onto a double bond   308
radical cyclization onto a triple bond   315
radical cyclization reactions   307
radical cyclizations   307, 308
radical halogenations   192
radical polymerization of activated alkenes   234
radical relay chlorination   166
radical relay mechanism   164
radical relay process   163, 165
radical traps   331
radioimmunoassays   65
rapamycin   366, 367

rate-determining step   109
rate enhancement   22, 25, 30
RCM-based macrocyclizations   261
RCM-based ring closure reactions   259
reductive amination   7
reductive cleavage strategy   294
reductive coupling of carbonyl compounds   253
reductive desilylation   310
Reformatsky reactions   258
regio- and stereo-controlled glycosylation   347
regioselective bis-functionalization   209
regioselective enolate formations   258
regioselective functionalization   192
regioselective hydroxylation   181
regioselective IMDA reaction   291
relay chlorine transfer   173
relay model   261, 263
remote functionalization   162
remote glycosylation reactions   345, 350
remote glycosylation strategy   349, 354
remote intramolecular glycosylation   345
removal of the tether   213
replication cycle   151
replication fidelity   135
resorcinarene   112
resorcinarene-based cages   127
retro aldol fission   300
retro Diels–Alder process   294
retro-Diels–Alder reaction   195
reversed-phase HPLC   52
reversible covalent templates   189
(L)-rhamnal-derived silyl ether   312
β-L-rhamnoside   348
β-L-rhamnoside linkage   350
β-L-rhamnoside linkages   347
β-L-rhamnosides   354
rhizoxin   332
ribonucleoside   329
ribose modifications   137
ribose-modified nucleic acids   138
ribozymes   134, 136, 152
rigid matrix   43
ring closing metathesis (RCM)   259, 380
ring closure reactions   249
ring-opening and ring-closing methathesis
   (RORCM)   80
RNA   2
RNAse P   134
RNA synthesis   14
RNA world   152
„RNA-world"   136
roseophilin   263, 266
[2]rotaxane   77, 87

rotaxane   18, 27
[2]rotaxanes   79, 88
rotaxanes   11, 75
rSNA   141
ruthenium(II)carbonyl porphyrins   14

saccharide template   208
Saegusa oxidation   332
Sakurai coupling   288
salicyl alcohol template   375
salt bridge formation   146
samarium iodide   258
scavenger templates   19
selection methods   136
selective hydroxylation   180
self-assembled block copolymer aggregates   127
self-assembling calix[4]resorcinarene
   oligomers   127
self-assembling reversible capsules   119
self-assembly   8, 204
self-association of amyloid peptides   145
self-complementary hexamer   147
self-complementary sequences   137
self-replicating system   17, 29
self-replication   15
self-replicator   27
semi-empirical PM3 calculations   193
sensor   48
separation factor α   46
serine proteases   166
β-sheet conformation   56
β-sheet template   152
shell-closure reactions   106, 111, 115
sila-hexenyl radicals   314
silicon anomeric effects   328
silicon-tethered ene cyclization   362, 363
silicon-tethered ene precursors   363
silicon-tethered radical cyclization   313
silicon tethers   278, 296
silicon transfer oligomerization   235
siloxane bonds   44
siloxane template   367
silyl acetals   278, 296, 327, 328
silyl acetals as the tether-directing group   339
silyl acetal tether   280, 330
silylacetate   239
silyl ethers   296
silyl ether tethers   321, 325
silylformylation   368
silyl tether   301
silyl-tethered radical cyclization   312
silyl-tethered radical cyclization of nucleosides
   326

silyl-tethered radical cyclizations 315, 323
singlet oxygen ene reaction 193
singlet oxygen photosensitization 193
slippage processes 89, 90
SmI$_2$-mediated Reformatsky reaction 258
S,N-acyl rearrangement 148
sodium dodecylsulfate (SDS) 175
„softball" 121
„softball" dimers 122
„softballs" 119
solid-phase synthesis 141
solid-phase synthesis of amide-linked oligo-
    nucleotides 142
solid-state template effect 192
solvation requirements 153
solvent effects 110, 239
solvent-induced steric effect 243
solvent–solute interactions 241
solvophobic effect 243
space occupancy factor 125
specificity constant 135
square root law 18
[$\pi\cdots\pi$] stacking 84, 91
statine 323, 324
stationary phases 58
stationary phases in chromatography 64
statistical reaction 22
Steglich/Keck conditions 235
stereochemical descriptors ^{f}C and ^{f}A 203
stereochemical transcription 276
stereocontrolled synthesis of 22-hydroxylated
    steroids 314
stereocontrol on the IMDA reaction 289
stereodefined quaternary centers 383
stereoelectronic restraints 212
stereoisomerism 113
in–in, in–out, and *out–out* stereoisomerism 200
stereoisomer separation 231
stereoregular copolymer 40
stereoselective 1,5 C–H insertion reactions 376
stereoselective C–H insertion reactions 376
stereoselective glycosylation 337, 342, 351
stereoselective intramolecular glycosylation 343
stereoselective macroolefination 372
stereospecific methylation of steroids 310
steroid hydroxylations 180
steroids 49, 309
stoechospermol 299
A^{1,3}-strain 305
strand association energetics 140
β-strand peptides 144
strand–strand association 152
subphthalocyanine 5

substituted vinylsilyl ether 322
substrate assembly 4
substrate-directed stereoselective synthesis 276
substrate selectivity 44
substrate–substrate bonding 8
substrate–template ternary complex 135
succinyl ester tethers 347
sugar pucker 147
sugars 48
sugar template 207
sulfide tether 303
sulfonate tether 293
sulfonate tethers 292
sulfoxide and sulfone tethers 304
sulfur templates 167
sulfur tethers 303
superphthalocyanine 6
supramolecular advanced materials 193
supramolecular aggregates 144
supramolecular chemistry 76, 105
supramolecular complex 93
surface imprinting method 52
surface of silica 44
surfactant mesophases 4
suspension polymerization 52
synthesis of *C*-glycosides 333

talaromycin A 312, 313, 386
Tamao oxidation 308, 312, 316, 325
tandem [4+2]/[3+2] cycloaddition 302
tandem [4+2]/[3+2] nitroalkene cycloaddition
    reactions 300
tandem [4+2]/[3+2] reaction 300
Taxol 258, 259, 292
Tebbe methylenation 284, 335
telomer 220, 226
telomer assay 225, 229, 245
telomer distribution histogram 222
telomerizations 230
telomerization terminator 220
telomer length 223
template
– for cyclization 10
– substrate interactions 8
template autonomy 135
template backbone 230
template-bound pMMA 239
template-bound radical 242
template-bound silyl ketene acetal initiator 239
template controlled free-radical oligomerization
    238
template controlled free-radical oligomerization
    of MMA 243

template controlled oligomerizations   219, 237
template controlled oligoselective free-radical
    polymerization of methyl methacrylate
    (MMA)   237
templated bis-functionalization   206
templated cyclization reactions   253
templated Diels–Alder reaction   26
templated *exo* Diels–Alder reaction   27
template-directed chlorination   170
template-directed epoxidations   172
template directed ligation   134, 148, 153
template directed ligation polymerization   152
template directed polymerization   133
template-directed reaction   193
templated radical macrocyclization   220, 221
template effect in ring closure reactions   258
template effect of low-valent-titanium [Ti]   257
template effects   3, 114
template efficiency   29
template efficiency parameters   26
template-free methyl acrylate telomerization
    224, 225
template mediated free-radical oligomerization
    of MMA   238
template mediated oligoselective polymerization
    of MMA mechanism   240
template monomer   65
template polymerization   39
template ratios   107
templates
– for macrocyclization   10
template–substrate complex   149
templating saccharide   207
temporary boronate tether   292
temporary connection   275
temporary sulfur connection   377
temporary tethering strategies   276
„tennis balls"   119, 120
terminator   231
terminator challenge experiments   241
terminator performance   236
ternary complex   14, 15, 23
terpene polyenes   172
terpenoid, marine natural product   298
tertiary structural folds   153
tether-directed allylation   360
tether-directed functionalization methods   192
tether-directed intramolecular glycosylation
    351
tether-directed β-O-mannoside syntheses   335
tether-directed radical cyclization   323, 333
tether-directed remote functionalization   189
tethered acceptor   338

tethered acetylene   382
tethered bis-azides   210, 211
tethered bis(β-keto ester)   201
tethered bis(buta-1,3-diene)   205
tethered bis(buta-1,3-diene)s   205
tethered radical cyclization   319
tethered reactions   8, 29
tethered reagents   163
tethered vinylcarbenes   209
tether-mediated nucleophile delivery   356
tether-reactive group conjugate   193, 212
tetraazacyclononane–Cu^{2+} chelates   55
tetraazamacrocycles   3
tetraethynylated C_{60} hexakis-adduct   196
tetrahedral transition state   62
C_{60} tetrakis-adducts   195
tetrameric assembly   124
tetramethylene dimethacrylate   51
tetrapyridylporphyrin   13, 20
tetrapyridylporphyrin template   26
tetrarylmethane-based stoppers   88
tetrol bowl   107, 116
tetronolide   285
theoretical plates   59
thermally initiated polymerization   66
thermal [2+2+2] retro-cycloaddition   205
thermodynamic control   81, 95
thermodynamic cycle   148
thermodynamic template   5, 7
thexylborane   291
thiamine pyrophosphate   184
thiazolium ions   184
thiophenyl-bearing template   238
thiopyridyl esters   249
thioxanthone tethered template   168
Thorpe–Ingold effect   303
Thorpe–Ziegler dinitrile method   253
three-dimensional molecular scaffolding   211
three-dimensional transfer of information   45
three-electron bond   169
three-point binding   56
thymidine nucleosides   324
titanium-induced intramolecular carbonyl
    coupling reactions   253
titanium mediated coupling of allylic alcohols
    256
topochemical solid-state synthesis   192
topochemistry   46
topological information   18
topological stereoisomers   82
topological templates   28
tosyliminoiodobenzene   183
transannular interactions   277, 328

transesterification   204, 213
transition state analogue   61
transition state imprinted polymer   63
transition state models   107, 119
transition state solvation   113
transition state stabilization   5
translational entropy   166
trefoil knot   18, 19
triazole stoppers   91
tribenzyl-levoglucosan   342
trichloroacetate template moiety   236
tricolorin A   263
tricolorin G   263
tricyclic β-lactam antibiotics   358, 359, 360
tricyclo-DNAs   137
trifluoroacetate   238
trile-helical structures   142
trimer carceplex   110
1,2,3-trimethoxybenzene   118
trimethylenemethane   316
trimethylol trimethacrylate-based beads   52
2-[(trimethylsilyl)methyl]prop-2-en-1-ol   316
2-(trimethylsilyl)oct-1-en-3-one   331
2,4,6-trinitrophenyl stoppers   83
methyl 2,3,6-tri-*O*-benzyl-4-*O*-(3,4,6-tri-*O*-
    benzyl-β-D-manno-pyranosyl)-α-D-gluco-
    pyranoside   387
1,2,4-triols   369, 370
triple helix   149
triply tethered template   169
tripyridyltriazine   11
triquinane   318
Tröger's base   384
tropylidenes   377
Tsuji–Trost allylation   265
tunicamycin V   328, 329, 330
turnover number   134
turnover rates   152
two-dimensional information transfer   42
two-point binding   56, 58, 180
type II IMDA reactions   295

Ullmann coupling   373
unsymmetrical biaryls   375
uranyl templates   6
urea-based calixarene dimers   126
UV hyperchromism   140

van der Waals contacts   108
van der Waals interactions   9, 121
van der Waals surface   140
vapor-phase osmometry   113
vesicles   175
vinyl acetate   238
vinylcarbenes   209
vinylene carbonate   238
4-vinyl-*N,N'*-diethylphenylamidine   56
4-vinylphenylboronic acid   46
4-vinylphenylboronic acids   40
vinyl polymerizations   40
vinyl silanes   281
vinyl silyl ether IMDA reactions   283
vinyl silyl ethers   281
vinyl sulfonates   292
vitamin B$_{12}$ 377, 378
Vorbrüggen coupling   356

Wadsworth–Emmons homologation   331
Watson–Crick base-pairing   140
Watson–Crick base pairing rules   137
Watson–Crick duplexes   137
Weinreb amide   366
wide-pore silica   42
Wieland–Miescher ketone   296
„wiffle balls"   119, 121, 122
Wilkinson's catalyst   364
Wittig reaction   332

p-xylylenediamine   80

zearanol   263
zeolites   4
Ziegler–Ruggli dilution principle   254
zwitterionic tetrahedral intermediate   122